FUNDAMENTALS OF
ELECTROCHEMICAL ANALYSIS

ELLIS HORWOOD SERIES IN ANALYTICAL CHEMISTRY
General Editor: Dr. R. A. CHALMERS

Founded as a library of fundamental books on important and growing subject areas in analytical chemistry, this series will serve chemists in industrial work and research, and in teaching or advanced study

Published or in active preparation:

Automatic Methods in Chemical Analysis
J. K. FOREMAN
P. S. STOCKWELL } Laboratory of the Government Chemist, London

Electroanalytical Chemistry
G. F. REYNOLDS, University of Reading

Analysis of Organic Solvents
V. ŠEDIVEC
J. FLEK } Institute of Hygiene and Epidemiology, Prague

Handbook of Process Stream Analysis
K. J. CLEVETT, Crest Engineering (U.K.) Inc.

Methods of Catalytic Analysis
G. SVEHLA, Queen's University of Belfast
H. THOMPSON, University of New York

Organic Reagents in Inorganic Analysis
Z. HOLZBECHER ET AL., Institute of Chemical Technology, Prague

Analysis of Synthetic Polymers
J. URBANSKI ET AL., Warsaw Technical University

Spectrophotometric Determination of the Elements
Z. MARCZENKO, Warsaw Technical University

Particle Size Analysis
Z. K. JELINEK, Organic Synthesis Research Institute, Pardubice, Czechoslovakia

Gradient Liquid Chromatography
C. LITEANU and S. GOCAN, University of Cluj, Rumania

Analytical Applications of Complex Equilibria
J. INCZEDY, University of Chemical Engineering, Veszprem, Hungary

Electrochemical Stripping Analysis
F. VYDRA, K. STULIK, E. JULÁKOVÁ, Charles University, Prague

Operational Amplifiers in Chemical Instrumentation
R. KALVODA, J. HEYROVSKY, Institute of Polarography Czechoslovak Academy of Science

FUNDAMENTALS OF ELECTROCHEMICAL ANALYSIS

Z. GALUS
Professor of Chemistry
Warsaw University

Translation Editor:
G. F. REYNOLDS
Director of Research Unit for Instrument Physics,
Department of Engineering and Cybernetics,
University of Reading

ELLIS HORWOOD LIMITED
Publisher Chichester

Halsted Press: a division of
JOHN WILEY & SONS Inc.
New York ● London ● Sydney ● Toronto

English Edition first published in 1976 by
ELLIS HORWOOD Ltd., Coll House, Westergate,
Chichester, Sussex, England
and
PWN—POLISH SCIENTIFIC PUBLISHERS
Warsaw, Poland

The Publisher's colophon is reproduced from James Gillison's drawing of the ancient Market Cross, Chichester

Translated by Stefan Marcinkiewicz

Distributed in:

Europe, Africa by
John Wiley & Sons Limited
Baffins Lane, Chichester, Sussex, England

Australia, New Zealand, South-east Asia by
JOHN WILEY & SONS AUSTRALASIA PTY LIMITED
1–7 Waterloo Road, North Ryde, N.S.W., Australia

Albania, Bulgaria, Chinese People's Republic, Czechoslovakia, Cuba, German Democratic Republic, Hungary, Korean People's Democratic Republic, Mongolia, Poland, Rumania, Socialist Republic of Vietnam, the U.S.S.R., Yugoslavia by
ARS POLONA
Krakowskie Przedmieście 7, 00-068 Warszawa, Poland

The Americas and all remaining areas by
Halsted Press: a division of
JOHN WILEY & SONS Inc.
605 Third Avenue, New York, N.Y. 10016, U.S.A.

Library of Congress Cataloging in Publication Data:
Galus, Zbigniew.
　　　Fundamentals of electrochemical analysis.
Translation of Teoretyczne podstawy
elektroanalizy chemicznej.
1.　Electrochemical analysis.　I. Title.
QD115.G3413　　　543′.087　　　76-5838
ISBN　0-470-15080-7　　(Halsted Press)
ISBN　85312-036-6　　(Ellis Horwood, Publisher)

© 1976 PWN—Polish Scientific Publishers, Warszawa

All rights reserved. No part of this publication may be reproduced, stored in retrieval system or transmitted in any form or by any means, electronic, mechanical, photocopying, recording or otherwise without prior permission of Państwowe Wydawnictwo Naukowe.

Printed in Poland by D.R.P.

Foreword to the English Edition

by G. F. REYNOLDS,
Department of Engineering & Cybernetics, University of Reading

It was a great pleasure to accept the invitation to edit and review critically the English translation of Professor Galus' book "Teoretyczne podstawy elektroanalizy chemicznej". It was evident that this book had made a major contribution to the literature of fundamental and theoretical electrochemistry and an edition in English was therefore highly desirable.

Professor Galus has concentrated his attention on polarography, stationary electrode voltammetry, chronopotentiometry and rotating disc electrode methods, but the theoretical and practical aspects discussed are of much wider significance and are applicable, generally, to the whole field of electrochemistry. Although primarily intended to present, discuss and elucidate the fundamentals and basic theory of electroanalytical chemistry, the book will have a wide appeal to practical analysts and electrochemists, as well as to theoreticians and physical chemists. The theory has been collected together from many authorities, discussed comparatively and elucidated, with the aid of more than 1300 mathematical equations and supported by more than 1200 references. The evidence relating to many important concepts is critically examined and important conclusions are presented. In addition, however, the practical aspects of the theory are given prominence, and the possibilities and limitations of practical exploitation in measurement, elucidation of reaction mechanisms, or analysis, are explained in detail.

A number of chapters and sections are of special interest and importance. In particular, reference should be made to the extensive use of diffusion in all its aspects, including diffusion controlled by transport and by charge transfer, the rate of electrode processes, kinetic parameters, catalytic electrode processes and electrode processes preceded, or followed, by chemical reactions. The final chapter on the theoretical aspects of some new polarographic methods is a useful indication of the lines on which the subject is progressing.

In editing the manuscript I have been at pains to make as few changes as possible and to preserve the author's style and enthusiasm, which transcends the translation. In general I have confined myself to stylistic emendation and to those changes necessary to improve clarity. Additions have been made only when it appeared necessary to augment the information given.

The book should commend itself to post-graduate students and research workers in electrochemistry and will be an important reference work for theoretical, physical and analytical chemists in the many fields in which these techniques have a place.

G. F. REYNOLDS
February 1976

Author's Preface

The rapid development of science in the modern world is accompanied by the appearance of new methods for the investigation of scientific problems. Old methods are revised and often rejected, but some of them, after years of neglect, become once more of interest to scientists. After modifications and introduction of improved apparatus they find a wide application in both theoretical and practical work.

The history of electroanalysis can be traced back to the classical electrogravimetric method, which was described in 1864. Although this method involved the flow of current through an electrolytic cell, no electrical measurements were made and results were calculated from gravimetric determinations of substances liberated at the electrodes. At the end of the last century microelectrodes which could be polarized to a required potential became available and, as a result, measurements of electric current during experiments became possible. This method did not find a wide application. Chronopotentiometry also was not recognized as an important analytical method although it had been described theoretically at the beginning of this century.

The rapid development of electroanalysis commenced in the third decade of the 20th century, following the introduction of polarography by Heyrovský. Although the importance of this method was quickly recognized by analytical and research chemists, its real development did not take place until about 30 years afterwards when many of the theoretical problems of polarography had been solved, mainly by the investigators of the Heyrovský school in Prague. Through their work the nature of phenomena observed during electrolysis at the electrode–electrolyte phase boundary became known and the results of this work stimulated research on new electroanalytical methods. As a result the theoretical principles of stationary electrode voltammetry, in which a stationary electrode is polarized with a potential linearly variable in time, were developed. Thanks to the theoretical research carried out by Levich, the rotating disc method became of importance, and due to the work of Gierst and Delahay and his collaborators at Baton Rouge, the interest in chronopotentiometry revived.

All these methods were applied to quantitative analysis of solutions.

The elucidation of various kinetic problems and the discovery of many rapid electrode and chemical reactions stimulated the search for methods suitable for investigation of such processes.

In the mid nineteen-fifties the potential pulse method was worked out. In this method a small potential disturbing the electrochemical equilibrium is rapidly applied to an electrode which is at equilibrium with the solution in which it is immersed. The resulting current is used as a measure of the electrode reaction rate. A modification of this technique, called the voltage pulse method, has also found practical application.

In addition to the methods in which the magnitude of the current is due to changes in electrode potential, a galvanostatic method was developed. In this method the equilibrium existing at the electrode is disturbed by a current pulse and the changes of the examined electrode potential are recorded. The difficulties connected with charging the double layer were partly overcome by means of the double pulse method. Soon after 1960 Reinmuth and Delahay introduced the coulostatic method.

In addition to the above, a method for the investigation of electrode processes kinetics, based on measurements of faradaic impedance, was under development after the end of the nineteen-forties. In recent years various variants of chronocoulometry have appeared and found practical application as auxiliary methods in studies of adsorption on electrodes.

Early in the nineteen-sixties Barker introduced a method of polarography using a square-wave alternating current. This was superior to the polarographic method using sinusoidal polarography which had been the subject of much work by Breyer at an earlier date. At approximately the same time as the pulse polarographic method, introduced by Barker and Gardner, came the faradaic rectification method, which has particular importance in kinetic studies, and the hanging drop electrode, developed mainly through the research carried out by Kemula and Kublik, and Shain and his group.

All methods mentioned above have become, or could become, the basis of volumetric analysis techniques similar to the well known amperometric titration method which was developed from classical polarography.

The coulometric methods called controlled potential coulometry and controlled current coulometry, including coulometric titration with constant current, are of considerable importance in analytic chemistry and research. In addition to these, another large group of electrometric techniques, consisting of potentiometric methods based on measurements of electro-

Author's Preface

motive forces of suitable cells, has been in use and the subject of development for many years. Conductometric methods are also of importance.

In my approach to this book on the electroanalytical methods I was faced with a difficult choice. Should I describe all the known methods briefly, or should I concentrate on some of them? I have chosen the second alternative and I have therefore concentrated on polarography, chronopotentiometry, stationary electrode voltammetry with linearly changing potential, and the rotating disc method. These methods were introduced into electroanalytical practice a relatively long time ago and they are in general use in analysis and research today. In my opinion an understanding of the theoretical principles of these four methods can facilitate the understanding of most other techniques.

This book is not intended as a collection of theoretical discussions of individual methods. I have tried to present the selected methods by comparing them in the context of a suitable problem. In the final conclusions I have tried to represent in the form of general equations the relationships describing the problems under discussion. I believe that this approach facilitates the understanding of the subject and the choice of a suitable method for the solution of a definite problem. This form of presentation of electroanalytical problems has been suggested by my experience gained in the course of lectures on electroanalytical techniques given in the Chemistry Department of Warsaw University.

I would like to thank Professor W. Kemula for encouragement and for discussions which helped me in the final selection of the form in which electroanalytical problems and methods are presented in this book.

Finally I would like to thank my wife Margaret who not only endured but also encouraged, assisted and inspired my work.

Zbigniew Galus

Preface to the English Edition

The book "Fundamentals of Electrochemical Analysis" appeared as a result of consideration of various problems, and experience gained in the course of monographic lectures on chemical electroanalysis given in the Chemistry Department of the University of Warsaw.

Electroanalytical methods are widely used in chemical analysis and in investigations of physical and chemical problems. Their development is particularly pronounced in the English speaking countries, where in recent years a rapid progress has been achieved, as can be seen from the lists of references given at the end of each Chapter of my book. For this reason I have accepted with great pleasure the proposition of translating my book into English and I hope that my method of presenting electroanalytical problems will find a favourable reception by the readers of the English edition.

During the preparation of the English edition I have tried to bring the book up to date by adding the material which was published after the end of the work on the Polish edition which appeared in 1971. Many of these additions and also certain changes in the presentation of the material have been suggested by my collaborators and friends from the Chemistry Department of Warsaw University, to whom I am deeply grateful.

I would also like to thank Dr. I. Ružić and the translator of the Russian edition of my book Mr. B. Ya. Kaplan for their valuable remarks and advice. Some of the changes have been introduced as a result of discussions with Prof. R. N. Adams who for two years directed my studies in his laboratory. I would also like to thank Prof. P. T. Kissinger, Prof. P. Zuman and Dr. S. W. Feldberg for discussions and advice. I am particularly grateful to the editors of the English edition of my book, Dr. G. F. Reynolds and Dr. R. A. Chalmers, for careful reading of the text and for many critical and valuable remarks, as well as to the translator of this book, Dr. S. Marcinkiewicz. I also wish to thank the authors and publishers who have so kindly given permission to reprint figures and tables.

Contents

Foreword to the English Edition V
Author's Preface VII
Preface to the English Edition XI

Chapter 1   Structure of the Electrode Double Layer 1
1.1   Structure in the Absence of Specific Adsorption 1
1.2   Structure in the Presence of Specific Adsorption 13
     References 21

**Chapter 2   Development and General Characterization of Electro-
analytical Techniques** 24
2.1   Chronoamperometry and Chronocoulometry 24
2.2   Polarography 26
2.3   Stationary Electrode Voltammetry 33
2.4   Rotating Disc Voltammetry 38
2.5   Chronopotentiometry 43
     References 47

Chapter 3   Rates of Electrode Processes 51
     References 76

Chapter 4   Diffusion to the Electrode 78
4.1   Linear Diffusion 82
4.2   Spherically Symmetrical Diffusion 83
4.3   Cylindrically Symmetrical Diffusion 86
4.4   Linear Diffusion to Expanding Drop Electrode 88
4.5   Convection Diffusion to Rotating Disc Electrode . . . 92
     References 94

**Chapter 5   Electrode Processes Controlled by Transport Rate:
Diffusion Currents** 95
5.1   Linear Diffusion Conditions 95

Chronoamperometry and Chronocoulometry 97
Chronopotentiometry 105
Stationary Electrode Voltammetry 110
Polarography 118
The Rotating Disc Method 124
Generalization of the Relationships 136
5.2 Spherical Diffusion Conditions 138
Chronoamperometry 139
Polarography: Corrections to the Ilkovič Equation 141
Stationary Electrode Voltammetry 150
Chronopotentiometry 153
General Consideration of Relationships 155
5.3 Cylindrical Diffusion Conditions 157
Chronoamperometry 157
Stationary Electrode Voltammetry 159
Chronopotentiometry 160
5.4 Limited Diffusion Field Conditions 164
Chronoamperometry 164
Stationary Electrode Voltammetry 166
Chronopotentiometry 169
5.5 Multi-Stage Electrode Processes 171
Polarography 172
Rotating Disc Method 172
Stationary Electrode Voltammetry 172
Chronopotentiometry 173
5.6 Electrode Processes Involving Several Depolarizers 174
Polarography 174
Stationary Electrode Voltammetry 174
Rotating Disc Method 175
Chronopotentiometry 175
References 176

Chapter 6 Electrode Processes Controlled by the Rate of Charge Transfer 183
6.1 Linear Diffusion Conditions 184
Chronoamperometry 185
Stationary Electrode Voltammetry 188
Chronopotentiometry 193
Polarography 196

		Rotating Disc Method	199
		Discussion of the Relationships Derived	203
	6.2	Spherically Symmetrical Diffusion Conditions	204
		Chronoamperometry	205
		Stationary Electrode Voltammetry	206
		Chronopotentiometry	208
	6.3	Cylindrically Symmetrical Diffusion Conditions	209
		References	209

Chapter 7 Equations of Electroanalytical Methods 211

7.1	Polarography		212
	Curves for Reversible Electrode Processes		212
	Wave Equations of Irreversible Electrode Processes		222
	Determination of Kinetic Parameters		224
7.2	Stationary Electrode Voltammetry		231
	Reversible Electrode Processes		231
	Irreversible Electrode Processes		235
	Determination of Kinetic Parameters		238
7.3	Chronopotentiometry		239
	Reversible Electrode Processes		239
	Irreversible Electrode Processes		242
	Determination of Kinetic Parameters		243
7.4	The Rotating Disc Method		245
	Reversible Electrode Processes		245
	Irreversible Electrode Processes		247
	Determination of Kinetic Parameters		249
	References		252

Chapter 8 Electrode Processes Preceded by First-Order Chemical Reactions 255

8.1	Chronoamperometry	262
8.2	Polarography	264
8.3	Stationary Electrode Voltammetry	268
8.4	Chronopotentiometry	275
8.5	Rotating Disc Method	278
8.6	General Discussion	282
8.7	Pseudo-First Order Chemical Reactions	285
8.8	Two First-Order Chemical Reactions	286
8.9	Effect of the Double Layer Structure on the Kinetics	288

8.10	Utility of the Methods for Kinetic Studies	292
	References	293

Chapter 9 Electrode Processes Followed by First-Order Chemical Reactions ... 296

9.1	Polarography	297
9.2	Stationary Electrode Voltammetry	299
9.3	Chronopotentiometry	303
9.4	The Rotating Disc Method	305
9.5	Generalization of the Relationships Discussed	306
9.6	Examples	308
	References	309

Chapter 10 Catalytic Electrode Processes ... 311

10.1	Chronoamperometry	313
10.2	Stationary Electrode Voltammetry	315
	Reversible Electrode Processes	315
	Irreversible Electrode Processes	319
10.3	Chronopotentiometry	321
10.4	Polarography	322
10.5	The Rotating Disc Method	324
10.6	General Discussion	325
10.7	Second-Order Catalytic Reactions	327
10.8	Examples of Catalytic Currents	328
	References	329

Chapter 11 Electrode Processes Preceded by Higher-Order Reactions ... 331

11.1	Polarography	332
11.2	Stationary Electrode Voltammetry	335
11.3	Chronopotentiometry	336
11.4	The Rotating Disc Method	337
	References	339

Chapter 12 Electrode Processes Followed by Dimerization ... 340

12.1	Polarography and Chronoamperometry	341
12.2	Stationary Electrode Voltammetry	344
12.3	Chronopotentiometry	347
12.4	The Rotating Disc Method	348
12.5	Discussion of General Relationships	349
	References	351

Contents XVII

Chapter 13 Electrode Processes Followed by Disproportionation . 353
13.1 Polarography 353
13.2 Stationary Electrode Voltammetry 354
13.3 Chronopotentiometry 357
13.4 The Rotating Disc Method 357
13.5 General Discussion 358
References 359

Chapter 14 Electroanalytical Investigations of Complexes 360
14.1 Reversibly Reducible Complexes 360
14.2 Irreversibly Reducible Complexes 369
14.3 The Mechanism of Reduction of Complex Compounds . 375
References 379

Chapter 15 ECE Processes 382
15.1 Chronoamperometry and Chronocoulometry 383
15.2 Polarography 386
15.3 Stationary Electrode Voltammetry 388
15.4 Chronopotentiometry 391
15.5 The Rotating Disc Method 393
References 394

Chapter 16 Influence of Adsorption on Electrode Processes . . . 396
16.1 Specifically Adsorbable Substances 396
Chronoamperometry and Chronocoulometry 397
Polarography 400
Stationary Electrode Voltammetry 407
Chronopotentiometry 411
16.2 Processes in the Presence of Adsorbed Electrode-Inactive Substances 413
References 417

Chapter 17 Cyclic Methods. Diffusion Processes 420
17.1 Cyclic Chronoamperometry and Double Step Chronocoulometry 422
17.2 Cyclic Stationary Electrode Voltammetry 427
17.3 Chronopotentiometry with Change of Direction of Current 434
17.4 Cyclic Chronopotentiometry 438

17.5	The Rotating Ring-Disc Electrode Method	440
	References	444

Chapter 18 Cyclic Methods. Kinetic Processes 446
- 18.1 Cyclic Chronoamperometry and Double Potential Step Chronocoulometry. 446
- 18.2 Cyclic Stationary Electrode Voltammetry 451
- 18.3 Current Reversal Chronopotentiometry. 454
- 18.4 The Rotating Ring-Disc Electrode Method 456
- References 458

Chapter 19 Deviations from Normal Depolarizer Transport . . . 459
- 19.1 Polarography 459
- 19.2 Stationary Electrode Voltammetry 466
- 19.3 Chronopotentiometry 467
- 19.4 The Rotating Disc Method 467
- References 468

Chapter 20 Some New Developments in Polarography 470
- 20.1 Principles of New Polarographic Methods 472
  - Alternating Current Sinusoidal Polarography. 473
  - Square-Wave Alternating Current Polarography . . . 476
  - Pulse Polarography 477
  - Triangular-Wave Alternating Current Polarography . . 479
- 20.2 Theoretical Aspects of New Polarographic Methods . . 480
- 20.3 Potential Usefulness of New Polarographic Methods . . 491
  - Investigation of the Kinetics of Electrode Processes and Accompanying Chemical Reactions 491
  - New Polarographic Techniques as Analytical Methods . 498
  - A.C. Polarography as a Method for Investigation of Double Layer Structure 501
- 20.4 Discussion of Some Other Electroanalytical Methods . . 504
- References 510

Index 515

Chapter 1

Structure of the Electrode Double Layer

Various kinds of electrode processes taking place under a variety of experimental conditions, have at least one property in common: they are electrolytic processes taking place in the layer of the solution adjacent to the electrodes. Therefore the discussion of these processes should start with the examination of this layer.

When a metal is introduced into a solution of an electrolyte, an electrical double layer appears at the phase boundary between the metal and the liquid. It consists of electrical charge present on the surface of the metal and of opposite electrical charges of the ions present in the immediate neighbourhood of the metal surface. The whole region of the phase boundary must be electrically neutral, which means that the total charge density excess on the liquid side of the double layer q_r must be neutralized by the numerically equal (but of opposite sign charge) density excess on the metal side q_M.

This condition of neutrality of the double layer can be expressed as:

$$-q_r = q_M. \tag{1.1}$$

We assume that the metal is an ideally polarized electrode, i.e. that the charged particles in practice do not cross the phase boundary and that changes in the electrode potential are equal to those in the externally applied potential. For example such an electrode is a mercury electrode in an oxygen-free supporting electrolyte solution. This has a finite range of potential values, limited on one side by oxidation of mercury and on the other by reduction of the cations of the supporting electrolyte.

1.1 Structure of the Electrode Double Layer in the Absence of Specific Adsorption

In order to discuss the distribution of ions in the electrical double layer it is necessary to propose a model of this layer and it is assumed that only electrostatic forces interact between the component of the solution and the electrode.

The first such model, proposed by Quincke [1] and Helmholtz [2], was a kind of flat condenser consisting of free ions present in the metal and oppositely charged ions present in the solution. In this model the potential drop across the double layer is linear (see Fig. 1.1). Theoretical considerations based on this model show some agreement with practical results. For instance the experimentally determined capacity of the double layer usually has a value between 20 and 40 $\mu F/cm^2$. Assuming that the distance between the plates of the condenser model is equal to the diameter of a water molecule and that the dielectric constant in the double layer in aqueous solutions is 10, the calculated values, obtained from the expression for the capacity of a flat condenser, are in fairly good agreement with the above experimental data. However, the Helmholtz theory does not explain the dependence of the capacity on electrolyte concentration and the dependence of the double layer capacity on the potential.

The Quincke and Helmholtz theories are nearer to reality in the case of concentrated solutions of electrolytes and large charges on electrodes: in such cases the potential drop across the double layer is similar to that shown in Fig. 1.1.

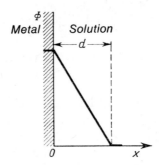

Fig. 1.1 The distribution of potential in the double layer according to the Helmholtz theory.

Quincke and Helmholtz considered electrostatic forces only, neglecting the thermal motion of ions. Their theory was further developed by Gouy [3, 4] and independently by Chapman [5], who showed that the thermal motion of ions should cause some diffusion of the double layer. According to this theory the concentration of ions in the double layer C_i^x can be described by the Boltzmann equation:

$$C_i^x = C_i^p \exp\left[-\frac{z_i F(\Phi_x - \Phi_r)}{RT}\right], \qquad (1.2)$$

1.1] Structure in the Absence of Specific Adsorption

where C_i^p is the concentration of ion i in the bulk of solution, z_i is the charge on this ion, Φ_x is the internal potential in the double layer at the distance x from the electrode, Φ_r is the internal potential in the bulk of solution, F and R are Faraday's constant and the gas constant, respectively, T is the absolute temperature.

Therefore the charge density ϱ in the layer is given by the relationship:

$$\varrho_x = \sum_i z_i F C_i^x = \sum_i z_i F C_i^p \exp\left[-\frac{z_i F(\Phi_x - \Phi_r)}{RT}\right]. \quad (1.3)$$

In the case of a large flat electrode it can be assumed that the charge density in a plane parallel to the electrode is constant, and consequently the general Poisson equation can be written in the form:

$$-\frac{\partial}{\partial x}\left[D_x^0 \frac{\partial \Phi_x}{\partial x}\right] = 4\pi \varrho_x, \quad (1.4)$$

where D_x^0 is the dielectric constant at distance x from the electrode.

Substituting this expression for ϱ_x in expression (1.4) we obtain:

$$\frac{\partial}{\partial x}\left[D_x^0 \frac{\partial \Phi_x}{\partial x}\right] = -4\pi F \sum_i z_i C_i^p \exp\left[-\frac{z_i F(\Phi_x - \Phi_r)}{RT}\right]. \quad (1.5)$$

Assuming that the dielectric constant has the average constant value D_s^0 in the region between x and the bulk of solution, and taking into account the identity:

$$\frac{\partial^2 \Phi_x}{\partial x^2} \equiv (1/2)\frac{\partial}{\partial x}\left(\frac{\partial \Phi_x}{\partial x}\right)^2 \quad (1.6)$$

we obtain the relationship:

$$\frac{\partial}{\partial \Phi_x}\left(\frac{\partial \Phi_x}{\partial x}\right)_x^2 = \frac{8\pi F}{D_s^0} \sum_i z_i C_i^p \exp\left(-\frac{z_i F(\Phi_x - \Phi_r)}{RT}\right), \quad (1.7)$$

which on integration gives:

$$\left(\frac{\partial \Phi_x}{\partial x}\right)_x = \pm\left[\frac{8\pi RT}{D_s^0}\sum_i C_i^p\left\{\exp\left[-\frac{z_i F(\Phi_x - \Phi_r)}{RT}\right] - 1\right\}\right]^{1/2}. \quad (1.8)$$

In the derivation of this equation, it has been assumed that $(\Phi_x - \Phi_r)$ and $\partial \Phi_x/\partial x$ tend to zero as $x \to \infty$.

If x_2 is the distance between the ions and the electrode at maximum approach, then according to the Gauss theorem the electrode surface charge density q_M is given by the relationship:

$$4\pi q_M = -D^0_{x_2}\left(\frac{d\Phi_x}{dx}\right)_{x=x_2}. \tag{1.9}$$

For $x = x_2$ equation (1.8) can be re-written in the following form:

$$\pm D^0_{x_2}\left[\frac{8\pi RT}{D^0_s}\sum_i C^p_i\left\{\exp\left[-\frac{z_i F(\Phi_2-\Phi_r)}{RT}\right]-1\right\}\right]^{1/2} = -D^0_{x_2}\left(\frac{\partial \Phi_x}{\partial x}\right)_{x=x_2}, \tag{1.10}$$

where $\Phi_2 = \Phi_x$ for $x = x_2$.

Equating the left-hand sides of equations (1.9) and (1.10) we obtain the expression for the charge of the electrode:

$$q_M = \pm\left[\frac{RTD^0_s}{2\pi}\sum_i C^p_i\left\{\exp\left[-\frac{z_i F(\Phi_2-\Phi_r)}{RT}\right]-1\right\}\right]^{1/2}, \tag{1.11}$$

from which q_M can be calculated when the potential difference $\Phi_2 - \Phi_r$ is known.

In the case of a solution containing only an electrolyte of the $z-z$ type a simplification may be achieved, since we now have $z_c = z_a = z$ and $C^p_c = C^p_a = C^p$, where indices c and a correspond to cations and anions, respectively. Thus:

$$\left[\sum_i C^p_i\left\{\exp\left[-\frac{z_i F(\Phi_2-\Phi_r)}{RT}\right]-1\right\}\right]^{1/2} =$$

$$= \left[C^p\left\{\exp\left[\frac{zF(\Phi_2-\Phi_r)}{RT}\right]+\exp\left[-\frac{zF(\Phi_2-\Phi_r)}{RT}\right]-2\right\}\right]^{1/2} =$$

$$= (C^p)^{1/2}\left\{\exp\left[\frac{zF(\Phi_2-\Phi_r)}{2RT}\right]-\exp\left[\frac{-zF(\Phi_2-\Phi_r)}{2RT}\right]\right\} =$$

$$= 2(C^p)^{1/2}\sinh\left[\frac{zF(\Phi_2-\Phi_r)}{2RT}\right]. \tag{1.12}$$

From equations (1.11) and (1.12) we obtain:

$$q_M = \left(\frac{2RTC^p D^0_s}{\pi}\right)^{1/2}\sinh\left[\frac{zF(\Phi_2-\Phi_r)}{2RT}\right]. \tag{1.13}$$

Using equations (1.13) and (1.9) and assuming that $D^0_{x_2} = D^0_s$ it is possible to derive an expression for the dependence of potential on the

1.1] Structure in the Absence of Specific Adsorption

distance from the electrode:

$$\left(\frac{\partial \Phi_x}{\partial x}\right)_{x_2} = -\left(\frac{32\pi RTC^p}{D_s^0}\right)^{1/2} \sinh\left[\frac{zF(\Phi_2-\Phi_r)}{2RT}\right]. \quad (1.14)$$

This equation gives the dependence of the field intensity at the inner border of the diffuse layer, on the potential in the plane Φ_2.

An analogous equation can be written for the electrical field intensity at any given point in the diffuse double layer:

$$\left(\frac{\partial \Phi_x}{\partial x}\right)_x = -\left(\frac{32\pi RTC^p}{D_s^0}\right)^{1/2} \sinh\left[\frac{zF(\Phi_x-\Phi_r)}{2RT}\right]. \quad (1.15)$$

Equation (1.15) can also be written in the form:

$$\left(\frac{\partial \Phi_x}{\partial x}\right)_x = -\frac{2RTk}{zF} \sinh\left[\frac{zF(\Phi_x-\Phi_r)}{2RT}\right], \quad (1.16)$$

where

$$k = \left[\frac{8\pi z^2 F^2 C^p}{D_s^0 RT}\right]^{1/2}. \quad (1.17)$$

Transformation and integration of equation (1.16) leads to the relationship:

$$-kx = \ln\tanh\left[\frac{zF(\Phi_x-\Phi_r)}{4RT}\right] - a. \quad (1.18)$$

Hence the potential at any given point in the double layer is given by the equation:

$$\Phi_x - \Phi_r = \frac{4RT}{zF} \tanh^{-1}[\exp(a-kx)]. \quad (1.19)$$

The integration constant a can be determined if it is assumed that $\Phi_x = \Phi_2$ when $x = x_2$. Then we obtain from equation (1.18):

$$a = \ln\tanh\left[\frac{zF(\Phi_2-\Phi_r)}{4RT}\right] + kx_2. \quad (1.20)$$

Introducing this expression for a into equation (1.19) we obtain an expression for dependence of potential on the distance from the electrode. For large values of x the exponential factor of equation (1.20) is small and in such cases the expression $\tanh^{-1}[\exp(a-kx)]$ approaches the expression $\exp(a-kx)$. Thus the distribution of potential has exponential character. The work of Parsons [6] has been used as a basis from which the above equations have been taken.

According to the theory of Gouy and Chapman the potential distribution can be expressed schematically as in Fig. 1.2.

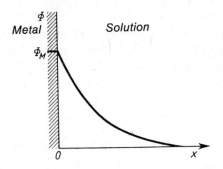

Fig. 1.2 The distribution of potential in the double layer according to the Gouy and Chapman theory.

Using expression (1.13) and the Lippmann equation we have:

$$-q_M = \left(\frac{\partial \sigma}{\partial E}\right)_{T,p,c_i}, \quad (1.21)$$

where E is the electrode potential with respect to any reference electrode. The equation is valid when the temperature, the pressure and the composition of the solution are constant. It is possible to obtain a relationship for the surface tension σ:

$$\sigma = \text{const} - \frac{4RT}{zF}\left(\frac{D_s^0 RTC^p}{2\pi}\right)^{1/2} \cosh\left[\frac{zF(\Phi_2-\Phi_r)}{2RT}\right]. \quad (1.22)$$

By differentiating equation (1.13) with respect to $(\Phi_2-\Phi_r)$ we obtain the differential capacity (C_d):

$$C_d = \frac{dq_M}{d(\Phi_2-\Phi_r)} = \frac{zF}{RT}\left(\frac{D_s^0 RTC^p}{2\pi}\right)^{1/2} \cosh\left[\frac{zF(\Phi_2-\Phi_r)}{2RT}\right]. \quad (1.23)$$

According to Gouy and Chapman the potential Φ_2 can be the same as that of the metal. This implies that the ions are point charges, and that the distance of their maximum approach to the electrode is not limited. If this assumption were correct, the differential capacity of the double layer calculated from equation (1.23) should be in agreement with results of experiments giving the values determined as $dq_M/d(\Phi_M-\Phi_r)$, where Φ_M is the inner potential of the electrode metal. However, considerable differences are observed, indicating that the double layer model proposed by Gouy and Chapman is not fully satisfactory. A further shortcoming

1.1] Structure in the Absence of Specific Adsorption 7

of this model is that it does not take into account the specific adsorption of ions on the electrode.

The theory proposed by Stern [7] in 1924 was free from these disadvantages. According to Stern the model proposed by Quincke and Helmholtz would be valid at a temperature of absolute zero at which the thermal motion has ceased. At higher temperatures some of the ions pass from the electrode surface to the diffuse region of the double layer. Stern suggested that the double layer should be divided into a rigid part having a thickness depending on the ionic radius and a diffuse part extending towards the bulk of solution from a plane situated at the distance x_2 from the electrode. This plane is called the outer Helmholtz plane. According to the Stern theory the potential drop across the double layer can be represented schematically as in Fig. 1.3.

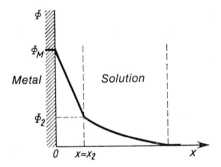

Fig. 1.3 The distribution of potential in the double layer according to the Stern theory.

According to the Stern model the potential difference between the electrode and the solution can be expressed as:

$$\Phi_M - \Phi_r = (\Phi_M - \Phi_2) + (\Phi_2 - \Phi_r). \tag{1.24}$$

Thus the total potential drop consists of the potential drop in the inner part of the double layer $(\Phi_M - \Phi_2)$ and in the diffuse layer $(\Phi_2 - \Phi_r)$.

Differentiating equation (1.24) with respect to the charge of the electrode we obtain the equation

$$\frac{d(\Phi_M - \Phi_r)}{dq_M} = \frac{d(\Phi_M - \Phi_2)}{dq_M} + \frac{d(\Phi_2 - \Phi_r)}{dq_M}, \tag{1.25}$$

which can also be written in the form:

$$\frac{1}{C_t} = \frac{1}{C_r} - \frac{1}{C_d}, \tag{1.26}$$

where C_t is the total capacity of the double layer, C_r is the capacity of the rigid layer, and C_d is the capacity of the diffuse layer.

The differential capacity of the double layer is equal to the total capacity of the two parts of the double layer connected in series:

$$C_t = \frac{C_r C_d}{C_r + C_d}. \tag{1.27}$$

When $C_r \gg C_d$ we have:

$$C_t \cong C_d = \frac{zF}{RT} \left(\frac{D_s^0 RTC^p}{2\pi} \right)^{1/2} \cosh \left[-\frac{zF(\Phi_2 - \Phi_r)}{2RT} \right] \tag{1.28}$$

[see equation (1.23)].

When $C_r \gg C_d$ we have:

$$C_t \cong C_r. \tag{1.29}$$

The theory of Gouy and Chapman enables only the capacity of the diffuse layer to be calculated. This is the main cause of differences between results of calculations based on this theory and experimental results.

In order to check the validity of the theory proposed by Gouy and Chapman, as modified by Stern, it is necessary to determine the value of C_r. This cannot be done experimentally. The validity of Stern's assumption was proved by Frumkin [8] in 1940, but a stricter proof was obtained several years later by Grahame [9], who assumed that the capacity of the rigid layer depends only on the charge of the electrode and not on the electrolyte concentration. Unlike Frumkin, he did not assign a constant value to C_r. Instead he calculated it from the experimentally determined value of C_t and the value of C_d calculated on the basis of the Gouy and Chapman equation [(1.28)].

Experiments carried out by Grahame [10, 11] showed that the contribution of the diffuse layer capacity to the total diffusion is small, particularly in the case of potentials considerably different from the zero charge potential, and that the rigid layer potential largely depends on the charge of the electrode. The assumption that it is independent of the charge could lead to serious errors. The curves in Fig. 1.4 show that the agreement between the calculated and the experimentally determined capacity values was good, although at positive potentials certain differences were observed.

At low electrolyte concentrations a minimum appears on the curve of potential against differential capacity. It is not observed in the case of concentrated solutions. Its potential is equal to that of zero charge potential for symmetrical electrolytes. The differential capacity curve obtained for

0.001 M NaF with a pronounced capacity minimum at the potential -0.5 V is shown in Fig. 1.5. The minimum is due to that fact that in these cases the total capacity is determined mainly by the diffuse layer capacity.

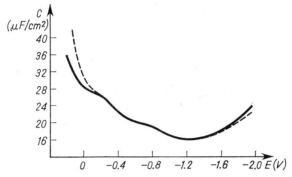

Fig. 1.4 Experimental (full line) and theoretical (broken line) curves of the dependence of the differential capacity of a mercury electrode in 0.1 M NaF at 25°C on the electrode potential (from [10] by permission of the copyright holders, the American Chemical Society).

The modified theory of Gouy and Chapman can be checked also by another method, based on the plot of $1/C_t$, determined experimentally, against $1/C_d$ which can be calculated. According to equation (1.26), at constant surface charge, the plot should be linear and its slope should be equal

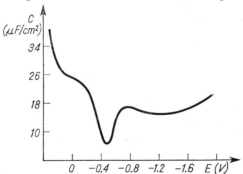

Fig. 1.5 Experimental curve of the dependence of the differential capacity of a mercury electrode in 0.001 M NaF on the electrode potential (from [10] by permission of the copyright holders, the American Chemical Society).

to one. The value of $1/C_r$ is obtained from the intercept on the $1/C_t$ axis. It is a condition of linearity that the system must not contain substances which are specifically adsorbed on the electrode.

The results of the above discussion can be used for calculating potential Φ_2 from experimental data. This potential is described by a relationship which can be readily derived from equation (1.19) or (1.13).

According to equation (1.13), at a constant charge and at potentials considerably different from the zero charge potential, the potential Φ_2 is given by the equation:

$$\Phi_2 = \mathrm{const} \pm \frac{RT}{zF} \ln C^p, \qquad (1.30)$$

where the signs $+$ and $-$ correspond to $\Phi_2 < 0$ and $\Phi_2 > 0$, respectively; it is assumed that $\Phi_r = 0$. The dependence of Φ_2 on electrode potential at several supporting electrolyte concentrations is illustrated by Fig. 1.6.

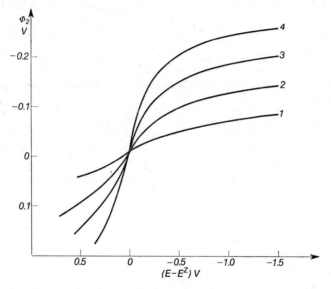

Fig. 1.6 Dependence of Φ_2 on electrode potential referred to the zero charge potential. Mercury electrode at 25°C in NaF solutions: curve 1—1.0 M; 2 — 0.1 M; 3 — 0.01 M and 4—0.001 M NaF (from [46] by permission of the copyright holders, Interscience Publishers).

Equation (1.30) was used earlier by Frumkin [12] for explaining the relation between the kinetics of hydrogen ion discharge on mercury and the structure of the double layer.

The theories of the double layer structure discussed above were later modified by various investigators. One objective of their work was the need to take into account the dependence of the dielectric constant of the medium, in which the double layer is formed, on the electrical field intensity and on the nature of ions present in the solution. It could be expected that the electrical field of the double layer would be significant, since it can be readily shown that it is very large in the zone adjacent to the electrode.

For the zone of an aqueous solution of an electrolyte at 25°C situated at the distance x_2 from the electrode equation (1.15) can be written as follows:

$$\left(\frac{\partial \Phi_x}{\partial x_x}\right)_{x=x_2} = -1.44 \times 10^5 q_M \text{ V/cm}, \quad (1.31)$$

where q_M is expressed in microcoulombs per cm².

Introducing into the equation $q_M = 10$ μC/cm² (this electrode charge is often observed) we come to the conclusion that the electrical field intensity in the double layer can readily exceed 10^6 V/cm.

In the modification of the Gouy and Chapman theory the finite volume of the ions present in the double layer was also taken into account.

In his studies on these problems Sparnaay [13], instead of solving the Poisson and Boltzmann equation, solved an equation involving the dependence of the dielectric constant on the electrical field intensity. Brodowsky and Strehlow [14] took into account the dependence of the dielectric constant on the electrolyte concentration, in addition to the dielectric saturation. Their work was based on the Wicke and Eigen [15] equation.

In recent years the structure of the inner layer has been studied by many authors, but knowledge of this structure and forces acting in this area is still far from complete. In several publications [16–25] various structures have been suggested for the layer formed at the mercury electrode–electrolyte solution phase boundary. The authors of these papers have usually assumed two possible orientations of water dipoles in the inner layer. Levine et al. [26] pointed out that the principal fault of this approach is either the neglect [16–18], or the inadequate account [19–25] of the lateral interaction of dipoles in the inner region. Another unsatisfactory feature is the use of the so-called effective dielectric constant.

Bockris and co-workers [24] used, for example, $D = 6$, whereas Macdonald and Barlow [19] stressed that a proper microscopic theory requires no such factor.

The nature of the humps on the capacity curves of the inner region and the corresponding orientation of water dipoles at the mercury–electrolyte solution interface is still attracting the attention of investigators. Hills and Payne [27] have proposed a dielectric model of the inner region which differs significantly from those of previous authors. On the basis of changes in excess surface entropy and volume they have come to the conclusion that the humps are due to adsorption of water. As a result of this adsorption the thickness of the inner region increases when the electrode charge becomes more positive. It was suggested that the hump observed at an electrode charge of about $4\,\mu C/cm^2$ corresponds to a maximum ratio of the inner region dielectric constant to the thickness of this region. Certain results of this work are however still open to doubt. This has been pointed out by Levine and co-workers [26] who developed a statistically-based theory describing the inner region capacity in cases where specific adsorption is absent. Their model of the region is a plane monolayer of polarizable water dipoles at fixed surface density. These are sandwiched *in vacuo* between two continuous media, the aqueous phase and the mercury phase, which carries a surface charge.

Two orientation states are assigned to the water dipoles, with the magnitudes of the dipole moment components perpendicular to the surface differing in the two states. The dipoles are situated at each site on a two-dimensional hexagonal lattice and random distribution over the sites of dipoles in the two states is assumed.

Dipole–dipole interactions and also dipole imaging, due to the two continuous media, were considered.

The coordination number for nearest-neighbour interactions in a plane hexagonal lattice is normally 6, but as a result of the appreciable dipole–dipole interactions at larger than nearest-neighbour distances, and also as a result of imaging, the effective coordination number reaches a value of about 15. In consequence the mean dipole moment per molecule is very much reduced and there is no need to introduce an effective dielectric constant to diminish the potential drop across the dipole layer to a "reasonable" value.

Fairly good values of capacities and their temperature coefficients are obtained in the neighbourhood of the zero charge potential.

Krylov [28] proposed a statistical theory of the double layer and introduced corrections to the classical theory.

The Gouy and Chapman theory is valid at low electrolyte concentration, but at higher concentration the drop in potential with increasing distance from the electrode is steeper than that indicated by classical theory. The thickness of the diffusion layer is thus negligible in comparison with that of the rigid layer.

Hurwitz, Sanfeld and Steinchen-Sanfeld [29] made a new approach to the treatment of the above factors affecting the Gouy and Chapman theory. This was based on the work of Prigogine, Mazur and Defay [30]. They also took into account the effect of the electric field on the acidity of the substance [31]. Nürnberg [32, 33] produced a further solution to the problem, while Jenard and Hurwitz [34] discussed this in relation to ionic association in the double layer field.

1.2 Structure of the Double Layer in the Presence of Specific Adsorption

Other forces, in addition to electrostatic interactions, can exist at the electrode–electrolyte interface. As a result, non-electrostatic adsorption of ions or molecules on the electrode can take place. This adsorption is often referred to as specific adsorption. The nature of forces connected with this phenomenon has been interpreted in various ways by a number of authors.

Grahame [35] postulated that formation of covalent bonds between adsorbed anions and the mercury electrode plane is of primary importance. However, Bockris and co-workers [24] pointed out that this postulate does not explain the fact that, in the case of halide ions, the specific adsorption on mercury increases with increasing atomic number, whereas the opposite relationship should be observed. Therefore they assigned the principal role to the type of hydration of the adsorbed species. They assumed that adsorption decreases with increasing "primary" hydration. This hypothesis explains the adsorption of halide ions, and the fact that in the alkali metals group only caesium ions are adsorbed on mercury electrodes.

Andersen and Bockris [36] postulated that the adsorption of halide and alkali metal ions is of the physical type, and Levine, Bell and Calvert [37] assigned the principal role in the specific adsorption to recoil forces. Specific adsorption has been observed in the case of inorganic anions [3, 4, 38, 39] and certain cations [40, 41].

Proof for the existence of specific adsorption of ions is the dependence of zero charge potential, measured with respect to a standard reference electrode, on the concentration of a surface active electrolyte. The effect of adsorption on the zero charge potential is reflected in the shapes of electrocapillary curves and the results of capacity determinations. Since the positive parts of electrocapillary curves show a decrease of surface tension resulting from adsorption of anions, which are desorbed at negative potentials, the shift of the zero charge potential in the negative direction increases with increasing specific interaction between anions and the electrode. This is shown in Table 1.1, in which the potentials are given with respect to the normal calomel electrode. Esin and Markov [43] found that zero charge potentials measured with respect to a standard reference electrode increase linearly with increasing log of electrolyte activity, or with increasing concentration of the adsorbate. This phenomenon, called the Esin and Markov effect, is a good criterion of existence of specific adsorption.

Table 1.1 ZERO CHARGE POTENTIALS E^0 OF MERCURY IN VARIOUS ELECTROLYTES (FROM [42] BY PERMISSION OF THE COPYRIGHT HOLDERS, THE AMERICAN CHEMICAL SOCIETY)

Electrolyte	E (V)	Electrolyte	E (V)
0.1 M NaF	−0.472	0.01 M KCN	−0.582
1.0 M NaF	−0.472	0.1 M KCN	−0.684
0.1 M KF	−0.471	1.0 M KCN	−0.790
0.001 M KCl	−0.474	0.1 M KSCN	−0.626
0.01 M KCl	−0.485	1.0 M CsCl	−0.556
0.1 M KCl	−0.505	0.1 M KNO$_3$	−0.516
1.0 M KCl	−0.555	0.1 M KOH	−0.476
0.1 M KBr	−0.573	0.05 M K$_2$CO$_3$	−0.473
0.1 M KI	−0.731	0.05 M K$_2$SO$_4$	−0.470
1.0 M KI	−0.830	0.1 M CH$_3$COOK	−0.488

The potential shift accompanying a change of activity of non-adsorbable electrolyte has been expressed by Parsons [44] by means of the following relationship:

$$\left(\frac{\partial E^+}{\partial \ln a_\pm^2}\right)_{T, p, q_M} = -\frac{RT}{2zF}\left\{1 + \frac{q_M/2A}{[1+(q_M/2A)^2]^{1/2}}\right\}, \quad (1.32)$$

where $A = (RTD^0C^p)^{1/2}$, $a_\pm$ is the activity of the electrolyte, E^+ is the electrode potential measured with respect to a reference electrode, which is reversible with respect to the cation.

It is possible to obtain simpler relationships from equation (1.32) which are also valid in the cases when specific adsorption is absent.

When $q_M \gg 1$, (1.32) becomes:

$$\left(\frac{\partial E^+}{\partial \ln a_\pm^2}\right)_{T, p, q_M} = -\frac{RT}{zF}. \quad (1.33)$$

When $q_M = 0$, we have:

$$\left(\frac{\partial E^+}{\partial \ln a_\pm^2}\right)_{q_M=0} = -\frac{RT}{2zF}, \tag{1.34}$$

and when $q_M/2A \ll -1$ we obtain:

$$\left(\frac{\partial E^+}{\partial \ln a_\pm^2}\right)_{q_M} = 0. \tag{1.35}$$

Parsons [44] compared the results obtained by means of equation (1.32) with those obtained experimentally by Grahame [10] for sodium fluoride solutions and found that the agreement was good.

Thus the Esin and Markov coefficient described by equations (1.32)–(1.35) can be used as a criterion of specific adsorption. When an ideally polarized electrode in a symmetrical electrolyte z–z gives good agreement with equation (1.32) we have the proof that specific adsorption is absent.

It can be shown that zero charge potential measured with respect to a standard reference electrode is independent of the concentration of an electrolyte consisting of non-adsorbable ions.

Since:

$$dE_r = dE^\pm + \frac{RT}{2zF} d\ln a_\pm^2, \tag{1.36}$$

we have:

$$\left(\frac{\partial E_r}{\partial \ln a_\pm^2}\right)_{q_M} = \left(\frac{\partial E^\pm}{\partial \ln a_\pm^2}\right)_{q_M} + \frac{RT}{2zF}. \tag{1.37}$$

Combining this equation with equation (1.34) we obtain:

$$\left(\frac{\partial E_r}{\partial \ln a_\pm^2}\right)_{q_M=0} = \frac{RT}{2zF} - \frac{RT}{2zF} = 0. \tag{1.38}$$

In the case of specific adsorption of anions the Esin and Markov coefficient is given by the relationship:

$$\left(\frac{\partial E^+}{\partial \ln a_\pm^2}\right)_{q_M} = \frac{RT}{zF}\left\{\left(\frac{\partial q_1}{\partial q_M}\right)_{a_\pm} + \left(\frac{\partial q_1}{\partial q_d}\right)_{a_\pm}\left(\frac{\partial q_d}{\partial q_M}\right)_{a_\pm}\right\}, \tag{1.39}$$

where q_d is the total charge in the diffusion layer and q_1 is the charge resulting from the presence of specifically adsorbed anions in the layer.

Analysis of this equation and comparison with Parsons' experimental data [44] show that in the case of strong specific adsorption the Esin and Markov coefficient does not depend on charge density.

Esin and Markov [43] explained that a shift of zero charge potential considerably larger than that expected on the basis of increase of concentration of specifically adsorbable electrolyte is due to the structure of the double layer, which consists of layers of discrete charges. This model was elaborated by Esin and Shikhov [45] and was discussed by Parsons [46].

In the case of specific adsorption of cations the zero charge potential is shifted in the direction of positive values. The existence of specific adsorption of certain cations was demonstrated by Frumkin and co-workers [40, 41, 47], who studied the adsorption of thallium and caesium ions.

In order to explain the observed facts it is necessary to obtain a suitable model of the double layer for the case of specific adsorption on the electrode. The first such model was described by Stern [7], but it was not satisfactory and therefore Ershler [48] proposed an improved model which took into account the effect of discrete charges in the adsorbed ions layer, suggested by Frumkin [12] and discussed by Esin and Shikhov [45]. Ershler assumed continuous distribution of the charge in the diffuse layer, which implies an infinite number of recoils for each ion on both sides of this layer at equipotentiality of the metal surface. Various other models proposed by later workers [9, 37, 49, 50, 51] can be regarded as modifications of the Ershler model. Here we will describe briefly one of them, proposed by Grahame [9].

Let us assume that ions which are not specifically adsorbable are at the distance x_2 from the electrode surface. Then the surface active substances approach the electrode up to the distance x_1, which is smaller than x_2. In this case the diffuse layer of the Stern model remains unchanged, but the inner layer is divided into two parts. The part situated nearer the electrode is separated from the rest of the solution by the inner Helmholtz plane at potential Φ_1 and occupied by specifically adsorbable ions. The other part is separated from the diffuse layer by the outer Helmholtz plane at potential Φ_2. The double layer model taking into account the specific adsorption is shown in Fig. 1.7.

The total potential difference across the inner layer $(\Phi_M - \Phi_2)$ depends on the electrode charge and the number of specifically adsorbed ions. It can be formally and generally described [52, 53] by the following equation:

$$\Phi_M - \Phi_2 = {}_{q_1}(\Phi_M - \Phi_2) + {}_{q_M}(\Phi_M - \Phi_2). \tag{1.40}$$

The first term of the right hand side of equation (1.40) determines the potential difference between the metal and the outer Helmholtz plane due to the presence of charge q_1 of the specifically adsorbable anions. The second term of this equation describes the potential drop in the layer due to the presence of charge q_M on the metal. It is assumed that the charge on the metal causes the appearance of an equal charge with opposite sign in the outer Helmholtz plane.

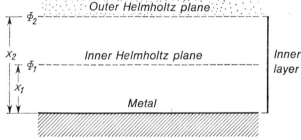

Fig. 1.7 Scheme of the double layer on the boundary between the electrode and the solution.

According to Parsons [46], if the metal surface and the outer Helmholtz plane is considered as a flat condenser, the components of equation (1.40) can be expressed by means of the following relationships:

$$_{q_1}(\Phi_M - \Phi_2) = \frac{4\pi q_1(x_2 - x_1)}{D_s^0} \tag{1.41}$$

and

$$_{q_M}(\Phi_M - \Phi_2) = \frac{4\pi q_M x_2}{D_s^0}. \tag{1.42}$$

From equations (1.41) and (1.42) it is possible to derive expressions describing two partial capacities of the inner layer. The first is connected with the change of the charge on the metal, accompanied by retention of a constant number of adsorbed ions:

$$_{q_M}C_r = \left[\frac{\partial q_M}{\partial(\Phi_M - \Phi_2)}\right]_{q_1} = \frac{dq_M}{d_{q_M}(\Phi_M - \Phi_2)}. \tag{1.43}$$

The second partial capacity of the inner layer is connected with the change of charge q_1 accompanied by the retention of charge q_M on the metal:

$$_{q_1}C_r = \left[\frac{\partial q_1}{\partial(\Phi_M-\Phi_2)}\right]_{q_M} = \frac{dq_1}{d_{q_1}(\Phi_M-\Phi_2)}. \qquad (1.44)$$

Charge q_1 and q_M can be determined experimentally and hence the capacities $_{q_M}C_r$ and $_{q_1}C_r$ can be calculated.

Thus the components of the potential difference $\Phi_M-\Phi_2$ described by equations (1.41) and (1.42) can be calculated from the relationships:

$$_{q_M}(\Phi_M-\Phi_2) = \int_0^{q_M} (_{q_M}C_r)^{-1}dq_M \qquad (1.45)$$

and

$$_{q_1}(\Phi_M-\Phi_2) = \int_0^{q_1} (_{q_1}C_r)^{-1}dq_1; \qquad (1.46)$$

the former is valid for a constant number of adsorbed ions and the latter for a constant charge of the electrode.

The methods of calculation of potential Φ_1, corresponding to the inner Helmholtz plane, have also been elaborated.

The relationships:

$$\frac{x_2-x_1}{x_2} = \frac{_{q_1}(\Phi_M-\Phi_2)q_M}{_{q_M}(\Phi_M-\Phi_2)q_1} \qquad (1.47)$$

or

$$\frac{x_2-x_1}{x_2} = \frac{_{q_M}C_r}{_{q_1}C_r} \qquad (1.48)$$

can be derived from equations (1.41) and (1.42), assuming that the dielectric constant in the inner layer remains unchanged. The right-hand sides of these equations can be evaluated. Since the ratio $(x_2-x_1)/x_2$ is connected with $_{q_1}(\Phi_1-\Phi_2)$ by means of the equation:

$$\frac{x_2-x_1}{x_2} = \frac{_{q_1}(\Phi_1-\Phi_2)}{_{q_1}(\Phi_M-\Phi_2)}, \qquad (1.49)$$

the potential Φ_1 can be calculated according to Parry and Parsons' proposition [54] by factorization of the difference $\Phi_1-\Phi_2$ into two components depending on q_1 and q_M:

$$\Phi_1-\Phi_2 = {_{q_M}(\Phi_M-\Phi_2)}\frac{x_2-x_1}{x_2} + {_{q_1}(\Phi_M-\Phi_2)}\lambda \qquad (1.50)$$

where λ determines the contribution of $(\Phi_1-\Phi_2)$ to the total potential difference ($\Phi_M - \Phi_2$). According to this expression the condition $0 < \lambda < 1$ is of course retained. For small values of q_1 it can be assumed that λ is approximately equal to $(x_2 - x_1)/x_2$, since the assumption of the linear distribution of potential is justified. For large values of q_1, $\lambda > (x_2 - x_1)/x_2$.

The relationship (1.50) can also be expressed in the following form:

$$\Phi_1 = \frac{q_M + \lambda q_1}{q_1 C_r} + \Phi_2. \qquad (1.51)$$

The potential Φ_2 can be expressed as a function of q_M, and hence Φ_2 is a function of q_1 and q_M.

The discussion has so far been limited to the structure of the double layer in the case of ionic adsorption on the surface of the electrode. In the case of adsorption of neutral substances maximum interaction with the electrode usually takes place at the zero charge potential. An increase of the potential (irrespective of its sign) is accompanied by desorption.

The shapes of experimental curves, electrocapillary curves or graphs of the dependence of the capacity on the potential, depend on the isotherm describing the dependence of the surface concentration on the concentration in the bulk of solution and also on the character of the dependence of the isotherm parameters on the electrode potential. In the case of electrocapillary curves there is a significant decrease in surface tension, as compared with that observed in the absence of surface active substances.

Various isotherms have been used for analysis of adsorption on the phase boundary between mercury electrodes and electrolyte solutions. In the case of very small surface concentrations of adsorbable ions or molecules, the Henry isotherm, showing the linear relation between the surface concentration (Γ) and the adsorbate concentration in the bulk of solution, is valid:

$$\Gamma = \beta C, \qquad (1.52)$$

where β is the adsorption equilibrium constant.

In this case the interactions between the adsorbed molecules are neglected. They are also neglected by the Langmuir isotherm

$$\frac{\Theta}{1 - \Theta} = \beta C, \qquad (1.53)$$

which, however, implies that the number of adsorption sites for the ions is limited. In the Langmuir isotherm Θ is the degree of surface saturation defined as the ratio $\frac{\Gamma}{\Gamma_M}$ where Γ_M is the maximum value of Γ corresponding to the maximum number of molecules or ions completely covering one square centimetre of the surface.

Frumkin modified the Langmuir isotherm by introducing a constant corresponding to the interaction between the adsorbed species:

$$\frac{\Theta}{1-\Theta} = \beta C \exp(2a\Theta), \qquad (1.54)$$

where a is the interaction constant. When $a = 0$ this isotherm becomes identical with the Langmuir isotherm.

The interactions are also taken into account in a simple manner by the virial isotherm.

The Flory–Huggins isotherm is often used in descriptions of adsorption from mixed solvents. It takes into account the difference between the sizes of the substances competing for the adsorption sites on the electrode surface:

$$\frac{\Theta}{r(1-\Theta)^r} = \beta C, \qquad (1.55)$$

where r is defined by the number of solvent molecules which must be desorbed in order to make room on the surface for the adsorption of one molecule of another substance.

Lawrence and Parsons [55] modified this isotherm in order to take into account the interaction between the adsorbed molecules.

Other isotherms have been proposed by Bockris and co-workers [56–58] and Levine and co-workers [59]. Levich and Krylov [60] derived an isotherm theoretically, which was based on the Ershler model, with hexagonal arrangement of adsorbed ions.

The presence of organic surface active substances has a particularly significant effect on the curves of dependence of differential capacity on potential. In addition to the considerable decrease of capacity in the potential region corresponding to the adsorption of the inert substance under study pseudocapacity peaks are usually observed connected with desorption of this substance at potentials negative and positive with respect to the maximum adsorption potential. The heights of these peaks, caused by changes of the electrode surface charge due to the desorption, depend on the frequency of the alternating current used in the determinations. Their shapes are also influenced by the nature of interactions between the adsorbed molecules, e.g. [61]. An example, the potential dependence of the differential capacity in the presence and absence of octyl alcohol in $1M$ KNO_3 solution, is shown in Fig. 1.8.

Published studies on the adsorption of various substances on electrode surfaces have been reviewed in several monographs [62–66].

In our discussion of the double layer we have deliberately omitted the problem of thermodynamic electrocapillarity theory, since it does not give any information regarding the structure of the double layer. This theory has been discussed in detail by Mohilner [67].

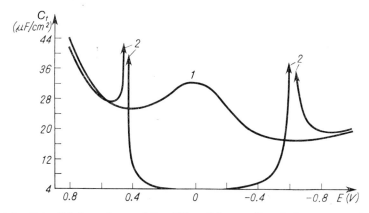

Fig. 1.8 Potential dependence of the differential capacity of the mercury electrode. Curve 1—1 M KNO$_3$; curve 2—1 M KNO$_3$ saturated with octyl alcohol; the potentials are given with respect to the zero charge potential, which is taken as zero.

Knowledge of the structure of the double layer plays an important part in the studies of the mechanisms of electrode processes. The discussion presented in this Chapter should be regarded as an introduction to this subject. The relation between the properties of the double layer and the kinetics of electrode processes has been discussed by Delahay [68].

References

[1] Quincke, G., *Pogg. Ann.*, **113**, 513 (1861).
[2] Helmholtz, H. L. F., *Ann. Physik*, **7**, 337 (1879).
[3] Gouy, G., *J. phys. radium*, **9**, 457 (1910).
[4] Gouy, G., *Compt. rend.*, **149**, 654 (1910).
[5] Chapman, D. L., *Phil. Mag.*, **25**, 475 (1913).
[6] Parsons, R., "*Modern Aspects of Electrochemistry*", Ed. Bockris, J. O'M., Butterworth, London 1954, Chapter 3.
[7] Stern, O., *Z. Elektrochem.*, **30**, 508 (1924).
[8] Frumkin, A. N., *Trans. Faraday Soc.*, **36**, 117 (1940).
[9] Grahame, D. C., *Chem. Revs.*, **41**, 441 (1947).

[10] Grahame, D. C., *J. Am. Chem. Soc.*, **76**, 4819 (1954).
[11] Grahame, D. C., *J. Am. Chem. Soc.*, **79**, 2093 (1957).
[12] Frumkin, A. N., *Z. physik. Chem.*, **164**, 121 (1933).
[13] Sparnaay, M., *Rec. trav. chim.*, **77**, 382 (1958).
[14] Brodowsky, H. and Strehlow, H., *Z. Elektrochem.*, **63**, 262 (1959).
[15] Wicke, E. and Eigen, M., *Z. Elektrochem.*, **56**, 551 (1952).
[16] Watts-Tobin, R. J., *Phil. Mag.*, **6**, 133 (1961).
[17] Mott, N. F. and Watts-Tobin, R. J., *Elektrochim. Acta*, **4**, 79 (1961).
[18] Macdonald, J. R. and Barlow, C. A., *J. Chem. Phys.*, **36**, 3062 (1962).
[19] Macdonald, J. R. and Barlow, C. A., *J. Chem. Phys.*, **39**, 419 (1963).
[20] Macdonald, J. R. and Barlow, C. A., in *"Electrochemistry"*—Proceedings of the 1st Australian Conference 1963, Ed. Friend, J. A. and Gutmann, F., Pergamon Press, London 1964, p. 199.
[21] Macdonald, J. R. and Barlow, C. A., *J. Chem. Phys.*, **44**, 202 (1966).
[22] Macdonald, J. R. and Barlow, C. A., *Surface Sci.*, **4**, 381 (1966).
[23] Barlow, C. A. and Macdonald, J. R., *Adv. Electrochem. Electrochem. Eng.*, **6**, 1 (1967).
[24] Bockris, J.O'M., Devanathan, M. A. V. and Müller, K., *Proc. Roy. Soc.*, **A274**, 55 (1963).
[25] Bockris, J. O'M., Green, M. and Swinkels, D. A. J., *J. Electrochem. Soc.*, **111**, 743 (1964).
[26] Levine, S., Bell, G. M. and Smith, A. L., *J. Phys. Chem.*, **73**, 3534 (1969).
[27] Hills, G. J. and Payne, R., *Trans. Faraday Soc.*, **61**, 326 (1965).
[28] Krylov, V. S., *Electrochim. Acta*, **9**, 1274 (1964).
[29] Hurwitz, H., Sanfeld, A. and Steinchen-Sanfeld, A., *Electrochim. Acta*, **9**, 929 (1964).
[30] Prigogine, I., Mazur, P. and Defay, R., *J. chim. phys.*, **50**, 146 (1953).
[31] Sanfeld, A. and Steinchen-Sanfeld, A., *Trans. Faraday Soc.*, **62**, 1907 (1966).
[32] Nürnberg, H. W., in *"Polarography 1964, Proc. 3rd Intern. Congr. Polarography, 1964"*, Ed. Hills G. J., Macmillan, London 1966, p. 149.
[33] Nürnberg, H. W., *Disc. Faraday Soc.*, **39**, 136 (1965).
[34] Jenard, A. and Hurwitz, H. D., *J. Electroanal. Chem.*, **19**, 441 (1968).
[35] Grahame, D. C., *J. Am. Chem. Soc.*, **74**, 4422 (1952).
[36] Andersen, T. N. and Bockris, J. O'M., *Electrochim. Acta*, **9**, 4 (1964).
[37] Levine, S., Bell, G. M. and Calvert, D., *Can. J. Chem.*, **40**, 518 (1962).
[38] Frumkin, A. N., *Z. physik. Chem.*, **103**, 43, 55 (1923).
[39] Frumkin, A. N., *Z. physik. Chem.*, **109**, 34 (1924).
[40] Frumkin, A. N. and Titevskaya, A. G., *Zhurn. Fiz. Khim.*, **31**, 485 (1957).
[41] Frumkin A. N., in *"Transactions of the Symposium on Electrode Processes"*, Ed. Yeager, E., J. Wiley, New York 1961, p. 1.
[42] Grahame, D. C., Coffin, E. M., Cummings, J. I. and Poth, M. A., *J. Am. Chem. Soc.*, **74**, 1207 (1952).
[43] Esin, O. A. and Markov, B. V., *Acta physicochim. U.R.S.S.*, **10**, 353 (1939).
[44] Parsons, R., in *"Proc. 2nd Intern. Congr. Surface Activity"*, Butterworth, London 1957, Vol. 3, p. 38.
[45] Esin, O. A. and Shikhov, V., *Zhurn. Fiz. Khim.*, **17**, 236 (1943).
[46] Parsons, R., *Adv. Electrochem. Electrochem. Eng.*, **1**, 1 (1961).
[47] Frumkin, A. N. and Polyanovskaya, N. S., *Zhurn. Fiz. Khim.*, **32**, 157 (1958).

References

[48] Ershler, B. W., *Zhurn. Fiz. Khim.*, **20**, 679 (1946).
[49] Grahame, D. C., *Z. Electrochem.*, **62**, 264 (1958).
[50] Levich, V. G., Kiryanov, V. A. and Krylov, V. S., *Dokl. Akad. Nauk SSSR*, **135**, 1193, 1425 (1960).
[51] Barlow, C. A. and Macdonald, J. R., *J. Chem. Phys.*, **40**, 1535 (1964).
[52] Grahame, D. C. and Parsons, R., *J. Am. Chem. Soc.*, **83**, 1291 (1961).
[53] Grahame, D. C., *J. Am. Chem. Soc.*, **80**, 4201 (1958).
[54] Parry, J. M. and Parsons, R., *Trans. Faraday Soc.*, **59**, 241 (1963).
[55] Lawrence, J. and Parsons, R., *J. Phys. Chem.*, **73**, 3577 (1969).
[56] Blomgren, E. and Bockris, J. O'M., *J. Phys. Chem.* **63**, 1475 (1959).
[57] Blomgren E., Bockris, J. O'M. and Jesch, C., *J. Phys. Chem.*, **65**, 2000 (1961).
[58] Wróblowa, H., Kovac, Z. and Bockris J. O'M., *Trans. Faraday Soc.*, **61**, 1523 (1965).
[59] Levine, S., Mingins, J. and Bell, G. M., *J. Electroanal. Chem.*, **13**, 280 (1967).
[60] Levich, V. G. and Krylov, V. S., *Dokl. Akad. Nauk SSSR*, **142**, 123 (1962).
[61] Sathyanarayana, S. and Baikerikar, K. G., *J. Electroanal. Chem.*, **21**, 449 (1969).
[62] Frumkin, A. N. and Damaskin, B. B., Plenary Lecture presented at the International Congress of Polarography in Kyoto, 1966, Butterworth, London 1967, p. 263.
[63] Frumkin, A. N. and Damaskin B. B., in *"Modern Aspects of Electrochemistry"*, Ed. Bockris J. O'M., Conway, B. E., Vol. 3, Butterworth, London 1964, Chapter 3.
[64] Damaskin, B. B., *Elektrokhimya*, **5**, 771 (1969).
[65] Payne, R., *J. Electroanal. Chem.*, **41**, 277 (1973).
[66] Damaskin, B. B., Petrii, O. A. and Batrakhov, V. V., *"Adsorption of Organic Compounds on Electrodes"*, Nauka, Moscow 1968 (in Russian).
[67] Mohilner D. M., in *"Electroanalytical Chemistry"*, Ed. Bard, A. J., Dekker, New York 1966, Vol. 1, p. 241.
[68] Delahay, P., *"Double Layer and Electrode Kinetics"*, Wiley–Interscience, New York 1965.

Chapter 2

Development and General Characterization of Electroanalytical Techniques

In this chapter the principles of chronoamperometry, polarography, chronopotentiometry, stationary electrode voltammetry and the rotating disc method are briefly discussed. A detailed discussion of various electroanalytical problems will be given later in this book.

Discussion of these topics in chronological order would necessitate commencing with chronopotentiometry, the theoretical principles of which had already been formulated at the beginning of the 20th century [1–4]. However, the importance of polarography in chemical analysis and physical and chemical research, as well as the fact that the principles of polarography are better known than those of the other methods, make it a more suitable subject for the start of the discussion. The introduction to polarography will be preceded by description of chronoamperometry, which is a closely related subject.

All the methods discussed in this book are based on electrolytic processes, which are carried out in conditions specific for each of the methods, and which involve the use of a suitable indicator electrode. The second electrode is a reference electrode giving a constant potential, such as the saturated calomel electrode. In the methods discussed, the migration of ions to the electrodes is practically eliminated as a result of introduction of a strong electrolyte into the sample solution at concentrations usually exceeding the concentration of the reacting substance by several orders of magnitude.

2.1 Chronoamperometry and Chronocoulometry

In chronoamperometry the current is recorded after the application (to the indicator electrode) of a potential at which the electrode reaction of the substance present in the solution can take place in the electrode phase (an amalgam electrode), or on the surface of the electrodes. The sample solution should be unstirred since diffusion should be the factor governing the transport of the reactants and the products of the electrode reaction.

During the measurements the electrode surface should remain unchanged. Any electrode mentioned in section 2.3 can be used.

When the potential in an electrolytic reduction process is sufficiently negative (or sufficiently positive in an electrolytic oxidation process) the concentration of the electrolysed substance on the surface of the electrode decreases to zero at the moment of application of the potential to the electrode.

In the case of a flat electrode and linear diffusion the rate of change of the current is given by the following equation:

$$i_g = \frac{nFAD^{1/2}C^0}{(\pi t)^{1/2}}, \tag{2.1}$$

where C^0 is the concentration of the depolarizer in the bulk of the solution, D is the diffusion coefficient, t is the electrolysis time, A is the electrode surface area, F is Faraday's constant, and n is the number of electrons exchanged between one ion or molecule and the electrode.

According to this equation, which is often called the Cottrell equation [5, 6], the current of chronoamperometric electrolysis tends to zero when the time tends to infinity. This is the result of the progressive decrease of the depolarizer concentration in the region close to the surface.

The relationships expressing the rate of current change are more complex when the diffusion of electroactive substance to the surface of the electrode is not linear.

The chronoamperometric method is now frequently used in electrochemical research. In recent years a variant of this method in which the potential is changed twice has been introduced into kinetic studies. The first potential change initiates, as before, the electrode process. After a brief electrolysis the potential is again changed to the initial or some other value, at which the product of the primary electrode process can be converted back into its original form.

Recently the measurements of current have been replaced by measurements of charge transferred during the electrolysis at constant potential. This method has been called chronocoulometry [7, 8]. It is carried out under experimental conditions identical with those used in the chronoamperometric method.

In chronocoulometry the equivalent of equation (2.1) is the relationship expressing the rate of change of charge Q:

$$Q = \frac{2nFAD^{1/2}C^0 t^{1/2}}{\pi^{1/2}}. \qquad (2.2)$$

Chronocoulometry is particularly useful in studies on adsorption of electroactive substances [9]. It is also used in kinetic studies on electrode processes and chemical reactions connected with electrode reactions. Chronocoulometric methods have been recently reviewed by several authors [10–12].

2.2 Polarography

In Heyrovský's polarographic method [13] the electrode employed is the dropping mercury electrode (D.M.E.) which was first introduced by Kučera [14] for electrocapillary studies. The diameter of the capillaries used in this method is about 0.04–0.08 mm. When the difference in hight between the end of the capillary and the mercury level is sufficiently large the mercury flows from the capillary in the form of small drops. Their mass depends on the radius of the capillary and on the surface tension, i.e. it also depends on the potential. Such an electrode is placed in the solution under study and is connected to the reference electrode. A potential difference is applied to this system of electrodes and the current in the resulting electrolytic circuit is measured. A simple polarographic circuit is shown schematically in Fig. 2.1. In polarography the potential of the dropping electrode is usually negative with respect to the reference electrode.

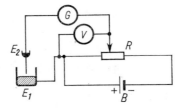

Fig. 2.1 Scheme of a simple polarographic circuit. B—voltage source; R—resistance; V—voltmeter; G—galvanometer; E_1—reference electrode; E_2—dropping electrode.

If the substance present in the solution is reducible at the dropping electrode, then at a characteristic potential difference reduction takes place and current flows in the circuit. At the same time the equivalent amount of another substance is oxidized at the reference electrode. The oxidized substance is usually the reference electrode material itself, e.g. mercury in the case of a calomel electrode. Since the surface area of the dropping electrode is small, and that of the reference electrode is large, the current intensity in these conditions depends on the former. This small current has only a slight effect on the distribution of concentration of substances constituting the reference electrode. This is one of the major reasons why the potential of this electrode is constant during polarographic measurements.

It should be realized that during an electrolysis, taking place under polarographic conditions, at a constant potential, the recorded current is not constant, but changes periodically (instantaneous current). At the moment of detachment of the drop the surface area of the electrode is practically equal to zero, although strictly speaking, it is equal to the cros-sectional area of the capillary. As a new drop grows, the surface area gradually increases to a maximum at the moment preceding the detachment of the drop. In general the surface area is virtually constant over the last 30 per cent of drop life [14a, 14b]. The dependence of current, i, on time, t, at constant potential is shown in Fig. 2.2, where t_1 is the lifetime of a single drop.

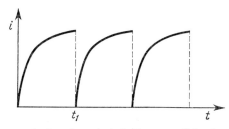

Fig. 2.2 Current changes during electrolysis in the case of the dropping mercury electrode.

Thus the current also depends on the lifetime of the mercury drop and the rate of flow of mercury from the capillary, since the average surface area of the drop electrode increases with increasing flow rate of mercury.

The above qualitative observations may be expressed more exactly. The mercury dropping from the capillary is under total pressure P which can be expressed as follows:

$$P = hdg, \qquad (2.3)$$

where h is the height of the column of mercury (from the end of the capillary to the mercury level in the resevoir), g is the gravitational constant and d is the density of mercury.

During the growth of the drop in the solution surface tension opposes the increase of its surface area, and as a result a back-pressure, P_b, acts on the drop.

$$P_b = \frac{2\sigma}{r} \tag{2.4}$$

where r is the radius of the drop at a given time, t, and σ is the surface tension. The pressure P_b acts in the direction opposite to that of the total pressure P.

Assuming that the growing drop of mercury is spherical, its volume, v, at any time, t, can be expressed by the equation:

$$v = \frac{mt}{d} = \frac{4}{3}\pi r^3 \tag{2.5}$$

where m is the mass of the drop flowing out of the capillary in unit time.

From equation (2.5)

$$r = \left(\frac{3mt}{4\pi d}\right)^{1/3}. \tag{2.6}$$

The average radius $\bar{r}$ at time t can be expressed by the integral:

$$\bar{r} = \frac{1}{t_1}\int_0^{t_1} r\,dt. \tag{2.7}$$

On the basis of equations (2.6) and (2.7) the following is obtained on integration:

$$\bar{r} = \frac{3}{4}\left(\frac{3\overline{m}t_1}{4\pi d}\right)^{1/3}, \tag{2.8}$$

where $\overline{m}$ is the average rate of mercury flow during the time t_1 of existence of the drop, known as the efficiency of the capillary.

Substituting equation (2.8) in equation (2.4) we obtain:

$$P_b = \frac{2\sigma}{\dfrac{3}{4}\left(\dfrac{3\overline{m}t_1}{4\pi d}\right)^{1/3}} = 4.31\,\frac{d^{1/3}\sigma}{(\overline{m}t_1)^{1/3}}. \tag{2.9}$$

P_b appears in the expression:

$$P_b = h_b dg, \tag{2.10}$$

where h_b is the height of mercury column corresponding to the total pressure P_b which depends on the capillary forces opposing the outflow of mercury.

From equations (2.9) and (2.10)

$$h_b = \frac{3.1}{(\overline{m}t_1)^{1/3}}. \tag{2.11}$$

The numerical constant in equation (2.11) was obtained assuming $d = 13.53$ g/cm^3 and $\sigma = 400$ dynes/cm.

Using Poiseuille's equation an expression for the average mercury flow rate is obtained:

$$\overline{m} = \frac{\pi r_c^4 dP}{8\eta l} \tag{2.12}$$

where r_c and l are the radius and the length of the capillary, respectively, and η is the viscosity of the solution.

Since:
$$P \sim h - h_b = h_r \tag{2.13}$$
then:
$$\overline{m} = k(h - h_b) = kh_r. \tag{2.14}$$

The weight of the detaching drop of mercury depends on the surface tension and the radius of the capillary

$$\overline{m} t_1 g = 2\pi r_c \sigma. \tag{2.15}$$

When the surface tension and the radius of the capillary are constant, the product $\overline{m}t_1$ is also constant. Therefore the lifetime of the drop is inversely proportional to h_r

$$t_1 = \frac{k'}{\overline{m}} = \frac{k''}{h_r} \tag{2.16}$$

where k' and k'' are constants.

Thus the outflow rate and the drop time, but not their product (which is the mass of the drop), will depend on h_r.

The capillary efficiency $\overline{m}$ does not depend on the medium into which the flow of the mercury takes place or on the electrode potential, but to a certain degree on surface tension, which is included in the correction term for the back pressure [see equation (2.4)].

The lifetime of the drop depends on the properties of the medium and on the electrode potential. Since, however, the duration of the drop depends on the surface tension according to equation (2.15), the dependence of drop lifetime on the electrode potential is expressed by the electrocapillary curve.

In various calculations it is necessary to know the electrode surface area. It may be assumed that the drop of mercury is spherical, as it has been found experimentally [15] that this assumption is almost strictly correct. Its surface area A at any time t is:

$$A = 4\pi r^2. \tag{2.17}$$

Combining equations (2.17) and (2.6) the following expression is obtained:

$$A = 4\pi \left(\frac{3mt}{4\pi \times 13.6}\right)^{2/3} = 0.85 m^{2/3} t^{2/3} \tag{2.18}$$

The average surface area of the drop during the drop time t_1 is given by the equation:

$$\bar{A} = \frac{1}{t_1} \int_0^{t_1} A\, dt; \tag{2.19}$$

$$\bar{A} = \frac{1}{t_1} \int_0^{t_1} 0.85 m^{2/3} t^{2/3} dt = 0.51\, m^{2/3} t_1^{2/3}. \tag{2.20}$$

The rate of increase of the surface area of the drop at time t can be expressed as:

$$\frac{d\bar{A}}{dt} = \frac{2}{3} 0.85\, m^{2/3} t^{-1/3}. \tag{2.21}$$

The maximum surface area of the drop at time t_1 is:

$$A_m = 0.85\, m^{2/3} t_1^{2/3}. \tag{2.22}$$

The above discussion has been concerned with changes of current caused by changes of the surface area of the drop during a polarographic process taking place at a constant potential. However, in polarography the relation between the current and the applied potential is investigated. The initial potential is so chosen that the rate of the electrode process is negligibly small. The flow of current in the circuit is observed only when the voltage applied to the electrodes, which changes at a constant rate, raises the potential of the dropping electrode to a value close to the standard potential of a depolarizer present in the solution. This condition applies when the electrode reaction of the depolarizer takes place without overvoltage. Over a certain potential range the current increases with increasing negative potential of the electrode, and during this time the concentration of the depolarizer on the surface of the electrode decreases. When this concentration decreases to zero the current reaches a constant value depending on the rate of depolarizer transport to the surface of the electrode. In these conditions we have the maximum current, which is often called the limiting current, or diffusion current. The last name is however not fully justified, since it suggests that the value of the current is limited by the rate of diffusion transport alone.

Convection caused by the drop growth also plays an important part in the transport of depolarizer to the surface of the electrode. It will be discussed in Chapters 4 and 6.

A polarographic curve expressing the relation between the current and the potential of a dropping electrode is a step-shaped wave and is often called the polarographic wave (or step). Such curves are recorded by means of a polarograph of which many are available, the majority being motor-driven instruments.

If the solution contains another substance which is reducible at the dropping electrode, and if the potential of this reduction is sufficiently different from that of the first wave, a new polarographic wave is formed at its characteristic potential and a further increase in current takes place. This is shown in Fig. 2.3. The height of the wave is usually proportional to the concentration of the reducible substance and, therefore, polarography is widely used in quantitative analysis.

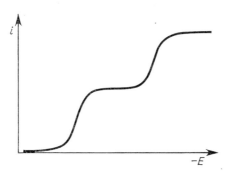

Fig. 2.3 Polarographic waves when two depolarizers are present in the solution.

The limiting current i_g is described by the equation derived by Ilkovič [6]:

$$i_g = 605 \, nm^{2/3} t_1^{1/6} D^{1/2} C^0 \tag{2.23}$$

where C^0 is the depolarizer concentration in the bulk of the solution, D is the diffusion coefficient of the depolarizer, n is the number of electrons involved in the elementary process, m is the efficiency of the capillary, and t_1 is the drop time.

Equation (2.23) is called the Ilkovič equation.

Since in reversible electrode processes polarographic waves are formed at potentials very close to the standard potentials of the reacting species, polarography also provides a method of qualitative determination of the composition of solutions.

The periodic regeneration of the surface of the dropping mercury electrode gives a high degree of reproducibility to polarographic determinations. Dropping electrodes are easily prepared, and the electrical circuit used in polarography is readily assembled according to the scheme in Fig. 2.1. However, in practice, polarographs which automatically record the current–voltage curves are used. The first such apparatus was constructed in 1925 by Heyrovský and Shikata. In their instrument the current–voltage curve was recorded on photographic paper by means of a ray of light reflected from the moving mirror of the galvanometer.

Polarographs based on this principle are no longer manufactured. They give good results, but they require photographic treatment of the exposed film, and for this reason more recent polarographs record the current–voltage curves directly on paper in suitable pen-chart recorders.

Automatic polarographs such as those mentioned above are very convenient, and various types, some specially adapted for specific research and analytical work, are in commercial manufacture. Modern polarographs include potentiostats which make it possible to study solutions having a high resistance, such as solutions in organic solvents. Since polarographs are also used for recording stationary electrode voltammetric curves, modern instruments generally have facilities for selecting from several rates of potential sweep. It is also possible to change the direction of the polarization rapidly.

All the modern polarographs contain devices for compensation of capacity currents, caused by changes of electrode surface area during polarography. Such devices make it possible to improve polarographic analysis of dilute solutions.

Polarography has been described and discussed in various reviews and monographs published by numerous authors in many languages.

Many reviews of polarography have been published and space does not permit the mention of those of an early date. However, mention must be made of the second edition of "*Polarography*" by Kolthoff and Lingane [17] and of works published later by von Stackelberg [18], Meites [19], Milner [20], Tsfasman [21], Kryukova, Sinyakova and Arefeva [22], Zagórski [23], Chodkowski [23], Proszt, Cieleszky and Györbiró [24].

A complete and very lucid description of the principles of polarography has been given by Heyrovský and Kuta [25].

Polarography of electrode processes combined with chemical reactions has been discussed by Guidelli [26].

Papers read at several conferences on polarography have appeared in the form of books [27–32], and various branches of polarography have been reviewed in monographs [33–36].

2.3 Stationary Electrode Voltammetry

Stationary electrode voltammetry was adopted as an electroanalytical technique much later than polarography. The first theoretical considerations of the reversible process were published by Randles [37] and Ševčik [38] in 1948. No significant use was made of this method, until several years later when hanging mercury electrodes were applied to electroanalysis [39–41].

In stationary electrode voltammetry, as in polarography, the variation of current with the potential applied to the indicator electrode is recorded but, contrary to polarography, the relationship gives a graph in the form of peaks, as it is shown schematically in Fig. 2.4. During measurements

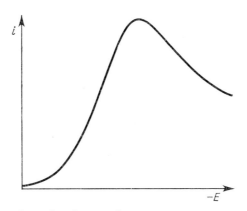

Fig. 2.4 Stationary electrode voltammetric curve.

the electrolytic cell should be protected from external shock, so that movement of the substances to the electrode surface and away from it occurs only as a result of diffusion forces.

Apparatus used in stationary electrode voltammetry may be very similar to that used in polarography. The only difference consists in substituting a constant surface microelectrode for the dropping mercury electrode. The latter can however be used in particular cases, when the rate of change in the potential applied to the electrode is so fast that practically no increase in electrode surface takes place during the recording of the stationary electrode voltammetric curve.

Mercury electrodes in the form of hanging drops are commonly used [39–41]. Other forms of mercury electrode are also acceptable, provided that their surface is small, compared with that of the reference electrode.

A number of other materials in addition to mercury can also be used for constructing indicator electrodes for stationary electrode voltammetry. Usually they are chosen from the point of view of analysis of redox systems which react with the electrode in the positive potential region. Platinum electrodes of various shapes, such as flat plates and wires, are suitable; diffusion transport to such electrodes has been described in detail. In recent years such electrodes have frequently been used in studies on mechanisms of oxidation of a number of organic compounds. Other noble metals, such as those of the gold and platinum group, can be used as electrode materials. Graphite electrodes are also frequently used in stationary electrode voltammetry. Since appreciable residual currents are observed when electrodes made of pure graphite are employed, various organic impregnating agents, inactive over a wide range of positive potentials, are used in order to decrease them.

Electrodes made of carbon paste have been successfully used. Usually they consist of a mixture of spectrally pure graphite and paraffin oil. Such electrodes were used for the first time by Adams [42], and their usefulness was later shown in a number of papers [43–46]. Usually the carbon paste fills cavities cut in a piece of Teflon. One side of the paste layer is in contact with the solution, and the other with a metallic conductor, usually a copper wire. The construction of such electrodes is presented schematically in Fig. 2.5. A great advantage of carbon paste electrodes is the very small residual current in the positive potential region, up to about -0.3 V. At this potential the reduction current of oxygen adsorbed at the electrode is observed. The small residual currents observed when carbon paste electrodes are used is due to the small capacitance of the double layer at the electrode–electrolyte interface, the small capacitance being in turn due to the presence of large particles of the substance joining the carbon particles. When other liquids such as carbon tetrachloride were used instead of paraffin oil the results were less satisfactory. Good results have been obtained with carbon paste electrodes containing silicone oil. It was found that the capacitance of electrodes, and consequently the residual current decreases with increasing density of the silicone oil (i.e. the length of its molecular chains). However, this is accompanied by a decrease in the rate of electrode processes [47].

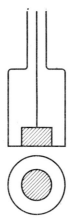

Fig. 2.5 Scheme of electrode made of carbon paste (horizontal and vertical sections). The shaded part—carbon paste; the remaining part of the electrode is made of Teflon.

It is noteworthy that carbon paste electrodes frequently make it possible to carry out measurements at much higher potentials than in the case of platinum electrodes, since the latter exhibit a much lower oxygen evolution potential.

Among other electrodes which have only recently found application, mention may be made of mercury covered with a layer of sparingly soluble mercury(I) salts [48, 49] or mercury oxide [50]. Mercury dissolves when the electrode is polarized to positive potentials, but this process is markedly inhibited by the layer of sparingly soluble substances covering the surface of the metal. Due to this inhibition, the mercury oxidation current rarely exceeds 1 μA/cm^2. Redox processes of various substances can sometimes take place at such electrodes at potential more positive than the redox potential of Hg/HgX (where X is halogen or oxygen ion). Since appreciable residual currents accompany the electrode processes, concentrations of the substances investigated must be high.

The layer of solid substance at the electrode surface usually decreases the electrode process rate (as compared with that taking place at a pure platinum electrode). Using passivated mercury electrodes, it is sometimes possible to achieve more positive potentials than those observed when platinum electrodes are used. At the same time the residual current due to oxidation of water at the electrode is kept at a low level.

Boron carbide electrodes [51] can be used both in the positive and the negative potential regions, provided that the negative potentials do not exceed −1.0 V. It was found that due to passivation, the activity of boron carbide electrodes changes during the process and therefore their usefulness in electroanalytical practice is limited. The solid products of oxidation of organic compounds, which are frequently formed during the electrode processes, can be easily removed from the surface of carbon paste electrodes, by simply substituting fresh paste for the used portion. It is difficult to remove such oxidation products from the surface of boron carbide electrodes without changing the state of the surface.

The potential applied to the electrodes, which changes linearly with time, is often obtained by means of a polarograph. The sweep rate is usually 0.1–0.8 V/min. Modern polarographs in which the range of sweep rates is much wider than usual (0.05–6 V/min), e.g. the Hungarian Radelkis OH-102 type, are very useful in stationary electrode voltammetry.

Most polarographs are designed for a potential sweep of 0.05 V/min. This is the minimum rate that should be used in stationary electrode voltammetry. Lower rates result in erroneous results, due to a considerable convection effect.

Although the lower limit of useful potential sweep can be obtained by use of a polarograph, special voltage generating devices are necessary for obtaining the upper limit. Voltage generated by such devices as a function of time is presented in Fig. 2.6. With such generators the po-

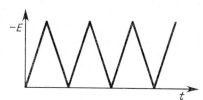

Fig. 2.6 Scheme of voltage change used in stationary electrode voltammetry.

tential sweep rate can be changed by hundreds of volts per second, and the voltage amplitude can be controlled at will, although in practice it seldom exceeds 2 V. Furthermore, cyclic polarization of the indicator electrode is possible. In the case when a reducible substance is present in the solution in addition to the supporting electrolyte, a change of the indicator electrode potential in the negative direction results in a change of the reduction current, the function having a peak, shaped as shown in Fig. 2.4.

The substance formed at the electrode surface in the reduction process can be reoxidized when the direction of polarization is reversed. If the reaction is fast enough, the oxidation is observed at the same potential region as that of the formation of the reduction peak.

Consider the mechanism of formation of such curves. In the cathodic part the reduction process starts at a potential close to the standard potential of the reducible substance present in the solution (provided that the electrode process is fast). The initial part of the curve of reduction current against electrode potential is very similar to the polarogram of the solution. The divergence between the polarogram and the stationary electrode voltammetric curve becomes apparent when the negative potential of the cathode increases. A current peak is formed on the stationary electrode voltammetric curve at a potential less than that at which the plateau of the polarographic wave is reached. At this moment the concentration of the depolarizer at the electrode surface is not yet at zero level. Further negative polarization of the electrode results in a decrease of the current, due to the decrease in depolarizer concentration in the region of the electrode. In cases where the reduction takes place at moderately negative potentials, and where more negative potentials are necessary to decompose the supporting electrolyte, the decrease of current becomes appreciable, after passing the peak. Frequently the solution contains another species which is reduced at a more negative potential than the first species; moreover, the difference between the reduction potential of the second depolarizer and the decomposition potential of the supporting electrolyte may not be very great. Only a moderate decrease in current is observed after the peak current in such cases.

The theory of fast electrode processes was presented by Randles [37] and Ševčik [38] in 1948. The Randles–Ševčik equation describes the peak current i_p of fast electrode processes:

$$i_p = 2.72 \times 10^5 n^{3/2} D^{1/2} A V^{1/2} C^0. \qquad (2.24)$$

It follows from equation (2.24) that the peak current depends on the following factors: concentration of the depolarizer in the bulk of the solution (C^0), diffusion coefficient of the substance being reduced or oxidized (D), the area of the electrode surface (A) and the number of electrons taking part in the elementary electrode process (n). Furthermore, the current increases with increasing polarization rate (V). This is understandable since the thickness of the diffusion layer decreases and the concentration gradients increase with increasing polarization rate. In polarography the limiting current is proportional to the concentration of the depolarizer in the bulk of the solution, but in the case of stationary electrode voltammetry it is the peak current which is proportional to concentration. The linear dependence of the peak current on the concentration of the reacting substance makes this method useful in quantitative analysis.

The development of stationary electrode voltammetry has been markedly enhanced by the introduction of hanging mercury drop electrodes [52–56]. Initially, this method was used mainly in electroanalysis. Metal ions to be determined are reduced at the electrode, usually by electrolysis at an appropriate constant potential. The liberated metal dissolves in the mercury. When the electrolysis lasts for a sufficiently long period, the concentration of the metals in the resulting amalgams is much higher then that of the corresponding ions in the solution. In order to obtain high and reproducible concentration of the amalgam, the solution should be vigorously stirred, at constant rate, during the concentrating electrolysis. The oxidation currents of the amalgamated metals recorded in stationary electrode voltammetry conditions are much higher than those of reduction of the corresponding ions present in the solution. This technique is discussed in detail by Neeb [57].

The stationary electrode voltammetric technique has also been employed in studies of mechanisms of electrode processes. It has been found to be particularly useful in investigations of mechanisms of oxidation and reduction of a number of organic substances [58–62]. In this case, in contrast to polarography, the substance produced during cathodic reduction remains in the neighbourhood of the electrode surface and can be reoxidized when the direction of polarization is reversed. From the intensity of the recorded oxidation currents and from the potentials at which the currents are observed, information on reversibility or irreversibility of the system and on chemical reactions accompanying electrode processes can be obtained. Extensive treatment of the theory of kinetic and adsorption currents has been presented by Shain *et al.* [63–65]. A considerable contribution to the theory of stationary electrode voltammetry has also been made by Savéant and Vianello [66–69].

A factor limiting the application of this technique is the appreciable capacity current observed at high polarization rates. In cases when the high capacity current exists, however, stationary electrode voltammetry can be used for investigation of the structure of the double layer [70, 71].

2.4 Rotating Disc Voltammetry

In both of the techniques discussed above, the solution analysed is not stirred. On the contrary, precautions are taken to ensure that only diffusion forces act upon the transport of substance electrolysed. This condition can only be fully satisfied in stationary electrode voltammetry. In polarography diffusion is accompanied by some transport due to convection which unavoidably results from the expansion of the drop into the bulk of the solution.

In the rotating disc technique, the basic principles of which are briefly outlined below, the convection transport of the depolarizer to the electrode plays a very important part, and its rate can be controlled. In this technique electrodes having various shapes are rotated at a constant rate in the solution under examination. Alternatively fixed electrodes in uniformly stirred solutions are used.

Rotating disc electrodes are now extensively employed in physicochemical research. A great advantage of such electrodes—as compared with electrodes having other shapes—is the reproducibility of results and the possibility of precise specification of the rates at which the reacting substances are transported to the electrode.

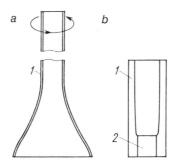

Fig. 2.7 Disc electrode. *1*—insulator layer; *2*—electrode material.

A disc electrode is shown diagramatically in Fig. 2.7. Only the lower surface of the electrode is in electrical contact with the solution. Its axis, side walls and the upper surface are insulated with non-conducting materials.

An alternative method of construction, frequently employed, is to seal the electrode material into a piece of soft glass (Fig. 2.7*b*). The lower part of the glass, with the electrode material sealed in, is ground till a uniform plane surface is formed [72]. Plastics such as Teflon may be substituted for glass. In this case the electrode material is pressed into a cavity formed in a cylindrical piece of the plastic [73]. Rotating electrodes are usually made of platinum, although satisfactory results are obtained with electrodes made of other materials, such as carbon paste [74], similar to that used in stationary electrode voltammetry. The paste is placed in a pit cut in a Teflon cylinder and is appropriately smoothed. Then it is connected to the electrical circuit by means of a piece of copper wire. Disc electrodes made of copper are also employed [75]. When amalgamated, they can be used in the negative potential region (disc cathode).

Proper functioning of rotating disc electrodes requires the fulfilment of several conditions. First of all, the electrode surface must be sufficiently smooth. Furthermore, relatively large dimensions are essential. It follows from the theoretical principles that the diameter of the active surface must be infinitely large, compared with the thickness of the diffusion layer. Since the latter is of the order of 10^{-3} cm, electrodes from 1–2 mm to several cm in diameter are employed in practice.

The electrode must be firmly mounted on its shaft in order to minimize vibrations in the horizontal plane. Two bearings placed at an appropriate distance from each other provide a sufficiently stable system. Good results are also obtained by fixing the electrode on an extention of the motor shaft. The required constant velocities can be obtained by use of an appropriate synchronized motor.

The arrangements for measurement may be practically the same as those for polarography, or stationary electrode voltammetry at low rates of polarization. The rotating electrode and the standard reference electrode are both connected into the electric circuit. Any type of polarograph is suitable for recording current–voltage curves. A moderate rate of potential sweep is necessary, especially with slow rotation of the electrode. The current-potential curves are very similar to polarograms but the limiting current is described by parameters somewhat different from those used in polarography.

Electroanalytical techniques involving convection transport were developed towards the end of the last century and therefore attempts to describe the rate of transport of the substance to the electrode were undertaken comparatively early. Initially, heterogeneous processes at the solid–liquid interface were considered in a general manner. Assuming that the rate of the process is defined by the transport rate of the substance, proper conclusions concerning the electrode process can be drawn, provided that the transport by migration is kept to a minimum.

Noyes and Whitney [76] came to the conclusion that it is always the diffusion rate that limits the dissolution rate of solids. They assumed that a very thin layer of saturated solution is formed at the solid–liquid interface and therefrom the diffusion into the bulk of the solution takes place. Nernst [77], who further developed the concept of Noyes and Whitney, introduced the idea of a diffusion layer and described the current i by the equation:

$$i = \frac{nFAD(C^0 - C)}{\delta}, \qquad (2.25)$$

where δ is the thickness of the diffusion layer, C is the depolarizer concentration at the electrode surface, F is the Faraday constant, A is the electrode area and C^0 is the depolarizer concentration in the bulk of the solution.

Nernst's theory and his equation were qualitatively correct but they neglected the hydrodynamics of the process: equation (2.25) does not contain the experimental factor determining the transport rate—the rotation velocity, but only the thickness of the diffusion layer, which depends on the rotation frequency.

It is not possible to evaluate the thickness of the diffusion layer solely on the basis of Nernst's theory. Calculations carried out on the basis of equation (2.25), in which experimental data were also considered, led to the relationship:

$$\delta = \frac{K}{N^x} \qquad (2.26)$$

where K is a constant and N is the number of electrode rotations per minute. The values found for x were between 0.3 and 1.

The first attempts to solve the problem of material transport on the basis of general principles of hydrodynamics were made by Eucken [78]. Assuming diffusion layer variability, he took into account not only diffusion, but also convection transport of the reagents. This solution of the problem, involving not a disc but a plate, was not very precise, due to a number of simplifying assumptions.

The full and correct equation describing the current in the case of a rotating disc electrode was given by Levich [79, 80]:

$$i = \frac{nFAD^{2/3}\omega^{1/2}(C^0 - C)}{1.62\nu^{1/6}} \qquad (2.27)$$

where ω is the angular velocity of disc electrode rotation and ν the kinematic viscosity of the solution.

Equation (2.27) is, in general, similar to equation (2.25) and can be obtained from it by substituting $\delta = 1.62 D^{1/3} \nu^{1/6} / \omega^{1/2}$.

The rotating disc technique was not widely employed for a long period, and most of the work done was by Russian electrochemists. A more rapid development of the method commenced about 1960. By that time the first theoretical considerations on kinetic processes at disc electrodes were presented by Koutecký and Levich [81]. Moreover, the rotating ring-disc electrode was introduced into electroanalytical research [82, 83]. In this modification an additional ring electrode closely surrounds the basic disc, as shown schematically in Fig. 2.8. The distance between the electrodes

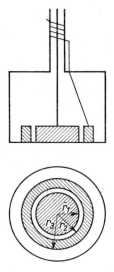

Fig. 2.8 Scheme of ring-disc electrode. Shaded part—conductor; white surface—insulator.

should be small. Such electrodes can be employed in detecting and studying unstable products of the process taking place at the inner electrode. Provided that the potential of the ring differs sufficiently from that of the disc, the unstable product is transferred from the latter electrode into the field of the former, where it can be either oxidized or reduced.

Since 1960 the rotating disc technique has been employed in the study of the kinetics of electrode processes [84], as well as that of chemical reactions taking place at the electrode surface [85]. In the last few years further theoretical studies concerning single disc and ring-disc electrodes have been published.

Nowadays the rotating disc technique is generally accepted as an important tool in the studies of the kinetics of moderately fast reactions. Its widespread analytical application is mainly due to the fact that substances which react with the electrode at positive potentials can be analysed. Since in this case the currents are much greater than those obtained in polarography, the rotating disc technique makes it possible to analyse fairly dilute solutions (of the order of $10^{-5}\,M$ or even lower).

A number of monographs on the rotating disc technique have been published. Theoretical problems concerning this subject were discussed by Levich [86] in his book on physico-chemical hydrodynamics and also in monographs by Albery and Hitchman [87] and Pleskov and Filinovskii [88]. An earlier extensive study of rotating disc electrodes was published by Riddiford [89]. The problems have also been discussed by Adams [12].

2.5 Chronopotentiometry

The three techniques discussed above: polarography, stationary electrode voltammetry and rotating disc voltammetry differ from each other substantially, though in each of them there is the common factor that the recorded currents are proportional to the concentration of the depolarizer in the bulk of the solution. On the other hand the dependence on time is different for each method (the duration of drop, polarization rate, rotation velocity). In each technique similar electric circuits are employed; a polarograph is suitable for both potential-sweep and current recording. The recorded curves represent the dependence of current on the electrode potential.

In the technique considered in this section a different electric circuit arrangement is employed and the time dependence of the electrode potential, at constant current intensity, is recorded. A typical curve is shown

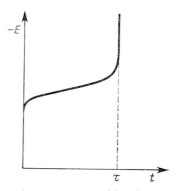

Fig. 2.9 Chronopotentiometric curve; τ—transition time.

in Fig. 2.9. Since the reduced form of the depolarizer is practically absent before the electrolysis commences the electrode potential is more positive than the reduction potential. As soon as the electrical circuit is completed, the direct current starts to flow, and—if the indicator electrode is connected to the negative pole of the current source—the reduction process starts. As the concentration of the reduced form becomes measurable at the very beginning of electrolysis and is higher at higher densities of the current in the electric circuit, the positive potential of the indicator electrode changes rapidly, becoming close to the standard potential, E^0, of the substance being studied (for reversible processes). In the course of electrolysis the concentration of the reduced form at the electrode surface increases at the expense of the concentration of the oxidized form.

As a result of these changes, the potential of the indicator electrode is shifted towards more negative values. In the case of a reversible electrode process, the potential is described by Nernst's equation.

$$E = E^0 + \frac{RT}{nF} \ln \frac{C_{\text{Ox}}(0, t)}{C_{\text{Red}}(0, t)} \tag{2.28}$$

where E^0 is the standard potential of the substance analysed, and $C_{\text{Ox}}(0, t)$ and $C_{\text{Red}}(0, t)$ are time-dependent concentrations of the oxidized and reduced forms at the electrode surface.

After a certain time the concentration of the oxidized form becomes close to zero, and according to equation (2.28) the electrode potential should be very high and negative. In practice it changes rapidly, reaching the value of the supporting electrolyte potential or that of some other species present, which is reducible at potentials more positive than that at which decomposition of the supporting electrolyte takes place.

The time elapsing from the beginning of electrolysis till the potential drop caused by the decrease in concentration of the analysed substance to zero at the electrode surface is called the transition time τ. It depends on current density and on concentration of the substance electrolysed. An equation combining the three parameters was derived at the beginning of this century and is known as Sand's equation:

$$\tau^{1/2} = \frac{\pi^{1/2} n F D^{1/2} C^0}{2i_0} \tag{2.29}$$

where i_0 is the current density.

Equation (2.29) was derived on the assumption that the substance is transported to the electrode by diffusion only. For this reason it is necessary to use solutions of sufficiently high supporting electrolyte concentrations to ensure that this is virtually the case. Measurements should be carried out under conditions of undisturbed diffusion. Protection of the solution from even the smallest disturbance is essential.

As already mentioned, the electrical circuit employed in chronopotentiometry differs from that used in the techniques previously discussed. The chronopotentiometric circuit is shown diagrammatically in Fig. 2.10,

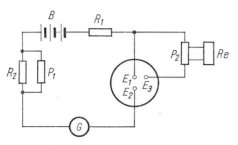

Fig. 2.10 Scheme of chronopotentiometric circuit.

where B is a battery or some other source of voltage of 200–300 V and R_1 is a large resistance limiting the current in the circuit. It is usually a resistor which is only roughly calibrated and therefore another, accurately calibrated, resistor, R_2, is added to the circuit. The voltage drop during the direct current flow through the resistor is precisely measured with the potentiometer P_1. Galvanometer G is used for checking that the circuit is always closed during the measurements.

When the indicator electrode, E_1, and supporting electrode, E_2, are connected into the above circuit, the potential of E_1 will change in the manner already described. The electrode process being studied takes place at E_1 and its potential is measured by means of potentiometer P_2 and can be recorded on the recorder Re.

Disturbances in C_{Ox} and C_{Red} at the indicator electrode are avoided by making the resistance of the potential recording circuit very large. This is chosen so that the current in the recording circuit is not higher than about 0.1% of that in the electrolytic circuit. The best results are obtained when the input resistance of the recording circuit is of the order of 10^{12} ohms.

Transition times may vary depending on current density and depolarizer concentration. Very short transition times may be observed at low concentrations and high current densities, but chronopotentiometric, experiments are usually so planned that the transition times are not shorter than one millisecond, since at shorter times the curves are considerably deformed and the measured transition times do not agree exactly with Sand's equation. Long transition times, exceeding one minute, are also not recommended, since in such conditions it is difficult to protect solutions from disturbances which affect the dependence of the concentration gradients on diffusion alone. The recording apparatus chosen depends on the transition time. Pen recorders are used at transition times usually exceeding one second; at shorter transition times $E-t$ curves are obtained on an oscillograph screen and can be photographed.

Electrodes made of various materials may be employed. In the negative potential region hanging mercury electrodes of different types, as well as mercury electrodes having larger areas are suitable. In order to avoid the creeping of solution along the walls of the vessel, the mercury is placed in Teflon or glass spoons, coated with a hydrophobic silicone layer.

In the positive potential region platinum electrodes are usually employed, as in stationary electrode voltammetry, but electrodes made of gold, boron carbide and carbon paste, as well as mercury covered with sparingly soluble calomel [48, 49] can be employed. Such electrodes are useful in studies on oxidation of organic compounds.

The importance of chronopotentiometry as an analytical method is limited. This is largely due to the non-linear concentration–transition time relationship. On the other hand kinetics and chemistry of electrode processes are fields in which chronopotentiometry is widely used and its importance is growing. However, the development of chronopotentiometry has been slow. Although the theoretical foundations were established at the beginning of this century [2, 3], it was rediscovered only during the nineteen fifties, through the published work of Gierst and Juliard [90, 91] and that of Delahay et al. [92–95]. Since that time a rapid development of this branch of electrochemistry has been observed. This includes the theoretical work on various types of kinetic current [92–98].

In recent years chronopotentiometry with reverse of direction of current [99, 100] and cyclic chronopotentiometry [101] have attracted much interest.

In the former technique, the direction of current is reversed before the transition time is reached or at the moment when it is reached, and the transition time of electrolysis of the primary electrode reaction product is recorded. In cyclic chronopotentiometry such changes are repeated many times.

Particular problems which can be studied by the above methods will be discussed in detail in later chapters. Chronopotentiometry has been described and discussed in several monographs [6, 102, 103].

References

[1] Weber, H. F., *Wied. Ann.* **7**, 536 (1879).
[2] Sand, H. J. S., *Phil. Mag.*, **1**, 45 (1901).
[3] Karaoglanoff, Z., *Z. Elektrochem.*, **12**, 5 (1906).
[4] Rosebrugh T. R. and Miller W. L., *J. Phys. Chem.*, **14**, 816 (1910).
[5] Cottrell, F. G., *Z. physik. Chem.*, **42**, 385 (1902).
[6] Delahay, P., *"New Instrumental Methods in Electrochemistry"*, Interscience, New York 1954, Chapter 3.
[7] Anson, F. C., *Anal. Chem.*, **36**, 932 (1964).
[8] Christie, J. H., Lauer, G. and Osteryoung, R. A., *J. Electroanal. Chem.*, **7**, 60 (1964); Osteryoung, R. A. and Anson, F. C., *Anal. Chem.*, **36**, 975 (1964).
[9] Anson, F. C., *Anal. Chem.* **38**, 54 (1966).
[10] Murray, R. W., *"Chronopotentiometry, Chronoamperometry and Chronocoulometry"*, in *Physical Methods of Organic Chemistry*, Ed. Weissberger, A. and Rossiter, B. W., 4th Ed., Part II, Interscience, New York 1969, Chapter VIII.
[11] Piekarski, S. and Adams, R. N., *"Voltammetry with Stationary and Rotated Electrodes in Chemistry"*, Vol. I, Part II, Ed. Weissberger, A., Wiley, New York 1971.
[12] Adams, R. N., *"Electrochemistry at Solid Electrodes"*, Dekker, New York 1969.
[13] Heyrovský, J., *Chem. Listy*, **16**, 256 (1922).
[14] Kučera, B., *Ann. Physik.*, **11**, 529, 698 (1903).
[14a] Reynolds, G. F., *Research*, **9**, 170 (1956).
[14b] Reynolds, G. F., *Z. analyt. Chem.*, **173**, 65 (1960).
[15] MacNevin, W. M. and Balis, E. W., *J. Am. Chem. Soc.*, **65**, 660 (1943); Smith, G. S., *Trans. Faraday Soc.*, **47**, 63 (1951).
[16] Ilkovič, L., *Coll. Czechoslov. Chem. Communs.*, **6**, 498 (1934).
[17] Kolthoff, I. M. and Lingane, J. J., *"Polarography"*, Interscience, New York 1952.
[18] von Stackelberg, M., *"Polarographische Arbeitsmethoden"*, de Gruyter, Berlin 1960.
[19] Meites, L., *"Polarographic Techniques"*, 2nd Ed., Interscience, New York 1965.
[20] Milner, G. W. C., *"The Principles and Applications of Polarography and other Electroanalytical Processes"*, Longmans, London 1957.
[21] Tsfasman, T. S., *"Electronic Polarographs"*, Metallurgizdat, Moscow 1960 (in Russian).
[22] Kryukova, T. A., Sinyakova, S. J. and Arefeva, T. V., *"Polarographic Analysis"*, Goskhimizdat, Moscow 1959 (in Russian).
[23] Zagórski, Z., *"Polarographic Analysis"*, WNT, Warsaw 1972 (in Polish); Chodkowski, J., *"Introduction to Theoretical Polarography"*, PWN, Warsaw 1958 (in Polish).
[24] Proszt, J., Cieleszky, V. and Györbiro, K., *"Polarographie mit besonderer Berücksichtigung der klassischen Methoden"*, Akadémiai Kiadó, Budapest 1967.
[25] Heyrovský, J. and Kuta, J., *"Principles of Polarography"*, Academic Press, New York 1966.
[26] Guidelli, R., *"Chemical Reactions in Polarography"*, in *Electroanalytical Chemistry*, Ed. Bard, A. J., Vol. 5, Dekker, New York 1971.

[27] Proceedings of the First International Polarographic Congress in Prague, Prirodovědecké Vydavatelstvi, Prague 1952.
[28] *Advances in Polarography*, Ed. Longmuir, 1. S., Pergamon Press, London 1960.
[29] *Progress in Polarography*, Ed. Zuman, P. and Kolthoff, I. N., Interscience, New York 1962.
[30] *Polarography 1964*, Ed. Hills, G. J., MacMillan, London 1965.
[31] "*Modern Aspects of Polarography*", Ed. Kambara, T., Plenum Press, New York 1966.
[32] *International Congress of Polarography in Kyoto (1966)*, Butterworth, London 1967.
[33] Březina, M. and Zuman, P., "*Die Polarographie in der Medizin, Biochemie und Pharmazie*", Akademische Verlagsgesellschaft, Leipzig 1956.
[34] Břeyer, B. and Bauer, H. H., "*Alternating Current Polarography and Tensammetry*", Interscience, New York 1963.
[35] Crow, D. R., "*Polarography of Metal Complexes*", Academic Press, London, New York 1969.
[36] Zuman, P., "*Substituent Effects in Organic Polarography*", Plenum Press, New York 1967.
[37] Randles, J. E. B., *Trans. Faraday Soc.*, **44**, 327 (1948).
[38] Ševčik, A., *Coll. Czechoslov. Chem. Communs*, **13**, 349 (1948).
[39] Kemula, W. and Kublik, Z., *Anal. Chim. Acta*, **18**, 104 (1958).
[40] Kemula, W. and Kublik, Z., *Roczniki Chem.*, **32**, 941 (1958).
[41] Ross, J. W., DeMars, R. D. and Shain, I., *Anal. Chem*, **12**, 1768 (1956); Frankenthal, R. P. and Shain, I., *J. Am. Chem. Soc.*, **78**, 2969 (1956).
[42] Adams, R. N., *Anal. Chem.*, **30**, 1576 (1958).
[43] Olson, C. and Adams, R. N., *Anal. Chim. Acta*, **22**, 582 (1960).
[44] Olson, C., Lee, H. Y. and Adams, R. N., *J. Electroanal. Chem.*, **2**, 396 (1961).
[45] Galus, Z. and Adams, R. N., *J. Phys. Chem.*, **67**, 862 (1963).
[46] Galus, Z. and Adams, R. N., *J. Phys. Chem.*, **67**, 866 (1963).
[47] Adams, R. N., *Rev. Polarography Japan*, **11**, 71 (1963).
[48] Kuwana, T. and Adams, R. N., *J. Am. Chem. Soc.*, **79**, 3609 (1957).
[49] Kuwana, T. and Adams, R. N., *Anal. Chim. Acta*, **20**, 51, 60 (1959).
[50] Kublik, Z., D. Sc. Thesis, Warsaw University Press, Warsaw 1968 (in Polish).
[51] Mueller, T. R., Olson, C. and Adams, R. N., in "*Advances in Polarography*", Ed. Longmuir, J. S., Pergamon Press, Oxford 1960, Vol. 1, p. 198; Mueller, T. R. and Adams, R. N., *Anal. Chim. Acta*, **23**, 467 (1960).
[52] Kemula, W., Kublik, Z. and Głodowski, S., *J. Electroanal. Chem.*, **1**, 91 (1959/60).
[53] Kemula, W., Galus, Z. and Kublik, Z., *Roczniki Chem.*, **23**, 1431 (1959).
[54] Kemula, W. and Galus, Z., *Bull. Acad. Polon. Sci., cl. III*, **7**, 729 (1959).
[55] Kemula, W., Rakowska, E. and Kublik, Z., *J. Electroanal. Chem.*, **1**, 205 (1959/60).
[56] Kemula, W., Galus, Z. and Kublik, Z., *Bull. Acad. Polon. Sci., cl. III*, **6**, 661 (1958).
[57] Neeb, R., "*Inverse Polaragraphie und Voltammetrie*", Academie-Verlag, Berlin 1969.
[58] Kemula, W. and Kublik, Z., *Bull. Acad. Polon. Sci., cl. III*, **6**, 653 (1958).
[59] Kemula, W., Kublik, Z. and Cyrański, R., *Roczniki Chem.*, **36**, 1349 (1960).
[60] Kemula, W., Grabowski, Z. R. and Kalinowski, M. K., *Naturwiss.*, **22**, 1 (1960).
[61] Galus, Z., Lee, H. Y. and Adams, R. N., *J. Electroanal. Chem.*, **5**, 17 (1962); Alden, J. R. A. and Chambers, J. Q., *J. Electroanal. Chem.*, **5**, 152 (1962).

References

[62] Buck, R. P. and Griffith, L. P., *J. Electrochem. Soc.*, **109**, 1005 (1962).
[63] Nicholson, R. S. and Shain, I., *Anal. Chem.*, **36**, 706 (1964).
[64] Nicholson, R. S. and Shain I., *Anal. Chem.*, **37**, 178 (1965).
[65] Wopshall, R. H. and Shain, I., *Anal. Chem.*, **39**, 1514 (1967).
[66] Savéant, J. M. and Vianello, E., in *"Advances in Polarography"*, Ed. Longmuir, J. S., Pergamon Press, Oxford 1960, Vol. 1, p. 367.
[67] Savéant, J. M. and Vianello, E., *Electrochim. Acta*, **12**, 629 (1967).
[68] Savéant, J. M., *Electrochim. Acta*, **12**, 753 (1967).
[69] Savéant, J. M. and Vianello, E., *Electrochim. Acta*, **8**, 905 (1963).
[70] Loveland, J. W. and Elving, P. J., *J. Phys. Chem.*, **56**, 250, 255, 935, 945 (1952).
[71] Gokhshtein, Ya. P., *Zhurn. Fiz. Khim.*, **32**, 1481 (1958).
[72] Frumkin, A. N. and Tedoradze, G., *Z. Elektrochem.*, **62**, 251 (1958).
[73] Vielstich, W. and Jahn, D., *Z. Elektrochem.*, **64**, 43 (1960).
[74] Galus, Z., Olson, C., Lee, H. Y. and Adams, R. N., *Anal. Chem.*, **34**, 164 (1962).
[75] Bagotskaya, I. A., *Dokl. Akad. Nauk SSSR*, **85**, 1057 (1952).
[76] Noyes, A. A. and Whitney, W. R., *Z. physik. Chem.*, **22**, 689 (1897).
[77] Nernst, W., *Z. physik. Chem.*, **47**, 52 (1904).
[78] Eucken, A., *Z. Elektrochem.*, **38**, 341 (1932).
[79] Levich, W. G., *Acta Physicochim. U.R.S.S.*, **17**, 257 (1942).
[80] Levich, W. G., *Acta Physicochim. U.R.S.S.*, **19**, 117, 133 (1944).
[81] Koutecký, J. and Levich, V. G., *Zhurn. Fiz. Khim.*, **32**, 1565 (1958).
[82] Frumkin, A. N. and Nekrasov, L. N., *Dokl. Akad. Nauk SSSR*, **126**, 115 (1959).
[83] Frumkin, A. N., Nekrasov, L. N., Levich, V. G. and Ivanov, J., *J. Electroanal. Chem.*, **1**, 84 (1959/60).
[84] Jahn, D. and Vielstich, W., *J. Electrochem. Soc.*, **109**, 849 (1962).
[85] Vielstich, W. and Jahn, D., *Z. Elektrochem.*, **64**, 43 (1960).
[86] Levich, V. G., *"Physicochemical Hydrodynamics"*, Prentice-Hall, Englewood Cliffs, N. J. 1962.
[87] Albery, W. J. and Hitchman, H. L., *"Ring-Disc Electrodes"*, Clarendon Press, Oxford 1971.
[88] Pleskov, Ya. V. and Filinovskii, V. Yu., *"Rotating Disc Electrode"*, Nauka, Moscow 1972 (in Russian).
[89] Riddiford, A. C., in *"Advances in Electrochemistry and Electrochemical Engineering"*, Ed. Delahay, P., Vol. 4, Interscience, New York 1966.
[90] Gierst, L. and Juliard, A., *"Proc. Intern. Comm. Electrochem. Thermodynam. and Kinet."*, Tamburini, Milano 1950, p. 117, 279.
[91] Gierst, L. and Juliard, A., *J. Phys. Chem.*, **57**, 701 (1953).
[92] Berzins, T. and Delahay, P., *J. Am. Chem. Soc.*, **75**, 4205 (1953).
[93] Mamantov, G. and Delahay, P., *J. Am. Chem. Soc.*, **76**, 5323 (1954).
[94] Delahay, P., Mattax, C. C. and Berzins, T., *J. Am. Chem. Soc.*, **76**, 5319 (1954).
[95] Delahay, P., *Disc. Faraday Soc.*, **17**, 205 (1954).
[96] Reinmuth, W. H., *Anal. Chem.*, **33**, 233 (1961).
[97] Testa, A. C. and Reinmuth, W. H., *Anal. Chem.*, **33**, 1320 (1961).
[98] Reinmuth, W. H., *Anal. Chem.*, **32**, 1514 (1960).
[99] Furlani, C. and Morpurgo, G., *J. Electroanal. Chem.*, **1**, 351 (1960); King, R. M. and Reilley, C. N., *J. Electroanal. Chem.*, **1**, 434 (1960); Dračka, O., *Coll. Czechoslov. Chem. Communs.*, **25**, 338 (1960).

[100] Testa, A. C. and Reinmuth, W. H., *Anal. Chem.*, **33**, 1324 (1961); Murray, R. W. and Reilley, C. N., *J. Electroanal. Chem.*, **3**, 64, 182 (1962).
[101] Herman, H. B. and Bard, A. J., *Anal. Chem.*, **35**, 1121 (1963).
[102] Davis, D. C., "*Applications of Chronopotentiometry to Problems in Analytical Chemistry*" in "*Electroanalytical Chemistry*", Vol. 1, Ed. Bard, A. J., Dekker, New York 1965.
[103] Paunovic, M., *J. Electroanal. Chem.*, **14**, 447 (1967).

Chapter 3

Rates of Electrode Processes

In discussions of the nature of electrode processes the reversible reactions are usually considered separately from those which are irreversible, since different equations describe the two groups. In chemical electroanalysis the criteria of reversibility are not clearly defined, although the concept of reversibility is often used. The definitions and criteria of reversibility depend on the kind of electroanalytical technique and hence expressions such as "polarographic reversibility" are sometimes used by investigators of electrochemical phenomena.

The general definition of reversibility of electrode processes can be formulated of the basis of the thermodynamics of irreversible processes. For a reversible process we have:

$$dS = \frac{Q_{el}}{T} \qquad (3.1)$$

where dS is the entropy change, Q_{el} is the elementary heat and T is the absolute temperature.

For an irreversible process:

$$dS > \frac{Q_{el}}{T}. \qquad (3.2)$$

Though the total energy of the system and the surroundings is always constant, the entropy in irreversible processes increases and:

$$dS + dS^* > 0 \qquad (3.3)$$

where dS is the entropy of heat transfer from the surroundings to the system, and dS^* is the entropy of heat transfer from the system to the surroundings.

The inequality (3.2) can be written as:

$$TdS - Q_{el} > 0. \qquad (3.4)$$

The difference between TdS and Q_{el} will be called the uncompensated elementary heat $\bar{Q}_{el}$:

$$TdS - Q_{el} = \bar{Q}_{el}. \qquad (3.5)$$

In irreversible processes $\bar{Q}_{el}$ is not equal to zero and is always positive. Obviously, according to equation (3.1), $\bar{Q}_{el}$ of reversible processes is equal to zero.

De Donder [1, 2] defined chemical affinity A as the ratio of $\bar{Q}_{el}$ to the increase of the reaction progress number $d\varepsilon$:

$$A = \frac{\bar{Q}_{el}}{d\varepsilon}. \tag{3.6}$$

If the work during the process is limited to a change of volume we can write the inequality (3.2) as:

$$TdS - dU - pdV > 0 \tag{3.7}$$

and hence:

$$dU = TdS - pdV - \bar{Q}_{el} \tag{3.8}$$

and

$$dU = TdS - pdV - Ad\varepsilon \tag{3.9}$$

In these equations dU and dV are the changes of intrinsic entropy and the change of volume, respectively, and p is the pressure.

Utilizing the definition of chemical affinity [1, 2]:

$$A = -\sum_\gamma \mu_\gamma \nu_\gamma \tag{3.10}$$

and, bearing in mind that:

$$d\varepsilon = \frac{dm_\gamma}{\nu_\gamma}, \tag{3.11}$$

the Gibbs equation is obtained:

$$dU = TdS - pdV + \sum_\gamma \mu_\gamma dm_\gamma, \tag{3.12}$$

where ν_γ is the stoichiometric coefficient of substance γ, μ is the chemical potential, and dm_γ is the change of the number of moles of substance γ.

Now consider, following Prigogine [3], a closed system consisting of two components having potentials E^I and E^{II}, respectively. Electrically charged particles are transferred from one component of this system to another and this process is identical to an electrode reaction. If the valence of the transferred ion is n_γ the following relation between the current and the rate of reaction is obtained:

$$i = n_\gamma F \frac{d\varepsilon_\gamma}{dt} \tag{3.13}$$

where F is the Faraday constant and $d\varepsilon_\gamma = -dm_\gamma^I = dm_\gamma^{II}$.

For such processes the term corresponding to electrical work must be added to the equation expressing the first law of thermodynamics:

$$dU = Q_{el} - pdV + (E^I - E^{II})idt. \qquad (3.14)$$

Equation (3.12) for the two-phase system being considered can be transformed to:

$$dS = \frac{dU}{T} + \frac{pdV}{T} - \sum_\gamma \left(\frac{\mu_\gamma^I dm_\gamma^I}{T} + \frac{\mu_\gamma^{II} dm_\gamma^{II}}{T} \right). \qquad (3.15)$$

Combining equations (3.13) and (3.14) with equation (3.15) we obtain the expression for entropy change:

$$dS = \frac{Q_{el}}{T} + \frac{\bar{A}_\gamma d\varepsilon_\gamma}{T} \qquad (3.16)$$

where $\bar{A}_\gamma$ is given by the equation:

$$\bar{A}_\gamma = A_\gamma + n_\gamma F(E^I - E^{II}) = (\mu_\gamma^I + n_\gamma FE^I) - (\mu_\gamma^{II} + n_\gamma FE^{II}). \qquad (3.17)$$

$\bar{A}_\gamma$ is the chemical affinity corresponding to the transfer of component γ from phase I to phase II.

The terms $\mu_\gamma + n_\gamma FE$ of equation (3.17) are called electrochemical potentials [4] and are usually represented by the symbol $\bar{\mu}_\gamma$.

In this discussion it will be more convenient to write equation (3.5) in the form:

$$dS = \frac{Q_{el}}{T} + \frac{\bar{Q}_{el}}{T}. \qquad (3.5a)$$

In order to simplify the equations put: $dS_e = Q_{el}/T$ and $dS_i = \bar{Q}_{el}/T$. Then equation (3.5a) becomes

$$dS = dS_e + dS_i. \qquad (3.18)$$

It follows from this equation that the entropy change can be considered as the sum of entropy changes corresponding to heat exchange with the surroundings dS_e and to the irreversible process taking place in the system dS_i.

It follows from equation (3.16) that:

$$dS_i = \frac{\bar{A}_\gamma d\varepsilon_\gamma}{T}. \qquad (3.19)$$

Thus the rate of entropy increase is given by the expression:

$$\frac{dS_i}{dt} = \frac{\bar{A}_\gamma d\varepsilon_\gamma}{T dt}. \qquad (3.20)$$

In order to express the entropy increase by means of electrochemical parameters the electrochemical affinity $\bar{A}_\gamma$ must be expressed by the equation:

$$\bar{A}_\gamma = A_\gamma + n_\gamma FE, \tag{3.21}$$

where A_γ is the chemical affinity.

Since in the state of electrochemical equilibrium the electrochemical affinity equals zero, we have:

$$A_\gamma + n_\gamma FE^e = 0 \tag{3.22}$$

where E^e is the equilibrium potential of the system being studied.

Equation (3.21) can be written in the following form:

$$\bar{A}_\gamma = n_\gamma F(E-E^e) = n_\gamma F\eta_\gamma \tag{3.23}$$

where η_γ is the overvoltage of the electrochemical process taking place in the system. It is given by the relationship:

$$\eta_\gamma = E - E^e. \tag{3.24}$$

Thus the overvoltage is the difference between the irreversible potential of the given system and the equilibrium potential.

According to equation (3.24) the overvoltage of cathodic processes has a minus sign and that of anodic processes has a plus sign.

Combining equations (3.23) and (3.13) with equation (3.20):

$$\frac{dS_i}{dt} = \frac{i_\gamma \eta_\gamma}{T}. \tag{3.25}$$

In a series of processes taking place in the system studied the entropy increase is equal to the sum of entropies of the individual processes:

$$\frac{dS_i}{dt} = \frac{1}{T}\sum_\gamma i_\gamma \eta_\gamma. \tag{3.26}$$

This equation was derived on the basis of thermodynamics of irreversible processes by van Rysselberghe [5] and by Zakharov [6]. Later these problems were further discussed by Guidelli [7].

It follows from equations (3.25) and (3.26) that the rate of entropy increase depends on the overvoltage of each process and on the current intensity. Thus the reversibility of electrode processes depends on two factors: the nature of processes taking place in the system, i.e. the overvoltage which, as will be shown, increases with decreasing process rate, and the experimental conditions which determine the current flowing in the system.

Overvoltages of different processes must be compared under identical conditions, since overvoltage depends on the intensity of current flowing in the system. For this reason the relationship (3.26) is complex.

It follows from the equations discussed above that under electrochemical conditions an electrode process can be reversible when the overvoltage equals zero.

In practice the current need not be very small for reversibility of the electrode process to be observed, but it must be small enough for the corresponding rate of transport of the depolarizer to the electrode to be considerably slower than the rate of electron transfer. This is illustrated in Fig. 3.1,

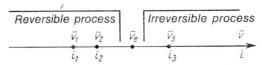

Fig. 3.1 Effect of depolarizer transport rate on the reversibility of an electrode process. $\bar{v}_e$—charge transfer rate; $\bar{v}_1, \bar{v}_2, \bar{v}_3$—transport rates.

which shows the straight line on which currents i and the corresponding depolarizer transport rates $\bar{v}$ are marked. In a solution at a constant concentration C the increase of the current from i_1 to i_3 can take place only as a result of increased rate of depolarizer transport from $\bar{v}_1$ to $\bar{v}_3$. When the charge transfer rate is much faster than that of the depolarizer transport, the ratio of concentrations of the oxidized form Ox to the reduced form Red on the electrode surface is expressed by the Nernst equation:

$$E = E^0 + \frac{RT}{nF} \ln \frac{C_{\text{Ox}}(0,t)}{C_{\text{Red}}(0,t)}. \quad (3.27)$$

Since this equation was derived for thermodynamically reversible processes the overvoltage of such a process always equals zero, because the system reaches the equilibrium potential as a result of the fast charge transfer rate. If by changing the experimental conditions the transport rate were increased to a value comparable with that of the charge transfer rate the system would not reach the equilibrium state and the ratio of concentrations of the Ox to Red forms on the electrode surface would differ from the value predicted by the Nernst equation for a given potential. This "non-nernstian" system of concentrations on the electrode surface causes the electrode potential to become different from the equilibrium potential. The difference between the electrode potential and that calculated from the Nernst equation determines the overvoltage.

In cases where the transport rates are similar to charge transfer rates (they can be slightly slower or faster) the electrode potential is close to the equilibrium potential and the overvoltage is small; such processes are called *quasi*-reversible. When the transport rates are much faster than the charge transfer rates the ratio of concentrations is almost independent of the charge transfer kinetics. The charge transfer may be so slow that the potential practically never reaches the value of the equilibrium potential given by the Nernst equation. Such processes are called totally irreversible.

A set of current vs. voltage curves may be obtained for any process studied, corresponding to various conditions of depolarizer transport. These curves are shown in Fig. 3.2. They could be recorded by using a rotating

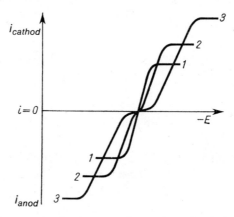

Fig. 3.2 Schematic current–voltage curves for various depolarizer transport rates. The transport rate increases from curve *1* to curve *3*.

electrode if it were possible in practice to make a considerable increase in the rate of rotation of the electrode without causing the turbulent flow. Curve *1* in Fig. 3.2 represents the *i–E* relationship for a reversible electrode process. An increase of the transport rate leading to an increase of the limiting current makes the process *quasi*-reversible (curve *2*). Any further increase of the transport rate increases the degree of irreversibility of the process.

Thus the reversibility of an electrode process is a relative concept and there are no processes that could be called reversible in the absolute sense. However, this term is useful in electrochemical studies carried out by means of a method with strictly defined transport rate values. It follows from the above reasoning that a process can have the properties of reversible processes when it is examined by some methods, but it can have the properties of irreversible processes when it is examined by other methods, in which the transport of the substance to the electrode is faster.

The depolarizer transport rates are so important in the determination of reversibility that they should be examined in detail in relation to the polarographic, stationary electrode voltammetric and chronopotentiometric methods, and the rotating disc method.

In the above methods the depolarizer transport rate depends on kinetic parameters. In polarography it depends on the drop time (t_1), in chronopotentiometry on the transition time (τ), in stationary electrode voltammetry on the reciprocal of the rate of change of potential ($1/\Delta V$), and in the rotating disc method on the reciprocal of the disc electrode rotation rate ($1/\omega$). The time of equilibrium disturbance on the electrode surface decreases and the transport rate increases with decreasing value of the time parameter.

The equations describing the transport rate $\bar{v}$ in various methods can be derived by suitably modifying the basic equations of those methods discussed in Chapter 2 (i.e. the Ilkovič, Sand, Randles and Ševčik, and Nernst and Levich equations). These equations show the relations between current depolarizer concentration, electrode surface area, and depolarizer transport rate. By elimination of the first two factors and the quotient nF the equations describing the transport rate [8] are obtained.

In the case of polarography this rate is given by the equation:

$$\bar{v}_p = \frac{2D^{1/2}}{(\pi t_1)^{1/2}} \tag{3.28}$$

where D is the diffusion coefficient and t_1 is the drop time.

In chronopotentiometry:

$$\bar{v}_c = \frac{\pi^{1/2} D^{1/2}}{2\tau^{1/2}}. \tag{3.29}$$

In stationary electrode voltammetry:

$$\bar{v}_{sv} = 2.82 n^{1/2} D^{1/2} V^{1/2}. \tag{3.30}$$

For the rotating disc electrode method:

$$\bar{v}_d = \frac{D}{\delta} \tag{3.31}$$

where δ is the thickness of the diffusion layer.

The applicability of these methods to the determination of the reversibility of electrode processes, or strictly speaking, to kinetic studies of electrode processes, depends on the possibility of controlling the transport rates in each.

Comparison of equations (3.28)–(3.31) readily shows that classical polarography is the least suitable for this purpose, since in this case the drop time can vary within narrow limits usually from 1 to 8 seconds. It is possible, however, that at transport rate values corresponding to these drop times, certain electrode processes will have reached the equilibrium state; when the rate constant of the electrode process is large and the process is reversible, or when the rate constant is small and the process is irreversible.

In the remaining three methods a different situation obtains. If under certain conditions the process reaches the equilibrium state (i.e. in a reversible process), its kinetics can be measured at a faster mass transport rate by suitably increasing the electrode rotation rate in the rotating disc method, or the potential change rate in stationary electrode voltammetry. In chronopotentiometry the current density should be increased to such an extent that the transition time is shortened considerably.

Theoretically the values of τ, V and ω can be changed at will, but in practice the useful range of time parameters is considerably restricted.

In stationary electrode voltammetry it is not advisable to use very slow polarization rates, since the process is then considerably affected by convection. For this reason the practical lower limit of the useful potential change rate is 0.1 V/min. The use of very fast polarization rates is limited by the effect of the double layer charging on the faradaic electrode process. Results of our experiments indicate that it is difficult to use in kinetic studies polarization rates exceeding 100 V/s when conventional apparatus is employed. Therefore it will be assumed in our further discussions that this is the fastest useful polarization rate, although faster rates have sometimes been used in investigations of the kinetics of electrode processes.

In chronopotentiometry the time parameter, i.e. the transition time, is limited by similar factors. At transition times exceeding 60 s the convection effect is considerable and leads to results which are not reproducible and which are contrary to theory. Also in this method the effect of charging the double layer is large when the time parameter is small. Therefore it appears that a transition time of the order of 10^{-3} s is the shortest time which can be used without causing considerable distortion, due to the effects of the double layer.

In the case of rotating disc methods the electrode rotation rate cannot be very slow, since in such a case, using a continuous change of electrode potential by means of a polarograph, it is possible to obtain curves showing the current peak. Also in the part of the current–voltage curve preceding the limiting current the current would be increased by the diffusion current. This can be better explained on the basis of analysis of the Nernst and Levich equations, according to which the current tends to zero when the electrode rotation rate tends to zero. This conclusion is obviously wrong, since at rotation rate close to zero the current is still flowing and the current vs. voltage curves would be similar to stationary electrode voltammetric curves.

The limitations of the use of very fast electrode rotation rates are not due to technical difficulties in constructing the suitable apparatus, but to the disappearance of laminarity and the appearance of turbulent flow.

According to Levich [9], when a smooth and well centred electrode is used, the transport ceases to be laminar at Reynolds number values (Re) between 10^4 and 10^5. The dimensionless Reynolds number is defined by the equation

$$Re = \frac{r^2 \omega}{\nu} \tag{3.32}$$

where r is the radius of the disc electrode, and ν is the kinematic viscosity of the solution.

It follows from equation (3.32) that it is easier to maintain laminar flow when the electrode radius is small (the dimensions of the working electrode and also any outer planes surrounding the electrode and constructed from an insulator must be taken into account), but the electrode must not be too small, since in such cases considerable boundary effects appear. However, disc electrodes having a radius even smaller than 1 mm have been used [10]. If it is assumed that the smallest radius of the electrode is 1.4 mm (conductor plus the insulating screen), and further that $\nu = 10^{-2}$ cm^2 s^{-1}, and the limiting value of Re is 2×10^4, we find that the fastest electrode rotation rate still leading to laminar flow is 10^4 rad/s.

On the basis of equations (3.28)–(3.30), using the above range of useful time parameters and assuming that the diffusion coefficient $D = 9 \times 10^{-6}$ cm^2 s^{-1}, the ranges of transport rates characteristic of the methods discussed have been calculated. The results of these calculations are shown in Table 3.1, from which it can be seen that in the classical polarographic method the range of change of mass transport rate is narrow. The other methods, and in particular chronopotentiometry and stationary electrode voltammetry, are much more suitable.

The data in Table 3.1 show that polarographic processes are reversible when their charge transfer rate constant is much larger (e.g. ten times larger) than the maximum material transport rate, i.e. when the constant is larger than 2.4×10^{-2} cm/s. The data in this Table also show that when the rate constant is 2.4×10^{-2} cm/s (polarographically reversible), it is chronopotentiometrically practically irreversible at transition times of the order of 10^{-3} s, since in this case the transport rate is several times faster than the charge transport rate. It is also practically irreversible in stationary electrode voltammetry at high rates of potential change.

Table 3.1 MINIMUM AND MAXIMUM VALUES OF KINETIC PARAMETERS

Method	Kinetic parameter X	Minimum value X (s)	Maximum transport rate (cm/s) corresponding to X minimum	Maximum value X (s)	Minimum transport rate (cm/s) corresponding to X maximum
Polarography	t_1	2	2.4×10^{-3}	8	1.5×10^{-3}
Chronopotentiometry*	τ	2×10^{-3}	6×10^{-2}	60	3.4×10^{-3}
Stationary electrode voltammetry*	$1/V$	10^{-2}	1.2×10^{-1}	590	4.8×10^{-4}
Rotating disc method	$1/\omega$	1.3×10^{-3}	1.6×10^{-2}	8.0×10^{-2}	2×10^{-3}

* V expressed in Hz

It follows from this discussion that the rate of depolarizer transport to the electrode, which depends on experimental conditions, plays an important part in the description of reversibility, which also depends on the nature of the process and on the rate of charge transfer between the oxidized and the reduced species. It is necessary to consider how the process rate is defined and what factors affect this rate. Let us consider reduction of a substance Ox, which can be expressed schematically by means of the following equation:

$$Ox + ne \rightarrow Red. \tag{3.33}$$

Let us assume that the substrate and the product of the electrode reaction are soluble in the solution and that the product dissolves in the electrode material, such as mercury, giving an amalgam, and further that the electrode process is not accompanied by chemical generation of the depolarizer, or by deactivation of the product, and that specific adsorption of Ox, Red and the supporting electrolyte ions does not take place on the electrode. If these assumptions are made the electrode process (3.33) can be divided into several stages. This was done by Delahay [11] who followed an idea of Parsons [12]. He divided the electrode process into the following stages:

(a) substance Ox is present in the bulk of solution beyond the borders of the diffusion double layer, while n electrons are present in the electrode;

(b) substance Ox is present in the inner plane of maximum approach, and n electrons are still present in the electrode;

(c) the transition state is reached; according to the generally accepted terminology the parameters of this state are marked with the symbol $\ddagger$;

(d) after the electron transfer the resulting substance Red is present in the plane of maximum approach;

(e) substance Red diffuses away from the diffusion double layer.

The standard free energies G^0 of these reaction stages can be expressed as follows:

$$G_a^0 = \mu_{Ox}^0 + n\mu_e^0 - nF\Phi_M, \tag{3.34}$$

$$G_b^0 = \mu_{Ox}^0 + n\mu_e^0 + zF\Phi_2 - nF\Phi_M, \tag{3.35}$$

$$G_d^0 = \mu_{Red}^0 + z'F\Phi_2, \tag{3.36}$$

$$G_e^0 = \mu_{Red}^0, \tag{3.37}$$

where Φ_2 is the potential in the maximum approach plane, Φ_M is the inner potential of the metal, referred to the inner potential of the solution, which is arbitrarily assumed to be equal to zero, z and z' are the charges of ions of Ox and Red, the signs being retained, μ_{Ox}^0, μ_{Red}^0, and μ_e are the standard chemical potentials of Ox, Red, and electrons in the metal, respectively.

In order to calculate the rate of reaction (3.33) by means of the theory of absolute reaction rates [13] it is necessary to know the values of the standard free energies of activation of the oxidation and reduction reactions. In the case of reduction this energy is equal to the difference between the free energies of the transition and the initial (a) states: $\overrightarrow{\Delta G_{\ddagger}^0} = G_{\ddagger}^0 - G_a^0$; similarly the energy of activation of the reverse reaction is equal to the difference between the energies of the transition and the final (e) states: $\overleftarrow{\Delta G_{\ddagger}^0} = G_{\ddagger}^0 - G_e^0$.

Theoretical calculation of the value of $\Delta G_{\ddagger}^0$ would require the postulation of a detailed model of the transition state which, in the present state of our knowledge, is not always possible. Therefore, in order to overcome this difficulty it is assumed that the potential-dependent part of the function $G_{\ddagger}^0 - G_b^0$ is a certain fraction α depending on the potential of a part of the function $G_d^0 - G_b^0$. Thus we introduce into the reasoning the coefficient α called the cathodic transition coefficient, or the coefficient of electron transfer of the cathodic process. It was first used in the discussion of the kinetics of electrode processes by Erdey-Gruz and Volmer [14] in 1930, although the problem of the effect of potential on electrode kinetics was earlier considered by Butler [15] and Audubert [16].

The relation between $G_{\ddagger}^0 - G_b^0$ and $G_d^0 - G_b^0$ is expressed by the following equation:

$$_{el}G_{\ddagger}^0 - {}_{el}G_b^0 = \alpha({}_{el}G_d^0 - {}_{el}G_b^0) \tag{3.38}$$

where the subscript el marks the potential-dependent parts of the standard free energies.

On the basis of (3.36) and (3.35) equation (3.38) can be written in the form:

$$_{el}G_{\ddagger}^0 - {}_{el}G_b^0 = \alpha[(z'-z)F\Phi_2 + nF\Phi_M] = \alpha nF(\Phi_M - \Phi_2), \tag{3.39}$$

since $n = z' - z$.

In the derivation of equations (3.38) and (3.39) it was assumed that the potential-dependent part of the standard free energy changes smoothly on passing from state (b) to state (c).

The correctness of equations (3.38) and (3.39) has been questioned by van Rysselberghe [17].

The standard free energy of activation of reduction $\overrightarrow{\Delta G_{\ddagger}^0}$ can be resolved into two components; the chemical part $_{ch}\overrightarrow{\Delta G_{\ddagger}^0}$ and the electrochemical part $_{el}\overrightarrow{\Delta G_{\ddagger}^0}$:

$$\overrightarrow{\Delta G_{\ddagger}^0} = G_{\ddagger}^0 - G_a^0 = {}_{ch}\overrightarrow{\Delta G_{\ddagger}^0} + {}_{el}\overrightarrow{\Delta G_{\ddagger}^0} = {}_{ch}\overrightarrow{\Delta G_{\ddagger}^0} + {}_{el}G_{\ddagger}^0 - {}_{el}G_a^0. \tag{3.40}$$

Using equation (3.39) the following is obtained from equation (3.40):

$$\overrightarrow{\Delta G_{\ddagger}^0} = {}_{ch}\overrightarrow{\Delta G_{\ddagger}^0} + [{}_{el}G_{\ddagger}^0 - {}_{el}G_b^0] + [{}_{el}G_b^0 - {}_{el}G_a^0]$$

$$= {}_{ch}\overrightarrow{\Delta G_{\ddagger}^0} + \alpha nF(\Phi_M - \Phi_2) + zF\Phi_2. \tag{3.41}$$

Similarly the standard free energy of activation of oxidation is expressed as:

$$\overleftarrow{\Delta G_{\ddagger}^0} = G_{\ddagger}^0 - G_d^0 = {}_{el}\overleftarrow{\Delta G_{\ddagger}^0} + {}_{ch}\overleftarrow{\Delta G_{\ddagger}^0} = {}_{ch}\overleftarrow{\Delta G_{\ddagger}^0} + {}_{el}G_{\ddagger}^0 - {}_{el}G_d^0 \tag{3.42}$$

and after further transformations:

$$\overleftarrow{\Delta G_{\ddagger}^0} = {}_{ch}\overleftarrow{\Delta G_{\ddagger}^0} + [{}_{el}G_{\ddagger}^0 - {}_{el}G_e^0] + [{}_{el}G_e^0 - {}_{el}G_d^0]$$

$$= {}_{ch}\overleftarrow{\Delta G_{\ddagger}^0} + (\alpha - 1)nF(\Phi_M - \Phi_2) + z'F\Phi_2. \tag{3.43}$$

Using the theory of absolute reaction rates and assuming that the transmission coefficient equals 1 the expressions describing the currents of reduction (i_{cathod}) and of oxidation (i_{anod}) are obtained:

$$i_{cathod} = nFA \frac{kT}{h} a_{Ox} \exp\left(\frac{-\overrightarrow{\Delta G^0_{\ddagger}}}{RT}\right) \quad (3.44)$$

and:

$$i_{anod} = nFA \frac{kT}{h} a_{Red} \exp\left(\frac{-\overleftarrow{\Delta G^0_{\ddagger}}}{RT}\right) \quad (3.45)$$

where k is the Boltzmann constant, h is Planck's constant, and a_{Ox} and a_{Red} are the activities of substances Ox and Red, respectively, in the bulk of solution.

Equations (3.44) and (3.45) can be suitably combined with equations (3.41) and (3.43). Putting $f = F/RT$:

$$i_{cathod} = nFA \frac{kT}{h} a_{Ox} \exp\left(-\frac{_{ch}\overrightarrow{\Delta G^0_{\ddagger}}}{RT}\right) \exp(-\alpha n f \Phi_M) \exp(\alpha n - z) f \Phi_2$$

$$(3.46)$$

and:

$$i_{anod} = nFA \frac{kT}{h} a_{Red} \exp\left(-\frac{_{ch}\overleftarrow{\Delta G^0_{\ddagger}}}{RT}\right) \exp[(1-\alpha) n f \Phi_M] \exp[(\alpha n - z) f \Phi_2]$$

$$(3.47)$$

Strictly speaking these equations should contain the activity coefficients of active complexes, which may depend on the potential.

Assuming that the reduction current has positive sign and the oxidation current is negative, the total current is:

$$i = i_{cathod} - i_{anod}. \quad (3.48)$$

Since, at the equilibrium potential, $i_{anod} = i_{cathod}$, the total current is equal to zero, although the exchange of electrons between the Ox and Red form takes place. The rate of this exchange depends on the kinetics of the electrode process and is often characterized by the exchange current. The concept of exchange current was introduced in 1936 by Butler [18] and it gained rapid acceptance. The term "exchange current" was first used by Dolin and Ershler [19] in 1940.

The exchange current observed is usually given the symbol i^0, whereas the symbol i_t^0 stands for the true exchange current which is measured when $\Phi_2 = 0$, i.e. when the double layer effect is eliminated. The relation between these two currents is described by the equation:

$$i^0 = i_t^0 \exp[(\alpha n - z)f\Phi_2]. \tag{3.49}$$

Combining equations (3.46), (3.47) and (3.49) we obtain:

$$i = i_t^0 \exp[(\alpha n - z)f\Phi_2]\{\exp[-\alpha nf(\Phi_M - \Phi_M^e)] - \exp[(1-\alpha)nf(\Phi_M - \Phi_M^e)]\} \tag{3.50}$$

where Φ_M^e is the equilibrium potential.

Until recently, exchange current values uncorrected for the double layer effect were quoted in the literature, and for this reason it is convenient to transform equation (3.50) by introducing $i^0 = i_t^0 \exp[(\alpha n - z)f\Phi_2]$:

$$i = i^0 \{\exp[-\alpha nf(\Phi_M - \Phi_M^e)] - \exp[(1-\alpha)nf(\Phi_M - \Phi_M^e)]\}. \tag{3.51}$$

Since overvoltage is given by the formula

$$\eta = \Phi_M - \Phi_M^e = E - E^0 \tag{3.52}$$

where E stands for potentials determined with respect to a definite reference electrode, equation (3.51) can be expressed in a simpler form:

$$i = i^0[\exp(-\alpha nf\eta) - \exp(1-\alpha)nf\eta]. \tag{3.53}$$

When the overvoltage of an electrode process is small, $|\eta| \ll 1/\alpha nf$ or $|\eta| \ll 1/(1-\alpha)nf$, equation (3.53) can be made linear. The following simple expression is obtained:

$$i = -i^0 nf\eta \tag{3.54}$$

which is often used in electrochemical relaxation methods.

When the overvoltages are very large as compared with $1/\alpha nf$ and $1/(1-\alpha)nf$, equation (3.53) is reduced to a simpler form, since it is possible to neglect one of the terms in square brackets. Then for cathodic processes we obtain:

$$i_{cathod} = i^0 \exp(-\alpha nf\eta), \tag{3.55}$$

and for anodic processes:

$$i_{anod} = i^0 \exp(1-\alpha)nf\eta. \tag{3.56}$$

In general these equations can be used when the overvoltages exceed 0.1 V. Then the electrode process is completely irreversible and consequently one of the terms of equation (3.53) can be neglected. This is equivalent to the assumption that, at large cathodic overvoltages, the anodic process has only a small effect on the current. Similarly, at large anodic overvoltages the cathodic process has only a minor effect on the current.

These expressions are illustrated by the curves shown in Fig. 3.3. It may be seen that in an irreversible electrode process (curve b) at reduction potentials the oxidation rate is very slow. In a reversible process (curve a) the total current is affected by the anodic and the cathodic processes according to equation (3.48).

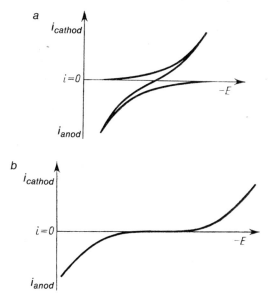

Fig. 3.3 Schematic anodic-cathodic current-voltage curves. a—reversible process; b—irreversible process.

Equations (3.55) and (3.56) can be written as follows:

$$\eta = \frac{2.303}{\alpha nf} \log i^0 - \frac{2.303}{\alpha nf} \log i, \qquad (3.55a)$$

$$\eta = -\frac{2.303}{(1-\alpha)nf} \log i^0 + \frac{2.303}{(1-\alpha)nf} \log i, \qquad (3.56a)$$

i.e. they have the form of the Tafel equation [20]:

$$\eta = a \pm b \log i. \qquad (3.57)$$

Comparison of equations (3.55a), (3.56a) and (3.57) allows the physical meaning of coefficients a and b of the Tafel equation to be readily seen. It also follows from these equations that for large overvoltages (i.e. in irreversible processes) the relation between overvoltage and the logarithm of current intensity is linear.

Apart from exchange currents, rate constants are used in descriptions of rates of electrode processes in a manner analogous to chemical kinetics. In electrode kinetics the dimensions of the constants are cm s^{-1}. The real standard rate constant k_s^t of an electrode process is described by the following equation:

$$k_s^t = \frac{kT}{h} \exp\left(-\frac{_{ch}\overrightarrow{\Delta G_{\ddagger}^0}}{RT}\right) \exp(-\alpha n f \Phi_M^0)$$

$$= \frac{kT}{h} \exp\left(-\frac{_{ch}\overleftarrow{\Delta G_{\ddagger}^0}}{RT}\right) \exp(1-\alpha) n f \Phi_M^0 \qquad (3.58)$$

where Φ_M^0 stands for the standard potential of the system studied. This is connected with the equilibrium potential according to the Nernst equation:

$$\Phi_M^{e^-} = \Phi_M^0 + \frac{1}{nf} \ln \frac{a_{Ox}}{a_{Red}}. \qquad (3.59)$$

In electroanalytical practice the observed standard rate constants of electrode processes, k_s, are often used. These are somewhat dependent on experimental conditions at variable potential Φ_2. The relation between the two constants can be expressed as follows:

$$k_s = k_s^t \exp(\alpha n - z) f \Phi_2. \qquad (3.60)$$

For the reasons discussed above, when $\Phi_2 = 0$, $k_s = k_s^t$.

Since some authors give the results of kinetic studies in the form of standard exchange current values, while others quote standard rate constants, it should be remembered that these two kinetic constants are related as follows:

$$i^0 = nFk_s a_{Ox}^{(1-\alpha)} a_{Red}^{\alpha} \qquad (3.61)$$

whereas i_t^0 and k_s^t are related according to the equation:

$$i_t^0 = nFk_s^t a_{Ox}^{(1-\alpha)} a_{Red}^{\alpha}. \qquad (3.62)$$

The exchange current and the standard rate constant are related according to equations (3.61) and (3.62). Therefore the results of kinetic studies can, in principle, be reported by quoting k_s or i^0 values. However, it should be remembered that the exchange current increases with increasing concentration of the Ox and Red forms. For this reason standard exchange current values referred to unit concentration of the Ox and Red forms are often used.

Constants k_s indicate the rates of electrode processes directly.

The values of standard rate constants of various electrode processes are shown in Table 3.2.

A large collection of rate constants has been published by Tanaka and Tamamushi [21] and, more recently, by Tamamushi [22].

In addition to standard rate constants which characterize the kinetics of electrode processes, other constants are used in practical work. They are usually called k_{fh}^0 and k_{bh}^0 for cathodic and anodic processes, respectively. The subscripts fh and bh indicate that the constants characterize heterogeneous forward and reverse processes respectively.

These constants characterize electrode process rates at electrode potentials equal to zero, usually referred to the normal hydrogen electrode. Comparison of such constants for different electrode processes cannot lead to the true comparison of kinetics of these processes, since standard potentials of the systems compared are different.

Table 3.2 STANDARD RATE CONSTANTS OF SOME ELECTRODE PROCESSES

Electrode reaction	Supporting electrolyte	Electrode	k_s (cm/s)	References
$Bi^{3+} + 3e \rightarrow Bi$	$1\,M\,HClO_4$	mercury	3.0×10^{-4}	[23]
$Cd^{2+} + 2e \rightarrow Cd$	$1\,M\,KNO_3$	mercury	~ 1.0	[24]
$Ce^{4+} + e \rightarrow Ce^{3+}$	$1\,M\,H_2SO_4$	platinum	3.7×10^{-4}	[25]
$Cr^{3+} + e \rightarrow Cr^{2+}$	$1\,M\,KCl$	mercury	1.0×10^{-5}	[26]
$Cr(CN)_6^{3-} + e \rightarrow Cr(CN)_6^{4-}$	$0.1\,M\,KCN + 0.1\,M\,KCl$	mercury	5.2×10^{-2}	[27]
$Cr(CN)_6^{3-} + e \rightarrow Cr(CN)_6^{4-}$	$0.1\,M\,KCN + 0.7\,M\,KC$	mercury	9.1×10^{-1}	[27]
$Cs^+ + e \rightarrow Cs$	$1\,M\,N(CH_3)_4OH$	mercury	2×10^{-1}	[23]
$Fe^{3+} + e \rightarrow Fe^{2+}$	$1\,M\,H_2SO_4$	platinum	5.3×10^{-3}	[28]
$Fe(CN)_6^{3-} + e \rightarrow Fe(CN)_6^{4-}$	$1\,M\,KCl$	platinum	9×10^{-2}	[26]
$Hg^+ + e \rightarrow Hg$	$0.2\,M\,HClO_4$	mercury	3.5×10^{-1}	[29]
$Ni^{2+} + 2e \rightarrow Ni$	$2.5\,M\,Ca(ClO_4)_2$	mercury	1.6×10^{-7}	[30]
$Pb^{2+} + 2e \rightarrow Pb$	$1\,M\,HClO_4$	mercury	2.0	[31]
$Tl^+ + e \rightarrow Tl$	$1\,M\,HClO_4$	mercury	1.8	[31]
$V^{3+} + e \rightarrow V^{2+}$	$1\,M\,HClO_4$	mercury	3.2×10^{-3}	[32]
$V^{3+} + e \rightarrow V^{2+}$	$1\,M\,H_2SO_4$	mercury	1.03×10^{-3}	[33]
$V^{3+} + e \rightarrow V^{2+}$	$0.5\,M\,K_2C_2O_4$	mercury	1.4×10^{-3}	[26]
$Zn^{2+} + 2e \rightarrow Zn$	$1\,M\,KCl$	mercury	6×10^{-3}	[34]
$Zn^{2+} + 2e \rightarrow Zn$	$1\,M\,KBr$	mercury	8×10^{-3}	[23]
$Zn^{2+} + 2e \rightarrow Zn$	$1\,M\,KI$	mercury	7×10^{-2}	[23]
$Zn^{2+} + 2e \rightarrow Zn$	$1\,M\,KSCN$	mercury	1.7×10^{-2}	[23]

The constants k_{fh}^0 and k_{bh}^0 are related to the standard rate constant k_s according to the following equations:

$$k_s = k_{fh}^0 \exp\left[-\frac{\alpha n F E^0}{RT}\right] \tag{3.63}$$

and

$$k_s = k_{bh}^0 \exp\left[\frac{(1-\alpha)nFE^0}{RT}\right] \tag{3.64}$$

where E^0 is the electrode potential with respect to a standard reference electrode, such as the saturated calomel electrode.

It follows from equations (3.63) and (3.64) that when E^0 and α are known k_s can be readily calculated from k_{fh}^0 and k_{bh}^0 values. These constants can be calculated for any potential by means of the following equations:

$$k_{fh} = k_{fh}^0 \exp\left[-\frac{\alpha n F E}{RT}\right] \tag{3.65}$$

$$k_{bh} = k_{bh}^0 \exp\left[\frac{(1-\alpha)nFE}{RT}\right] \tag{3.66}$$

where k_{fh} and k_{bh} are the heterogeneous rate constants of the electrode process at the potential E.

It follows from equations (3.65) and (3.66) that when $E = 0$, $k_{fh} = k_{fh}^0$ and $k_{bh} = k_{bh}^0$.

Equations (3.65) and (3.66) express the dependence of rate constants of electrode processes on the electrode potential.

In recent years the mechanisms and kinetics of electron transfer between the electrode and the reacting ion or molecule have been studied. The reactions investigated are relatively simple and they do not involve rupture and formation of bonds, but the electron transfer causes changes in bond lengths in the reacting molecules, or solvated ions, and also certain changes in the solvation sheath.

Work leading to the above results was carried out by Marcus [35] in the period since 1956. The results of this work have been improved and ex ended in further work [36–41] and have been the subject of a review [42].

Theoretical aspects of electrode kinetics have also been studied by Levich and Dogonadze and their co-workers. These authors have published several papers [43–51] on the theory of kinetics of homogeneous redox electrode reactions taking place on metal electrodes and on electrodes consisting of semiconductors. These studies have been reviewed partly by Levich [52, 53] and by Dogonadze and Kuznetsov [54].

Earlier in this chapter we introduced the term transition coefficient in order to describe the electrochemical activation energy of electrode processes. In modern theories of electrode processes electrochemical activation energy is described in detail.

In the description of activation energy on the basis of the generally accepted theories developed by Marcus, Levich and Dogonadze, which have been elaborated in greater detail than other theories, the transition coefficient does not appear explicitly in the equations, but its value can be obtained by comparing the expressions describing the activation energy or current with those obtained by classical methods and involving α or β.

We will present briefly the principles and the fundamental equations obtained by Marcus [40–42]. His theory and also that of Levich and Dogonadze concerns in principle adiabatic electrode processes in which, during the electron transfer, new bonds are not formed and the existing ones are not ruptured. For such processes we can assume that both Red and Ox forms remain in solution. Examples of such processes are electron exchanges in $Fe(H_2O)_6^{3+}/Fe(H_2O)_6^{2+}$, MnO_4^-/MnO_4^{2-} and $Fe(CN)_6^{3-}/Fe(CN)_6^{4-}$ systems.

Assuming that the electron exchange is the rate-determining step and the other steps are either equilibrium states or are fast, according to Marcus the free electrochemical activation energy of the cathode process can be described by the equation:

$$\overline{\Delta G_{\ddagger}^0} = \frac{\lambda}{4} + \frac{w^{Red}+w^{Ox}}{2} + \frac{nF(E-E_f^0)}{2} + \frac{[nF(E-E_f^0)+(w^{Red}-w^{Ox})]^2}{4\lambda} \quad (3.67)$$

where w^{Ox} and w^{Red} stand for the electrical work of transfer of one mole of substance Ox and Red from the bulk of solution to the outer Helmholtz plane, which is regarded as identical with the charge transfer plane. When specific adsorption of Ox and Red is absent

$$w^{Ox} = z_{Ox}F\Phi_2 \quad (3.68)$$

and

$$w^{Red} = z_{Red}F\Phi_2; \quad (3.69)$$

since $z_{Ox} = z_{Red}+n$ we have

$$w^{Red} = (z_{Ox}-n)F\Phi_2. \quad (3.69a)$$

When potential Φ_2 is equal to zero, w^{Ox} and w^{Red} in equation (3.67) can be neglected. These parameters are also absent when the substances reacting with the electrode are electrically neutral.

E_f^0 in equation (3.67) is the formal potential, whereas λ can be derived theoretically from Marcus's equations and reasoning:

$$\lambda = \lambda_0 + \lambda_i. \qquad (3.70)$$

λ_0 depends on the sizes and shapes of the reacting substances. For spherical molecules participating in the electrochemical reaction λ_0 is given by the expression

$$\lambda_0 = \frac{1}{2}(ne)^2 \left(\frac{1}{a} - \frac{1}{r}\right)\left(\frac{1}{D_0} - \frac{1}{D_s}\right) \qquad (3.71)$$

where a is the radius of the reacting molecule, r is twice the distance from the centre of the molecule to the electrode surface, D_0 is the square of the refractive index and D_s is the static dielectric constant.

The value of λ_i depends on the magnitude of changes in bond lengths and bond angles in the inner coordination sphere of reacting substances and on the values of the force constants of vibrations.

Equation (3.67) can be formally written as:

$$\overrightarrow{\Delta G^0_\ddagger} = {}_{ch}\overrightarrow{\Delta G^0_\ddagger} + {}_{el}\overrightarrow{\Delta G^0_\ddagger}, \qquad (3.72)$$

$$_{ch}\overrightarrow{\Delta G^0_\ddagger} = \frac{\lambda}{4}, \qquad (3.73)$$

and

$$_{el}\overrightarrow{\Delta G^0_\ddagger} = \frac{(2z_{Ox}-n)F\Phi_2}{2} + \frac{nF(E-E_f^0)}{2} + \frac{n^2F^2[(E-E_f^0)-\Phi_2]^2}{4\lambda}. \qquad (3.74)$$

When the effect of potential Φ_2 can be neglected ($\Phi_2 = 0$) and the electrode potential is equal to the formal potential $\lambda/4$ is equal to the standard free activation energy of the electrode process and its connection with the standard rate constant is given by the equation

$$k_s = \varkappa \varrho Z_h \exp\left(-\frac{\lambda}{4RT}\right) \qquad (3.75)$$

where Z_h is the frequency of collision.

The process of oxidation at the electrode can be considered similarly:

$$\overleftarrow{\Delta G^0_\ddagger} = \frac{\lambda}{4} + \frac{w^{Ox}+w^{Red}}{2} - \frac{nF(E-E_f^0)}{2} + \frac{[nF(E-E_f^0)+(w^{Red}-w^{Ox})]^2}{4\lambda}. \qquad (3.76)$$

Utilizing equations (3.68) and (3.69) we can express the electrical part of the activation energy of the oxidation process by means of the equation:

$$_{el}\overrightarrow{\Delta G^0_{\ddagger}} = \frac{(2z_{ox}-n)F\Phi_2}{2} - \frac{nF(E-E^0_f)}{2} + \frac{n^2F^2[(E-E^0_f)-\Phi_2]^2}{4\lambda}. \quad (3.77)$$

These relationships lead to two important conclusions. The first one is that the Tafel equations which were discussed earlier are a simplified form of more complex relationships and the second one that the electrical transition coefficient is potential dependent. These problems have been discussed by Mohilner [55] on the basis of Marcus's ideas.

Utilizing the above relationships we can express the rate constants of anodic and cathodic processes as follows:

$$k_{fh} = \varkappa_\varrho Z_h \exp\left(-\frac{\overrightarrow{\Delta G^0_{\ddagger}}}{RT}\right) = \varkappa_\varrho Z_h \exp\left(-\frac{\lambda}{4RT}\right) \exp\left(-\frac{_{el}\overrightarrow{\Delta G^0_{\ddagger}}}{RT}\right), \quad (3.78)$$

$$k_{bh} = \varkappa_\varrho Z_h \exp\left(-\frac{\overleftarrow{\Delta G^0_{\ddagger}}}{RT}\right) = \varkappa_\varrho Z_h \exp\left(-\frac{\lambda}{4RT}\right) \exp\left(-\frac{_{el}\overleftarrow{\Delta G^0_{\ddagger}}}{RT}\right). \quad (3.79)$$

In the case of an irreversible electrode process the anodic and cathodic currents are given by the following equations:

$$i_{cathod} = nFAk_{fh}C^0_{Ox} = nFAk_sC^0_{Ox}\exp\left(-\frac{(2z_{ox}-n)F\Phi_2}{2RT}\right) \times$$

$$\exp\left(-\frac{nF\eta}{2RT}\right) \exp\left(-\frac{n^2F^2(\eta-\Phi_2)^2}{4\lambda RT}\right), \quad (3.80)$$

$$i_{anod} = nFAk_{bh}C^0_{Red} = nFAk_sC^0_{Red}\exp\left(-\frac{(2z_{ox}-n)F\Phi_2}{2RT}\right) \times$$

$$\exp\left(\frac{nF\eta}{2RT}\right) \exp\left(-\frac{n^2F^2(\eta-\Phi_2)^2}{4\lambda RT}\right). \quad (3.81)$$

Using the observed exchange current introduced earlier we can write equations (3.80) and (3.81) in a shorter form:

$$i_{cathod} = i^0 \exp\left(-\frac{nF\eta}{2RT}\right) \exp\left(-\frac{n^2F^2(\eta-\Phi_2)^2}{4\lambda RT}\right), \quad (3.80a)$$

$$i_{anod} = i^0 \exp\left(\frac{nF\eta}{2RT}\right) \exp\left(-\frac{n^2F^2(\eta-\Phi_2)^2}{4\lambda RT}\right). \quad (3.81a)$$

These relationships can be compared with equations (3.55) and (3.56) which were derived on the basis of formal considerations.

It follows from equations (3.80a) and (3.81a) that potential dependences of log i_{cathod} and log i_{anod} are not linear as appears from equations (3.55) and (3.56). However it should be stressed that so far deviations from the linear dependence of log $i-E$ which would confirm equations (3.80a) and (3.81a) have not been observed. As mentioned earlier, this problem is connected with the potential-dependence of the transition coefficient.

Assuming that $\Phi_2 = 0$ it is possible to derive the expression for the transition coefficient from equations (3.55) and (3.80a):

$$\alpha = \frac{1}{2} + \frac{nF(\eta - \Phi_2)}{2\lambda}. \tag{3.82}$$

This equation shows that in the region of small overvoltages and when neither Ox nor Red forms are specifically adsorbed on the electrode surface the transition coefficient should be equal to 1/2 and should depend on overvoltage at least in a certain region. This was demonstrated in a simple way by Oldham [56].

Combining in pairs equations (3.63) and (3.65) and (3.64) and (3.66) we obtain the expressions:

$$k_{fh} = k_s \exp\left[-\frac{\alpha nF(E-E_f^0)}{RT}\right] \tag{3.83}$$

and

$$k_{bh} = k_s \exp\left[\frac{(1-\alpha)nF(E-E_f^0)}{RT}\right]. \tag{3.84}$$

Equation (3.83) shows that log k_{fh} constantly increases with increasing negative potential values, but this increase is limited by the frequency of collisions of the reacting substance with the surface of the electrode. The existence of the upper limit of rate constants of electrode processes, k_{fh}^m, was foreseen earlier [57, 58].

Thus when $E \to -\infty$

$$k_{fh} = k_{fh}^m. \tag{3.85}$$

Utilizing the results of Reiss's work [59] Oldham described the upper limit of the rate constants by means of the equation:

$$k_{fh}^m = \frac{3D}{3L} \tag{3.86}$$

where D is the diffusion coefficient of the reacting substance and L is the mean jump length, which in the case of aqueous solutions is equal to about 1.5 Å. Assuming that $D = 10^{-5}$ cm^{2}s^{-1} we obtain from equation (3.86) the maximum value of $k_{fh}^m = 10^3$ cm s^{-1}.

Rates of Electrode Processes

Considering similarly equation (3.84) it can be shown that k_{bh} will reach a similar extreme value (slightly different due to a different diffusion coefficient) when the electrode potential $E \to +\infty$.

The non-dependence of the rate constants on the potential means that the transition coefficients of these processes become equal to zero.

Combination of equations (3.83) and (3.84) with the elimination of k_s gives the equation:

$$\frac{k_{bh}}{k_{fh}} = \exp\frac{[nF(E-E_f^0)]}{RT} \tag{3.87}$$

which shows that when k_{fh} reaches a maximum value at very high positive potential values the potential dependence of k_{fh} is given by the relationship

$$k_{fh}(E \to +\infty) = k_{bh}^m \exp\left[-\frac{nF(E-E_f^0)}{RT}\right]. \tag{3.88}$$

This equation shows that in these conditions the transition coefficient of the cathodic process is equal to unity, since in this region the whole change of the electrode potential and not only a part of this potential modifies the cathodic rate constant.

Thus the consideration of the potential dependence of the cathodic rate constants leads to three equations describing this dependence as functions of overvoltage, i.e. equations (3.85), (3.88) and (3.83). The last one describes the potential dependence of k_{fh} for potentials close to the formal potential. This equation is usually employed.

Oldham described the general potential dependence of k_{fh} by means of the equation:

$$k_{fh} = \frac{k_s k_{fh}^m k_{bh}^m}{k_s k_{bh}^m + k_{fh}^m k_{bh}^m \exp\left[\dfrac{\alpha nF(E-E_f^0)}{RT}\right] + k_s k_{fh}^m \exp\left[\dfrac{nF(E-E_f^0)}{RT}\right]}. \tag{3.89}$$

A similar equation expresses the changes of k_{bh}:

$$k_{bh} = \frac{k_s k_{fh}^m k_{bh}^m}{k_s k_{bh}^m \exp\left[-\dfrac{nF(E-E_f^0)}{RT}\right] + k_{fh}^m k_{bh}^m \exp\left[-(1-\alpha)\dfrac{nF(E-E_f^0)}{RT}\right] + k_s k_{fh}^m}. \tag{3.90}$$

These equations are valid for any potential.

On the basis of these equations Oldham derived the overvoltage dependence of $\log k_s$ which is shown in Fig. 3.4.

The values used in the calculations were as follows:

$T = 298$ K, $n = 2$, $k_{fh}^m = 10^3$ cm s^{-1}, $k_s = 10^{-1}$ cm s^{-1} and $\alpha = 0.30$. The last two values are close to the data characterizing the electrode reaction of cadmium on mercury electrodes.

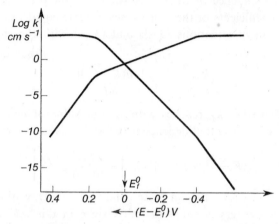

Fig. 3.4 Dependence of rate constants k_{fh} and k_{bh} on electrode potential according to Oldham (from [56] by permission of the copyright holders, Elsevier Publishing Co.).

The relationships shown in Fig. 3.4 indicate that equation (3.83) is valid in a wide potential range of 400 mV. This range would be even wider if the electrode reaction were a one-electron reaction with lower standard rate constant.

Equations (3.89) and (3.90) are obviously approximations. They show the transition of α from 0 to 1 in the form of jumps (straight sections of curves shown in Fig. 3.4). In reality the potential dependence between the extreme values should correspond to a smooth curve.

The problems considered by Oldham were also discussed by Reinmuth [60] who showed that the limitation due to the frequency of collisions should be reflected not in the impedance of charge transfer of the Faraday reaction but in the impedance of transport. He also proved the possibility of existence of electrode processes in which the rate is not limited by the frequency of collisions with the electrode. An example of such processes is electron exchange between adsorbed reagents.

So far the dependence of the transition coefficient on electrode potential has not been convincingly demonstrated. Parsons and Passeron [61] claimed in their paper published in 1966 that they observed this dependence in the case of the electron exchange Cr(III)+$e \rightarrow$ Cr(II) in perchlorate media but later Anson and co-workers [62] came to the conclusion that this relationship is doubtful, since the accuracy of existing methods used for studying electrode kinetics is insufficient for convincing demonstration of such relationships.

It should be stressed that the potential dependence of α is foreseen also by the theory of Levich and Dogonadze and by the concept of Schmidt and Mark [63].

Recently the potential dependence of α was considered by Japanese workers [64] who discussed the conditions in which it can be observed (these conditions were also discussed by Mohilner [55]). They observed quadratic dependence of the logarithm of rate constant of electron exchange in policyclic aromatic hydrocarbons on the free energy of reaction and on this basis they came to the conclusion that the Marcus theory is the correct one. However it has been shown in several experimental studies that the Tafel relationships are linear over wide potential ranges [65, 66], which suggests that α is constant.

It is probable that further improvements in experimental methods of investigation of electrode kinetics will make it possible to confirm the conclusions regarding α resulting from these theories or to reveal simplifications and errors made in the theoretical considerations.

Various published concepts regarding the classical transition coefficient have been reviewed by Bauer [67]. The nature of this coefficient has also been discussed by Parsons [68].

It should be remembered that when more than one electron is exchanged during an electrode process the slopes of Tafel relationships and other equations of this type have a more complex sense.

In such cases parameter αn determined experimentally from the slope of the plot of log $i(k_{fh}, k_{bh})$ vs. E is described by the following relationship:

$$\alpha n = i - 1 + \alpha_i \qquad (3.91)$$

and

$$\beta n = n - i + \beta_i \qquad (3.92)$$

where i stands for the the number of the slowest step and α_i and β_i are the transition coefficients of this step, which according to the previous reasoning should be equal to 0.5.

When, for example, $n = 3$ and $i = 2$, $\alpha n = 1.5$.

In further chapters of this book parameters αn and βn will often appear in kinetic equations. In the case of polyelectron processes they should be understood in the sense following from equations (3.91) and (3.92).

References

[1] de Donder, Th., *"L'affinité"*, Ed. van Rysselberghe, P., Paris 1936.
[2] de Donder, Th. and van Rysselberghe, P., *"Affinity"*, Stanford University Press, Stanford 1936.
[3] Prigogine, I., *"Introduction to Thermodynamics of Irreversible Processes"*, 2nd Ed., Wiley, Interscience, New York 1962.
[4] Guggenheim, E. A., *"Thermodynamics"*, Amsterdam 1949.
[5] van Rysselberghe, P., *J. Phys. Chem.*, **57**, 275 (1953).
[6] Zakharov, E. M., *Zhurn. Fiz. Khim.*, **38**, 2950 (1964).
[7] Guidelli, R., *Trans. Faraday Soc.*, **66**, 1185, 1194 (1970).
[8] Galus, Z., *Chem. Anal. (Warsaw)*, **10**, 803 (1965).
[9] Levich, V. G., *"Physicochemical Hydrodynamics"*, Prentice-Hall, Englewood Cliffs 1962, p. 86.
[10] Azim, S. and Riddiford, A. C., *Anal. Chem.*, **34**, 1023 (1962).
[11] Delahay, P., *"Double Layer and Electrode Kinetics"*, Interscience, New York 1965; Mohilner, D. M. and Delahay, P., *J. Phys. Chem.*, **67**, 558 (1963).
[12] Parsons, R., *Trans. Faraday Soc.*, **47**, 1331 (1951).
[13] Eyring, H., Glasstone, S. and Laidler, K. J., *J. Chem. Phys.*, **7**, 1053 (1939).
[14] Erdey-Gruz, T. and Volmer, M., *Z. physik. Chem.*, **150A**, 203 (1930).
[15] Butler, J. A. V., *Trans. Faraday Soc.*, **19**, 729, 734 (1924).
[16] Audubert, R., *J. chim. phys.*, **21**, 351 (1924).
[17] van Rysselberghe, P., *Electrochim. Acta*, **8**, 583, 709 (1963).
[18] Butler, J. A. V., *Proc. Roy. Soc.*, **157A**, 423 (1936).
[19] Dolin, P. I. and Erszler, B. B., *Acta Physicochim. URSS*, **13**, 747 (1940); ibid. **14**, 886 (1940).
[20] Tafel, J., *Z. physik. Chem.*, **50**, 641 (1905).
[21] Tanaka, N. and Tamamushi, R., *Electrochim. Acta*, **9**, 963 (1964).
[22] Tamamushi, R., *Kinetic Parameters of Electrode Reactions* (Supplement to Ref. [21], 1972).
[23] Randles, J. E. B. and Somerton, K. W., *Trans. Faraday Soc.*, **48**, 951 (1952).
[24] Bauer, H. H., Smith, D. L. and Elving, P. J., *J. Am. Chem. Soc.*, **82**, 2094 (1960).
[25] Galus, Z. and Adams, R. N., *J. Phys. Chem.*, **67**, 866 (1963).
[26] Randles, J. E. B. and Somerton, K. W., *Trans. Faraday Soc.*, **48**, 937 (1952).
[27] Delahay, P., *"Progress in Polarography"*, Interscience, New York 1962, Vol. 1, page 72.
[28] Anson, P. C., *Anal. Chem.*, **33**, 939 (1961).
[29] Imai, H. and Delahay, P., *J. Phys. Chem.*, **66**, 1108 (1962).
[30] Galus, Z., *Roczniki Chem.*, **42**, 783 (1968).
[31] Randles, J. E. B., *Trans. Symp. Electrode Processes, Philadelphia 1959*, Wiley, New York 1961, page 209.
[32] Joshi, K. M., Mehl, W. and Parsons, R., *Trans. Symp. Electrode Processes, Philadelphia 1959*, Wiley, New York 1961, page 249.
[33] Randles, J. E. B., *"Progress in Polarography"*, Interscience, New York 1962, Vol. 1, page 123.
[34] Randles, J. E. B., *Disc. Faraday Soc.*, **1**, 11 (1947).
[35] Marcus, R. A., *J. Chem. Phys.*, **24**, 966, 979 (1956).

[36] Marcus, R. A., *Can. J. Chem.*, **37**, 155 (1959).
[37] Marcus, R. A., *J. Chem. Phys.*, **26**, 867, 872 (1957).
[38] Marcus, R. A., *Disc. Faraday Soc.*, **29**, 21 (1960).
[39] Marcus, R. A., *J. Phys. Chem.*, **67**, 853, 2889 (1963).
[40] Marcus, R. A., *J. Chem. Phys.*, **38**, 1335 (1963); **39**, 1734 (1963); **43**, 679 (1965).
[41] Marcus, R. A., *Electrochim. Acta*, **13**, 995 (1968).
[42] Marcus, R. A., *Ann. Rev. Phys. Chem.*, **15**, 155 (1964).
[43] Levich, V. G. and Dogonadze, R. R., *Dokl. Akad. Nauk SSSR*, **124**, 123 (1959); **133**, 158 (1960).
[44] Levich, V. G. and Dogonadze, R. R., *Coll. Czechoslov. Chem. Communs.*, **26**, 193 (1961).
[45] Dogonadze, R. R., *Elektrokhimiya*, **1**, 1434 (1965).
[46] Dogonadze, R. R. and Tchizmardsev, Yu. A., *Dokl. Akad. Nauk SSSR*, **144**, 1077 (1962); **145**, 849 (1962).
[47] Dogonadze, R. R., Kuznetsov, A. M. and Tchizmardsev, Yu. A., *Zhurn. Fiz. Khim.*, **38**, 1195 (1964).
[48] Dogonadze, R. R. and Kuznetsov, A. M., *Elektrokhimiya*, **1**, 742 (1965); *Izv. Akad Nauk SSSR, ser. khim.*, 1885, 2140 (1964).
[49] Dogonadze, R. R. and Kuznetsov, A. M., *Elektrokhimiya*, **3**, 380 (1967).
[50] Dogonadze, R. R., Kuznetsov, A. M. and Levich, V. G., *Electrochim. Acta*, **13**, 1025 (1968).
[51] Levich, V. G. and Kharkats, Y. I., *Elektrokhimiya*, **6**, 562 (1970).
[52] Levich, V. G., *Adv. Electrochem. Electrochem. Eng.*, **4**, 249 (1966).
[53] Levich, V. G., "*Itogi nauki. Elektrokhimiya 1965*", Moscow 1967, Vsesoj. Inst. Nauchnoy i Tekhnich. Inf. (in Russian).
[54] Dogonadze, R. R., and Kuznetsov, A. M., "*Itogi nauki. Elektrokhimiya 1967*", Moscow, Vsesoj. Inst. Nauchnoy i Tekhnich. Inf. (in Russian), 1969; Dogonadze, R. R. and Kuznetsov, A. M., *Progress Surf. Sci.*, **6**, 1 (1975).
[55] Mohilner, D. M., *J. Phys. Chem.*, **73**, 2652 (1969).
[56] Oldham, K. B., *J. Electroanal. Chem.*, **16**, 125 (1968).
[57] Randles, J. E. B., *Trans. Faraday Soc.*, **48**, 828 (1952).
[58] Rubin, H. and Collins, F. C., *J. Phys. Chem.*, **58**, 958 (1954).
[59] Reiss, H., *J. Chem. Phys.*, **18**, 996 (1950).
[60] Reinmuth, W. H., *J. Electroanal. Chem.*, **21**, 425 (1969).
[61] Parsons, R. and Passeron, E., *J. Electroanal. Chem.*, **12**, 524 (1966).
[62] Anson, F. C., Rathyen, N. and Frisbee, R. F., *J. Electrochem. Soc.*, **117**, 477 (1970).
[63] Schmidt, P. P. and Mark, H. B., Jr., *J. Chem. Phys.*, **58**, 4290 (1973).
[64] Suga, K., Mizota, H., Kanzaki, Y. and Aoyagui, S., *J. Electroanal. Chem.*, **41**, 313 (1973).
[65] Bockris, J. O'M., Mittal, K. L. and Sen, R. K., *Nat. Phys. Sci.*, **234**, 118 (1971).
[66] Yamaoka, H., *J. Electroanal. Chem.*, **25**, 381 (1970); **36**, 457 (1972).
[67] Bauer, H. H., *J. Electroanal. Chem.*, **16**, 419 (1968).
[68] Parsons, R., *Croat. Chem. Acta*, **42**, 281 (1970).

Chapter 4

Diffusion to the Electrode

In any detailed discussion of electrode processes the distribution of depolarizer concentration, which is a function of distance from the electrode and of the electrolysis time, must be exactly defined. This distribution depends on factors governing the transport of the depolarizer to the electrode.

It can be assumed (neglecting the surface effects) that before electrolysis commences the system is at equilibrium. The chemical potential a_s of component s is independent of its position in the system. During the electrolysis, when the activity of the depolarizer on the surface of the electrode is smaller than that in the bulk of the solution, there is a change of the value of chemical potential μ_s, which is defined as:

$$\mu_s = \mu_s^0 + RT\ln a_s \pm z_s F\Phi \qquad (4.1)$$

where μ_s^0 is the standard chemical potential, $\pm z_s F$ is the charge of one mole of the depolarizer s (the signs $+$ and $-$ correspond to cationic and anionic depolarizers respectively), and Φ is the electrical potential of the part of the system being studied.

If, as a result of electrolysis, a concentration gradient, and consequently also a chemical potential gradient appear in the system, it will tend to return to the equilibrium state. Since at equilibrium the chemical potential is the same in all parts of the system, the depolarizer will be transferred from the regions of higher concentration to those of lower concentration, i.e. in the direction of the electrode. The rate of this transfer will be proportional to the chemical potential gradient [1–3] along the x-axis, which is assumed to be the only direction in which the difference in chemical potential appears:

$$V_{s,x} = -K_s \frac{\partial \mu_s}{\partial x}. \qquad (4.2)$$

Differentiating equation (4.1) with respect to x and combining the result with equation (4.2) we obtain:

$$V_{s,x} = -K_s \left(\pm z_s F \frac{\partial \Phi}{\partial x} + RT \frac{\partial \ln a_s}{\partial x} \right) \qquad (4.3)$$

where $V_{s,x}$ is the rate of transfer of the depolarizer in the direction of decreasing chemical potential of the depolarizer, and K_s is the rate of transfer of component s under the influence of unit force per mole, i.e. the mobility of depolarizer s.

Multiplying both sides of equation (4.3) by the concentration of component s (C_s):

$$f_{s,x} = -C_s V_{s,x} = -K_s C_s \left(\pm z_s F \frac{\partial \Phi}{\partial x} + RT \frac{\partial \ln a_s}{\partial x} \right) \qquad (4.4)$$

where $f_{s,x}$ is the depolarizer flux, i.e. the amount of substance passing through plane x in mole s^{-1}cm^{-2}.

In the above expressions the minus sign appears, since the diffusion takes place in the direction opposite to that of the concentration increase.

In the case of transport of neutral molecules to the electrode, equations (4.3) and (4.4) become simpler, as the terms containing the electrical potential disappear. In the case of transport of ions there are two possibilities: (1) constant electrical potential Φ and variable activity, and (2) constant activity and variable electrical potential. The former corresponds to transport by diffusion in which the movement of ions is not affected by the electric field, and the latter to transport by migration which is connected with the effect of the field on electrically charged particles present in the solution.

The solution will contain various ions besides those that react with the electrode. If the values corresponding to the latter ions are indicated by subscript s, the flux of these ions is given by the equation:

$$f_s = -z_s K_s C_s F \frac{\partial \Phi}{\partial x} - RTK_s \frac{\partial C_s}{\partial x}. \qquad (4.5)$$

Assuming further that the diffusion is of linear character, i.e. that it takes place from a cylinder of liquid perpendicular to a flat electrode, if a depolarizer having the concentration C plays a part in the electrode process, its transport to the electrode takes place under the influence of both concentration and potential gradients.

The magnitude of the total current can be described by taking into account fluxes of ions of all species passing through a certain plane parallel to the electrode at the distance $x = x_1$ from the electrode.

$$i = AF \left(\sum z_c f_c - \sum z_a f_a \right) \qquad (4.6)$$

where A is the surface area of the plane which is also the surface of the electrode, and subscripts c and a correspond to cations and anions, respectively.

Expressing f_c and f_a by means of equation (4.4) we obtain from equation (4.6):

$$\frac{i}{AF} = F\frac{\partial \Phi}{\partial x}\left(-\sum z_c^2 K_c C_c - \sum z_a^2 K_a C_a\right) + RT\left(-\sum z_z K_c \frac{\partial C_c}{\partial x} + \sum z_a K_a \frac{\partial C_a}{\partial x}\right). \quad (4.7)$$

Substituting the expression for $\partial \Phi/\partial x$ obtained from equation (4.7) in equation (4.5) we obtain the following expression for the depolarizer flux:

$$f_s = \frac{z_s K_s C_s}{\sum z_c^2 K_c C_c + \sum z_a^2 K_a C_a}\left[\frac{i}{AF} + RT\left(\sum z_c K_c \frac{\partial C_c}{\partial x} - \sum z_a K_a \frac{\partial C_a}{\partial x}\right)\right] - RTK_s \frac{\partial C_s}{\partial x}. \quad (4.8)$$

Since the transport number t_s of ion s is defined as:

$$t_s = \frac{z_s^2 K_s C_s}{\sum z_c^2 K_c C_c + \sum z_a^2 K_a C_a} \quad (4.9)$$

equation (4.8) can be simplified to:

$$f_s = \frac{t_s}{z_s}\left[\frac{i}{AF} + RT\left(\sum z_c K_c \frac{\partial C_c}{\partial x} - \sum z_a K_a \frac{\partial C_a}{\partial x}\right)\right] - RTK_s \frac{\partial C_s}{\partial x}. \quad (4.10)$$

In these equations the summation applies to all the anions and cations, including the depolarizer ion.

Two extreme cases of application of this equation merit consideration:

(1) A very small ionic depolarizer concentration in comparison with the total concentration of other ions.

(2) A solution containing only one electrolyte which, on dissociation. gives the ionic depolarizer.

Consider the first case. If the concentration of the depolarizer is low in comparison with the total concentration of the ionic components of the solution, then according to equation (4.9) t_s is almost equal to zero, and the first term of equation (4.10) can be neglected. In that case

$$f_s = -RTK_s \frac{\partial C_s}{\partial x}. \quad (4.11)$$

The second case is simple when it is assumed that the electrode reaction of depolarizer s gives a neutral product and that the depolarizer is a cation. In such a case $z_s C_s = z_a C_a$ and equation (4.10) becomes:

$$f_s = \frac{t_s}{z_s}\left[\frac{i}{AF} + RT(K_s - K_a)z_s \frac{\partial C_s}{\partial x}\right] - RTK_s \frac{\partial C_s}{\partial x}. \qquad (4.12)$$

Combining Fick's first law of diffusion, applied to components s diffusing in the direction perpendicular to the surface of the electrode,

$$f_{s,x} = -D_s \frac{\partial C_s}{\partial x}, \qquad (4.13)$$

where D_s is the diffusion coefficient of depolarizer s, with the equation

$$f_{s,x} = -K_s RT \frac{\partial C_s}{\partial x} \qquad (4.14)$$

which is deduced from equation (4.4), assuming that there is no potential gradient and $a_s = C_s$, we obtain the expression describing the diffusion coefficient:

$$D_s = K_s RT. \qquad (4.15)$$

Using equation (4.15) and the relationships:

$$t_s + t_a = 1 \qquad (4.16)$$

and

$$\frac{t_s}{t_a} = \frac{z_s K_s}{z_a K_a} \qquad (4.17)$$

we obtain, instead of equation (4.12):

$$f_s = \frac{t_s i}{z_s AF} - D_s \frac{\partial C_s}{\partial x}\left[1 + \frac{z_s}{z_a}\right] t_a. \qquad (4.18)$$

In these equations t_a stands for the transport number of the anion (it has been assumed that the depolarizer is the cation).

Comparison of this equation with equation (4.11) shows that the electrical potential gradient causes a change of the value of depolarizer flux as compared with that defined by diffusion alone. This change depends on the transport numbers of the depolarizer and the conjugated anion, and on the ratio of the charges on these two ions.

It will be shown later that in the case of low concentrations of the principal electrolyte, the existence of an electrical potential gradient causing the migration of ions, leads to changes in the limiting current value.

4.1 Linear Diffusion

Consider the case of an excess of the principal electrolyte in the solution, which is so large that the flux of depolarizer to the surface of the electrode depends almost completely on diffusion. In this case, on the basis of equation (4.13), the basic equation of diffusion for an isotropic medium is obtained.

Consider an element of volume having the shape of a parallelepiped with the walls parallel to the three axes of coordinates and having the dimensions $2dx$, $2dy$ and $2dz$. The centre of this element is in point P, having the coordinates x, y, z, and the concentration at this point is C.

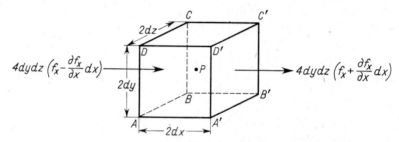

Fig. 4.1 Model of linear diffusion (by permission, from *The Matematics of Diffusion*, by T. Crank, published by the Oxford University Press).

Assume also that walls $ABCD$ and $A'B'C'D'$ are perpendicular to the x-axis, as shown in Fig. 4.1. The rate of penetration of the substance into the element of volume, through the wall $ABCD$, having the coordinate $x-dx$ is equal to $4dydz\left(f_x - \frac{\partial f_x}{\partial x}dx\right)$, where f_x is the flux along the x-axis at the point P.

The rate of escape of the substance from the element of volume $8dxdydz$ through the wall $A'B'C'D'$ is equal to $4dydz\left(f_x + \frac{\partial f_x}{\partial x}dx\right)$. The contribution of the amount of diffusing substance in the element contained between the walls $ABCD$ and $A'B'C'D'$ to the rate of increase is $-8dxdydz\frac{\partial f_x}{\partial x}$.

The contributions from the other pairs of walls: $-8dxdydz\frac{\partial f_y}{\partial y}$ and $-8dxdydz\frac{\partial f_z}{\partial z}$ can be similarly obtained.

The rate of increase of the amount of diffusing substance in the element under consideration can also be described by means of the expression $8dxdydz\, \frac{\partial C}{\partial t}$. Therefore we have:

$$8dxdydz\frac{\partial C}{\partial t} = -8dxdydz\left(\frac{\partial f_x}{\partial x}+\frac{\partial f_y}{\partial y}+\frac{\partial f_z}{\partial z}\right), \tag{4.19}$$

or after simplification:

$$\frac{\partial C}{\partial t} = -\left[\frac{\partial f_x}{\partial x}+\frac{\partial f_y}{\partial y}+\frac{\partial f_z}{\partial z}\right]. \tag{4.19a}$$

In equation (4.19) f_x, f_y and f_z are the fluxes of the substance diffusing along the axes x, y and z.

Combining equation (4.19) with the equation defining the flux, (4.13), we obtain the required equation of diffusion:

$$\frac{\partial C}{\partial t} = D\left[\frac{\partial^2 C}{\partial x^2}+\frac{\partial^2 C}{\partial y^2}+\frac{\partial^2 C}{\partial z^2}\right]. \tag{4.20}$$

When the diffusion is unidimensional, i.e. when the concentration gradient exists along the x-axis only, equation (4.20) is reduced to:

$$\frac{\partial C}{\partial t} = D\frac{\partial^2 C}{\partial x^2}. \tag{4.21}$$

This equation is known as Fick's second law of diffusion, since it was first obtained by Fick [4] in 1855.

Equation (4.21) will be used later, in order to describe the distribution of concentration of the depolarizer with respect to time and distance from the electrode for definite conditions dictated by the kind of process involved.

4.2 Spherically symmetrical diffusion

An electrode process does not always take place under conditions of linear diffusion. It has already been mentioned in Chapter 2 that spherical electrodes are often used. The basic diffusion equation for such cases differs slightly from the linear diffusion equation (4.20). Under the conditions of spherically symmetrical diffusion the substance diffuses in the direction of the centre of the sphere along the lines that are the extensions of the radii.

Let us consider a spherical electrode immersed in a solution containing the depolarizer and an amount of the principal electrolyte in sufficient concentration for it to be assumed that the diffusion takes place only as a result of the concentration gradient. Such an electrode having the radius r_0 is shown schematically in Fig. 4.2. It is, of course, an idealized electrode, since in practice it must be connected by a conductor to the rest of the electrical circuit. The presence of this connection, irrespective of its size, decreases the diffusion field, and therefore has an effect on the final result of our discussion, based on the ideal model.

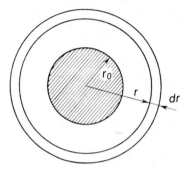

Fig. 4.2 Model of spherically symmetrical diffusion to spherical electrode; r_0—radius of spherical electrode.

Consider an infinitely thin layer having thickness dr and surrounding the electrode at the distance r from its centre (see Fig. 4.2). The layer is limited by two surfaces: at the distance r from the centre of the electrode (area $4\pi r^2$) and at the distance $r+dr$ [area $4\pi(r+dr)^2$].

As in the case of linear diffusion we evaluate the number of moles of the depolarizer which diffuse through these surfaces. During the time dt a quantity dN_r molecules diffuses through the surface nearer to the centre of the electrode. This value is obtained from an equation analogous to equation (4.13) expressing Fick's first law of diffusion:

$$dN_r = 4\pi r^2 D \left(\frac{\partial C}{\partial r}\right)_r dt. \qquad (4.22)$$

At this distance from the electrode the flux is given by the expression:

$$f_r = -\frac{dN_r}{4\pi r^2 dt} = -D\left(\frac{\partial C}{\partial r}\right)_r. \qquad (4.23)$$

Similarly for the surface situated at the distance $(r+dr)$ from the centre of the electrode the number of moles of diffusing depolarizer is given by the expression:

$$dN_{r+dr} = 4\pi(r+dr)^2 D \left(\frac{\partial C}{\partial r}\right)_{r+dr} dt, \qquad (4.24)$$

4.2] Spherically Symmetrical Diffusion

while the flux at the distance $(r+dr)$ is given by the expression:

$$f_{r+dr} = -D\left(\frac{\partial C}{\partial r}\right)_{r+dr}. \qquad (4.25)$$

Since the concentration gradient at the surface $(r+dr)$ depends on that present at the distance r from the centre of the electrode according to the equation

$$\left(\frac{\partial C}{\partial r}\right)_{r+dr} = \left(\frac{\partial C}{\partial r}\right)_r + \frac{\partial}{\partial r}\left(\frac{\partial C}{\partial r}\right)dr, \qquad (4.26)$$

equation (4.24) can be expressed as:

$$dN_{r+dr} = 4\pi(r+dr)^2 D dt \left[\left(\frac{\partial C}{\partial r}\right)_r + \frac{\partial^2 C}{\partial r^2}dr\right], \qquad (4.27)$$

or in an expanded form:

$$dN_{r+dr} = 4\pi D dt \left[r^2\left(\frac{\partial C}{\partial t}\right) + 2r\left(\frac{\partial C}{\partial r}\right)dr + r^2\left(\frac{\partial^2 C}{\partial r^2}\right)dr + \right.$$

$$\left. \left(\frac{\partial C}{\partial r} + 2r\frac{\partial^2 C}{\partial r^2}\right)(dr)^2 + \left(\frac{\partial^2 C}{\partial r^2}\right)(dr)^3\right]. \qquad (4.28)$$

The last two terms, which contain higher powers of very small quantities, can be neglected, and therefore equation (4.28) can be expressed in a simpler form:

$$dN_{r+dr} = 4\pi D dt \left[r^2\left(\frac{\partial C}{\partial r}\right) + 2r\left(\frac{\partial C}{\partial r}\right)dr + r^2\left(\frac{\partial^2 C}{\partial r^2}\right)dr\right]. \qquad (4.29)$$

The change of concentration in the spherical layer having thickness dr and surrounding the electrode, is equal to the ratio of the difference between the number of moles of depolarizer entering through the surface situated at the distance $(r+dr)$ from the centre of the electrode and the number of moles of depolarizer leaving through the surface situated nearer the electrode, to the volume of the layer, which is of course equal to $4\pi r^2 dr$.

This can be expressed mathematically as:

$$dC = \frac{dN_{r+dr} - dN_r}{4\pi r^2 dr}. \qquad (4.30)$$

The change of concentration, depending on the distance r from the electrode and the time t of the electrolysis, is described by the following relationship:

$$\frac{\partial C}{\partial t} = \frac{dN_{r+dr} - dN_r}{4\pi r^2 dr\, dt}. \qquad (4.31)$$

Combining equation (4.31) with relationships (4.22) and (4.29) we obtain:

$$\frac{\partial C}{\partial t} = D\left[\frac{\partial^2 C}{\partial r^2} + \frac{2}{r}\frac{\partial C}{\partial r}\right]. \qquad (4.32)$$

This equation is the equivalent of equation (4.20) describing linear diffusion processes.

4.3 Cylindrically Symmetrical Diffusion

Solid electrodes, made of platinum, gold, graphite and other materials, are often used in electrochemistry, especially in processes taking place at positive potentials. Such electrodes often consist of a suitable noble metal in the form of a wire, and from the point of view of geometry they are relatively small cylinders. This kind of electrode is shown schematically in Fig. 4.3.

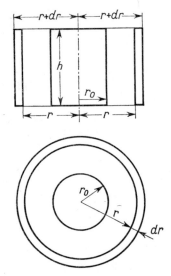

Fig. 4.3 Model of cylindrically symmetrical diffusion to cylindrical electrode (horizontal and vertical sections); r_0—radius of the cylindrical electrode

Considering this type of electrode, call the radius of the cylindrical electrode r_0, and its height h. The depolarizer diffuses from the bulk of solution in the direction of the centre of the electrode. In order to obtain an equation describing cylindrically symmetrical diffusion, a reasoning identical with that used in the case of spherical diffusion is employed.

4.3] Cylindrically Symmetrical Diffusion

Consider the number of moles of depolarizer passing through a surface surrounding the electrode at the distance r from its centre:

$$dN_r = 2\pi r h D \left(\frac{\partial C}{\partial r}\right)_r dt. \tag{4.33}$$

The number of moles of depolarizer diffusing through a surface situated at the distance $(r+dr)$ from the centre of the electrode is given by the equation:

$$dN_{r+dr} = 2\pi h(r+dr) D \left(\frac{\partial C}{\partial r}\right)_{r+dr} dt. \tag{4.34}$$

Substituting equation (4.26) in equation (4.34):

$$dN_{r+dr} = 2\pi h D\, dt \left[r\frac{\partial C}{\partial r} + \frac{\partial C}{\partial r} dr + r\frac{\partial^2 C}{\partial r^2} dr + \frac{\partial^2 C}{\partial r^2}(dr)^2 \right]. \tag{4.35}$$

The last term (the square of a very small quantity) can be neglected, hence:

$$dN_{r+dr} = 2\pi h D\, dt \left[r\frac{\partial C}{\partial r} + \frac{\partial C}{\partial r} dr + r\frac{\partial^2 C}{\partial r^2} dr \right]. \tag{4.36}$$

The change of depolarizer concentration in the thin layer having thickness dr and surrounding the cylindrical electrode is given as in the case of spherical diffusion, by the ratio of the difference between the number of moles of depolarizer diffusing to the electrode through the outer surface of the layer and the number of moles of depolarizer diffusing to the electrode through the surface situated nearer the electrode at the distance r from its centre, to the volume of the layer, equal to $2\pi r h dr$.

Thus:

$$dC = \frac{dN_{r+dr} - dN_r}{2\pi r h dr}. \tag{4.37}$$

The change of concentration, depending on the distance from the electrode and the electrolysis time, is given by the formula:

$$\frac{\partial C}{\partial t} = \frac{dN_{r+dr} - dN_r}{2\pi h dr dt}. \tag{4.38}$$

Combining equation (4.38) with equations (4.33) and (4.36) the required relationship is obtained:

$$\frac{\partial C}{\partial t} = D\left[\frac{\partial^2 C}{\partial r^2} + \frac{1}{r}\frac{\partial C}{\partial r}\right]. \tag{4.39}$$

This equation of cylindrical diffusion is formally similar to equation (4.32), which was deduced for spherically symmetrical diffusion.

In the discussion of diffusion to the cylinder it has been assumed that the base and the upper plane of the cylinder, having the surface areas πr^2, do not act as part of the electrode. This is true only in the case of the upper plane which acts as the electrical connection and is usually sealed in glass or some other insulating material.

On the other hand, the base of the cylinder is usually not insulated and therefore it acts as part of the electrode. However it is usually very small in comparison with the total surface of the electrode and as a result equation (4.39) describes with acceptable accuracy the distribution of depolarizer concentration depending on the distance from the electrode and the electrolysis time.

Equations (4.32) and (4.39) can also be obtained from the equation of linear diffusion by transformation of coordinates.

4.4 Linear Diffusion to an Expanding Drop Electrode

In the types of diffusion discussed previously it has been assumed that an electrical potential gradient was absent and that the solution studied was not disturbed. The depolarizer transport was therefore due to diffusion alone.

However, in certain electrochemical techniques the transport of depolarizer to the electrode by processes other than diffusion is stipulated. The description of such transport is more difficult than that of the transport by diffusion since in this case convection must also be taken into consideration.

An important electrochemical technique in which depolarizer transport to the electrode takes place by both diffusion and convection, is the polarographic method. The surface of the dropping mercury electrode used in this method increases with time and the movement of the surface of the electrode causes a larger time dependence of concentration than that existing in the absence of convection. This can be expressed schematically as:

$$\frac{\partial C}{\partial t} = \left(\frac{\partial C}{\partial t}\right)_{\text{diff.}} + \left(\frac{\partial C}{\partial t}\right)_{\text{conv.}}. \tag{4.40}$$

The growth of each drop of mercury causes a movement of the substance in the direction of the surface of the drop, and as a result the total depolarizer flux consists of two components: diffusion $f_{\text{diff.}}$ described by equation (4.13), and convection by

$$f_{\text{conv.}} = SC \tag{4.41}$$

4.4] Linear Diffusion to an Expanding Drop Electrode 89

where S is the convection rate. In this case the expression for the total depolarizer flux f_s is:

$$f_s = f_{\text{diff.}} + f_{\text{conv.}} = -D\frac{\partial C}{\partial x} + SC. \qquad (4.42)$$

The interpretation of diffusion at a growing sphere is fairly complex and therefore we will first consider the linear diffusion model. This is equivalent to replacement of a spherical electrode expanding in the direction of the solution by a flat electrode moving in the same direction. Such a model, shown in Fig. 4.1, is also useful in the present discussion. The reasoning used in Section 4.1 is valid in this case also and only the description of the fluxes need be changed.

For the above reasons equation (4.19a) can be combined with equation (4.42) for fluxes in the directions of axes x, y, and z. The relationship obtained is:

$$\frac{\partial C}{\partial t} = D\left[\frac{\partial^2 C}{\partial x^2} + \frac{\partial^2 C}{\partial y^2} + \frac{\partial^2 C}{\partial z^2}\right] - S_x\frac{\partial C}{\partial x} - S_y\frac{\partial C}{\partial y} - S_z\frac{\partial C}{\partial z}. \qquad (4.43)$$

In the case of polarographic dropping mercury electrode the concentration difference exists principally along the axis perpendicular to the electrode surface, and not in the planes parallel to this surface. Therefore, both $\partial C/\partial y = 0$ and $\partial C/\partial z = 0$. Then from the general equation (4.43) the following relationship is obtained:

$$\frac{\partial C}{\partial t} = D\frac{\partial^2 C}{\partial x^2} - S_x\frac{\partial C}{\partial x}. \qquad (4.44)$$

Comparing this equation with equation (4.21) it can be seen that the second term of the right-hand side of equation (4.44) determines the effect of convection, due to the movement of the electrode in the direction of the solution. This term, appearing in equation (4.44) in the general form, must be so expressed that the concentration changes could be represented in the form of functions of the distance from the electrode and the duration of electrolysis only. For this purpose consider an imaginary layer surrounding the spherical drop of mercury. We will call the radius of the electrode r_1 and that of the outer boundary of the layer r_2. If we consider that the layer has a constant volume ΔV, this volume will be the difference between the volume of a sphere of radius r_2 and the volume of the electrode:

$$\Delta V = V_2 - V_1 = \tfrac{4}{3}\pi(r_2^3 - r_1^3). \qquad (4.45)$$

When the difference $r_2 - r_1 = x$ is very small, in comparison with the radius of the drop, equation (4.45) can be written as:

$$\Delta V = \tfrac{4}{3}\pi[(r_1+x)^3 - r_1^3] \cong 4\pi r_1^2 x = Ax = \text{const}, \qquad (4.46)$$

where A stands for the surface area of the spherical electrode.

Since ΔV is time independent:

$$\frac{d(\Delta V)}{dt} = 0. \qquad (4.47)$$

During the expansion of the drop its surface area A increases and, therefore, according to equation (4.46) x must decrease. For this reason the surface of the drop of mercury approaches the imaginary boundary and moves, with respect to the liquid situated at the distance x from the surface of the electrode, at the relative rate:

$$S_x = \frac{dx}{dt}. \qquad (4.48)$$

Differentiating equation (4.46) with respect to time:

$$\frac{d(\Delta V)}{dt} = \frac{d(Ax)}{dt} = A\frac{dx}{dt} + x\frac{dA}{dt} \qquad (4.49)$$

and combining equations (4.47) and (4.49) with equation (4.48):

$$S_x = \frac{dx}{dt} = -\frac{x}{A}\frac{dA}{dt}. \qquad (4.50)$$

The surface area of the drop electrode has been described in chapter 2 in terms of capillary efficiency and the drop time [equation (2.22)]. Differentiating this expression with respect to time gives:

$$\frac{dA}{dt} = \frac{2}{3} \times 0.85 \, m^{2/3} t^{-1/3}. \qquad (4.51)$$

Combining equations (2.30) and (4.51) with equation (4.50), the expression for the convection rate S_x in terms of time and of distance from the surface of the electrode is obtained:

$$S_x = -\frac{2}{3}\frac{x}{t}. \qquad (4.52)$$

Substituting the expression for S_x in equation (4.44) we finally obtain [5, 6]:

$$\frac{\partial C}{\partial t} = D\frac{\partial^2 C}{\partial x^2} + \frac{2x}{3t}\frac{\partial C}{\partial x}. \qquad (4.53)$$

This equation takes into account the effect of the rate of increase of the drop on the depolarizer transport to the electrode, but it does not take into account the spherical character of diffusion. Such a simplification can be justified when the spherical electrode is large, or when the drop time is short.

The equation describing the diffusion up to an expanding spherical electrode, taking into account both convection transport and the spherical diffusion, is more complex, since it contains the term describing the spherical diffusion $\left(\dfrac{2}{r}\dfrac{\partial C}{\partial r}\right)$. Hence the general equation of convection diffusion in this case is:

$$\frac{\partial C}{\partial t} = D\frac{\partial^2 C}{\partial r^2} + \frac{2}{r}\frac{\partial C}{\partial r} - S_r\frac{\partial C}{\partial r} \qquad (4.54)$$

where S_r is the rate of movement of the liquid along the radius of the electrode. Analogously to equation (4.48) this rate is expressed by

$$S_r = \frac{dr}{dt}. \qquad (4.55)$$

The radius of a drop electrode increases with increasing time according to the relationship:

$$r_0 = \left(\frac{3mt}{\pi \times 13.6}\right)^{1/3} = at^{1/3} \qquad (4.56)$$

where a stands for a constant value equal to $(3m/13.6\pi)^{1/3}$ and m is the efficiency of the capillary.

It may be assumed that the origin of the system of coordinates is in the middle of the drop and that the distance from the origin is r. Then for a non-compressible solution the position of a certain point in the solution can be calculated by using the equation:

$$r^3 = a^3 t + \text{const.} \qquad (4.57)$$

The rate of expansion of drop S_{r_0} can be determined by differentiating equation (4.56) with respect to time:

$$S_{r_0} = -\frac{dr_0}{dt} = -\frac{a^3}{3r_0^2}. \qquad (4.58)$$

The equation of the rate of movement of solution S_r is obtained by differentiating equation (4.57):

$$S_r = \frac{dr}{dt} = \frac{a^3}{3r^2}. \qquad (4.59)$$

Introducing this expression for S_r into equation (4.54) we finally obtain the relationship [7, 8]:

$$\frac{\partial C}{\partial t} = D\frac{\partial^2 C}{\partial r^2} + \frac{2}{r}\frac{\partial C}{\partial r} - \frac{a^3}{3r^2}\frac{\partial C}{\partial r}. \qquad (4.60)$$

4.5 Convection Diffusion to Rotating Disc Electrode

In the cases discussed previously, material approaches the electrode by diffusion. Only in the case of transport to an expanding sphere was the diffusion accompanied by convection, as the second factor taking the depolarizer to the surface of the electrode.

The role of convection is much greater in the case of transport of material to a disc electrode. In the discussion of such transport, however, the diffusion transport should be taken into account. The general equation describing such a case has already been given by equations (4.43) and (4.44).

When the solution is stirred with sufficient vigour the stationary state is established soon after the start of electrolysis. Neglecting the decrease of depolarizer concentration due to electrolysis, we can assume that the concentration is time independent and hence $\partial C/\partial t = 0$. Equations (4.43) and (4.44) therefore become simpler:

$$S_x\frac{\partial C}{\partial x} + S_y\frac{\partial C}{\partial y} + S_z\frac{\partial C}{\partial z} = D\left[\frac{\partial^2 C}{\partial x^2} + \frac{\partial^2 C}{\partial y^2} + \frac{\partial^2 C}{\partial z^2}\right] \qquad (4.61)$$

and

$$S_x\frac{dC}{dx} = D\frac{d^2 C}{dx^2}. \qquad (4.62)$$

These equations, like equations (4.43) and (4.44), have a general character. The equation for the particular case of the rotating disc is obtained by expressing the rate of movement of depolarizer by convection by means of values characteristic of the rotating electrode.

The solution of the problem of movement of a liquid caused by a disc rotating around an axis perpendicular to its plane has been given by von Karman [9] and in greater detail by Cochran [10]. They assumed that the volume of the solution in which the disc is present is infinite, the disc is large, and the flow of the liquid is a laminar flow. This solution of the problem leads to the model of movement of the liquid shown in Fig. 4.4.

At greater distances from the disc the liquid moves in a direction perpendicular to the surface. Near the disc, in the thin layer directly adhering to the surface of the electrode, the liquid also gains a centrifugal velocity.

4.5] Convection Diffusion to Rotating Disc Electrode

The thickness of this thin layer of liquid δ_0 is expressed by

$$\delta_0 = 2.6\left(\frac{\nu}{\omega}\right)^{1/2} \quad (4.63)$$

where ν is the kinematic viscosity of the liquid and ω is the angular velocity of the rotating disc. Hence S_x from equation (4.62) can be expressed as:

$$S_x = (\nu\omega)^{1/2} H(\xi) \quad (4.64)$$

where $H(\xi)$ is a function of ξ and ξ is defined as:

$$\xi = \left(\frac{\omega}{\nu}\right)^{1/2} x. \quad (4.65)$$

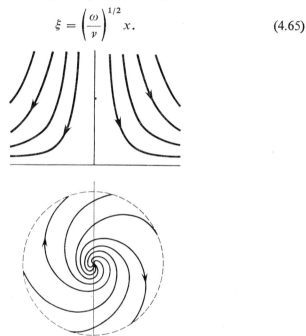

Fig. 4.4 Model of the movement of liquid in the vicinity of a disc electrode. Upper part—vertical section; lower part—horizontal section (from [11] by permission of the copyright holders, Prentice-Hall Inc.).

The simple analytical equations describing this function can be used only for large and small values of ξ. Thus for $\xi \ll 1$ (in the layer situated in the region close to the electrode):

$$H = -0.51\,\xi^2, \quad (4.66)$$

whereas for large values of ξ the following asymptotic expression is valid:

$$H \to -0.89 + 2.10\mathrm{e}^{-0.89\xi}. \quad (4.67)$$

Since for large values of ξ the second term of the right-hand side of equation (4.67) can be regarded as small in comparison with -0.89, we have:

$$H = -0.89. \tag{4.67a}$$

For these two different areas of the diffusion layer, one of which is situated near the surface of the disc and the other farther removed from it, we have two equations of diffusion by convection but they are valid only in the case of transport to a rotating disc. In these equations, instead of the generally expressed velocity S_x, there is the parameter determining the rate of convection transport to the disc, i.e. its angular velocity ω. Combining equation (4.62) with equations (4.64), (4.65) and (4.66) we obtain for very small values of x:

$$-0.51 \frac{\omega^{3/2} x^2}{\nu^{1/2}} \frac{dC}{dx} = D \frac{d^2 C}{dx^2}. \tag{4.68}$$

Combination of equation (4.62) with equations (4.64), (4.65) and (4.67a) leads to the expression:

$$-0.89 (\nu\omega)^{1/2} \frac{dC}{dx} = D \frac{d^2 C}{dx^2}. \tag{4.69}$$

Equations (4.68) and (4.69) are the two complete forms of the equation of diffusion by convection to a rotating disc, which are valid for various areas of the diffusion layer.

These problems have been discussed in detail by Levich [11] in a monograph dealing with physical and chemical hydrodynamics. They are also discussed in detail by Pleskov and Filinovskii [12].

References

[1] Guggenheim, E. A., *J. Phys. Chem.*, **33**, 842 (1929).
[2] Onsager, L. and Fuoss, R. M., *J. Phys. Chem.*, **36**, 2689 (1932).
[3] Hartley, G. S., *Phil. Mag.*, **12**, 473 (1931).
[4] Fick, A., *Pogg. Ann.*, **94**, 59 (1855).
[5] Ilkovič, D., *J. chim. phys.*, **35**, 129 (1939).
[6] MacGillavry, D. and Rideal, E. K., *Rec. trav. chim.*, **56**, 1013 (1937).
[7] Koutecký, J., *Českoslov. čas. fys.*, **2**, 117 (1952); *Czechoslov. J. Phys.*, **2**, 50 (1953).
[8] Matsuda, H., *Bull. Chem. Soc. Japan*, **26**, 342 (1953).
[9] von Karman, T., *Z. angew. Math. u. Mech.*, **1**, 244 (1921).
[10] Cochran, W. G., *Proc. Cambridge Phil. Soc.*, **30**, 365 (1934).
[11] Levich, V. G., "*Physicochemical Hydrodynamics*", Prentice-Hall, Englewood Cliffs, N. J. 1962, Chapter 2.
[12] Pleskov, Yu. V. and Filinovskii, V. Yu., "*Rotating Disc Electrode*", Nauka, Moscow 1972 (in Russian).

Chapter 5

Electrode Processes Controlled by Transport Rate: Diffusion Currents

This chapter commences the systematic description of various types of electrode processes which are encountered in electroanalytical practice.

The problems discussed here are the simplest, since they are limited to exchange of electrons between the electrode and the electrolysed substance, which reaches the electrode by diffusion, or (in polarography and in particular in the rotating disc method) by convection. The possibility of transport by migration is virtually eliminated by introduction of a supporting electrolyte to the solution at a concentration exceeding that of the depolarizer by at least two orders of magnitude. We assume that the substance Ox which can be reduced on the electrode according to the equation

$$\text{Ox} + ne \rightleftharpoons \text{Red} \qquad (5.1)$$

is present in the solution before start of the electrolysis. Therefore, although the reduction is discussed the mathematical treatment is the same as for the anodic oxidation of the substance Red. Substance Ox reacts with the electrode and yields Red according to equation (5.1). The progress of this reaction depends on the potential. When it is sufficiently positive the reduction is insignificant, and when it is sufficiently negative the concentration of Ox on the electrode surface equals zero. In the intermediate range of potentials the ratio of Ox to Red on the electrode surface is given by the Nernst equation:

$$E = E^0 + \frac{RT}{nF} \ln \frac{C_{\text{Ox}}(0, t)}{C_{\text{Red}}(0, t)}. \qquad (5.2)$$

Therefore it is assumed that the electrode reaction (5.1) is fast and that it is controlled by diffusion alone.

5.1 Electrode Processes Under Linear Diffusion Conditions

In order to describe currents due to electrode reactions it is necessary to solve the equation expressing Fick's second law of linear diffusion for the Ox form:

$$\frac{\partial C_{Ox}(x, t)}{\partial t} = D_{Ox} \frac{\partial^2 C_{Ox}(x, t)}{\partial x^2} \quad (5.3)$$

where D_{Ox} is the diffusion coefficient of the Ox form.

In some cases, when the electrode is spherical or cylindrical, or when the depolarizer also reaches the electrode by convection, equation (5.3) has additional terms (see the previous chapter).

In order to solve equation (5.3), or a related equation, it is necessary to define the initial conditions describing the depolarizer concentration before electrolysis, the boundary conditions describing the mode of change of the depolarizer concentration on the electrode surface during the electrolysis and the change of the depolarizer concentration during the electrolysis at the other end of the region under consideration (usually a point at an infinite distance from the electrode is considered theoretically).

For the case discussed in this chapter some of these conditions can be formulated jointly for all the methods discussed in this book, since they are all identical. Thus the initial conditions are identical and can be expressed as follows:

$$t = 0, \quad x \geqslant 0, \quad C_{Ox} = C_{Ox}^0. \quad (5.4)$$

This means that irrespective of the method, the depolarizer concentration in the solution is the same in the entire system before the start of the electrolysis. We call the initial concentration C_{Ox}^0. At the same time, irrespective of the method, it can be assumed that the concentration due to electrolysis changes only on the electrode surface and in its immediate neighbourhood. At a sufficient distance from the electrode the concentration C_{Ox} is practically equal to C_{Ox}^0 even after a prolonged electrolysis. This can be expressed as:

$$t \geqslant 0, \quad x \to \infty, \quad C_{Ox} \to C_{Ox}^0. \quad (5.5)$$

The boundary conditions describing the change of the depolarizer concentration on the electrode surface during the electrolysis are specific for each method discussed in this book. As a result of differences between these conditions and of differences between the initial equations expressing Fick's second law of diffusion (due to the spherical or cylindrical diffusion and to convection) the final equations, obtained after solving the initial equations, are different.

Later in this chapter the solutions of the diffusion equation (5.1) for the electrochemical methods discussed in this book will be considered consecutively.

5.1.1 CHRONOAMPEROMETRY AND CHRONOCOULOMETRY

In chronoamperometry the potential at which the process (5.1) takes place is applied to the electrode and changes in current in the course of the electrolysis are recorded. The exact description of the time-dependence of the current requires the solution of equation (5.3), including conditions (5.4) and (5.5) and the additional condition expressing the change of concentration C_{Ox} at the surface of the electrode. This condition can be formulated on the basis of the Nernst equation determining concentration $C_{Ox}(0, t)$ for the given electrode potential. The final equations are simplified, however, by assuming that the electrode potential is so negative that the concentration of substance Ox at the electrode surface is equal to zero. Under polarographic conditions there are potentials at which the "plateau" of the limiting current is already formed. In such conditions the second boundary condition is:

$$t > 0, \quad x = 0, \quad C_{Ox} = 0. \tag{5.6}$$

Equation (5.3) with conditions (5.4)–(5.6) can be solved by the Laplace transformation. Multiplying both sides of equation (5.3) by e^{-st} and integrating with respect to time t from 0 to ∞:

$$\int_0^\infty e^{-st} \frac{\partial C_{Ox}(x, t)}{\partial t} dt = D \int_0^\infty e^{-st} \frac{\partial^2 C_{Ox}(x, t)}{\partial x^2} dt. \tag{5.7}$$

Assuming that the order of integration and differentiation can be interchanged, the right-hand side of equation (5.7) can be expressed as:

$$D \int_0^\infty e^{-st} \frac{\partial^2 C_{Ox}(x, t)}{\partial x^2} dt = D \frac{\partial^2}{\partial x^2} \int_0^\infty e^{-st} C_{Ox}(x, t) dt$$

$$= D \frac{\partial^2 \bar{C}_{Ox}(x, s)}{\partial x^2} \tag{5.8}$$

where $\bar{C}_{Ox}(x, s)$ is the transform of $C_{Ox}(x, t)$; it depends on the new variables.

Integrating the left-hand side of equation (5.7):

$$\int_0^\infty e^{-st} \frac{\partial C_{Ox}(x, t)}{\partial t} dt = [C_{Ox}(x, t) e^{-st}]_0^\infty$$

$$+ s \int_0^\infty C_{Ox}(x, t) e^{-st} dt. \tag{5.9}$$

Taking into account the initial condition (5.4), equation (5.9) can be rewritten as:

$$\int_0^\infty e^{-st} \frac{\partial C_{Ox}(x, t)}{\partial t} \, dt = -C_{Ox}^0 + s\bar{C}_{Ox}(x, s). \tag{5.10}$$

This gives, instead of equation (5.3), the simple differential equation:

$$D \frac{d^2 \bar{C}_{Ox}(x, s)}{dx^2} - s\bar{C}_{Ox}(x, s) + C_{Ox}^0 = 0. \tag{5.11}$$

Using a similar transformation of the boundary conditions (5.5) and (5.6) the following expression is obtained instead of condition (5.5):

$$x \to \infty, \quad \bar{C}_{Ox}(x, s) \to \frac{C_{Ox}^0}{s}, \tag{5.12}$$

and instead of (5.6):

$$x = 0, \quad \bar{C}_{Ox}(x, s) = 0. \tag{5.13}$$

The general solution of equation (5.11) is:

$$\bar{C}_{Ox}(x, s) = \frac{C^0}{s} + C_1 \exp\left(-\frac{s^{1/2} x}{D^{1/2}}\right) + C_2 \exp\left(\frac{s^{1/2} x}{D^{1/2}}\right). \tag{5.14}$$

It follows from the boundary condition (5.12) that $C_2 = 0$, since otherwise the term $C_2 \exp[(s^{1/2} x)/D^{1/2}]$ would tend to infinity for $x \to \infty$. Hence

$$\bar{C}_{Ox}(x, s) = \frac{C_{Ox}^0}{s} + C_1 \exp\left(-\frac{s^{1/2} x}{D^{1/2}}\right). \tag{5.15}$$

Utilizing condition (5.13):

$$\bar{C}_{Ox}(x, s) = \frac{C_{Ox}^0}{s} + C_1 \exp\left(-\frac{s^{1/2} 0}{D^{1/2}}\right) = 0, \tag{5.16}$$

and hence

$$C_1 = \frac{C_{Ox}^0}{s}. \tag{5.17}$$

Combining equation (5.17) with equation (5.15):

$$\bar{C}_{Ox}(x, s) = \frac{C_{Ox}^0}{s} \left[1 - \exp\left(-\frac{s^{1/2} x}{D^{1/2}}\right)\right]. \tag{5.18}$$

Consideration should now be turned from function $\bar{C}_{Ox}(x, s)$ to the primary function $C_{Ox}(x, t)$. This can be done by re-transformation of expression (5.18), but the final result, i.e. the time-dependence of the current, can be obtained in a simpler way. The derivative $\left[\dfrac{\partial C_{\mathrm{Ox}}(x, t)}{\partial x}\right]_{x=0}$ should be calculated and the result should be combined with the general equation describing the magnitude of the current:

$$i = nFDA\left[\frac{\partial C_{\mathrm{Ox}}(x, t)}{\partial x}\right]_{x=0}. \tag{5.19}$$

In order to calculate the derivative we differentiate expression (5.18) with respect to x and obtain:

$$\frac{\partial \overline{C}_{\mathrm{Ox}}(x, s)}{\partial x} = \frac{C^0_{\mathrm{Ox}}}{s}\left[\frac{s^{1/2}}{D^{1/2}}\exp\left(-\frac{s^{1/2}x}{D^{1/2}}\right)\right]. \tag{5.20}$$

For $x = 0$ we have:

$$\left[\frac{\partial \overline{C}_{\mathrm{Ox}}(x, s)}{\partial x}\right]_{x=0} = \frac{C^0_{\mathrm{Ox}}}{(sD)^{1/2}}. \tag{5.21}$$

This expression is readily re-transformed, and as a result of re-transformation:

$$\left[\frac{\partial C_{\mathrm{Ox}}(x, t)}{\partial x}\right]_{x=0} = \frac{C^0_{\mathrm{Ox}}}{(\pi Dt)^{1/2}}. \tag{5.22}$$

Combining equation (5.22) with equation (5.19) we obtain the required relationship:

$$i_g = \frac{nFD^{1/2}AC^0_{\mathrm{Ox}}}{\pi^{1/2}t^{1/2}} \tag{5.23}$$

where i_g is the limiting current, i.e. the maximum current that can be obtained in the given conditions. Its value depends on the depolarizer concentration C^0_{Ox} and on its diffusion coefficient D, also on the electrode surface area A and electrolysis time t. F is the Faraday constant, and n is the number of electrons exchanged in the fundamental electrode process (the number of electrons exchanged between one ion or one molecule and the electrode).

This equation was derived by Cottrell [1], and its agreement with experimental data was checked by Laitinen and Kolthoff [2–4], who used platinum electrodes to ensure the linearity of diffusion. In this work they used electrodes constructed in various ways, and obtained a good agreement with theory in the case of anodic oxidation of ferrocyanide. Similar studies were carried out by Adams and Zimmerman [5, 6] who employed electrodes consisting of platinum and carbon paste.

Equation (5.23) describes the time-dependence of the limiting current for a flat electrode, provided that during the reduction the electrode potential is sufficiently negative. According to this equation the time-dependence of the current corresponds to the curve shown in Fig. 5.1. The current tends to zero as the time tends to infinity.

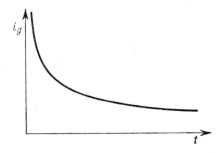

Fig. 5.1 Change of the limiting current (i_g) as a function of time (t) for a flat electrode at constant potential.

Using equation (5.23) it is easy to obtain the equation approximately describing the limiting current observed under polarographic conditions. For this purpose the electrode surface area should be expressed by means of parameters m and t.

It is known that the surface area of the mercury drop electrode, at a given moment of the drop time t, is:

$$A_t = 0.85 m^{2/3} t^{2/3}. \qquad (5.24)$$

Putting this expression into equation (5.23):

$$i_t = \frac{0.85}{\pi^{1/2}} nFD^{1/2} m^{2/3} t^{1/6} C_{Ox}^0. \qquad (5.25)$$

Expressing C_{Ox}^0 in mmole l^{-1}, m in mg s^{-1} and i_t in μA:

$$i_t = 463 n D^{1/2} m^{2/3} t^{1/6} C_{Ox}^0. \qquad (5.26)$$

This expression is very similar to the Ilkovič equation in which only the numerical coefficient is larger. This will be discussed in detail later in this chapter.

Equation (5.23) describes the limiting current, i.e. the maximum current that can be observed under these conditions, because in formulating the boundary condition we have assumed that all the molecules or ions of substance Ox are reduced at the electrode as soon as they reach it. This, in fact, happens in the case when the potential applied to the electrode is sufficiently negative, i.e. more negative than that at which the formation of the "plateau" of the limiting polarographic current takes place. A different situation exists in the case when the potential applied to a stationary flat electrode immersed in a non-stirred solution is so positive that the concentration of the oxidized form at the surface of the electrode is greater than zero. In such case the concentration of the oxidized form depends on the potential.

If the reduced form is soluble in the solution or in mercury it diffuses away from the surface of the electrode during its formation in the electrode process. It is assumed that the distribution of the concentration of reduced substance also obeys the linear diffusion equation.

In order to solve this new problem the boundary condition (5.6) should be replaced by a new condition which, according to the previous arguments, is the Nernst equation. However, this equation contains the concentration of the Red form, and therefore a system of two equations, consisting of equation (5.3) and the analogous equation for the Red form, must be solved. The solution of these two differential equations gives the functions $C_{Ox}(x, t)$ and $C_{Red}(x, t)$, describing the concentrations of the substances Ox and Red, as functions of the distance from the electrode and of time.

The initial conditions are as follows:

$$t = 0, \quad x \geqslant 0, \quad C_{Ox} = C_{Ox}^0, \quad C_{Red} = 0. \qquad (5.27)$$

It is assumed that the reduced substance is formed as a result of the electrode process only.

During the process, the concentrations at a very large distance from the electrode are:

$$t > 0, \quad x \to \infty, \quad C_{Ox} \to C_{Ox}^0, \quad C_{Red} \to 0. \qquad (5.28)$$

It has been mentioned that the first boundary condition is formulated on the basis of the Nernst equation. Since it is assumed that the electrode process is fast, the distribution of concentrations on the electrode surface depends on the electrode potential according to the following equation:

$$E = E^0 + \frac{RT}{nF} \ln \frac{C_{Ox}(0, t)}{C_{Red}(0, t)}. \qquad (5.29)$$

This equation can be written in a shorter form:

$$\Theta = \frac{C_{Ox}(0, t)}{C_{Red}(0, t)} \qquad (5.30)$$

where

$$\Theta = \exp\left[\frac{nF(E-E^0)}{RT}\right]. \qquad (5.31)$$

The second boundary condition can be found by considering the fluxes of the oxidized and the reduced forms. The sum of these fluxes is equal to zero, which can be expressed as

$$D_{Ox}\left(\frac{\partial C_{Ox}(x,t)}{\partial x}\right)_{x=0} + D_{Red}\left(\frac{\partial C_{Red}(x,t)}{\partial x}\right)_{x=0} = 0 \qquad (5.32)$$

where D_{Red} is the diffusion coefficient of the reduced form. Condition (5.32) means that one mole of substance Red is formed by electrolytic reduction of one mole of substance Ox.

Using the Laplace transformation once more the following expressions for $C_{Ox}(x,t)$ and $C_{Red}(x,t)$ are obtained:

$$C_{Ox}(x,t) = C_{Ox}^0 \frac{\Theta\left(\frac{D_{Ox}}{D_{Red}}\right)^{1/2} + \mathrm{erf}\left(\frac{x}{2D_{Ox}^{1/2}t^{1/2}}\right)}{1 + \Theta\left(\frac{D_{Ox}}{D_{Red}}\right)^{1/2}}, \qquad (5.33)$$

$$C_{Red}(x,t) = C_{Ox}^0 \frac{\left(\frac{D_{Ox}}{D_{Red}}\right)^{1/2} \mathrm{erfc}\left(\frac{x}{2D_{Red}^{1/2}t^{1/2}}\right)}{1 + \Theta\left(\frac{D_{Ox}}{D_{Red}}\right)^{1/2}}, \qquad (5.34)$$

where erfc is defined by the relationship

$$\mathrm{erfc}\,\gamma = 1 - \mathrm{erf}\,\gamma \qquad (5.35)$$

and erfγ is the integral of the error function which is defined by the equation

$$\mathrm{erf}\,\gamma = \frac{2}{\pi^{1/2}} \int_0^\gamma \exp(-z^2)\,dz. \qquad (5.36)$$

The dependence of function erfγ on parameter γ is shown in Fig. 5.2.

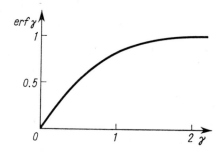

Fig. 5.2 Dependence of function erf γ on parameter γ

When the argument of the function is equal to zero the value of the function is also equal to zero, whereas for values of the argument greater than two the value of the error function is practically equal to one.

Returning to functions $C_{Ox}(x, t)$ and $C_{Red}(x, t)$ expressed by equations (5.33) and (5.34), it is possible from these to obtain expressions for the concentrations of substances Ox and Red on the surface of the electrode:

$$C_{Ox}(0, t) = C_{Ox}^0 \frac{\Theta \left(\frac{D_{Ox}}{D_{Red}}\right)^{1/2}}{1+\Theta \left(\frac{D_{Ox}}{D_{Red}}\right)^{1/2}}, \qquad (5.37)$$

$$C_{Red}(0, t) = C_{Ox}^0 \frac{\left(\frac{D_{Ox}}{D_{Red}}\right)^{1/2}}{1+\Theta \left(\frac{D_{Ox}}{D_{Red}}\right)^{1/2}}. \qquad (5.38)$$

It follows from equation (5.37) that the concentration at the surface of the electrode is equal to zero when $\Theta = 0$, i.e. when the electrode potential is much more negative than the normal potential of the Ox/Red system.

Using equation (5.37) it is possible to derive the expression describing the current intensity. Differentiating $C_{Ox}(x, t)$ given by equation (5.37) with respect to x and taking the value obtained for $x = 0$, we have the expression describing the flux of substance Ox on the surface of the electrode. This differentiation can be carried out using the following expression for the error function:

$$\frac{d\,\text{erf}[\gamma(u)]}{du} = \frac{2 d\gamma(u)}{\pi^{1/2} du} \exp - [\gamma(u)]^2, \qquad (5.39)$$

$$\left(\frac{\partial C_{Ox}}{\partial x}\right)_{x=0} = \frac{C_{Ox}^0}{(\pi D_{Ox} t)^{1/2} \left[1+\Theta \left(\frac{D_{Ox}}{D_{Red}}\right)^{1/2}\right]}. \qquad (5.40)$$

Substituting $\left(\frac{\partial C_{Ox}}{\partial x}\right)_{x=0}$ expressed by equation (5.40) in equation (5.19) we obtain the expression describing the current:

$$i = \frac{nFAD_{Ox}^{1/2} C_{Ox}^0}{\pi^{1/2} t^{1/2} \left[1+\Theta \left(\frac{D_{Ox}}{D_{Red}}\right)^{1/2}\right]}. \qquad (5.41)$$

Equation (5.41) becomes identical with equation (5.23) when $\Theta = 0$, i.e., according to equation (5.30), when the concentration of Ox on the surface of the electrode is equal to zero.

The derivation of equation (5.41) is not necessary for obtaining the expression describing the limiting current, since this current is obtained when the depolarizer concentration on the surface of the electrode is equal to zero. Therefore the Ilkovič equation can also be derived for the simple boundary condition that $C_{Ox}(0, t) = 0$.

In recent years the chronocoulometric method [7–10], having little in common with chronoamperometry, has been developed. In this method the determination of charge flowing in the circuit is carried out instead of the current determination. The time-dependence of the charge can be described in a simple way by means of equation (5.42), provided that all the conditions under which equation (5.23) was derived remain unchanged.

$$Q = \frac{2nFAD^{1/2}C^0 t^{1/2}}{\pi^{1/2}}. \tag{5.42}$$

It follows from equation (5.42) that the relationship between the charge and the square root of electrolysis time should be linear and that the straight line should pass through the origin of the coordinates. This is not observed in practice, since the change of the electrode potential from the definite value before the potential jump to the new value at which the electrode process takes place involves charging or discharging of the electrode double layer.

Let us call the charge considered Q_{dl}. It should appear as a constant, time-independent term on the right-hand side of equation (5.42):

$$Q = \frac{2nFAD^{1/2}C^0 t^{1/2}}{\pi^{1/2}} + Q_{dl}. \tag{5.42a}$$

This makes it possible to determine the capacity of the electrode double layer examined.

The chronocoulometric method is particularly useful in studies on adsorption of depolarizers, which will be discussed in chapter 16.

A new electroanalytical method was proposed in recent years by Oldham [11, 12]. It is based on the semi-integral of the current which is the parameter recorded in the experiment. The derivation of this parameter can be written in the form of the operator $d^{-1/2}/dt^{-1/2}$ since it is equivalent to differentiation to the power $-1/2$. It leads to a function having properties intermediate between those of the original function and its integral with respect to time.

When the potential is such that the concentration on the electrode surface is equal to zero, if τ is the transition time, then

$$\frac{d^{-1/2}}{dt^{-1/2}} i(\tau) = m(\tau) = nFAC^0 \sqrt{D}.$$

This equation, showing the proportionality between $m(\tau)$ and concentration, is the basis of Oldham's electroanalytical method. After measuring $m(\tau)$ it is easy to calculate the depolarizer concentration in the bulk solution.

The theoretical aspects of this method have been considered by Grenness and Oldham [13]. The method of calculation of parameter $m(\tau)$ has been also discussed. The dependence of $m(\tau)$ on potential has a shape identical with that of polarographic curves recorded in the case of the dropping electrode. A characteristic feature of this voltammetric method is the lack of coupling with time or frequency dependence. The complex method of obtaining the parameter $m(\tau)$ is a disadvantage of the method.

In another publication [14] the curves obtained in the case when an electrode having surface area independent of time is polarized by a potential have been described and an apparatus facilitating the derivation of parameter $m(\tau)$ has been proposed [15].

5.1.2 Chronopotentiometry

Derivation of the equivalent of chronoamperometric equation (5.23) for the chronopotentiometric method requires the solution of equation (5.3). The conditions (5.4) and (5.5) are also valid in this case, whereas the boundary condition describing the changes of depolarizer concentration on the surface of the electrode is different. Since classical chronopotentiometry is a method of electrolysis at constant current in the circuit, the flux of substance reducible at the electrode surface is also constant and is given by the expression:

$$D_{Ox} \left(\frac{\partial C_{Ox}(x, t)}{\partial x} \right)_{x=0} = \frac{i}{nFA}. \tag{5.43}$$

This equation is the boundary condition necessary for the solution of equation (5.3) for chronopotentiometric conditions.

Using the Laplace transformation as in the case of chronoamperometry, and utilizing the initial condition we obtain:

$$D_{Ox} \frac{d^2 \bar{C}_{Ox}(x, s)}{dx^2} - s\bar{C}_{Ox}(x, s) + C_{Ox}^0 = 0. \tag{5.44}$$

Transformation of the boundary conditions leads to the expressions:

$$x \to \infty, \quad \overline{C}_{Ox}(x, s) \to \frac{C^0_{Ox}}{s}, \qquad (5.45)$$

$$x = 0, \quad \frac{d\overline{C}_{Ox}(x, s)}{dx} = \frac{\lambda}{s}, \qquad (5.46)$$

where

$$\lambda = \frac{i}{nFAD_{Ox}}.$$

The general solution of equation (5.44) can be expressed by equation (5.14), and because of the boundary condition (5.45) it can also be expressed by equation (5.15).

Constant C_1 in equation (5.15) is calculated from the boundary condition (5.46).

Differentiating $\overline{C}(x, s)$ given by equation (5.15) with respect to x gives:

$$\frac{d\overline{C}_{Ox}(x, s)}{dx} = -\frac{s^{1/2}}{D^{1/2}_{Ox}} C_1 \exp\left(-\frac{s^{1/2}x}{D^{1/2}_{Ox}}\right). \qquad (5.47)$$

Re-transformation of this equation is easy, since it consists of expressions given in tables of transforms [16, 17].

Reversed transformation leads to the expression:

$$C_{Ox}(x, t) = C^0_{Ox} - 2\lambda \left(\frac{D_{Ox}t}{\pi}\right)^{1/2} \exp\left(-\frac{x^2}{4D_{Ox}t}\right)$$

$$+ \lambda x \, \text{erfc}\left(\frac{x}{2D_{Ox}^{1/2} t^{1/2}}\right). \qquad (5.48)$$

This equation describes the dependence of the change of concentration of the reducible substance on the electrolysis time and the distance from the electrode. It has been derived by several workers [18–21].

The purpose of this discussion is an attempt to obtain the time-dependence of the current, and therefore it is necessary to study the processes taking place at the electrode surface for $x = 0$. In the case of electrolysis in chronopotentiometric conditions the time-dependence of the concentration at the electrode surface is given by the expression:

$$C_{Ox}(0, t) = C^0_{Ox} - 2\lambda \left(\frac{D_{Ox}t}{\pi}\right)^{1/2} \qquad (5.49)$$

which is obtained by simplification of equation (5.48).

Although the limiting current is measured in polarography and peak current in stationary electrode voltammetry, it is the transition time which is measured in chronopotentiometry. The transition time is the electrolysis time τ after which the depolarizer concentration at the surface of the electrode decreases to zero. It can be expressed as:

$$C_{Ox}(0, \tau) = 0. \qquad (5.50)$$

Thus, we obtain the mathematical definition of the transition time by making the right-hand side of equation (5.49) equal to zero

$$C_{Ox}^0 - 2\lambda \left(\frac{D_{Ox}\tau}{\pi} \right)^{1/2} = 0. \qquad (5.51)$$

Equation (5.51) is usually written as:

$$\tau^{1/2} = \frac{\pi^{1/2} n F D_{Ox}^{1/2} C_{Ox}^0 A}{2i}. \qquad (5.52)$$

As in chronoamperometry, C_{Ox}^0 stands for depolarizer concentration in the bulk of solution, D is the depolarizer diffusion coefficient, A is the electrode surface area, i is the current intensity, F is the Faraday constant and n is the number of electrons exchanged in the elementary electrode process.

Equation (5.52) is often referred to as the Sand equation since it was first derived by Sand [19].

It follows from this equation that in the case of constant electrode surface area and constant depolarizer concentration C_{Ox}^0, the product $i\tau^{1/2}$

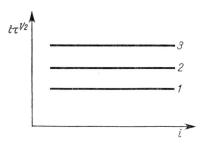

Fig. 5.3 Dependence of $i\tau^{1/2}$ on current intensity. The depolarizer concentration (or the electrode surface) increases from straight line *1* to straight line *3*.

is independent of the applied current density. This is shown in Fig. 5.3, in which straight lines *1*, *2* and *3* are the functions of depolarizer concentration increasing from *1* to *3* in the bulk of the solution or of the increasing electrode surface area. In general it is assumed that the constant values of the product $i\tau^{1/2}$ can be observed only at electrolysis times not exceeding several tens of seconds and on condition that the electrode process takes place under conditions of linear diffusion. At longer electrolysis times the convection resulting from density gradients caused by the electrode reaction increases the product $i\tau^{1/2}$. These problems have been studied by Bard [22, 23].

The determination of the transition time is simple in the case of reversible processes, provided that the chronopotentiometric curves are not distorted by capacity effects, or by effects of changes of the electrode surface taking place during the process. In other cases the determination of the transition time may be difficult.

Several methods of determination of τ have been proposed and discussed [24–29].

The effects of double layer charging during the recording of chronopotentiometric curves have also been analysed theoretically [29–31]. Olmstead and Nicholson have investigated experimentally the reduction of ferric iron on a mercury electrode in the presence of oxalates and found that the charge consumption by the double layer during the recording of chronopotentiograms is adequately described by the theoretical model which they were using [30].

Reinmuth [32] considered also a higher order electrode process taking place under chronopotentiometric conditions.

Chronopotentiometry is not as widely used in analytical practice as polarography and stationary electrode voltammetry, since the relation between the depolarizer concentration and the transition time is not linear. For this reason attempts were made to modify the chronopotentiometric method by simplifying this relationship. Senda [33], and later Kambara and Tachi [34] observed that when the current increases linearly with increasing $t^{1/2}$, the transition time is directly proportional to the depolarizer concentration. This kind of chronopotentiometry has been elaborated by Hurwitz and Gierst [35–37].

Using a current having the density:

$$i_0 = S\sqrt{t} \qquad (5.53)$$

where S is the so-called amplitude coefficient, we obtain, by the method previously described, the equation giving the concentration changes of the Ox form:

$$C_{\text{Ox}}(x, t) = C_{\text{Ox}}^0 - \frac{S}{2nFD_{\text{Ox}}} \left\{ \sqrt{\pi D_{\text{Ox}}} \left(t + \frac{x^2}{2D_{\text{Ox}}} \right) \right.$$
$$\left. \times \text{erfc} \frac{x}{2\sqrt{D_{\text{Ox}}t}} - x\sqrt{t} \exp \left(\frac{-x^2}{4D_{\text{Ox}}t} \right) \right\}. \tag{5.54}$$

The concentration at the electrode surface is given by the following simple expression:

$$C_{\text{Ox}}(0, t) = C_{\text{Ox}}^0 - \frac{St}{2nF} \sqrt{\frac{\pi}{D_{\text{Ox}}}}. \tag{5.55}$$

By substituting $C_{\text{Ox}}(0, t) = 0$ in equation (5.55) the expression describing the transition times is obtained:

$$\tau = \frac{2nFD_{\text{Ox}}^{1/2} C_{\text{Ox}}^0}{\pi^{1/2} S}. \tag{5.56}$$

Kambara and Tachi [34] gave the general integral equation for any time-dependence of the current impulse. An extensive and detailed discussion of these factors was given by Murray and Reilley [38, 39], who investigated theoretically chronopotentiometry in which the current impulses change according to the relationship $i_0 = St^r$, where r can have any value. For $r > -1$ the relation between the concentration and the transition time is:

$$\tau^{r+1/2} = \frac{nFD_{\text{Ox}}^{1/2} C_{\text{Ox}}^0 \Gamma(r+3/2)}{S\Gamma(r+1)} \tag{5.57}$$

where Γ stands for the gamma function. The ratio of gamma functions in equation (5.57) can be readily calculated from tables of this function.

Murray and Reilley also investigated chronopotentiometry in which the current changes according to the relationship $i_0 = Se^t$, as well as chronopotentiometry involving periodic current pulses.

Polarization of a dropping mercury electrode with constant intensity current during the drop lifetime has also been used [40–44]. In this case the current slowly increases like the voltage during the recording of the polarogram. The maintenance of a practically constant value of current intensity during the drop life time leads to considerable changes in current density, particularly during the first period of the drop lifetime, which considerably complicates the method.

Kies [45, 46] described a different procedure in which the dropping mercury electrode is polarized with a current having its intensity changing with time according to the relationship $i = at^{2/3}$ where the time is measured from the moment of detachment of the old drop from the capillary. Under such conditions the current density should remain constant although the electrode surface area changes.

Barański and Galus [47] considered theoretically chronopotentiometry in an internal electrolysis system and found a good agreement between the derived relationships and the experimental data.

5.1.3 Stationary Electrode Voltammetry

In the case of stationary electrode voltammetry, as in the two methods previously discussed, the solution of equation (5.3) with conditions (5.4) and (5.5) requires the definition of the boundary condition describing the concentration change of the reducible substance at the surface of the electrode. In the case of a reversible electrode process this condition is given by the Nernst equation, but in the case of an irreversible process its definition is different. In stationary electrode voltammetry this difference between the boundary conditions leads to different final relationships.

The introduction of the boundary condition defined by the Nernst formula makes it necessary to consider the dependence of concentration on time and distance from the electrode of both the oxidized and the reduced forms. Therefore in addition to equation (5.3) the analogous equation for the Red form must be used:

$$\frac{\partial C_{\text{Red}}(x, t)}{\partial t} = D_{\text{Red}} \frac{\partial^2 C_{\text{Red}}(x, t)}{\partial x^2}. \tag{5.58}$$

In additional to the initial condition (5.4) the initial condition for the reduced form must be defined:

$$t = 0, \quad x \geqslant 0, \quad C_{\text{Red}} = 0. \tag{5.59}$$

This expresses the fact that substance Red is formed in an electrolytic process, and that it is not present in the solution when this process is started.

The condition defining the concentration of substance Red on the border of the diffusion region:

$$t \geqslant 0, \quad x \to \infty, \quad C_{\text{Red}} \to 0 \tag{5.60}$$

must be added to the boundary condition (5.5). The former condition is obvious, bearing in mind that the reduced form is generated at the electrode. Therefore the concentration of substance Red decreases with increasing distance from the electrode and at infinite distance it is practically equal to zero.

The boundary conditions describing the time-dependence of concentrations of substances Ox and Red at the surface of the electrode are:

$$t > 0, \quad x = 0, \quad \frac{C_{Ox}}{C_{Red}} = \exp\left[\frac{nF(E-E^0)}{RT}\right], \quad (5.61)$$

and

$$D_{Ox}\frac{\partial C_{Ox}}{\partial x} = -D_{Red}\frac{\partial C_{Red}}{\partial x} = \frac{i(t)}{nFA}. \quad (5.62)$$

Condition (5.61) is the Nernst formula written in a slightly different form, and condition (5.62) expresses the fact that the sum of fluxes of substances Ox and Red at the surface of the electrode is equal to zero.

In stationary electrode voltammetry the potential in equation (5.61) is the following function of time:

$$E = E_i - Vt \quad (5.63)$$

where E_i is the initial potential at which the polarization starts, and V is the rate of change of polarization voltage (scan rate).

Utilizing equation (5.63) the boundary condition (5.61) can be written in the form:

$$\frac{C_{Ox}}{C_{Red}} = \Theta e^{-at} \quad (5.64)$$

where

$$\Theta = \exp\left[\frac{nF(E_i-E^0)}{RT}\right], \quad (5.65)$$

and

$$a = \frac{nFV}{RT}. \quad (5.66)$$

Using the Laplace transformation and initial conditions (5.4) and (5.59), from equations (5.3) and (5.58) we obtain equation (5.11) and the following relationship for substance Red:

$$D_{Red}\frac{d^2\overline{C}_{Red}}{dx^2} - s\overline{C}_{Red} = 0. \quad (5.67)$$

The general solution of equation (5.11) has been given above [equation (5.15)], and the solution of equation (5.67) is:

$$\overline{C}_{Red}(x, s) = C_2 \exp\left(-\frac{s^{1/2}x}{D_{Red}^{1/2}}\right). \quad (5.68)$$

Applying Laplace transformation to equation (5.62):

$$D_{Ox}\left(\frac{\partial \overline{C}_{Ox}}{\partial x}\right)_{x=0} = -D_{Red}\left(\frac{\partial \overline{C}_{Red}}{\partial x}\right)_{x=0} = \Delta(s) \quad (5.69)$$

where $\Delta(s)$ is the transform of the expression $i(t)/nFA$:

$$\Delta(s) = L\left\{\frac{i(t)}{nFA}\right\}. \quad (5.70)$$

Differentiating $\overline{C}_{Ox}(x, s)$ and $\overline{C}_{Red}(x, s)$ given by equations (5.15) and (5.68) with respect to x we obtain:

$$\frac{d\overline{C}_{Ox}(x, s)}{dx} = -C_1 \frac{s^{1/2}}{D_{Ox}^{1/2}} \exp\left(-\frac{s^{1/2}x}{D_{Ox}^{1/2}}\right), \quad (5.71)$$

$$\frac{d\overline{C}_{Red}(x, s)}{dx} = -C_2 \frac{s^{1/2}}{D_{Red}^{1/2}} \exp\left(-\frac{s^{1/2}x}{D_{Red}^{1/2}}\right). \quad (5.72)$$

Constants C_1 and C_2 can be readily determined by solving equations (5.71) and (5.72) for $d\overline{C}(0, s)/dx$ and combining the solutions with equation (5.69). Introducing the results into equations (5.15) and (5.68):

$$\overline{C}_{Ox}(x, s) = \frac{C_{Ox}^0}{s} - \frac{\Delta(s)}{s^{1/2}D_{Ox}^{1/2}} \exp\left(-\frac{s^{1/2}x}{D_{Ox}^{1/2}}\right) \quad (5.73)$$

and

$$\overline{C}_{Red}(x, s) = \frac{\Delta(s)}{s^{1/2}D_{Red}^{1/2}} \exp\left(-\frac{s^{1/2}x}{D_{Red}^{1/2}}\right). \quad (5.74)$$

Since it is required to obtain equations describing the current, the primary interest is in expressions giving the values of functions $\overline{C}_{Ox}$ and $\overline{C}_{Red}$ on the surface of the electrode:

$$\overline{C}_{Ox}(0, s) = \frac{C_{Ox}^0}{s} - \frac{\Delta(s)}{s^{1/2}D_{Ox}^{1/2}}. \quad (5.75)$$

$$\overline{C}_{Red}(0, s) = \frac{\Delta(s)}{s^{1/2}D_{Red}^{1/2}}, \quad (5.76)$$

from which it is possible obtain the integral equations [48]:

$$C_{Ox}(0, t) = C_{Ox}^0 - \frac{1}{\pi^{1/2}D_{Ox}^{1/2}} \int_0^t \frac{f(\tau)d\tau}{\sqrt{t-\tau}} \quad (5.77)$$

and

$$C_{\text{Red}}(0, t) = \frac{1}{\pi^{1/2} D_{\text{Red}}^{1/2}} \int_0^t \frac{f(\tau)d\tau}{\sqrt{t-\tau}} \tag{5.78}$$

where

$$f(t) = D_{\text{Ox}}\left(\frac{\partial C_{\text{Ox}}}{\partial x}\right)_{x=0} = \frac{i(t)}{nFA}. \tag{5.79}$$

In order to eliminate the concentrations and obtain the integral equation, the solution of which gives the flux due to substance Ox at the electrode surface, the boundary condition (5.64) is utilized, combined with equations (5.77) and (5.78):

$$\int_0^t \frac{f(t)d\tau}{\sqrt{t-\tau}} = \frac{C_{\text{Ox}}^0 \sqrt{\pi D_{\text{Ox}}}}{1+\gamma\Theta e^{-at}} \tag{5.80}$$

where

$$\gamma = \frac{D_{\text{Ox}}^{1/2}}{D_{\text{Red}}^{1/2}}. \tag{5.81}$$

Considering equation (5.66) it can be seen that the term at is dimensionless and is proportional to the potential:

$$at = \frac{nFVt}{RT} = \frac{nF(E_i - E)}{RT}. \tag{5.82}$$

The final purpose of this discussion is the calculation of the current–potential curves, not the current–time curves. Therefore it is useful to carry out all the calculations with reference to at.

Transforming the variables:

$$\tau = \frac{z}{a} \tag{5.83}$$

and

$$f(t) = g(at) \tag{5.84}$$

the following is obtained from equation (5.80):

$$\int_0^{at} \frac{g(z)dz}{\sqrt{a}\sqrt{at-z}} = \frac{C_{\text{Ox}}^0 \sqrt{\pi D_{\text{Ox}}}}{1+\gamma\Theta e^{-at}}. \tag{5.85}$$

The integral equation (5.85) can be expressed in dimensionless form. This is very important in its further treatment, in which numerical methods are used.

By substituting
$$g(at) = C_{Ox}^0 \sqrt{\pi D_{Ox} a}\, \chi(at) \tag{5.86}$$
the final form of the integral equations is obtained:
$$\int_0^{at} \frac{\chi(z)\,dz}{\sqrt{at-z}} = \frac{1}{1+\gamma\Theta e^{-at}}. \tag{5.87}$$

The solution of equation (5.87) gives the values of $\chi(at)$ depending on at for the given value of $\gamma\Theta$. According to equations (5.61) and (5.64) the values of at are related to the potential as follows:
$$E = E^0 - \frac{RT}{nF}\ln\gamma + \frac{RT}{nF}\ln\gamma\Theta = \frac{RT}{nF} at, \tag{5.88}$$
or
$$(E-E_{1/2})n = \frac{RT}{F}\ln\gamma\Theta + \frac{RT}{F} at. \tag{5.89}$$

For this reason the values of $\chi(at)$ can be considered as values of $\chi[(E-E_{1/2})n]$ and they determine the current as a function of the potential. From equations (5.79), (5.84) and (5.86):
$$i = nFAC_{Ox}^0 \sqrt{\pi D_{Ox} a}\, \chi(at). \tag{5.90}$$

The above reasoning, presented according to Nicholson and Shain [49], is at least partly similar to that published by other authors.

Equation (5.87) has been solved by several methods. As early as 1948 Ševčik [50], who first solved this problem simultaneously with Randles [51], expressed the right-hand side of equation (5.87) in the form of an exponential series, but in his final expression, relating the peak current to the concentration, the value of the constant was too low. This was due to erroneous evaluation of the value of the $\chi(at)$ function at the peak current potential.

Reinmuth [52] expressed the function $\chi(at)$ as follows:
$$\chi(at) = \frac{1}{\pi}\sum_{j=1}^{\infty}(-1)^{j+1}\sqrt{j}\exp\left[-\frac{jnF}{RT}(E-E_{1/2})\right]. \tag{5.91}$$

The analytical solution of equation (5.87) can be obtained in the form of the following equation:
$$\chi(at) = \frac{1}{\pi\sqrt{at}\,(1+\gamma\Theta)} + \frac{1}{4\pi}\int_0^{at} \frac{dz}{\sqrt{at-z}\,\cosh^2\left(\frac{\ln\gamma\Theta - z}{2}\right)}. \tag{5.92}$$

This equation was derived by Matsuda and Ayabe [53] and by Gokhsteyn [54]. De Vries and Van Dalen [55] expressed the function $\chi(at)$ by the equation

$$\chi(at) = -\sum_{n=1}^{\infty}(-1)^n \exp[-n(at^{1/2}-at)]\sqrt{n}\,\mathrm{erf}\sqrt{nat}. \quad (5.93)$$

Nicholson and Shain [49] applied numerical methods to the solution of equation (5.87). The results of their calculations lead to the determination of the value of function $\chi(at)$ as a function of the potential. The graph of this function is shown in Fig. 5.4. Its shape determines the shape of stationary electrode voltammetric curves in reversible electrode processes.

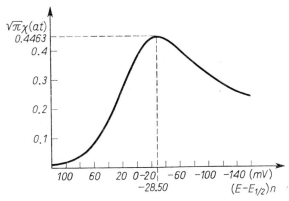

Fig. 5.4 Potential dependence of current function $\sqrt{\pi}\chi(at)$ for a reversible electrode process and linear diffusion.

In stationary electrode voltammetry the peak current is the characteristic quantity. The peak current equation can be readily obtained by using equation (5.90) and the graph of function $\chi(at)$, from which the maximum value of the function can be determined and then introduced into equation (5.90). The equation of the peak current i_p is:

$$i_p = 0.446 \frac{n^{3/2} F^{3/2}}{R^{1/2} T^{1/2}} A D_{Ox}^{1/2} V^{1/2} C_{Ox}^0. \quad (5.94)$$

At a temperature of 25°C this equation has the following simpler form:

$$i_p = 2.69 \times 10^5 n^{3/2} A D_{Ox}^{1/2} V^{1/2} C_{Ox}^0 \quad (5.94a)$$

where the peak current i_p is expressed in amperes, the electrode surface area A in cm^2, the depolarizer concentration in the bulk of solution C_{Ox}^0 in mole/ml, the depolarizer diffusion coefficient D_{Ox} in cm^2/s, and the scan rate V in V/s; n is the number of electrons exchanged in the elementary process. The last two equations show that the peak current is proportional to the depolarizer concentration, which makes the method suitable for analytical applications.

Equation (5.94) was first derived independently by Randles [51] and Ševčik [50]. Randles used the graphical method and Ševčik the Laplace transformation. The general validity of this equation was proved experimentally by Randles.

It has been mentioned that the value of the numerical coefficient of equation (5.94), determined by Ševčik, was too low. The correct value of this coefficient was determined by Randles and was experimentally checked by Delahay [56]. It was also confirmed by the results of all the above-mentioned theoretical investigations carried out by other authors.

The value of the Randles and Ševčik constant was determined experimentally by Mueller and Adams [57], who carried out stationary electrode voltammetric and chronoamperometric measurements, using the same solution and the same boron carbide electrode. They interpreted the results by means of the following equation, obtained by combining the Randles and Ševčik equation with the Cottrell equation:

$$K = \frac{i_p}{i_g t^{1/2} \pi^{1/2} V^{1/2} n^{1/2}} \tag{5.95}$$

where constant K should theoretically be equal to $0.452 \ F^{1/2} R^{1/2} T^{1/2}$.

These authors also showed that 2.72×10^5 is a satisfactory value for the constant in equation (5.94a).

Savéant and co-workers [58, 59] proposed a new variant stationary electrode voltammetry which they called convolution potential sweep voltammetry. It is based on calculation of a convolutive integral of the type:

$$I = \frac{1}{\pi^{1/2}} \int_0^t \frac{i(V)}{(t-V)^{1/2}} dV \tag{5.96}$$

directly from experimental current-time curves.

This method has much in common with the semi-integral electroanalysis proposed by Oldham [11].

5.1.3.1 *Stationary Electrode Voltammetric Process in which an Insoluble Product is Formed*

An example of this type of process is the reversible reduction of ions of certain metals on solid electrodes. The solution of the problem of a stationary electrode voltammetric process in which an insoluble product is formed is possible when it is assumed that the activity of the product is unity. This was done by Berzins and Delahay [60] who solved equation (5.3) with conditions (5.4) and (5.5) and with the second boundary condition defined on the basis of the Nernst equation:

$$t > 0, \quad x = 0,$$

$$C_{Ox}(0, t) = \exp\left[\frac{nF}{RT}(E_i - E^0) - \ln f_{Ox}\right]\exp(-at), \quad (5.97)$$

where E_i is the initial potential and f_{Ox} is the activity coefficient of the Ox form.

Using the Laplace transformation:

$$\left(\frac{\partial C_{Ox}}{\partial x}\right)_{x=0} = -C_{Ox}^0 i\left(\frac{a}{D_{Ox}}\right)^{1/2}\exp(-at)\,\text{erf}[i(at)^{1/2}] \quad (5.98)$$

where $i = \sqrt{-1}$.

Since [61]

$$\text{erf}[i(at)^{1/2}] = \frac{2i}{\pi^{1/2}}\int_0^{(at)^{1/2}} \exp z^2 dz, \quad (5.99)$$

i can be eliminated from equation (5.98).

Introducing the expression for a obtained from equation (5.66) into the resulting equation and combining it with equation (5.19) we obtain an expression describing the current i:

$$i = \frac{2n^{3/2}F^{3/2}}{\pi^{1/2}R^{1/2}T^{1/2}}\,AD_{Ox}^{1/2}C_{Ox}^0 V^{1/2}\Phi(at) \quad (5.100)$$

where $\Phi(at)$ is given by the expression:

$$\Phi(at) = \exp(-at)\int_0^{(at)^{1/2}} \exp z^2 dz. \quad (5.101)$$

The values of this function were tabulated by Miller and Gordon [62].
The function $\Phi(at)$ has its maximum equal to 0.541, which corresponds to the argument 0.924. The graph of this function is shown in Fig. 5.5.

Introducing the maximum value of the function into equation (5.100), an expression describing the peak current is obtained:

$$i_p = \frac{1.082 n^{3/2} F^{3/2}}{\pi^{1/2} R^{1/2} T^{1/2}} A D_{Ox}^{1/2} C_{Ox}^0 V^{1/2}. \tag{5.102}$$

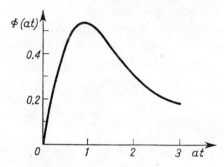

Fig. 5.5 Dependence of function $\Phi(at)$ on parameter at.

By substituting numerical values for the constants in the equation, at temperature 25°C:

$$i_p = 3.67 \times 10^5 n^{3/2} A D_{Ox}^{1/2} V^{1/2} C_{Ox}^0. \tag{5.103}$$

The dimensions of parameters in this equation are identical with those of equation (5.95).

It follows from equation (5.103) that, in the case of a reversible process leading to an insoluble product, the observed peak current is also proportional to the square root of the polarization rate and that the relation between the peak current and the concentration is linear. This provides suitable conditions for determination of metals by reduction of their ions at solid electrodes.

White and Lawson [63] continued the work of Berzins and Delahay, described above and, in the theoretical description of deposition and dissolution of metals at solid electrodes, they took into account the charge transfer kinetics, spherical diffusion and the uncompensated cell resistance.

5.1.4 POLAROGRAPHY

It has already been shown that, for the Ox form, the transport to a flat electrode moving in the direction of the solution is expressed by the following equation:

$$\frac{\partial C_{\text{Ox}}(x, t)}{\partial t} = D_{\text{Ox}} \frac{\partial^2 C_{\text{Ox}}(x, t)}{\partial x^2} + \frac{2x \partial C_{\text{Ox}}(x, t)}{3t \partial x}. \quad (5.104)$$

This equation with conditions (5.4)–(5.6), was first solved by Ilkovič [64, 65]. It takes into account the transport of the substance to the electrode due to convection caused by the growth of the drop, but it does not represent the diffusion to the growing mercury drop very accurately, since in its derivation the effect of the spherical diffusion was neglected. This equation will now be solved and its further modifications, leading to modification of the Ilkovič equation, will be discussed later in this chapter.

In order to reduce equation (5.104) to the equation describing linear diffusion in an unstirred solution, i.e. to equation (5.3), a new variable is introduced:

$$z = f(t)x \quad (5.105)$$

where $f(t)$ is an unknown function of time, which is so chosen that in equation (5.104) the term containing the first derivative of concentration with respect to distance disappears.

Since

$$\left(\frac{\partial C}{\partial t}\right)_x = \left(\frac{\partial C}{\partial t}\right)_z + \left(\frac{\partial C}{\partial z}\right)_t \frac{\partial z}{\partial t} = \left(\frac{\partial C}{\partial t}\right)_z + \frac{\partial C}{\partial z} xf'(t),$$

$$\left(\frac{\partial C}{\partial x}\right)_t = \left(\frac{\partial C}{\partial z}\right)_t \frac{\partial z}{\partial x} = f(t)\frac{\partial C}{\partial z},$$

the following relationship is obtained instead of equation (5.104):

$$\frac{\partial C_{\text{Ox}}}{\partial t} + \left[xf'(t) - \frac{2}{3}\frac{x}{t}f(t)\right]\frac{\partial C_{\text{Ox}}}{\partial z} = D_{\text{Ox}}[f(t)]^2 \frac{\partial^2 C_{\text{Ox}}}{\partial z^2}. \quad (5.106)$$

If

$$f(t) = t^{2/3}, \quad (5.107)$$

the expression preceding $\partial C/\partial z$ on the left-hand side of equation (5.106) becomes equal to zero. Then:

$$\frac{\partial C_{\text{Ox}}}{\partial t} = D_{\text{Ox}} t^{4/3} \frac{\partial^2 C_{\text{Ox}}}{\partial z^2}. \quad (5.108)$$

On introducing a new variable:

$$p = \tfrac{3}{7} t^{7/3} \quad (5.109)$$

the following expression is finally obtained:

$$\frac{\partial C_{Ox}(z, p)}{\partial p} = D_{Ox}\frac{\partial^2 C_{Ox}(z, p)}{\partial z^2}. \quad (5.110)$$

The initial and boundary conditions are defined by analogy with equations (5.4)–(5.6):

$$\tau = 0, \quad z \geqslant 0, \quad C_{Ox} = C^0, \quad (5.111)$$

$$\tau \geqslant 0, \quad z \to \infty, \quad C_{Ox} \to C^0, \quad (5.112)$$

$$\tau > 0, \quad z = 0, \quad C_{Ox} = 0. \quad (5.113)$$

Proceeding as in subsection 5.1.1 we obtain the following general solution describing the change of the concentration:

$$\bar{C}(z, s) = \frac{C^0}{s}\left[1 - \exp\left(-\frac{s^{1/2}z}{D_{Ox}^{1/2}}\right)\right]. \quad (5.114)$$

In this case also, it is desired to obtain the expression describing the current on the dropping electrode, using an equation analogous with (5.19). Therefore, it is possible to differentiate $\bar{C}_{Ox}(z, s)$ with respect to z instead of transforming in reverse function (5.114). Differentiating and then taking the value of the function for $z = 0$ we obtain:

$$\left[\frac{\partial \bar{C}_{Ox}(z, s)}{\partial z}\right]_{z=0} = \frac{C^0}{(sD_{Ox})^{1/2}}. \quad (5.115)$$

By means of reverse transformation of this function [11, 12] the following expression is obtained:

$$\left[\frac{\partial C_{Ox}(z, p)}{\partial z}\right]_{z=0} = \frac{C^0}{(\pi D_{Ox} p)^{1/2}}. \quad (5.116)$$

The current intensity is given by:

$$i = nFAD_{Ox}\left(\frac{\partial C_{Ox}}{\partial x}\right)_{x=0} = nFAD_{Ox}\left(\frac{\partial C_{Ox}}{\partial z}\right)_{z=0}\frac{\partial z}{\partial x}. \quad (5.117)$$

According to equations (5.105) and (5.107):

$$\frac{\partial z}{\partial x} = f(t) = t^{2/3}. \quad (5.118)$$

Combining equations (5.116) and (5.118) with equation (5.117):

$$i = nFAD_{Ox}\frac{C^0 t^{2/3}}{(\pi D_{Ox} p)^{1/2}}. \quad (5.119)$$

Substituting the value of p from expression (5.109) into equation (5.119), the final result is obtained:

$$i = nFAD_{Ox} \frac{C^0 t^{2/3}}{\left(\pi D_{Ox} \frac{3}{7} t^{7/3}\right)^{1/2}} = \sqrt{\frac{7}{3}} \frac{nFD_{Ox}^{1/2} A C^0}{(\pi t)^{1/2}}. \quad (5.120)$$

This equation differs from that previously derived, (5.26), only by the constant $\sqrt{7/3}$. The current observed in the case of a flat electrode moving in the direction of the solution is thus $\sqrt{7/3}$ times larger than that observed in the case of a stationary flat electrode having the same surface area. Therefore the factor $\sqrt{7/3}$ is the measure of the contribution of convection to the diffusion transport of an electrolysed substance to the electrode.

Proceeding as for a stationary flat electrode and inserting the term for the dropping electrode surface $A = 0.85\,(mt)^{2/3}$ the following relationship is obtained:

$$i_t = 706 n C^0 D_{Ox}^{1/2} m^{2/3} t^{1/6}. \quad (5.121)$$

In this equation which describes the instantaneous current (expressed in microamperes) at time t in the life of the drop, the concentration is expressed in millimoles per litre, and m in milligrams per second.

When t is equal to the drop time t_1 the current is the maximum current recorded at the moment at which the mercury drop detaches.

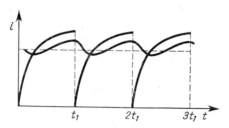

Fig. 5.6 Time dependence of instantaneous and average currents; t_1—drop time. The average current is shown by the wavy line.

The current changes during the drop time are shown in Fig. 5.6. These curves showing the large current changes during the drop life can be obtained by means of a low inertia galvanometer or by displaying the potential drop across a constant resistance with an oscillograph.

In practical polarography galvanometers of long period (4–8 seconds) are employed, or else an RC circuit (R is the resistance and C is the capacity of the circuit) is incorporated. Under these circumstances the recorded current corresponds approximately to the mean value, and oscillations of the current, corresponding to electrode surface changes, are observed around the mean current (drop wave oscillations).

The mean current during the drop time can be determined by integration of the current over the drop time:

$$\bar{i} = \frac{1}{t_1} \int_0^{t_1} i \, dt. \tag{5.122}$$

Since we know the dependence of i on t [equation (5.121)] the solution of equation (5.122) can be obtained:

$$\bar{i} = \frac{1}{t_1} \int_0^{t_1} 705 n C^0 D_{Ox}^{1/2} m^{2/3} t^{1/6} dt = \frac{6}{7} \times 706 n C^0 D_{Ox}^{1/2} m^{2/3} t^{1/6}$$

$$= 605 n D_{Ox}^{1/2} C^0 m^{2/3} t^{1/6}. \tag{5.123}$$

This equation is known as the Ilkovič equation, since it was first obtained by Ilkovič [64]. It does not take into account the spherical nature of diffusion at a growing mercury drop, the decrease of the depolarizer concentration in the neighbourhood of the electrode, the non-centric growth of the drop during the outflow of mercury from the capillary, and the screening effect of the capillary glass on the electrode. Modifications of equations (5.121) and (5.123) to take account of these effects will be discussed later in this chapter.

Equation (5.121) for the instantaneous current makes it possible to analyse polarographic current–voltage curves, obtained on individual mercury drops. According to this equation the dependence of log i on log t should be linear and the slope of the straight line should be equal to $\frac{1}{6}$. Deviations from this relationship make it possible to determine the character of electrode processes.

It follows from the Ilkovič equation that the limiting* current is proportional to the depolarizer concentration in the bulk of the solution, pro-

---

* It follows from equations (5.26) and (5.12) that the contribution of convection to the transport of the substance to the electrode is considerable. In the case of diffusion to a flat electrode moving in the direction of the solution the numerical coefficient 463 appearing in equation (5.26), describing the diffusion process only, is replaced by a new coefficient $\sqrt{7/3}$ times larger. This increase in the magnitude of the coefficient which is a measure of the contribution of convection to the depolarizer transport to the electrode, is so significant that the term "diffusion current" suggesting the current due to discharge

vided that parameters m and t_1 remain unchanged. This can be written in the form of the following equation:

$$i_g = K_{Il} c^0 \tag{5.124}$$

where K_{Il}, called the Ilkovič constant, is defined by the formula:

$$K_{Il} = 605 n D_{Ox}^{1/2} m^{2/3} t^{1/6}. \tag{5.125}$$

The Ilkovič equation allows the limiting current to be related to the height of mercury head. Any change of this height causes changes of m and t_1. These relationships can be expressed by the following simple equations:

$$m = k'h_r, \tag{5.126}$$

$$t_1 = \frac{k''}{h_r}, \tag{5.127}$$

where h_r stands for the height of the mercury reservoir corrected for the back pressure (i.e. the pressure opposing the capillary forces).

Writing the Ilkovič equation in the form:

$$i_g = \text{const}\, m^{2/3} t^{1/6}, \tag{5.128}$$

and combining equation (5.128) with equations (5.126) and (5.127) we obtain the relationship:

$$i_g = \text{const}(k'h_r)^{2/3} \left(\frac{k''}{h_r}\right)^{1/6} = \overline{\text{const}}\, h_r^{1/2} \tag{5.129}$$

where $\overline{\text{const}}$ stands for an expression combining constants const, k' and k''. It follows from this relationship that by lifting the mercury reservoir the limiting current can be increased. The height of the reservoir, appearing in expression (5.129), is decreased by a certain quantity related to the back pressure.

The uncorrected height can be used for investigation of dependence on (5.129); in this case the dependence is also linear, but the straight line does not pass through the origin and it has the coordinates i_g, $-h^{1/2}$.

The dependence of the limiting current on the square root of the height of the mercury reservoir is very often used for determining the type of current involved. For example, in the case of processes controlled by the rate of a preceding chemical reaction a different relationship for dependence is observed.

of the depolarizer supplied by diffusion, should be changed to "limiting current" [66]. Replacement of the term "diffusion current" with "limiting current" will inevitably cause confusion, but this ought to be done for the sake of accuracy of description.

In any further discussion of the Ilkovič equation the diffusion current constant should be mentioned. This concept, introduced by Lingane [67], is defined by the relationship:

$$I = \frac{i_g}{C^0 m^{2/3} t_1^{1/6}} = 605 n D^{1/2}. \tag{5.130}$$

Lingane and Loveridge [68] found that when the product $m^{2/3} t_1^{1/6}$ is varied the diffusion current constant passes through a minimum.

Using the known values of diffusion current constant which are tabulated in some text-books on polarography [69] the concentration can be determined, by measuring the limiting current when the capillary parameters are known.

We should also remember the effect of the potential on the recorded limiting current. It is known that the mercury outflow rate has almost no dependence on potential, but the drop time is potential-dependent. By measuring the drop time at various potentials it is possible to obtain the electrocapillary curve. Therefore at very negative potentials, when the surface tension is small, the drop time is shorter, and as a result the limiting current is decreased. Since in the Ilkovič equation t_1 appears as $t_1^{1/6}$, these changes are not large, but they should be taken into account in accurate measurements.

5.1.5 The Rotating Disc Method

The case of the rotating disc method is not approached by solving equation (5.3). Instead, an equation describing the changes of depolarizer concentration, depending on the distance from the electrode under conditions of laminar convection transport of the depolarizer to the disc electrode surface is used.

When the solution is vigorously stirred, which is usually the case in rotating disc electrolysis, the stationary state is established immediately after the start of electrolysis. Under these conditions the depolarizer concentration is the following function of distance (x) from the disc surface ($x = 0$):

$$S_x \frac{dC_{Ox}}{dx} = D \frac{d^2 C_{Ox}}{dx^2}. \tag{5.131}$$

This dependence is a simplified form of equation (4.61). It is obtained on the assumption that the concentration gradient in the neighbourhood of the rotating disc is formed only in the direction perpendicular to its surface.

In this method the boundary condition describing the mode of change of the depolarizer concentration at the electrode surface can be formulated in various ways.

(a) The applied potential can be so negative during the reduction (or so positive during the oxidation) that the depolarizer concentration on the surface of the disc is practically zero. This is the procedure commonly used, although in practice an increasing potential is applied continuously to the electrodes, starting (in the case of a reduction) with a negative value such that the initial rate of reaction is virtually zero. It will be recognized that this is akin to the starting condition in polarography.

It is convenient to accept the above condition, since it is the purpose to derive an expression for the limiting current. Therefore, the approach resembles that of subsection 5.1.1, although in that case transport of material to the electrode surface was by diffusion only.

(b) The rotating disc electrode can be applied in the chronopotentiometric method (in practice this method is not used). The electrolysis then takes place at constant current, and the condition is similar to condition (5.42).

(c) The rotating electrode can be rapidly polarized with a potential increasing at a constant rate. In this case the boundary condition determining the concentration of the depolarizer on the electrode surface is similar to that used in stationary electrode voltammetry with linear diffusion (5.64). When the polarization is sufficiently fast the curves recorded for a rotating disc electrode show the current peak which is more or less pronounced, depending on hydrodynamic conditions.

These three possible variants of the rotating disc electrode method will now be discussed in detail.

5.1.5.1 *The Limiting Current with Constant Potential Applied to the Rotating Disc Electrode*

The problem of the movement of a liquid due to the rotation of a disc placed perpendicularly to the rotation axis was studied by von Karman [70] and Cochran [71]. Using the results of this work Levich [72, 73] determined the dependence of the concentration on the distance from the rotating disc electrode and he also obtained the expression describing the limiting current for these electrodes.

This expression can be derived starting with equation (5.131) with the boundary conditions (5.5) and (5.6). It is assumed that the potential applied to the electrode during the experiment is such that the electrode process is faster than the transport of the substance to the electrode, and as a result, the depolarizer concentration on the electrode surface is equal to zero.

Integrating equation (5.131) gives the expression:

$$\frac{dC_{ox}}{dx} = C_1 \exp\left\{\frac{1}{D}\int_0^x S_x(z)\,dz\right\}. \tag{5.132}$$

A second integration gives the relationship:

$$C_{ox} = C_1 \int_0^x \exp\left\{\frac{1}{D}\int_0^t S_x(z)\,dz\right\} dt + C_2. \tag{5.133}$$

In order to calculate the integration constants C_1 and C_2 conditions (5.5) and (5.6) must be used. It follows from condition (5.6) that $C_2 = 0$, since the integral in expression (5.133) disappears for $x = 0$.

From condition (5.5):

$$C_{ox}^0 = C_1 \int_0^\infty \exp\left\{\frac{1}{D}\int_0^t S_x(z)\,dz\right\} dt. \tag{5.134}$$

In order to evaluate the integral in equation (5.134) we divide the integration range into two parts $0 \leqslant x \leqslant \delta$ and $\delta \leqslant x \leqslant \infty$ (i.e. the diffusion layer and the remainder of the solution situated further from the electrode).

Calling the integral W:

$$W = W_1 + W_2 = \int_0^\infty \exp\left\{\frac{1}{D}\int_0^t S_x(z)\,dz\right\} dt$$

$$= \int_0^\delta \exp\left\{\frac{1}{D}\int_0^t S_x(z)\,dz\right\} dt + \int_\delta^\infty \exp\left\{\frac{1}{D}\int_0^t S_x(z)\,dz\right\} dt. \tag{5.135}$$

In the layer adjoining the electrode S_x is described by the relationship:

$$S_x \cong -\frac{\omega^{3/2} x^2}{2\nu^{1/2}}, \tag{5.136}$$

and therefore

$$W_1 \cong \int_0^\infty \exp\left(-\frac{\omega^{3/2} t^3}{6D\nu^{1/2}}\right) dt \tag{5.137}$$

where ν is the kinematic viscosity of the solution and ω is the angular velocity of the electrode.

Since the integral W_1 is rapidly convergent:

$$W_1 \cong \int_0^\infty \exp\left(-\frac{\omega^{3/2}t^3}{6D\nu^{1/2}}\right) dt. \tag{5.138}$$

Introducing the new variable $u = \omega^{1/2}t/\sqrt[3]{6D\nu^{1/2}}$:

$$W_1 = \frac{\sqrt[3]{6D\nu^{1/2}}}{\omega^{1/2}} \int_0^\infty \exp(-u^3)\, du. \tag{5.139}$$

The integral in equation (5.139) is expressed as function Γ:

$$\int_0^\infty \exp(-u^3)\, du = \frac{1}{3}\int_0^\infty \frac{e^{-t}dt}{t^{2/3}} = \frac{1}{3}\Gamma\left(\frac{1}{3}\right) = \Gamma\left(1+\frac{1}{3}\right) = 0.89, \tag{5.140}$$

therefore equation (5.139) can be written in the following final form:

$$W_1 = 1.62\frac{D^{1/3}}{\nu^{1/3}}\sqrt{\frac{\nu}{\omega}}. \tag{5.141}$$

Similarly the second integral W_2 is equal to:

$$W_2 = \int_0^\infty \exp\left\{\frac{1}{D}\int_0^t S_x(z)\,dz\right\} dt = \int_0^\infty \exp\left\{-0.89\frac{\sqrt{\nu\omega}}{D}t\right\} dt$$

$$= \frac{D}{0.89\sqrt{\nu\omega}}\exp\left(\frac{-0.89\sqrt{\nu\omega}\delta_0}{D}\right) \cong \exp\left(\frac{-2.3\nu}{D}\right)\frac{D}{\nu}\sqrt{\frac{\nu}{\omega}}. \tag{5.142}$$

$D/\nu \cong 10^{-3}$ and $W_1 \gg W_2$. This means therefore that the principal change in concentration takes place in the layer adjoining the electrode.

C_1 can be readily determined from equations (5.134) and (5.141). Introducing the resulting expression into equation (5.133):

$$C_{Ox} = \frac{C_{Ox}^0}{1.62\left(\dfrac{D}{\nu}\right)^{1/3}\sqrt{\dfrac{\nu}{\omega}}}\int_0^x \exp\left\{\frac{1}{D}\int_0^t S_x(z)\,dz\right\} dt. \tag{5.143}$$

The concentration increases with increasing distance from the surface of the disc and at a distance of the order of δ it approaches the value of C_{Ox}^0. The dependence of concentration on the distance from the electrode surface is shown in Fig. 5.7.

Differentiating equation (5.143) with respect to x and considering the result for $x = 0$ gives the general expression for the intensity of the limiting current in the case of a rotating disc electrode:

$$i_g = nFAD\left(\frac{\partial C_{\text{Ox}}}{\partial x}\right)_{x=0} = nFAD \frac{C^0_{\text{Ox}}}{1.62\left(\frac{D}{\nu}\right)^{1/3}\sqrt{\frac{\nu}{\omega}}}$$

$$= 0.62nFAD^{2/3}C^0_{\text{Ox}}\nu^{-1/6}\omega^{1/2}, \tag{5.144}$$

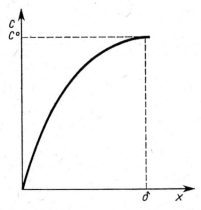

Fig. 5.7 Dependence of depolarizer concentration (C) on the distance from the disc electrode (x); δ—thickness of the diffusion layer.

in which the current i_g is expressed in amperes, C^0 in moles per litre, D and ν in cm²/s, and ω (the angular velocity of the rotating electrode) in radians per second.

Equation (5.144) can be expressed in the following general form:

$$i_g = \frac{nFADC^0}{\delta} \tag{5.145}$$

where δ is the thickness of the diffusion layer, shown schematically in Fig. 5.7.

By comparing equations (5.144) and (5.145) the following formula for δ is obtained:

$$\delta = 1.61D^{1/3}\nu^{1/6}\omega^{-1/2}. \tag{5.146}$$

Gregory and Riddiford [74] showed that convergence of integral (5.137) is not rapid enough to make it possible to pass from equation (5.137) to equation (5.138) without an error. Using the expansion given by these authors we obtain the following expression for the thickness of the diffusion layer:

$$\delta = 1.61 D^{1/3} v^{1/6} \omega^{-1/2} \left[1 + 0.3539 \left(\frac{D}{v} \right)^{0.36} \right]. \quad (5.146a)$$

Newman [75] carried out more detailed calculations for a wider interval of the ratio D/v and obtained the following description for δ:

$$\delta = 1.61 D^{1/3} v^{1/6} \omega^{-1/2} \left[1 + 0.2980 \left(\frac{D}{v} \right)^{1/3} + 0.14514 \left(\frac{D}{v} \right)^{2/3} \right]. \quad (5.146b)$$

Further work was carried out by Kassner [76], leading to the following equation:

$$\delta = 1.611 D^{1/3} v^{1/6} \omega^{-1/2} \left\{ 1.1203 A + 0.6977 \left(\frac{D}{v} \right)^{2/3} \exp\left[-3.11 \left(\frac{v}{D} \right) \right] \right\} \quad (5.146c)$$

which is valid when $0 < \frac{D}{v} < 0.25$.

The dependence of parameter A on the ratio $\frac{D}{v}$ is shown in Table 5.1.

Table 5.1 DEPENDENCE OF PARAMETER A ON $\frac{D}{v}$

$\frac{D}{v}$	A	$\frac{D}{v}$	A
0	0.8934	0.060	1.0143
0.001	0.9209	0.080	1.0268
0.003	0.9341	0.100	1.0368
0.005	0.9424	0.120	1.0451
0.007	0.9487	0.140	1.0521
0.009	0.9541	0.160	1.0580
0.020	0.9747	0.200	1.0675
0.040	0.9981	0.250	1.0762

When $\frac{D}{v} = 10^{-3}$ the value of δ given by Levich is about 3% higher than that obtained from equations (5.146a, b and c), but at higher values of $\frac{D}{v}$, of the order of 0.1, this difference is larger (up to 17%).

The correctness of equation (5.144) was checked by other authors. The linearity of the relation between i_g and $\omega^{1/2}$, predicted by this equation, was first checked experimentally by Kabanov and Siver [77, 78]. Investigating the reduction of oxygen and hydrogen ions they found that the discrepancies between experimentally determined and theoretical values of the limiting currents do not exceed 3%. The linearity of the relation between i_g and $\omega^{1/2}$ was also checked by Hogge and Kraichmann [79, 80] and by other authors.

After detailed investigations of these relationships Gregory and Riddiford [74] introduced a corrective term to equation (5.144).

This equation also leads to the conclusion that $i_g = 0$ when $\omega = 0$. Several authors [74, 81–83] showed that this conclusion is wrong, since in this case diffusion transport always takes place.

It should be stressed that equation (5.144) is valid for laminar flow of the liquid in the vicinity of the disc. This laminarity is preserved when the Reynolds number Re, defined by the expression:

$$Re = \frac{r^2 \omega}{\nu}$$

does not excess 10^5. In this equation r is the total radius of the disc, the conductor and the insulator surrounding it. It follows from this equation that it is easier to maintain the laminar flow when the disc is small, but in such cases the boundary effects can play an important part.

The limits of applicability of disc electrodes, due to the requirements for laminar flow, are valid when the disc surface is ideally smooth and the disc is well centred. The shape of the disc also affects the agreement between the experimental and the theoretical results. Blurton and Riddiford [84, 85] observed that electrodes in which the immersed parts were cylindrical can give results different from those expected on theoretical grounds. Similar electrodes were investigated by Adams and co-workers. Prater and Adams [86] compared the results obtained by means of electrodes in which the disc was placed in a cylindrical insulator with those in which it was placed in a bell-shaped insulator. The discrepancies between these results were within the experimental error in the range 1–40 revs. per second.

The results can also be affected by the shape of the vessel in which the electrode is placed and by the distance of the electrode from the bottom of the vessel when it is smaller than 0.5 cm [74].

Prater and Adams [86] showed that the results are not affected by changing the volume of the electrolytic vessel between 100 ml and 9 litres.

In many publications is was assumed that the same accessibility of disc electrode surface for the depolarizer would lead to a homogeneous distribution of current density on the surface of the disc. However Newman [87] showed theoretically that in this case the current density would change along the radius of the electrode as a result of ohmic decreases when the electrode potential is insufficient for reaching the limiting current. This was confirmed on the basis of distribution of changes of thickness of metal deposited on the electrode surface [88, 89]. Newman's conclusions have also been confirmed in an extensive work published by Miller and Bellavance [90].

5.1.5.2 Electrode Processes on Rotating Disc Electrodes at Constant Current

The theory of electrode processes on rotating disc electrodes at constant current intensity was first discussed in detail by Levich [73], who used the successive approximations method, and then by Siver [91]. Both these treatments give only approximate solutions over limited time intervals. The complete solution of the problem was obtained by Hale [92] and by Filinovskii and Kiryanov [93]. Hale obtained the solution by using the numerical method, while Filinovskii and Kiryanov employed the Laplace transformation method. The latter method will be briefly described.

The depolarizer concentration C_{Ox} can be expressed by the unidimensional equation of convective diffusion:

$$\frac{\partial C_{Ox}}{\partial t} = D_{Ox}\frac{\partial^2 C_{Ox}}{\partial x^2} - S_x\frac{\partial C_{Ox}}{\partial x} \qquad (5.147)$$

where S_x is the velocity vector, functionally dependent on position in space, but independent of time in the case of a hydrodynamic stationary state. For large values of the Schmidt number ($Sc = \nu/D$) S_x can be represented with sufficient accuracy by the relationship (5.136) if it is in the vicinity of the disc electrode surface.

In order to simplify further calculations dimensionless variables $y = x/\delta$, and $k = Dt/\delta^2$ will be used instead of variables x and t and a dimensionless concentration $g(y, k) = [C^0 - C(x, t)]/C^0$ will be introduced, where C^0 is the depolarizer concentration in the bulk solution, and δ is the thickness of the diffusion layer. In this case equation (5.147) becomes:

$$\frac{\partial g}{\partial k} = \frac{\partial^2 g}{\partial y^2} + \alpha y^2 \frac{\partial g}{\partial y} \qquad (5.148)$$

where $\alpha = 2.128$.

When the depolarizer flux on the disc electrode surface is described by the equation $D_{Ox}\left(\frac{\partial C_{Ox}}{\partial x}\right)_{x=0} = U(t)$, the boundary conditions are:

$$k \geqslant 0, \quad y \to \infty, \quad g \to 0, \qquad (5.149)$$

$$k > 0, \quad y = 0, \quad \left(\frac{\partial g}{\partial y}\right)_{y=0} = W(k), \qquad (5.150)$$

where

$$W(k) = -\frac{U(t)}{i_g}, \qquad (5.151)$$

and i_g is described by the relationship:

$$i_g = \frac{nFD_{Ox}C^0 A}{\delta}. \qquad (5.152)$$

Using the Laplace transformation method, employing the above specified conditions and introducing the substitution

$$\bar{g}(y, s) = \Psi(y, s)\exp\left(-\frac{\alpha y^3}{6}\right) \qquad (5.153)$$

where $\bar{g}(y, s)$ stands for the transform of function $g(y, t)$, we obtain the equation:

$$\Psi(y, s) = \frac{W(s)Ai[(s+\alpha y)/\alpha^{2/3}]}{\alpha^{1/3}Ai'(s/\alpha^{2/3})} \qquad (5.154)$$

where $Ai(z)$ is the Airy function which increases exponentially as $z \to \infty$, and $Ai'(z)$ is the derivative of this function.

It is rather difficult to return to the original function since it requires the application of the interpolation formula to the logarithmic derivative of the Airy function. Using this formula and the further necessary transformations the following relationship is obtained for the depolarizer concentration on the disc electrode surface in the case of electrolysis at constant current density:

$$C_{Ox}(0, t) = C^0\left\{1 - \frac{i_0}{i_g}\left[1.07\,\mathrm{erf}\sqrt{3.10\frac{D_{Ox}t}{\delta^2}} - 0.73\exp\left(-1.65\frac{D_{Ox}t}{\delta^2}\right)\mathrm{erf}\sqrt{1.45\frac{D_{Ox}t}{\delta^2}}\right]\right\} \qquad (5.155)$$

where i_0 is the current density.

Since the transition time τ is reached when the depolarizer concentration on the surface decreases to zero, the following relationship is obtained by making the right-hand side of equation (5.155) equal to zero:

$$\frac{nFD_{Ox}C°A}{\delta i_0} = 1.07 \operatorname{erf} \sqrt{3.10 \frac{D_{Ox}\tau}{\delta^2} - 0.73} \times$$

$$\exp\left(-1.65 \frac{D_{Ox}\tau}{\delta^2}\right) \operatorname{erf} \sqrt{1.45 \frac{D_{Ox}\tau}{\delta^2}}. \quad (5.156)$$

This equation is quite general and can be used for interpretation of processes involving both long and short transition times. It shows that the transition time becomes longer as a result of convection. In the initial stage of electrolysis the inequality $t \leqslant \delta^2/D_{Ox}$ is valid for low values of t, and the effect of convection on the concentration on the electrode surface is slight. In this case equation (5.155) becomes:

$$C_{Ox}(0, t) = C°\left[1 - \frac{i_0}{i_g} \frac{2}{\sqrt{\pi}} \sqrt{\frac{D_{Ox}t}{\delta^2}}\right]$$

$$= C° - \frac{2i_0 \sqrt{t}}{nF\sqrt{\pi D_{Ox}}}, \quad (5.157)$$

i.e. it is reduced to the equation characteristic of chronopotentiometry in unstirred solutions.

5.1.5.3 *Dependence of Current on Rate of Potential Change for Processes Taking Place on Rotating Disc Electrodes*

When the potential applied to the electrode increases rapidly and linearly it is sometimes difficult to observe a pronounced "plateau" on the limiting current.

This situation occurs when the time during which the potential changes is shorter than that in which the stationary diffusion conditions are established. The latter depends on the thickness of the diffusion layer and can be shortened by increasing the speed of rotation of the disc electrode, which of course decreases the thickness of the diffusion layer.

The problem of dependence of limiting current on potential under such conditions has been discussed by Fried and Elving [94] who used the concept of the Nernst diffusion layer. In 1967 Girina, Filinovskii and Feoktistov [95] published a study of this problem for reversible electrode processes. They used the approximate method of solving the problems of non-stationary convective diffusion to rotating disc electrodes, proposed by Filinovskii and Kiryanov [93]. Introducing dimensionless parameters $k = Dt/\delta^2$ and $y = x/\delta$ and assuming that $D_{Ox} = D_{Red} = D$, it is possible to write the convective diffusion equations for substances Ox and Red in the following form (as in subsection 5.1.4.2):

$$\frac{\partial C_{Ox}}{\partial k} = \frac{\partial^2 C_{Ox}}{\partial y^2} + \alpha y^2 \frac{\partial C_{Ox}}{\partial y}, \qquad (5.158)$$

$$\frac{\partial C_{Red}}{\partial k} = \frac{\partial^2 C_{Red}}{\partial y^2} + \alpha y^2 \frac{\partial C_{Red}}{\partial y}. \qquad (5.159)$$

For reversible electrode processes the boundary condition can be expressed by the equation:

$$\frac{C_{Ox}}{C_{Red}} = \Theta \exp(-\sigma k) \qquad (5.160)$$

where Θ is defined by equation (5.65), and C_{Ox} and C_{Red} are the concentrations of the substances on the surface of the electrode, and where:

$$\sigma = \frac{nFV\delta^2}{RTD} = 0.412 \frac{nF}{RT} \left(\frac{\nu}{D}\right)^{1/3} \frac{V}{N} \qquad (5.161)$$

where N is the speed of rotation of the electrode in revolutions per second.

If the condition is satisfied that the fluxes of substances Ox and Red on the electrode surface are equal:

$$D_{Red}\left(\frac{\partial C_{Red}}{\partial y}\right)_{y=0} = -D_{Ox}\left(\frac{\partial C_{Ox}}{\partial y}\right)_{y=0} = \frac{i(k)}{nFA} \qquad (5.162)$$

can be added to condition (5.160).

According to Filinovskii and Kiryanov [93] the concentrations of Ox and Red forms on the electrode surface are expressed by the formulae:

$$C_{Red}(0, k) = \frac{d}{nFdk} \int_0^k i(k-\lambda) R(\lambda) d\lambda, \qquad (5.163)$$

$$C_{Ox}(0, k) = C^0 - \frac{d}{nFdk} \int_0^k i(k-\lambda) R(\lambda) d\lambda, \qquad (5.164)$$

where C^0 is the concentration of Ox in the bulk of solution, and $R(\lambda)$ is the function characterizing the non-stationary diffusion process on the surface of the rotating disc.

By combining equations (5.163), (5.164) and (5.160) the expression describing the current is obtained:

$$i(k) = \frac{nFADC^0}{\delta} \frac{d}{dk} \int_0^k \frac{1 - \frac{C^0_{Red}}{C^0_{Ox}} \Theta \exp[-\sigma(k-\lambda)]}{1 + \Theta \exp[-\sigma(k-\lambda)]} \times$$

$$\left[\frac{\exp(-3.10\lambda)}{(\pi\lambda)^{1/2}} + 0.94 \operatorname{erf}(3.10\lambda)^{1/2} \right] d\lambda. \qquad (5.165)$$

When $k \ll 1$ equation (5.165) leads to the relationship:

$$\frac{i(k)}{i_g} = \frac{\sigma}{4}(1+1/4) \int_0^k \frac{d\lambda}{\sqrt{\pi(k-\lambda)} \cosh \frac{\sigma(\lambda - k_{1/2})}{2}} \qquad (5.166)$$

where i_g is the limiting current described by equation (5.152), and $k_{1/2} = \ln \Theta/\sigma$ is brought in to introduce the half-wave potential.

It follows from relationship (5.166) that when $k \ll 1$ the effect of convection on the current is slight, since this relationship is identical with the Randles and Ševčik equation, valid for unstirred solutions.

When $k > 1$ the current is given by the relationship:

$$\frac{i(k)}{i_g} = \frac{1 - \frac{C^0_{Red}}{C^0_{Ox}} \exp[-\sigma(k-k_{1/2})]}{1 + \exp[-\sigma(k-k_{1/2})]}. \qquad (5.167)$$

During the polarization of the electrode with rapidly changing potential ($\sigma \gg 1$) a maximum appears on the recorded current–potential curve provided that the speed of rotation of the electrode N is not large. The current of this maximum i_m is proportional to $\sqrt{\sigma}$, i.e. $\sqrt{V/N}$ and is expressed by the following relationship:

$$\frac{i_m}{i_g} = 1.86 \left(\frac{\nu}{D}\right)^{1/6} \left(\frac{V}{N}\right)^{1/2}. \qquad (5.168)$$

When the electrode polarization rate is not fast the "plateau" of the current is well formed and does not depend on the rate of application of the potential to the electrode.

The reasoning above is purely theoretical in character, but it permits the conditions which should be observed during experimental work to be determined. For example, the polarization rate at a given speed of electrode rotation which produces current–voltage curves with a well-formed "plateau" gives the limiting current.

In addition to the problem considered in this part of the chapter certain variants appear when condition (5.160) is changed. Reversible electrode processes have been discussed, but it is also possible to consider *quasi-reversible* and irreversible processes, leading to the formation of insoluble products. Such cases have not yet been treated theoretically, and it is unlikely that any future solutions of these problems will be of practical importance.

5.1.5.4 *Other Variants of Hydrodynamic Voltammetry*

Disc electrodes are generally used in electrochemical studies, but rotating microelectrodes, which are usually made of platinum wire, have often been used for analytical purposes. Although they are of importance in analytical studies it is difficult to describe exactly the current flowing through them, since the hydrodynamic conditions are complex.

In a series of theoretical investigations Matsuda [96–103] considered various systems of electrodes in hydrodynamic voltammetry.

The current flowing in a system in which the solution passes through a tube, the inner surface of which is the electrode, has also been considered from the theoretical point of view [104–106].

The results of these studies will not be discussed here. They contain solutions which are often complex and are difficult to express in the form of analytical relationships. They have been mentioned since they deal with systems of electrodes which either have found practical application or are likely to be used in practical work in the near future.

In recent years porous electrodes through which the test solutions are flowing have become of considerable practical importance. Such electrodes have been made of various conducting materials, e.g. finely divided silver [107–109], amalgamated particles of nickel [110], and platinum [111], tightly packed platinum gauze [112–113], powdered graphite [115], grains of glassy carbon [116–118], carbon fibres [119, 120] and amalgamated pieces of copper wire [121].

Such electrodes can be used for various purposes, both analytical and preparative.

Fujinaga [122] has discussed various problems connected with this method of electroanalysis. It is also the subject of several publications of general value [123–126].

5.1.6 GENERALIZATION OF THE RELATIONSHIPS

In this chapter, so far, equations relating the current to the depolarizer concentration have been discussed in the context of linear diffusion. Analysis of equations (5.23), (5.52), (5.95) and (5.144) indicates that the current is proportional to the depolarizer concentration, and that it depends on the factors determining the rates of transport to the electrode. These rates depend on the depolarizer diffusion coefficient and the kinetic parameter (see Chapter 3). Calling this parameter X allows the above four equations to be expressed by one general relationship:

$$i_g = KAC^0 X^{-1/2} \qquad (5.169)$$

where A is the electrode surface area, C^0 is the depolarizer concentration in the bulk of solution, and K is a constant, characteristic of a specific method. These constants are shown in Table 5.2.

Table 5.2 VALUES OF CONSTANT K IN EQUATION (5.169)

Method	Kinetic parameter X	Constant K
Chronoamperometry	t	$\dfrac{nFD^{1/2}}{\pi^{1/2}}$
Chronopotentiometry	τ	$\dfrac{\pi^{1/2}nFD^{1/2}}{2}$
Stationary electrode voltammetry	$\dfrac{1}{V}$	$2.69 \times 10^5 n^{3/2} D^{1/2}$
Rotating disc method	$\dfrac{1}{\omega}$	$0.61 nFD^{2/3}v^{-1/6}$

In chronopotentiometry the magnitude of the direct current appears in equation (5.169) instead of i_g.

It follows from equation (5.169) that the product $i_g X^{1/2}$ should be independent of the kinetic parameter (see Fig. 5.8), but it should depend on the electrode surface area and the depolarizer concentration. This dependence of the $i_g X^{1/2}$–X relationship is characteristic of processes controlled by the rate of linear diffusion of the depolarizer to the electrode.

Large currents can be recorded at a given concentration for a given electrode surface area, only when the kinetic parameter is small, then the analytical sensitivity of the method is higher. However, it has already been pointed out that a considerable decrease in the kinetic parameter has an unfavourable effect on the experimental results, since at low values of this parameter the capacitance currents become appreciable, as compared with the faradaic currents.

In the rotating disc method a large decrease in the kinetic parameter leads to a change from laminar to turbulent flow and, as a result, the reproducibility of results decreases.

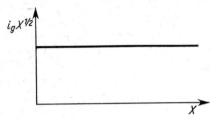

Fig. 5.8 Dependence of $i_g X^{1/2}$ on kinetic parameter X.

5.2 Electrode Processes under Spherical Diffusion Conditions

In the first part of this chapter electrode processes controlled only by the rate of diffusion have been discussed. Discussion was limited to cases in which the diffusion of the substance to the electrode is linear. In practice spherical electrodes are often used. Diffusion to such electrodes is spherically symmetrical, as in the case of dropping mercury electrodes, used in polarography.

Stationary electrode voltammetric and chronopotentiometric measurements are also often carried out by using hanging mercury drop electrodes. Therefore, the evaluation of the effect of spherical diffusion on recorded currents or on transition time is an important consideration.

Exact relationships describing this type of diffusion can be obtained by solving the equation of Fick's second law of diffusion for spherical diffusion conditions:

$$\frac{\partial C_{\text{Ox}}(r, t)}{\partial t} = D_{\text{Ox}} \left[\frac{\partial^2 C_{\text{Ox}}(r, t)}{\partial r^2} + \frac{2}{r} \frac{\partial C_{\text{Ox}}(r, t)}{\partial r} \right]. \tag{5.170}$$

As in the case of linear diffusion the solution of this equation requires the formulation of the initial conditions describing the depolarizer concentration before electrolysis, which determine the depolarizer concentration at the ends of the region ($r \to \infty$) and the mode of change of this concentration on the electrode surface during the electrolysis. These conditions are identical with those formulated for linear diffusion.

For all the methods discussed in this book the initial condition can be formulated as follows:

$$t = 0, \quad r \geqslant r_0, \quad C_{\text{Ox}} = C_{\text{Ox}}^0, \tag{5.171}$$

whereas the condition determining the concentration at the end of the region, also common for all the methods discussed, is as follows:

$$t \geq 0, \quad r \to \infty, \quad C_{Ox} \to C_{Ox}^0. \tag{5.172}$$

The solution of equation (5.171) also requires another boundary condition, but it is specific for each individual method.

5.2.1 Chronoamperometry

If it is assumed that a constant negative potential applied to a spherical electrode is such that, from the start of chronoamperometric electrolysis, the concentration of the oxidized form on the electrode surface is equal to zero, then the second boundary condition is:

$$t > 0, \quad r = r_0, \quad C_{Ox} = 0. \tag{5.173}$$

The solution of equation (5.170) can be simplified considerably by putting:

$$B = rC_{Ox}. \tag{5.174}$$

Then equation (5.170) becomes:

$$\frac{\partial B(r, t)}{\partial t} = D_{Ox} \frac{\partial^2 B(r, t)}{\partial r^2}, \tag{5.175}$$

i.e. a equation identical with that derived for linear diffusion is obtained.

Equation (5.175) can be solved [127–129] by a method similar to that used in the case of linear diffusion, utilizing the Laplace transformation.

The final result is obtained in the form of an equation:

$$C_{Ox}(r, t) = C_{Ox}^0 \left[1 - \frac{r_0}{r} \operatorname{erfc}\left(\frac{r - r_0}{2D_{Ox}^{1/2} t^{1/2}} \right) \right] \tag{5.176}$$

where function erfc (γ) is defined by relationship (5.35).

It follows from equation (5.176) that the concentration increases with increasing distance from the electrode, and at a sufficient distance it reaches the value of the initial concentration. The dependence of the concentration on the distance from the electrode for various electrolysis times is shown in Fig. 5.9. It is assumed that $r_0 = 0.1$ cm and $D_{Ox} = 10^{-5}$ cm^2/s.

In order to obtain the equation describing the current it is necessary to differentiate equation (5.176) with respect to r. The derivative of the concentration with respect to distance for $r = r_0$ is:

$$\left(\frac{\partial C_{Ox}(r, t)}{\partial r} \right)_{r=r_0} = C_{Ox}^0 \left[\frac{1}{\sqrt{\pi D_{Ox} t}} + \frac{1}{r_0} \right]. \tag{5.177}$$

Since the current is given by the general equation:

$$i = nFD_{Ox}A \left(\frac{\partial C_{Ox}}{\partial r} \right)_{r=r_0} \tag{5.178}$$

by combining equations (5.177) and (5.178):

$$i_g = nFD_{Ox}AC_{Ox}^0 \left[\frac{1}{\sqrt{\pi D_{Ox}t}} + \frac{1}{r_0} \right]. \tag{5.179}$$

This result differs significantly from expression (5.23), which was derived for linear diffusion. The difference is due to the appearance of the term $nFD_{Ox}AC_{Ox}^0/r_0$ in equation (5.179).

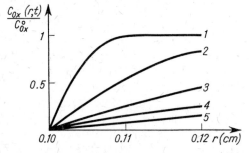

Fig. 5.9 Dependence of the ratio of concentration at the electrode to that in the bulk of solution on the distance from the electrode (r) for electrolysis at constant potential carried out on a spherical electrode having the radius 0.1 cm. Curves *1, 2, 3* and *4* illustrate concentration changes after 1, 10, 100 and 1000 seconds, respectively. Curve *5* represents the concentration system after infinitely long electrolysis time. It is assumed in the calculations that $D = 10^{-5}$ cm² s⁻¹ (from [129a] by permission of the copyright holders, Interscience Publishers).

In the case of spherical electrodes the recorded current is larger than that observed with flat electrodes having identical surface areas. The difference increases with increasing electrolysis time. In the case of linear diffusion the current tends to zero when the electrolysis time increases, but with spherical diffusion it tends to a constant value. This time dependence of the current is shown in Fig. 5.10.

Analysis of equation (5.179) makes it possible to determine, in the case of spherical electrodes, when the process can be described by the simplified equation derived for the linear diffusion conditions. This is possible under the following condition:

$$\frac{1}{\sqrt{\pi D_{Ox}t}} \gg \frac{1}{r_0}.$$

There are two possible ways of fulfilling this condition. The first way consists in using large spherical electrodes, since in this case $1/r_0$ is small. Alternatively electrolysis can be carried out for a short time, and in this case small electrodes can be used, since $1/\sqrt{\pi D_{Ox} t}$ is then relatively large.

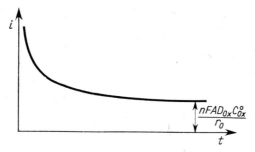

Fig. 5.10 Time (t) dependence of current (i) in electrolysis carried out at constant potential on a spherical electrode. The current tends to the constant value $nFAD_{Ox}C^0_{Ox}/r_0$.

In practice relatively small electrodes are used and the electrolysis is carried out for a comparatively long time.

Assuming that the radius of electrodes commonly used in experimental work $r_0 = 0.05$ cm and that the diffusion coefficient $D_{Ox} = 10^{-5}$ cm²/s, the spherical diffusion term can be neglected provided that the following equation is approximately valid:

$$\frac{1}{\sqrt{\pi D_{Ox} t}} = \frac{100}{r_0} = \frac{100}{0.05} = 2000 \text{ cm}^{-1}.$$

This equation gives the longest chronoamperometric electrolysis time for which the time dependence of the current can be described by the equation derived for conditions of linear diffusion. This time is 0.008 s. At longer electrolysis times, which are generally used in practice, it is necessary to take into account the effect of the spherical nature of the electrode on the recorded results.

It follows from the discussion above that, as a result of neglecting the spherical conditions of diffusion in derivation of the Ilkovič equation, certain discrepancies between the theoretical and the experimental results arise.

5.2.2 Polarography: Corrections to the Ilkovič Equation

The Ilkovič equation was derived for conditions differing from those actually existing in experimental work. Attempts were made by Lingane and Loveridge [130], Strehlow and von Stackelberg [131], and Kambara and Tachi [132, 133] in 1950 to introduce theoretically the effect of spherical diffusion to the equation describing the limiting current.

The treatment proposed by Lingane and Loveridge is perhaps the simplest. The Ilkovič equation can be obtained from that describing the rate of current change under potentiostatic conditions for linear diffusion, by expressing the surface area of the drop in terms of parameters m and t and introducing the coefficient $\sqrt{7/3}$. Therefore the Ilkovič equation, taking into account the spherical nature of diffusion, can be similarly derived from equation (5.179).

Taking into account the convection due to the drop growth by introducing the coefficient $\sqrt{7/3}$ into this equation:

$$i_g = nFD_{Ox}AC_{Ox}^0 \left(\sqrt{\frac{7}{3\pi D_{Ox}t}} + \frac{1}{r} \right) \tag{5.180}$$

where r is the radius of a drop at the given drop time t; it is related to m and t as follows:

$$r = \left(\frac{3mt}{4\pi d} \right)^{1/3}. \tag{5.181}$$

Expressing the surface area of the dropping electrode by means of equation (5.24) and introducing numerical values, the expression is obtained for instantaneous current at a given moment t in the life of the drop:

$$i_t = 709 n D_{Ox}^{1/2} C_{Ox}^0 m^{2/3} t^{1/6} + 31.56 n D_{Ox} C_{Ox}^0 m^{1/3} t^{1/3}. \tag{5.182}$$

Integration of this equation between 0 and t_1 leads to the expression for the mean current:

$$i_g = 607 n D_{Ox}^{1/2} C_{Ox}^0 m^{2/3} t_1^{1/6} \left(1 + \frac{39 D_{Ox}^{1/2} t_1^{1/6}}{m^{1/3}} \right). \tag{5.183}$$

It follows from equation (5.183) that when the spherical nature of the diffusion is allowed for the term $1 + 39 D_{Ox}^{1/2} t_1^{1/6}/m^{1/3}$ appears in the Ilkovič equation. The value of this term depends on the diffusion coefficient of the reacting substance and on the properties of the capillary. By substituting the values of D_{Ox}, m, and t_1 observed in practice it is found that the numerical value of this term is equal to about 1.1. This means that the mean current calculated, taking into account spherical diffusion, is about 10% larger than that calculated for linear diffusion.

Making allowances for the increased volume of solution from which the depolarizer diffuses to the electrode, compared with that of the flat electrode, Strehlow and von Stackelberg [131] obtained an equation similar to (5.183). In the case of a flat electrode the diffusion takes place from a column of liquid having a cross-sectional area equal to the electrode area, whereas in the case of a spherical electrode the column of liquid becomes wider on passing from the electrode to the bulk of solution (the electrode surface area is $4\pi r_0^2$ whereas the surface area at the distance Δr from the electrode is $4\pi(r_0+\Delta r)^2$, i.e. it increases with Δr).

On the basis of this reasoning Strehlow and von Stackelberg obtained equation (5.183), but their numerical coefficient in the corrective term, 17, was different. The value obtained for this coefficient by Kambara and Tachi was 39, and that obtained later by von Stackelberg [134] was 34.

The exact equation describing the diffusion current at a dropping electrode, taking into account the spherical nature of diffusion, was derived independently by Koutecký [135] and Matsuda [136]. The equation obtained by Koutecký for instantaneous current is:

$$i_t = 706 n D_{Ox}^{1/2} C_{Ox}^0 m^{2/3} t^{1/6} \left[1 + \frac{39 D_{Ox}^{1/2} t^{1/6}}{m^{1/3}} + 150 \left(\frac{D_{Ox}^{1/2} t^{1/6}}{m^{1/3}} \right)^2 \right], \quad (5.184)$$

that for the mean current is:

$$i_g = 605 n D_{Ox}^{1/2} C_{Ox}^0 m^{2/3} t^{1/6} \left[1 + \frac{34 D_{Ox}^{1/2} t^{1/6}}{m^{1/3}} + 100 \left(\frac{D_{Ox}^{1/2} t^{1/6}}{m^{1/3}} \right)^2 \right]. \quad (5.184a)$$

An equation similar to equation (5.184a) was obtained by Kambara and Tachi [137]. Matsuda [136] derived the equation of instantaneous current for commonly used polarographic capillaries. It is similar to equation (5.184) and it makes it possible to determine the spherical effect accurately:

$$i_t = 709 n D_{Ox}^{1/2} C_{Ox}^0 m^{2/3} t^{1/6} \left[1 + \frac{23.5 D_{Ox}^{1/2} t^{1/6}}{m^{1/3}} + 62.9 \left(\frac{D_{Ox}^{1/2} t^{1/6}}{m^{1/3}} \right)^2 + \ldots \right]. \quad (5.185)$$

Koutecký and von Stackelberg [138] have discussed the above corrections and their experimental verification in a detailed review.

Another method of solving this problem was proposed by Levich [139].

Newman [140], using a procedure similar to that proposed by Levich, found a new method of solving the problem and obtained a numerical constant equal to 34.626, which is very similar to that obtained by Koutecký and von Stackelberg [138], 34.7.

The allowance made for spherical diffusion is not the only possible correction to the original Ilkovič equation.

Various authors [141–149] have investigated the behaviour of the instantaneous current during the drop life. Some observed that, in the initial period of the growth of drops, the current observed is smaller than that calculated from the above equations. Airey and Smales [150] suggested that this is due to the decrease of the depolarizer content of the solution resulting from the electrolysis on the previous drop. It is fairly obvious that this effect should be most pronounced during the early stages of growth of the drop, since in later stages the larger drop reaches regions not affected by previous electrolysis.

Hans and Henne [151] observed differences between the time dependence of the instantaneous current measured on the first drop in the electrolysis and that measured on subsequent drops. Thus they confirmed the existence of the effect of decrease of depolarizer concentration, since the electrolysis on the first drop causes a decrease of the depolarizer concentration at the end of the capillary and subsequent drops grow in the solution containing the depolarizer at a lower concentration. This effect is much less pronounced when, according to Smoler's suggestion [152], the capillary is placed horizontally in the solution, since in this case during the detachment of the drop the composition of the solution at the end of the capillary is almost completely re-established (153).

Hans, Henne and Meurer [154] obtained an equation describing the instantaneous current, in which the above decrease in depolarizer concentration is taken into account:

$$i_t = 708nD_{Ox}^{1/2}C_{Ox}^0 m^{2/3}t^{1/6}\left[1 + \frac{\left(23.5 + \frac{t}{t_1} \times 15.5\right)D_{Ox}^{1/2}t^{1/6}}{m^{1/3}} + \frac{3000D_{Ox}t^{1/3}}{m^{2/3}}\right]. \quad (5.186)$$

The impoverishment effect was also considered theoretically by Markowitz and Elving [155, 156]. These authors assumed that it is the principal factor leading to the difference between the theoretical and experimental data. In their considerations they introduced a semi-empirical factor accounting for the impoverishment of depolarizer in the solution at the outlet from the capillary as compared with the initial concentration.

Several years later this effect was also considered by Duffey, Rahilly and Kidman [157]. On the basis of theoretical considerations and numerical data obtained by earlier investigators [144] they derived the following equation describing the maximum current:

$$i_g = 694.4 n D_{Ox}^{1/2} C_{Ox}^0 m^{2/3} t^{1/6} + 31.56 n D_{Ox} C_{Ox}^0 m^{1/3} t^{1/3}. \qquad (5.187)$$

This equation is simpler than that given by Markowitz and Elving [156].

Utilizing experimental results which were published earlier [143] the authors came to the conclusion that only about 6% of the impoverished solution remains after the detachment of the polarized drop. The effect of the solution impoverished in depolarizer at the outlet of the capillary was examined experimentally by O'Brien and Dieken [158, 159] who came to the conclusion that at the outlet of the capillary there is a cylindrical region extending for 0.35 mm below the capillary in which the impoverishment is significant. This region is transferred from drop to drop and only a small part of the impoverished solution neighbouring the upper part of the drop is taken down with the drop when it detaches itself. The new drop starts to grow in a layer of solution that has a depolarizer concentration only abont 60% of that in the bulk solution.

O'Brien and Dieken [159] have found a good agreement between the change of limiting currents with time on individual drops and the values predicted by the Duffey and co-workers equation and in particular by the Markowitz and Elving equation.

This effect can be decreased by using Smoler's capillaries with elongated ends, by controlled breaking off of the mercury drops, or by polarizing every other drop leaving the capillary. When traditional capillaries with flat ends are used the depolarizer transport to upper parts of the drops is hindered.

A device for automatic recording of current–time curves on the first drop in non-impoverished solution was described by Kirova-Eisner and co-workers [160].

The problem of screening of the mercury drops by the glass of the capillary was investigated by Mairanovskii and Neiman [161] and was taken into account by Matsuda [136] in the derivation of equation (5.185). If the screening effect were not observed the instantaneous current would be given by the equation derived by Matsuda:

$$i_t = 709 n D_{Ox}^{1/2} C_{Ox}^0 m^{2/3} t^{1/6} \left[1 + \frac{36.3 D_{Ox}^{1/2} t^{1/6}}{m^{1/3}} + 343 \left(\frac{D_{Ox}^{1/2} t^{1/6}}{m^{1/3}} \right)^2 + \ldots \right].$$

$$(5.188)$$

In the discussion of modifications of the Ilkovič equation it has been assumed so far that the drop increases with concentric symmetry as shown in Fig. 5.11a. In fact the drop grows from a definite point on its surface as shown in Fig. 5.11b.

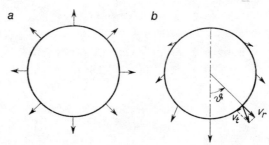

Fig. 5.11 Scheme of centric (*a*) and non-centric (*b*) growth of dropping electrode (from [66] by permission of the copyright holders, Elsevier Publishing Co.).

The growth of the drop is governed by radial expansion and by the downward displacement of the drop centre. The relative rate of movement of the solution to the electrode S_x is given by the equation:

$$S_x = \frac{m(1+\cos\vartheta)}{4\pi d}\left(\frac{1}{r^2} - \frac{1}{r_0^2}\right) \approx -\frac{2x}{3t}(1+\cos\vartheta) \qquad (5.189)$$

where r_0 is the radius of the drop, d is the density of the solution, and r is the radius of the sphere consisting of the mercury drop and a constant volume of the solution.

After making the allowance for relationship (5.189) de Levie [66] obtained the equation which correctly describes the diffusion and the convection at the growing mercury drop:

$$\frac{\partial C_{Ox}(x,t)}{\partial t} = D_{Ox}\frac{\partial^2 C_{Ox}(x,t)}{\partial x^2} - S_x\frac{\partial C_{Ox}(x,t)}{\partial x}$$

$$= D_{Ox}\frac{\partial^2 C_{Ox}(x,t)}{\partial x^2} + \frac{2x}{3t}(1+\cos\vartheta)\frac{\partial C_{Ox}(x,t)}{\partial x}. \qquad (5.190)$$

The solution of this equation, given in the form of concentration gradient on the electrode surface, is:

$$\left[\frac{\partial C_{Ox}(x,t)}{\partial x}\right]_{x=0} = \frac{C_{Ox}^0(7+4\cos\vartheta)^{1/2}}{(3\pi D_{Ox}t)^{1/2}}, \qquad (5.191)$$

whereas the current density depending on ϑ is:

$$i_0(\vartheta) = nFC_{Ox}^0\left[\frac{(7+4\cos\vartheta)D_{Ox}}{3\pi t}\right]^{1/2}. \qquad (5.192)$$

Evaluation of the expression describing the total current due to the process taking place on the whole surface of the electrode gives the numerical coefficient in the equation describing the current. This is almost identical with the Ilkovič coefficient $\sqrt{7/3} = 1.527$, whereas the value obtained by de Levie is 1.505. Thus the non-concentric increase of the drop affects the total current value to a small extent.

It follows from equation (5.192) that the electrode process is not equally intense over the whole surface of the drop. At the bottom of the drop, where the radial growth speed is twice as fast as that postulated by Ilkovič, $\vartheta = 0$ and we have:

$$(i_0)_{\vartheta=0} = nFC_{Ox}^0 \left(\frac{11 D_{Ox}}{3\pi t}\right)^{1/2} = 1.915 nFC_{Ox}^0 \left(\frac{D_{Ox}}{\pi t}\right)^{1/2}. \qquad (5.193)$$

At the top of the drop, near the end of the capillary, the surface of the drop is not moved, $\vartheta = \pi$ and the expression for the current density is:

$$(i_0)_{\vartheta=\pi} = nFC_{Ox}^0 \left(\frac{D_{Ox}}{\pi t}\right)^{1/2}. \qquad (5.194)$$

This expression is identical to the equation describing the current in chronoamperometry at flat stationary electrodes. Thus the current density at the bottom of the drop is almost twice as high as that at the top, and this uneven distribution of current density exists also in the case of electrodes having very thin walls.

When the resistance of the solution cannot be neglected, the uneven distribution of current density on the electrode surface always causes uneven distribution of potential on the solution side of the electrode–solution interphase. Therefore, in the region of faster transport of the substance, at the bottom of the drop, the effective electrode potential is lower than in the slower transport regions. When the current density depends on the electrode potential, as in the regions of polarographic wave formation, the effect of uneven transport rate can be enhanced, or partly compensated.

Surface tension depends on the potential and therefore uneven distribution of potential causes differences in magnitude of surface tension in different regions of the drop surface. The resulting additional movements of the mercury cause further movement of the solution. The surface tension is usually strongly potential dependent in the region of polarographic wave formation, when the reduced or the oxidized form is specifically adsorbed on the electrode surface, since in this region the concentrations of the two forms change rapidly with changing potential. Therefore in such cases a maximum of the first kind can be expected.

In the above discussion of the effect of spherical diffusion it has been assumed that the mercury drop is spherical.

However, it is possible it is pear-shaped, as a result of elongation downwards, due to gravity, but MacNevin and Balis [162], and Smith [163], have shown that mercury drops having a diameter not exceeding 0.1 mm are spherical.

In derivation of the Ilkovič equation it was assumed that the capillary yield m was constant. However, as a result of the action of capillary forces, the back pressure changes with increasing radius of the drop. As a result of this the value of m can change during the drop time. This effect can be decreased by increasing the distance between the end of the capillary and the level of mercury in the reservoir.

Duda and Vrentas [164, 165] obtained the solution for the case when the mercury outflow rate depends on time. It has been estimated [155] that at the beginning of drop formation the instantaneons efficiency of the capillary is about 20–40% lower than that at the end of the drop life. However, experiments carried out by O'Brien and Dieken [159] showed that the variable mercury outflow rate taken into account by Duda and Vrentas in the equations describing the current does not explain the difference between the results of experiments and the Ilkovič equation and its modifications.

The problem of the effect of variable mercury outflow rate on polarographic currents was considered earlier by Los and Murray [166], but the character of their work is less exact than that of the solution proposed by Duda and Vrentas.

The Ilkovič equation has been the subject of much experimental verification. It is difficult, however, to compare experimentally observed currents with those calculated from this equation, since the value of the depolarizer diffusion coefficient in the presence of the supporting electrolyte must be known. It has been found, in general, that the experimentally determined current is usually smaller than that calculated from the corrected Ilkovič equation.

Using horizontal capillaries Kuta and Smoler [167] showed that this difference is largely due to the consumption of the depolarizer during the polarization of the previous drops.

It has been mentioned that the time dependence of the instantaneous current during drop life has been studied by many authors. According to the original Ilkovič equation the dependence of log i on log t should be linear and the slope of the straight line should be $\frac{1}{6}$. It was found that this slope is not exactly constant but changes slightly during the experiment.

Instantaneous currents recorded at the first drop after the start of polarization and at subsequent drops were investigated in detail by Kuta and Smoler [152, 153, 168, 169]. They showed that the time dependence of the current obeys Koutecký's equation (5.189), when the current is recorded at the first drop after the start of polarization.

When the electrode is an amalgam flowing from a capillary and when this amalgam is oxidized, the equation describing the anodic current differs from the above equations, corrected for the effect of spherical diffusion. In this case the spherical effect decreases the rate of transport of the substance to the electrode and, as a result, a minus sign appears in the equation before the term corresponding to the effect of spherical diffusion.

Strehlow and von Stackelberg [131] obtained the following equation for the mean anodic limiting current, due to the oxidation of an amalgam:

$$\bar{i}_g = 605 n D_{\text{Red}}^{1/2} C_{\text{Red}}^0 m^{2/3} t_1^{1/6} \left(1 - \frac{A D_{\text{Red}}^{1/2} t_1^{1/6}}{m^{1/3}} \right). \quad (5.195)$$

Using the value of constant A given by Koutecký (equal to 34), von Stackelberg and Toome [170] obtained a good agreement between the experimental and theoretical current values.

It is recommended [171] that in such cases the mercury flow rates should be of the order of 0.5 mg/s.

The recorded limiting current is affected by the kind and concentration of the supporting electrolyte. This effect is usually observed as changes in t_1, D and m values. The change in t_1 can be large, but since it appears in the Ilkovič equation as $t^{1/6}$ the effect on the current is small; also, the drop time can be accurately measured and its effect on recorded currents can be calculated.

The changes of the diffusion coefficient can be evaluated on the basis of the Stokes and Einstein law written in the general form:

$$D = \frac{\text{const}}{\eta} \quad (5.196)$$

where η is the viscosity.

Combination of equation (5.196) with the original Ilkovič equation gives the general relationship:

$$i_g \eta^{1/2} = \text{const.} \quad (5.197)$$

The effect of viscosity on the limiting current has been investigated by many authors [172–175].

5.2.3 Stationary Electrode Voltammetry

A theoretical treatment of stationary electrode voltammetry under conditions of spherical diffusion, as in cases previously discussed, requires the formulation of the boundary condition, which on the basis of the Nernst equation is as follows:

$$t > 0, \quad r = r_0, \quad \frac{C_{Ox}(r_0, t)}{C_{Red}(r_0, t)} = \Theta \exp(-at), \quad (5.198)$$

where a and Θ are defined by equations (5.66) and (5.65).

The introduction of the concentration of the Red form into the boundary condition requires the solution of a system of equations consisting of equation (5.170) and the equation for the reduced form:

$$\frac{\partial C_{Red}(r, t)}{\partial t} = D_{Red} \left[\frac{\partial^2 C_{Red}(r, t)}{\partial r^2} + \frac{2}{r} \frac{\partial C_{Red}}{\partial r} \right]. \quad (5.199)$$

It is also necessary to formulate the second boundary condition for $r = r_0$;

$$t > 0, \quad r = r_0,$$

$$D_{Ox} \left[\frac{\partial C_{Ox}(r, t)}{\partial r} \right]_{r=r_0} + D_{Red} \left[\frac{\partial C_{Red}(r, t)}{\partial r} \right]_{r=r_0} = 0. \quad (5.200)$$

The initial condition for the Red form is:

$$t = 0, \quad r \geqslant r_0, \quad C_{Red} = 0. \quad (5.201)$$

whereas the boundary condition describing the concentration for $r \to \infty$ is:

$$t \geqslant 0, \quad r \to \infty, \quad C_{Red} \to 0. \quad (5.202)$$

It is assumed that both reduced and oxidized forms are soluble in the solution.

The solution of the problem, as presented, was first obtained by Frankenthal and Shain [176]. It was not obtained in the analytical form but in the form of tables which make it possible to plot current–potential curves for spherically symmetrical diffusion conditions. These curves will be discussed in Chapter 7.

The expression for the peak current, which is of importance to analytical chemists, is:

$$i_p = 8.81 \times 10^5 n^{3/2} A D_{Ox}^{1/2} V^{1/2} C_{Ox}^0 \varphi. \qquad (5.203)$$

All the symbols appearing in this equation have been defined with the exception of function φ which depends on n, r_0, V, and D_{Ox} and cannot be expressed in analytical form. The values of this function at the peak current are shown in Table 5.3, with respect to the contribution of spherical diffusion which is given by the expression $(1/r_0)(D_{Ox}/nV)^{1/2}$.

Table 5.3 DEPENDENCE OF φ ON $\dfrac{1}{r_0}\left(\dfrac{D_{Ox}}{nV}\right)^{1/2}$ AT THE PEAK OF CURRENT–POTENTIAL CURVE (FROM [176] BY PERMISSION OF THE COPYRIGHT HOLDERS, THE AMERICAN CHEMICAL SOCIETY).

$\dfrac{1}{r_0}\left(\dfrac{D_{Ox}}{nV}\right)^{1/2}$	φ	$\dfrac{1}{r_0}\left(\dfrac{D_{Ox}}{nV}\right)^{1/2}$	φ
0	0.311*	0.420	0.344
0	0.310**	0.632	0.360
0.003	0.312	1.000	0.388
0.032	0.314	2.000	0.460
0.316	0.336	3.000	0.527

* Extrapolated.
** According to Randles [51].

Equation (5.203), together with Table 5.3, make it possible to calculate the current observed in stationary spherical electrode voltammetry. The results of these calculations show that the spherical diffusion can, in some cases, cause an increase in the peak current, compared with that observed in the case of linear diffusion at the electrodes with identical surface area. When $(1/r_0)(D_{Ox}/nV)^{1/2} = 2.00$ the peak current is 1.5 times larger than that observed in the case of flat electrodes.

In practice spherical electrodes having a radius of about 0.04 mm are often used, together with convenient polarization rates (about 7×10^{-3} V/s) such that conventional polarograms are obtained. Assuming that $D_{Ox} = 10^{-5}$ cm^2/s for two-electron processes, then $(1/r_0)(D_{Ox}/nV)^{1/2} \cong 0.67$. The increase of the peak current due to spherical diffusion, as compared with the current recorded under linear diffusion conditions, is 18%. This proves that, in stationary electrode voltammetry, when small spherical electrodes are used and the polarization rates are slow, the current increase due to spherical diffusion should be taken into account.

The Randles andŠevčik equation for linear diffusion can be used in two-electron processes when the voltage is applied at a rate exceeding several volts per second and when the electrode radius is 0.04 cm and $D = 10^{-5}$ cm^2/s.

The non-analytical equation (5.203) is inconvenient in practical work: the value of $(1/r_0)(D_{Ox}/nV)^{1/2}$ must be calculated for each experiment, and the corresponding value of φ found from a φ vs. $(1/r_0)(D_{Ox}/nV)^{1/2}$ graph obtained on the basis of data given in Table 5.3. This value of φ must be introduced into equation (5.203). Therefore a more convenient form of this relationship was produced by Reinmuth [177] in which the observed current i is related to that which would be observed in the case of a flat electrode having the same surface area (i_{pl}):

$$i = i_{pl} + nFAC^0_{Ox}\frac{D_{Ox}}{r_0}\left[\frac{1-\exp(-at)}{1+\Theta\exp(-at)}\right]. \quad (5.204)$$

Thus the second term of the right-hand side of equation (5.204) determines the effect of spherical diffusion on the measured current. It can also be expressed in the form $nFAD_{Ox}(C^0_{Ox} - C_{Ox})/r_0$ where C_{Ox} is the instantaneous concentration of the oxidized form on the electrode surface.

The potential dependence of this term has the shape of a polarographic wave. Therefore, the part of the total current due to the spherical effect depends on the potential, according to an equation resembling that for polarographic waves:

$$E = E^{0'} + \frac{RT}{nF}\ln\frac{i_{s\infty}-i_s}{i_s} \quad (5.205)$$

where $E^{0'}$ is the formal potential, i_s is the contribution of the spherical effect to the observed current at a given potential, and $i_{s\infty}$ is the limiting value of this contribution.

While in the case of flat electrodes the observed current increases linearly with the square root of the rate of potential change, the term representing the contribution of the spherical effect does not show this dependence. Therefore, according to equation (5.204), at fast polarization rates this term becomes very small as compared with current i_{pl}.

Reinmuth found that equation (5.204) gives results which are, within experimental error, identical with those obtained on the basis of the work of Frankenthal and Shain and are in good agreement with experimental data.

The equation of peak current for spherical diffusion conditions can also be written in the following form:

$$i_p = 2.69 \times 10^5 n^{3/2} A D_{Ox}^{1/2} V^{1/2} C_{Ox}^0 + \frac{0.724 \times 10^5 n A C_{Ox}^0 D_{Ox}}{r_0}. \quad (5.206)$$

This form, given by Nicholson and Shain [49], is very suitable in practical work. It is very similar to relationship (5.179), which was derived for chronoamperometric processes taking place under conditions of spherical diffusion.

5.2.4 Chronopotentiometry

The problem of transition time in chronopotentiometry carried out under conditions of spherical diffusion was solved by Mamantov and Delahay [178]. In order to describe this time it was necessary to obtain an expression for the time dependence of the concentration of the oxidized form in the case of reduction on the electrode surface. This was obtained by solving equation (5.170) with the initial condition (5.171) and the first boundary condition (5.172).

The second boundary condition is similar to that used in the solution of the analogous problem in the case of linear diffusion:

$$t > 0, \quad r = r_0, \quad \left[\frac{\partial C_{Ox}(t, r)}{\partial r}\right]_{r=r_0} = \frac{i_0}{nFD_{Ox}}, \quad (5.207)$$

where $i_0 = i/A$.

The expression obtained by Mamantov and Delahay for the rate of change of concentration of the electrolysed substance on the electrode surface is as follows:

$$C_{Ox}(r, t) = C_{Ox}^0 - \frac{i_0 r_0}{nFD_{Ox}} \left\{1 - \exp\left(\frac{D_{Ox} t}{r_0^2}\right) \mathrm{erfc}\left[\frac{(D_{Ox} t)^{1/2}}{r_0}\right]\right\}. \quad (5.208)$$

From this equation the expression for the transition time can be readily obtained. Since $C_{Ox}(r_0, \tau) = 0$, we have:

$$\frac{nFD_{Ox} C_{Ox}^0}{i_0 r_0} = 1 - \exp\left(\frac{D_{Ox} \tau}{r_0^2}\right) \mathrm{erfc}\left[\frac{(D_{Ox} \tau)^{1/2}}{r_0}\right]. \quad (5.209)$$

Equation (5.208) was derived by a different method by Koutecký and Čižek [179].

Equation (5.209) is unsuitable for practical work and therefore it is desirable to express it in a more useful form. When $D_{Ox} \tau / r_0^2 \ll 1$, i.e. when the transition time is short or when the spherical electrode is large we have:

$$\exp\left(\frac{D_{Ox}\tau}{r_0^2}\right) \cong 1, \tag{5.210}$$

$$\mathrm{erf}\left[\frac{(D_{Ox}\tau)^{1/2}}{r_0}\right] \cong 1 - \frac{2}{\pi^{1/2}} \frac{(D_{Ox}\pi)^{1/2}}{r_0}. \tag{5.211}$$

By introducing expressions (5.210) and (5.211) into equation (5.209) we obtain the Sand equation for the transition time under linear diffusion conditions. This is the expected result, since when $D_{Ox}\tau/r_0^2$ is small the effect of spherical diffusion is slight. When $D_{Ox}\tau/r_0^2$ is large it is possible, by expanding in series the function from equation (5.209), to obtain [180] a simpler equation which is suitable for practical work:

$$i_0 \tau^{1/2} = \frac{\pi^{1/2} n F D_{Ox}^{1/2} C_{Ox}^0}{2} + \frac{\pi n F D_{Ox} C_{Ox}^0 \tau^{1/2}}{4 r_0}. \tag{5.212}$$

This equation is not accurate, but it simplifies the discussion of the relation between transition time and current density in chronopotentiometry at spherical electrodes.

The first term of the right-hand side of equation (5.212) gives the expression $i_0 \tau^{1/2}$ for linear diffusion, whereas the second represents the contribution of the spherical nature of the electrode. While the first term is independent of the current density, the second one increases with decreasing current density. At small current densities, when the spherical electrode is small, this term can have large values, and when the current density tends to zero it tends to infinity.

The results of chronopotentiometric measurements, obtained by means of a spherical electrode, when plotted as an $i_0 \tau^{1/2}$–i_0 graph do not give a straight line parallel to the current axis, as obtained in the case of linear diffusion. This difference is particularly pronounced at low current densities (Fig. 5.12, curve *1*). Straight line *2* in Fig. 5.12 represents the same relationship for an electrode process taking place on a flat electrode having the same surface area.

The results of these discussions and equation (5.212) indicate that the relationships for the spherical and the flat electrodes are similar for large current densities.

In the case of spherical electrodes the value of the product $i_0 \tau^{1/2}$ increases with decreasing current density, and as a result, particularly when the electrode is small, the transition time may not be reached. This can be readily explained by writing equation (5.212) in the form:

$$i_0 = \frac{\pi^{1/2} n F D_{Ox}^{1/2} C_{Ox}^0}{2\tau^{1/2}} + \frac{\pi n F D_{Ox} C_{Ox}^0}{4 r_0}. \tag{5.213}$$

It follows from equation (5.213) that the transition time is reached according to the Sand equation when $\pi^{1/2}nFD_{ox}^{1/2}C_{ox}^0/2\tau^{1/2} \gg \pi nFD_{ox}C_{ox}^0/4r_0$. When the values of the two terms are comparable the transition time is longer than that calculated for linear diffusion. The transition time is not reached when the current density i_0 is lower than $\pi nFD_{ox}C_{ox}^0/4r_0$.

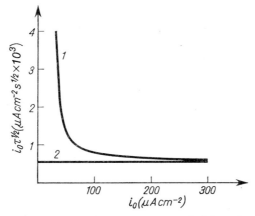

Fig. 5.12 Dependence of $i_0\tau^{1/2}$ (expressed in μA cm^{-2} s$^{1/2} \times 10^3$) on i_0 (expressed in μA cm^{-2}). Curve 1—spherical electrode having the radius $r_0 = 0.05$ cm; straight line 2—flat electrode. It is assumed in the calculations that $D = 10^{-5}$ cm^2 s^{-1}, $n = 2$, $C^0 = 10^{-3}$ M.

5.2.5 General Consideration of Relationships

In the discussions of limiting current, peak current and transition time in chronopotentiometry carried out in spherical diffusion conditions, equations (5.179), (5.206) and (5.212) describing these quantities were given. Utilizing the concept of the kinetic parameter they can be expressed in the general form:

$$i_0 X^{1/2} = G + \frac{BX^{1/2}}{r_0} \qquad (5.214)$$

where G and B are constants characteristic of the methods discussed and X is the kinetic parameter.

In order to facilitate the understanding of this problem and to facilitate the use of equation (5.214) the values of the above constants are given in Table 5.4.

In investigations involving spherical electrodes, when it is essential to evaluate the contribution of spherical diffusion or when it is desired to

Table 5.4 CONSTANTS G AND B IN EQUATION (5.214)

Method	Kinetic parameter	G	B
Chronoamperometry	t	$\dfrac{nFD^{1/2}C^0}{\pi^{1/2}}$	$nFDC^0$
Stationary electrode voltammetry	$\dfrac{1}{V}$	$2.79n^{3/2}FD^{1/2}C^0$	$0.75nFDC^0$
Chronopotentiometry	τ	$\dfrac{\pi^{1/2}nFD^{1/2}C^0}{2}$	$\dfrac{\pi nFDC^0}{4}$

determine the current which would be recorded under conditions of linear diffusion, on the basis of measured current values, the investigation of the dependence of $i_0 X^{1/2}$ on $X^{1/2}$ is probably the most suitable approach. In the case of linear diffusion a straight line parallel to the $X^{1/2}$ axis is obtained by this means (Fig. 5.13, curve 1), and in the case of a spherical electrode $i_0 X^{1/2}$ increases with increasing $X^{1/2}$ (Fig. 5.13, curve 2). The slope of line 2 depends on constant B and on the radius of the spherical electrode; line 2 approaches line 1 when the radius increases, provided that other parameters remain unchanged.

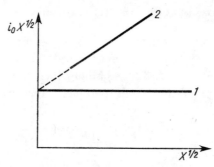

Fig. 5.13 Scheme of dependence of $i_0 X^{1/2}$ on $X^{1/2}$. Straight line 1—flat electrode; straight line 2—spherical electrode.

The dependence of $i_0 X$ on $X^{1/2}$ makes it possible to evaluate the effect of spherical diffusion in a more quantitative way. By extrapolating curve 2 to $X = 0$ we obtain the value of G, characteristic of the linear diffusion. By comparing this with an experimentally determined value of the product

$i_0 X^{1/2}$, we can ascertain if the investigated process can be analysed on the basis of the theory of processes carried out under linear diffusion conditions, and whether non-diffusional transport takes place during the experiment.

5.3 Electrode Processes under Cylindrical Diffusion Conditions

Cylindrical electrodes, which are often used in electroanalytical practice, are made of solid conductors such as platinum, and for this reason they are mainly used in investigations of electrode processes taking place in potential regions which are positive with respect to the hydrogen electrode. They are prepared by simply fixing a piece of platinum wire in soft glass or Teflon.

The principal equations describing the currents and chronopotentiometric transition time in the case of cylindrically symmetrical diffusion are now discussed. In order to obtain these equations it is necessary to solve the equation:

$$\frac{\partial C_{Ox}(r, t)}{\partial t} = D_{Ox}\left[\frac{\partial^2 C_{Ox}(r, t)}{\partial r^2} + \frac{1}{r}\frac{\partial C_{Ox}(r, t)}{\partial r}\right]. \quad (5.215)$$

The initial condition, which is common for all the methods discussed, is as follows:

$$t = 0, \quad r \geq r_0, \quad C_{Ox} = C_{Ox}^0. \quad (5.216)$$

The first boundary condition is also common:

$$t \geq 0, \quad r \to \infty, \quad C_{Ox} \to C_{Ox}^0, \quad (5.217)$$

where r_0 is the radius of the cylindrical electrode.

Cylindrical electrodes are used in chronoamperometry, stationary electrode voltammetry and chronopotentiometry. Therefore cylindrical diffusion for each method will be discussed separately.

5.3.1 Chronoamperometry

In the formulation of the second boundary condition for chronoamperometry it will be assumed that the potential applied to the electrode is so negative that the Ox form cannot exist on the electrode surface but is immediately reduced to the Red form. This can be expressed in an exact form as follows:

$$t > 0, \quad r = r_0, \quad C_{Ox} = 0. \quad (5.218)$$

The identical problem was solved earlier for heat transfer, as described by Carslaw and Jaeger [181]. Rius, Polo and Llopis [182] showed that the solution obtained can be adapted to the present diffusion problem.

The equation describing the dependence of the concentration of the Ox form on the distance from the electrode and on the time, is complex. In this case, however, it can be solved when the time dependence of the concentration at the electrode surface is known, since this makes it possible to formulate the following equation, describing changes of current in chronoamperometric processes taking place on cylindrical electrodes:

$$i_g = nFAD_{Ox}C_{Ox}^0 \frac{4}{\pi^2 r_0} \int_0^\infty \frac{\exp\left(-\frac{D_{Ox}t(u)}{r_0^2}\right)}{I_0^2(u) + Y_0^2(u)} \frac{du}{u} \qquad (5.219)$$

where u is an auxiliary variable, and I_0 and Y_0 are Bessel functions of the first and second kind, respectively. The integral of equation (5.219) was calculated by Jaeger and Clarke [183] for $D_{Ox}t/r_0^2$ values ranging from 0.01 to 10.

By developing equation (5.219) in a Bessel function series it can be written in the form of the following equations:

$$i_g = nFAD_{Ox}C_{Ox}^0 \frac{1}{r_0}\left\{\frac{1}{\pi^{1/2}Z^{1/2}} + \frac{1}{2} - \frac{1}{4}\left(\frac{Z}{\pi}\right)^{1/2} + \frac{1}{8}Z\ldots\right\} \qquad (5.220)$$

and

$$i_g = nFAD_{Ox}C_{Ox}^0 \frac{2}{r_0}\left\{\frac{1}{[\ln 4Z - 2\gamma]} - \frac{\gamma}{[\ln 4Z - 2\gamma]^2}\ldots\right\} \qquad (5.221)$$

where Z is a dimensionless parameter $D_{Ox}t/r_0^2$ and γ is the Euler constant, equal to 0.5772.

Equation (5.220) is used for small, and equation (5.221) for large values of parameter Z.

When Z is sufficiently small all the terms of equation (5.220), with the exception of the first, can be neglected. As a result we obtain the relationship derived at the beginning of this chapter for chronoamperometry carried out under linear diffusion conditions. This simplification is justifiable when $1/(\pi Z)^{1/2} \gg 1/2$, and it can be assumed that this inequality is satisfied when $Z \leqslant 0.01$. When the diameter of the electrode used in the measurements is 0.2 cm and $D_{Ox} = 10^{-5}$ cm^2/s, the equation derived for linear diffusion conditions can be used when the electrolysis time does not exceed 10 seconds.

It follows from equation (5.221) that, at longer electrolysis times, the current slowly decreases during the electrolysis. Analysis of equation (5.221) shows that for sufficiently large values of Z this relationship can be expressed in a simpler form:

$$i_g = nFAD_{Ox}C_{Ox}^0 \frac{2}{r_0} \frac{1}{\ln 4Z}. \qquad (5.222)$$

It can also be assumed, rather arbitrarily, that equation (5.222) gives approximately correct results when it is used instead of equation (5.221), provided that $Z > 10$. It follows from equation (5.222) that the current tends to zero, when the electrolysis time is very long (this is not observed in the case of spherical diffusion). Comparing the time dependence with that observed in chronoamperometry under linear diffusion conditions, in which the current tends to zero with increasing time, we can see that in the case of cylindrical electrodes the current tends to zero much more slowly. This is the result of the logarithmic time dependence of the current in the case of cylindrical diffusion. The determination of the time dependence for long electrolysis times is difficult, since it is difficult to avoid the effect of transport by convection.

5.3.2 STATIONARY ELECTRODE VOLTAMMETRY

In the case of stationary electrode voltammetry carried out under conditions by cylindrical diffusion, the second boundary condition is identical with that formulated for processes carried out on spherical electrodes. It is expressed by relationship (5.198) and in this case r_0 is the radius of the cylindrical electrode. Since this condition also introduces the concentration of the reduced form, it is necessary to solve not only equation (5.215), but also the analogous equation for the reduced form. In order to solve this system of two equations it is necessary to introduce the conditions defined by equations (5.200), (5.201) and (5.202). It is assumed that both Ox and Red forms are soluble in the solution.

This problem was solved by Nicholson [184], but, as in the case of spherical diffusion the solution is not obtained in an analytical form.

The peak current can be expressed as follows:

$$i_p = 8.88 \times 10^5 n^{3/2} AD_{Ox}^{1/2} V^{1/2} C_{Ox}^0 \Psi. \qquad (5.223)$$

This equation is valid for processes carried out at 25°C. It contains function Ψ which depends on parameter $(1/r_0)(D_{Ox}/nV)^{1/2}$. The expression for the peak current is shown in Fig. 5.14, which indicates that when the parameter $(1/r_0)(D_{Ox}/nV)^{1/2}$ tends to zero, function Ψ tends to 0.306.

By comparing the dependence of functions Ψ and φ [(see equation

(5.203)] on parameter $(1/r_0)(D_{ox}/nV)^{1/2}$ it is observed that function Ψ increases at a slower rate and therefore, the effect of cylindrical diffusion on the deviations of the results from those predicted on the basis of the

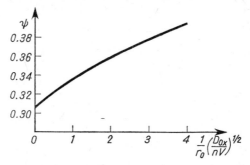

Fig. 5.14 Dependence of function Ψ on parameter $1/r_0(D_{ox}/nV)^{1/2}$.

theory of processes taking place under conditions of linear diffusion, is smaller than that of spherical diffusion. For example, for $(1/r_0)(D_{ox}/nV)^{1/2} = 0.648$, the value of function Ψ is 0.324, whereas that of function φ is 0.361. Since in the case of linear diffusion the value of the function at the maximum is 0.310, in the case of spherical diffusion the difference between the value of the current and that which would be observed under identical conditions if a flat electrode were used is 16.5%. In the case of cylindrical diffusion at the same value of parameter $(1/r_0)(D_{ox}/nV)^{1/2} = 0.648$ this difference is only 4.5%.

It has been shown that, when a polarograph is used as the source of linearly increasing voltage and when the electrode radius is 0.04 cm, the value of parameter $(1/r_0)(D_{ox}/nV)^{1/2}$ is about 0.65. Thus, under conditions frequently used in electroanalysis, the effect of the use of a cylindrical electrode on the deviation of results from those obtained with flat electrodes, is slight. It can be neglected when $(1/r_0)(D_{ox}/nV)^{1/2} < 0.2$, since the current increase due to cylindrical diffusion is smaller than 2%. Assuming that $D_{ox} = 10^{-5}$ cm²/s, $n = 2$, and $r_0 = 0.04$ cm we obtain $V = 0.28$ V/s. Therefore, in such cases the equation derived for linear diffusion can be used for interpretation of experimental results, provided that the polarization rate does not exceed the calculated value.

5.3.3 Chronopotentiometry

The boundary condition for the solution of equation (5.125) is identical with condition (5.207), but in this case r_0 is the radius of the cylindrical electrode. The problem of chronopotentiometric processes under conditions of spherical diffusion was solved by Peters and Lingane [185] who deduced an expression for the transition time. This expression has the form of a Bessel function series, similar to that describing stationary electrode voltammetry under cylindrically symmetrical diffusion conditions:

$$\frac{i_0 \tau^{1/2}}{C_{Ox}^0} = \frac{\pi^{1/2} n F D_{Ox}^{1/2}}{2} \times$$

$$\left[\frac{1}{1 - \frac{\pi^{1/2}}{4} Z^{1/2} + \frac{1}{4} Z - \frac{3\pi^{1/2}}{32} Z^{3/2} + \frac{21}{160} Z^2 - \frac{9\pi^{1/2}}{128} Z^{3/2} \ldots} \right] \quad (5.224)$$

where $Z = D_{Ox} \tau / r_0^2$.

Equation (5.224) differs from Sand's equation by the term in square brackets, which introduces the cylindrical effect into the equation describing the transition time. When $Z \to 0$, relationship (5.224) is reduced to the equation derived for conditions of linear diffusion. However, it has been found that in this case the description of the process is not correct when the dimensionless parameter Z is large [186]. For this reason Evans and Price [187] expanded the equation by addition of extra terms:

$$\frac{i_0 \tau^{1/2}}{C_{Ox}^0} = \frac{\pi^{1/2} n F D_{Ox}^{1/2}}{2} L \quad (5.225)$$

where L is a function of parameter Z.

Function L can be defined as a corrective coefficient for cylindrical electrodes. The value of $i_0 \tau^{1/2}/C_{Ox}^0$ for a flat electrode, multiplied by L, gives the value of $i_0 \tau^{1/2}/C_{Ox}^0$ for cylindrical electrodes.

The values of parameter L, shown in Table 5.5, correspond to transition time values from 0.09 to 25 seconds and $Z^{1/2}/\tau^{1/2}$ values varying from 0.004 to 0.4, which corresponds to changes in diffusion coefficient from 10^{-6} to 10^{-4} cm²/s and changes if electrode radius changes from 0.025 to 0.25 cm. As might be expected, L is very close to 1.000 when either τ or D_{Ox} is very small. In such conditions the diffusion layer is very thin and the diffusion process is almost linear. L is very close to unity in the case when r_0 is very large, since then the surface is almost flat.

Assuming $D_{Ox} = 10^{-5}$ cm²/s and $r_0 = 0.04$ cm, $Z^{1/2}/\tau^{1/2} = 0.08$. It follows from Table 5.5 that in this case the Sand equation can be applied, provided that the transition time is shorter than 0.1 s.

Equations (5.224) and (5.225) cannot be applied when the values of parameters $Z^{1/2}/\tau^{1/2}$ and τ are large. They should be used with care when $Z^{1/2}/\tau^{1/2}$ is greater than about 0.2.

Table 5.5 THE VALUES OF TERM L FOR VARIOUS VALUES OF D_{Ox}, r_0 AND τ (FROM [187] BY PERMISSION OF THE COPYRIGHT HOLDERS, ELSEVIER PUBLISHING CO.)

$z/\tau^{1/2}/t^{1/2}Z$	0.09	0.25	0.49	0.81	1.21	1.96	2.56	4.00	6.25	9.00	12.25	16.00	20.25	25.00	$t^{1/2}/\tau^{1/2}Z$
0.004	1.001	1.001	1.001	1.002	1.002	1.002	1.003	1.004	1.004	1.005	1.006	1.007	1.008	1.009	0.004
0.006	1.001	1.001	1.002	1.002	1.003	1.004	1.004	1.005	1.007	1.008	1.009	1.011	1.012	1.013	0.006
0.008	1.001	1.002	1.002	1.003	1.004	1.005	1.006	1.007	1.009	1.011	1.012	1.014	1.016	1.018	0.008
0.01	1.001	1.002	1.003	1.004	1.005	1.006	1.007	1.009	1.011	1.013	1.015	1.018	1.020	1.022	0.01
0.02	1.003	1.004	1.006	1.008	1.010	1.012	1.014	1.018	1.022	1.026	1.031	1.035	1.039	1.044	0.02
0.03	1.004	1.007	1.009	1.012	1.015	1.019	1.021	1.026	1.033	1.039	1.046	1.052	1.059	1.065	0.03
0.04	1.005	1.009	1.012	1.016	1.019	1.025	1.028	1.035	1.044	1.052	1.061	1.070	1.078	1.087	0.04
0.05	1.007	1.011	1.015	1.020	1.024	1.031	1.035	1.044	1.055	1.065	1.076	1.087	1.097	1.108	0.05
0.06	1.008	1.013	1.019	1.024	1.029	1.037	1.042	1.052	1.065	1.078	1.091	1.104	1.116	1.129	0.06
0.07	1.009	1.015	1.022	1.028	1.034	1.043	1.049	1.061	1.076	1.091	1.106	1.120	1.135	1.150	0.07
0.08	1.011	1.018	1.025	1.032	1.039	1.049	1.056	1.070	1.087	1.104	1.120	1.137	1.154	1.171	0.08
0.09	1.012	1.020	1.028	1.036	1.043	1.055	1.063	1.078	1.097	1.116	1.135	1.154	1.173	1.192	0.09
0.10	1.013	1.022	1.031	1.039	1.048	1.061	1.070	1.087	1.108	1.129	1.150	1.171	1.192	1.213	0.10
0.11	1.015	1.024	1.034	1.043	1.053	1.067	1.076	1.095	1.118	1.141	1.164	1.187	1.211	1.235	0.11
0.12	1.016	1.026	1.037	1.047	1.058	1.073	1.083	1.104	1.129	1.154	1.179	1.204	1.230	1.257	0.12
0.13	1.017	1.029	1.040	1.051	1.062	1.079	1.090	1.112	1.139	1.166	1.194	1.221	1.250	1.281	0.13
0.14	1.019	1.031	1.043	1.055	1.067	1.085	1.097	1.120	1.150	1.179	1.209	1.239	1.271	1.306	0.14
0.15	1.020	1.033	1.046	1.059	1.072	1.091	1.104	1.129	1.160	1.192	1.224	1.257	1.293	1.333	0.15
0.16	1.021	1.035	1.049	1.063	1.076	1.097	1.110	1.137	1.171	1.204	1.239	1.276	1.316	1.364	0.16
0.17	1.022	1.037	1.052	1.067	1.081	1.103	1.117	1.146	1.181	1.217	1.255	1.295	1.342	1.398	0.17
0.18	1.024	1.039	1.055	1.070	1.086	1.109	1.124	1.154	1.192	1.230	1.271	1.316	1.370	1.439	0.18
0.20	1.026	1.044	1.061	1.078	1.095	1.120	1.137	1.171	1.213	1.257	1.306	1.364	1.439	—	0.20
0.25	1.033	1.055	1.076	1.097	1.118	1.150	1.171	1.213	1.269	1.333	—	—	—	—	0.25
0.30	1.039	1.065	1.091	1.116	1.141	1.179	1.204	1.257	1.333	—	—	—	—	—	0.30
0.40	1.052	1.087	1.120	1.154	1.187	1.239	1.276	—	—	—	—	—	—	—	0.40

Cylindrical Diffusion Conditions

Dornfeld and Evans [188] derived an equation, using the method of Carslaw and Jaeger [189], which correctly describes the process when the value of Z is large:

$$\frac{i_0 \tau^{1/2}}{C_{Ox}^0} = \frac{\pi^{1/2} n F D_{Ox}^{1/2}}{2} R \qquad (5.226)$$

where R is the function:

$$R = \left[\pi^{1/2} \left\{ \frac{\ln \Delta}{4Z^{1/2}} + \frac{(1+\ln \Delta)}{8Z^{3/2}} + \frac{\frac{(3+\pi^2)}{2} - \ln \Delta - 3(\ln \Delta)^2}{64 Z^{5/2}} \right\} \right] \qquad (5.227)$$

$$\Delta = \frac{4Z}{e^\gamma} \qquad (5.228)$$

and γ is the Euler constant.

Equation (5.227) gives correct values of R when $Z^{1/2}$ is smaller than 1.6. Since relationship (5.225) is valid for $Z^{1/2}$ values smaller than 0.7, in a certain range of Z values the transition time is not precisely described.

Dornfeld and Evans calculated function R for parameter $Z^{1/2}$ values from 0.20 to 2.20, using a numerical method. The results of these calculations are shown in Table 5.6.

Table 5.6 DEPENDENCE OF THE CORRECTIVE FACTOR R ON PARAMETER Z (FROM [188] BY PERMISSION OF THE COPYRIHGT HOLDERS, ELSEVIER PUBLISHING Co.)

$Z^{1/2}$	R	$Z^{1/2}$	R	$Z^{1/2}$	R
0.20	1.086	0.90	1.368	1.60	1.628
0.25	1.108	0.95	1.388	1.65	1.646
0.30	1.128	1.00	1.407	1.70	1.664
0.35	1.149	1.05	1.426	1.75	1.682
0.40	1.170	1.10	1.444	1.80	1.700
0.45	1.190	1.15	1.463	1.85	1.718
0.50	1.210	1.20	1.482	1.90	1.735
0.55	1.230	1.25	1.501	1.95	1.753
0.60	1.250	1.30	1.519	2.00	1.770
0.65	1.270	1.35	1.538	2.05	1.788
0.70	1.291	1.40	1.556	2.10	1.805
0.75	1.310	1.45	1.574	2.15	1.822
0.80	1.330	1.50	1.592	2.20	1.840
0.85	1.349	1.55	1.610		

The results of theoretical predictions are in satisfactory agreement with the experimental results for reduction of hydrogen ions.

Chronopotentiometry under conditions of cylindrical diffusion has also been studied by Hurwitz [190].

The equations of currents or transition times under cylindrically symmetrical diffusion conditions cannot be presented in a simple analytical form. The relationships discussed above are only approximate and they can only be used under certain conditions. Generally, the effect of cylindrical diffusion is smaller than the spherical effect. Therefore, when spherical and cylindrical electrodes having the same radii and the same surface areas are used, the measured currents are larger and the transition times are longer in the case of spherical electrodes.

5.4 Electrode Processes under Limited Diffusion Field Conditions

It has been assumed so far that the substance Ox, which is reduced, diffuses to the electrode from the field extending to an infinite distance from the electrode surface; similarly the substance Red, formed in the electrode process could, at least theoretically, diffuse away from the electrode to the field extending also to an infinite distance. The situation is substantially different when the electrolysed substance is contained in a very thin solution layer at the electrode surface. In such a case, and under appropriately chosen experimental conditions Ox can be completely transformed into Red. From the mathematical point of view, the problem of electrolysis is similar to that in which the metal, contained in a small drop of mercury, is oxidized electrochemically. In analytical determinations oxidation of metals introduced into mercury drops is frequently employed and therefore a theoretical description of the recorded curves is important.

The restriction of the diffusion field makes the solution of diffusion equations more complex. The technique of thin layer electrolysis is rather new, but numerous papers dealing with this technique and its applications have been published. They have been reviewed by Hubbard and Anson [191, 192].

5.4.1 Chronoamperometry

Chronoamperometric electrode processes in limited diffusion field conditions have been considered first from the point of view of the description of current–time curves recorded during the oxidation of metals dissolved in a small drop of mercury.

Thus equation (5.199) has been solved at the initial condition:

$$t = 0, \quad r \geqslant r_0, \quad C_{\text{Red}} = C_{\text{Red}}^0, \tag{5.229}$$

and the boundary conditions:

$$t > 0, \quad r \to 0, \quad C_{\text{Red}} \text{ remains limited}, \tag{5.230}$$

$$t > 0, \quad r = r_0, \quad C_{\text{Red}} = 0. \tag{5.231}$$

A similar problem has been solved for heat transfer (see for instance [193]). Adapting the solutions to the given diffusion conditions an expression is obtained describing the distribution of concentration inside the mercury drop as a function of r and t.

$$C_{\text{Red}}(r, t) = C_{\text{Red}}^0 \sum_{k=1}^{\infty} 2\cos\nu_k \exp\left(-\frac{\nu_k^2 D_{\text{Red}} t}{r_0^2}\right) \frac{\sin \nu_k \dfrac{r}{r_0}}{\nu_k \dfrac{r}{r_0}} \tag{5.232}$$

where $\nu_1 = \pi$, $\nu_2 = 2\pi$, $\nu_3 = 3\pi$ …

Differentiating equation (5.232) with respect to r, calculating the derivative for $r = r_0$, and combining the relationship obtained with equation (5.178) (written for the Red form):

$$i_g = -\frac{2nFAD_{\text{Red}} C_{\text{Red}}^0}{r_0} \sum_{k=1}^{\infty} \exp\left(-\frac{\nu_k^2 D_{\text{Red}} t}{r_0^2}\right). \tag{5.233}$$

This form of the equation was put forward by Chovnik and Vashchenko [194]. After certain simplifications Stromberg and Zakharova [195] used it for the determination of the diffusion coefficients of metals in mercury.

Stevens and Shain [196] presented a different solution:

$$i_g = nFAD_{\text{Red}} C_{\text{Red}}^0 \left[-\frac{1}{(\pi D_{\text{Red}} t)^{1/2}} + \frac{1}{r_0} - \frac{2}{(\pi D_{\text{Red}} t)^{1/2}} \times \right.$$

$$\left. \sum_{n=1}^{\infty} \exp\left(-\frac{n^2 r_0^2}{D_{\text{Red}} t}\right) \right]. \tag{5.234}$$

When electrodes having $r_0 = 0.05$–0.1 cm are used and the duration of electrolysis does not exceed 30 seconds, the third term in the brackets is small and therefore, in this short time interval the i_g–$t^{1/2}$ relationship should be linear. The diffusion coefficient of a metal in mercury can be calculated from the slope of the straight line and from the intercept on the current axis.

The equations discussed above make it possible to determine diffusion coefficients of metals in mercury in a comparatively simple way. These equations are particularly useful for the determination of diffusion coefficients of metals which form unstable amalgams [197].

An expression describing the current of irreversible thin layer oxidation of amalgams has also been derived [198]. When $v \leqslant 0.16$ it has the following form:

$$i = -nFAC^0_{Red}k_s \exp\left[\frac{\beta nF}{RT}(E-E^0)\right] \exp(\Lambda^2 v) \operatorname{erfc}(\Lambda\sqrt{v}) \quad (5.235)$$

where $v = \dfrac{D_{Red}t}{l^2}$, $\Lambda = \dfrac{k_{bh}l}{D_{Red}}$, and l is the thickness of the electrolysed layer.

When $v = 0$ equation (5.232) is reduced to the relationship for electrolysis in semi-infinite zones; in this case $\exp(\Lambda^2 v)\operatorname{erfc}(\Lambda\sqrt{v})$ equals one.

The same authors [198] gave an expression describing irreversible oxidation of amalgams from small spherical electrodes.

5.4.2 Stationary Electrode Voltammetry

In the case of stationary electrode voltammetry the problem of electrolysis from a limited diffusion zone has been elucidated for a diffusion zone having the shape of a layer parallel to the electrode surface. In such conditions diffusion to the electrode is linear.

Hubbard and Anson [199] considered the cases in which polarization rates were slow. Adapting heat transfer equations [200] they derived an equation for the peak current:

$$i_p = \frac{n^2F^2AlVC^0_{Ox}}{4RT} \quad (5.236)$$

where A is the electrode surface area which is equal to that of the solution layer adjoining to the electrode, and l is the thickness of this layer.

In such conditions the complete i–E curve can be represented by the following equation:

$$i = \frac{n^2F^2AlVC^0_{Ox}}{RT} \frac{\Theta}{(1+\Theta)^2} \quad (5.236a)$$

where $\Theta = \exp\left[\dfrac{nF}{RT}(E-E^0)\right]$.

It follows from this equation that, under such conditions, the peak current (contrary to the relationship given by Randles and Ševčik) is proportional to the first power of the polarization rate. Equation (5.236) describes the process correctly when the reduction of substance Ox in the electrolysed solution is practically complete.

De Vries and Van Dalen [201] have further developed the theory of the stationary electrode voltammetric oxidation of metals from flat electrodes covered with a very thin mercury layer. The oxidized metal is dissolved in mercury and forms an amalgam. The preparation of such electrodes has been described by Moros [202] and by Ramaley, Brubaker and Enke [203].

De Vries and Van Dalen also solved the system of Fick equations written for the Ox and Red forms, with boundary conditions involving the effect of the limited diffusion zone. They solved the resulting non-linear integral Volterra equation by the numerical method.

The final results were presented by the authors in the form of current–potential curves calculated for definite conditions. It follows from these results that the differences between the curves recorded under conditions of very limited diffusion zone and those recorded under conditions of semi-infinite diffusion space, are appreciable. In the former case, when both the thickness of the mercury layer and the rate of potential change are small, the dependence of peak current value on polarization rate is in accord with that put forward by Hubbard and Anson. Since, in this case, the diffusion zone is very small, the oxidation of metal is rapid and therefore the current decreases rapidly to zero after the peak value is reached.

The above effect is rather important when several metals are dissolved in the mercury layer. In such cases the quantitative determination of the metals is fairly accurate, since the resolving power is high due to the small width of the current peaks. Figure 5.15 shows the course of the oxidation current for several metals dissolved in a very thin mercury layer and in a hanging mercury drop.

The initial electrolytic accumulation of metals is more effective in thin (less than 0.01 cm) mercury layers than in hanging mercury drops. When thin layer and hanging drop mercury electrodes of identical surfaces are used under identical conditions, more concentrated amalgams can be obtained in the former case. This is due to the smaller volume of mercury into which the reduced metal is introduced. This is very important from the analytical point of view since by means of such thin layer electrodes it is possible to analyse less concentrated solutions than would be possible if hanging drop mercury electrodes were used.

A more exact description of stationary electrode voltammetric oxidation of metals from thin mercury layers has been given by de Vries [204]. The earlier treatment had been found to give unsatisfactory results when thicker mercury layers (about 100 μm) and higher polarization rates are employed.

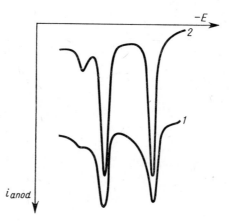

Fig. 5.15 Oxidation currents of metals dissolved in hanging mercury drop (curve *1*) and in a very thin layer of mercury (curve *2*).

On account of the fact that no analytical solutions of the problem had been produced, a simplified model was put forward by Roe and Toni [205].

Nigmatullin [206] solved the problem of irreversible electrode processes at spherical electrodes when the reduced form is dissolved in the drop, or the thin layer; for instance in the mercury layer covering a spherical solid electrode. Similar problems were discussed by Bucur et al. [207].

De Vries and Van Dalen [208] described exactly the effect of uncompensated ohmic drop on curves of anodic oxidation of metals at a thin flat mercury layer.

It has already been mentioned that the stationary electrode voltammetric technique with initial electrolytical accumulation into a hanging mercury drop electrode has found a wide analytical application.

In the initial stage of an analytical determination the voltage applied to the electrodes of the electrolytic cell is such that the potential of the hanging mercury drop electrode is sufficiently negative for the reduction of ions of metals present in the solution in trace amounts. The reduced metal is dissolved into the mercury drop, and as a result a gradual increase of the concentration of the metal is obtained, since its diffusion through the mercury in the narrow capillary tube is hindered. Electrolysis of very dilute solutions gives comparatively concentrated amalgams after several minutes, the oxidation currents of which can readily be recorded.

In order to obtain appreciably concentrated amalgams and to increase the sensitivity of the method the solution should be stirred in a reproducible way during the initial accumulating electrolysis, e.g. by means of synchronized magnetic stirrers.

Hanging drop electrodes of various types make it possible to analyse solutions containing metal ions in concentrations of the order of 10^{-8} M. The method was developed by Kemula and Kublik [209–211] and others (see for instance [211, 212]) and was utilized in the analysis of traces of metallic impurities in various substances. These problems are discussed by Neeb [212].

The hanging drop technique has been widely used by Stromberg *et al.* in analysis and other physico-chemical investigations. Some of these studies have been reviewed by the principal authors [213], and some of the later work was reviewed by Neeb [212].

The problem of concentration distribution of the reduced metal in the mercury drop has been discussed by several authors [214, 215] and a theoretical description of the oxidation of metals from hanging mercury drop has been given by Reinmuth [216].

Brainina *et al.* [217] discussed the problem of deposition of metals at solid electrodes with the purpose of subsequent oxidation and determination.

Keller and Reinmuth [218] proposed a new method of solving stationary electrode voltammetry problems. It is to a certain extent a development of earlier Reinmuth's work [48], but it has a more general character. In particular it makes it possible to elaborate electrolytic processes taking place in a finite diffusion region as well as *quasi*-reversible processes and electrode processes connected by chemical reactions of the first order. Corrections for spherical and cylindrical diffusion have been also discussed.

5.4.3 CHRONOPOTENTIOMETRY

Chronopotentiometric electrode processes with linear diffusion from a thin solution layer were considered by Christensen and Anson [219]. These authors adapted the solution of a similar problem relating to heat transfer [220] and derived an equation describing the time dependence of concentration of the electrolysed substance at the electrode surface:

$$C_{Ox}(0, t) = C_{Ox}^0 - \frac{il}{nFAD_{Ox}}\left[\frac{D_{Ox}t}{l^2} + \frac{1}{3} - \frac{2}{\pi^2}\sum_{k=1}^{\infty}\frac{1}{k^2}\exp\left(-\frac{D_{Ox}k^2\pi^2 t}{l^2}\right)\right] \quad (5.237)$$

where l, as previously, is the thickness of the solution layer. $C_{Ox}(t, 0)$ can be expressed by the equivalent equation:

$$C_{Ox}(0, t) = C_{Ox}^0 - \frac{2it^{1/2}}{nFA\pi^{1/2}D_{Ox}^{1/2}} - \frac{4it^{1/2}}{nFAD_{Ox}^{1/2}}\sum_{k=1}^{\infty}\text{erfc}\frac{kl}{(D_{Ox}t)^{1/2}}. \quad (5.238)$$

For high l values equation (5.238) is reduced to that derived for conventional chronopotentiometry.

Since the transition time is obtained when $C_{Ox}(0, \tau) = 0$, the relationships describing the transition time can be readily derived from equations (5.237) and (5.238). When $l^2 < 4\tau D_{Ox}$ the exponential term in equation (5.237) can be neglected and the transition time can be described by the simpler relationship:

$$\tau = \frac{nFAlC_{Ox}^0}{i} - \frac{l^2}{3D_{Ox}}. \quad (5.239)$$

It is also possible to choose such experimental conditions that $l^2/3D_{Ox}$ is small, as compared with the measured transition time. In such a case the transition time is described by a relationship according to which practically the whole amount of substance Ox contained in the solution layer is consumed in the electrode reaction:

$$\tau = \frac{nFAlC_{Ox}^0}{i}. \quad (5.240)$$

This equation suggests a simple method for the determination of the number of electrons n, even when the diffusion coefficient of the substance investigated is unknown.

The relationships derived were tested experimentally by Christensen and Anson. They found that the potential changes observed before the transition time are much more pronounced in thin layer chronopotentiometry than in the conventional technique.

Galus, Kemula and Sacha [221] considered the problem of chronopotentiometric oxidation of metals from the hanging mercury drop, when practically the whole amount of the metal introduced into the drop, or a considerable part of it, is oxidized in the initial concentrating electroreduction. Utilizing the results obtained by Crank [222] and adapting them to this problem they obtained an equation describing the concentration of the metal at the mercury drop surface:

$$C_{\text{Red}}(r_0, t) = C_{\text{Red}}^0 - \frac{ir_0}{nFAD_{\text{Red}}} \left\{ \frac{3D_{\text{Red}}t}{r_0^2} + \frac{1}{5} - 2\sum_{n=1}^{\infty} \frac{e^{-D_{\text{Red}}\alpha_n t}}{\alpha_n^2 r_0^2} \right\} \quad (5.241)$$

where $r_0 \alpha_n$ are the positive roots of the equation:

$$r_0 \alpha_n \cot r_0 \alpha_n = 1. \quad (5.242)$$

When the transition times are so long that the inequality $r_0^2 < 4D_{\text{Red}}t$ is fulfilled, the third term in brackets in equation (5.241) is small as compared with the others, and the transition time is then given by the equation:

$$\tau = \frac{nFAr_0 C_{\text{Red}}^0}{3i} - \frac{r_0^2}{15 D_{\text{Red}}}. \quad (5.243)$$

This equation is similar to equation (5.239), derived for conditions of linear diffusion.

In the case when the amount of the oxidized metal in the mercury drop is appreciable, the drop radius is small and the direct current is low, equation (5.243) is reduced to the simple form:

$$\tau = \frac{nFAr_0 C_{\text{Red}}^0}{3i}. \quad (5.244)$$

A different treatment of this problem has been described by Kemula and Strojek [223]. Zakharov *et al.* [224] derived similar equations. Barański [225] has given theoretical descriptions of chronopotentiometric oxidation of amalgams in an internal electrolysis system.

5.5 Multi-Stage Electrode Processes

In this chapter the theory and equations of one-stage reversible electrode processes have been discussed. In the case of reduction these processes can be formulated by the general equation:

$$\text{Ox} + ne \rightleftharpoons \text{Red}. \quad (5.245)$$

Some substances, however, are reduced in multi-stage processes. A two-stage process can be formulated as follows:

$$\text{Ox} + n_1 e \rightleftharpoons \text{Red}_1, \quad (5.246)$$

$$\text{Red}_1 + n_2 e \rightleftharpoons \text{Red}_2. \quad (5.247)$$

The electroactive substances Ox and Red_1 can be reduced at different potentials. When the reduction potential of Red_1 is more negative than that of Ox reduction, two steps should appear on the recorded curves, provided that the reduction potentials are sufficiently different from each other.

An example of such a process is the electroreduction of Cu(II) in a solution 1 M with respect to ammonia and ammonium chloride. The reaction takes place according to the equations:

$$Cu^{2+} + e \rightleftharpoons Cu^+, \qquad (5.248)$$

$$Cu^+ + e \rightleftharpoons Cu. \qquad (5.249)$$

The problem of multi-stage electrode processes in the context of specific analytical techniques will now be considered.

5.5.1 Polarography

When the reduction potentials of the substances participating in the multi-stage electrode process differ by more than $200/n$ mV, separate polarographic waves for each stage are obtained. For a two-stage reduction a polarogram with two waves is obtained, having the ratio of heights determined by $n_1 : n_2$. When $n_1 = n_2$ the waves are equal. The first of them corresponds to reduction of Ox to Red_1, n_1 electrons being exchanged in the elementary reaction, and the second corresponds to the reduction of Ox to Red_2; the number of electrons being exchanged in the elementary reaction is $n_1 + n_2$.

The analysis of overlapping polarographic waves of a two-stage process was carried out by Ružić [226].

5.5.2 Rotating Disc Method

The conclusions reached in the discussion of polarography are also approximately valid in the case of the rotating disc technique. In this case also two waves are recorded when the reduction potentials of Ox and Red_1 differ sufficiently from each other, and the ratio of their heights is $n_1 : n_2$.

5.5.3 Stationary Electrode Voltammetry

Multi-stage stationary electrode voltammetric processes have been discussed by Gokhshtein and Gokhshtein [227, 228]. This rather general discussion considered the multi-stage charge exchange in reversible, irreversible and *quasi*-reversible systems, but the results cannot be used directly for the interpretation of experimental curves. A two-stage electrode process has been discussed by Poltsyn and Shain [229].

In the case of a reversible process two independent cathodic peaks are observed, provided that the reduction potentials differ by at least $118/n$ mV. When the difference is smaller than $100/n$ mV the two peaks coalesce to form one broad peak. If Red_1 is more readily reduced than Ox one peak is observed, and its size corresponds to the direct reduction of Ox to Red_2.

For the determination of the second current peak height it is necessary to know the current which would be observed at the second peak potential if the first reduction process alone was taking place. This is the necessary condition for the determination of the zero line from which the second peak current must be measured. This problem has been considered by Gokhshtein and Gokhshtein [228, 230] and a more detailed discussion has been given by Poltsyn and Shain [229].

5.5.4 Chronopotentiometry

The problem of multi-stage electrode processes is the most complex in the case of chronopotentiometry. Whereas in the three methods previously discussed the values of the recorded multi-stage reduction currents are proportional to the number of electrons exchanged in each of the stages, the relationship between transition times measured in the case of a two-stage chronopotentiometric process is given by the formula:

$$\tau_2 = \tau_1 \left[2\frac{n_2}{n_1} + \left(\frac{n_2}{n_1}\right)^2 \right] \tag{5.250}$$

where τ_1 and τ_2 are the first and second transition times, respectively. Relationship (5.250) was derived by Berzins and Delahay [231].

In polarography, stationary electrode voltammetry and the rotating disc technique the measured currents were equal when n_1 was equal to n_2, whereas in chronopotentiometry $\tau_2 = 3\tau_1$ when $n_1 = n_2$, and $\tau_2 = 8\tau_1$ when $n_2 = 2n_1$, which follows from equation (5.250).

This relationship was tested by Berzins and Delahay on the example of the two-stage oxygen reduction. The experimental result was $\tau_2 = 2.97\tau_1$ which is in good agreement with the theory.

For a three-stage electrode process the following relationships are valid [232]:

$$\frac{\tau_3}{\tau_1} = \frac{n_3^2 + 2n_3(n_1+n_2)}{n_1^2}, \quad (5.251)$$

$$\frac{\tau_3}{\tau_2} = \frac{n_3^2 + 2n_3(n_1+n_2)}{n_2^2 + 2n_1 n_2}, \quad (5.252)$$

$$\frac{\tau_3}{\tau_1+\tau_2} = \frac{n_3^2 + 2n_3(n_1+n_2)}{(n_1+n_2)^2}. \quad (5.253)$$

Similar relationships have been discussed by Murray and Reilley [38, 39] and by Evans [233].

5.6 Electrode Processes Involving Several Depolarizers

In the preceding part of this chapter the case of a multi-stage electrode process with one depolarizer in the solution has been briefly discussed. Frequently, especially in analysis, solutions containing two or more depolarizers have to be investigated. The problem of limiting currents when several depolarizers are present in the solution is similar to that already described in the preceding section.

5.6.1 POLAROGRAPHY

Every depolarizer is characterized by its own independent polarographic wave, provided that the half-wave potentials of the individual processes are sufficiently separated from each other. The waves are adequately resolved when the potential differences are greater than $200/n$ mV. Since the waves are independent of each other they can be described by the Ilkovič equation.

Ružić and Branica [234, 235] presented a logarithmic analysis of overlapping reversible and totally irreversible polarographic waves. Ružić [236, 237] has dealt also with the problem of *quasi*-reversible waves.

5.6.2 STATIONARY ELECTRODE VOLTAMMETRY

When the difference between the potentials of the electrode processes of the individual depolarizers exceeds $120/n$ mV, independent current peaks are observed for each depolarizer. The determination of the second and further peaks is rather difficult since, at lower potential values, the intensities of peak currents corresponding to the depolarizer reacting with the electrode are smaller. Provided that the currents observed at the potentials of the measured peaks, as a result of electrode reactions of other depolarizers, are taken into account in the determination of the values of the peaks, each current peak may be considered independently. Its value can then be described by the Randles and Ševčik equation.

When the difference between the potentials of two substances reacting with the electrode is smaller than $120/n$ mV the system can be analysed by computer. Gutknecht and Perone [238] were able to make an accurate determination of the currents of In (II) and Cd (II) in 1 M HCl, although the difference between the peak potentials was only 40 mV.

5.6.3 ROTATING DISC METHOD

The remarks made in subsection 5.6.1 are also approximately valid for the rotating disc technique.

5.6.4 CHRONOPOTENTIOMETRY

As in the case of multi-stage electrode processes of one depolarizer, when several depolarizers are involved chronopotentiometry again presents the most complex problem. When two depolarizers are present in the solution, the transition time is described by the equation derived by Berzins and Delahay [231]:

$$(\tau_1 + \tau_2)^{1/2} - \tau_1^{1/2} = \frac{\pi^{1/2} n_2 F D_{Ox_2}^{1/2} C_{Ox_2}^0}{2i_0} \quad (5.254)$$

where τ_1 and τ_2 are the transition times of depolarizers Ox_1 and Ox_2 which react with the electrodes more and less readily, respectively. C_{Ox_2} is the concentration of Ox_2 in the bulk of the solution and the symbols D_{Ox_2} and n_2 also correspond to Ox_2.

It follows from equation (5.254) that τ_2 depends on the concentration of depolarizer Ox_1, which is reduced at more positive potentials. This relationship can be evaluated by considering a simple example. Assuming $C_{Ox_1}^0 = C_{Ox_2}^0$, $n_1 = n_2$ and $D_{Ox_1} = D_{Ox_2}$, we obtain $\tau_2 = 3\tau_1$.

Reilley, Everett and Johns [239] derived the following equation describing the case when N electroactive substances are present in the solution:

$$(\tau_1 + \tau_2 + \ldots \tau_N)^{1/2} - (\tau_1 + \tau_2 + \ldots \tau_{N-1})^{1/2}$$
$$= \frac{\pi^{1/2} n_N F D_{Ox_N}^{1/2} C_{Ox_N}^0}{2i_0} \quad (5.255)$$

where the subscript N corresponds to substance N.

Bos [240, 241] analysed in detail a chronopotentiometric process of two substances present in a solution, taking into account the effect of the number of the electrons exchanged, the potential difference at which the process took place, different concentrations of the substances and the type of electrode in finite and semi-infinite spaces. He discussed the effect of the above parameters on the transition time and the shape of the recorded chronopotentiometric curves.

The relationships discussed in this chapter have been concerned with electrode reduction processes but they are equally valid for anodic oxidation:

$$\text{Red} - ne \rightleftharpoons \text{Ox}. \quad (5.256)$$

In this case the index "Red" should be substituted for "Ox" in the symbols appearing in the equations, especially in those standing for the diffusion coefficient and the concentration.

References

[1] Cottrell, F. G., *Z. physik. Chem.*, **42**, 385 (1902).
[2] Laitinen, H. A. and Kolthoff, I. M., *J. Phys. Chem.*, **45**, 1062 (1941).
[3] Laitinen, H. A. and Kolthoff, I. M., *J. Am. Chem. Soc.*, **61**, 3344 (1939).
[4] Laitinen, H. A., *Trans. Electrochem. Soc.*, **82**, 289 (1942).
[5] Adams, R. N. and Zimmerman, J., unpublished work.
[6] Zimmerman, J., *Thesis*, University Kansas, Lawrence 1964.
[7] Anson, F. C., *Anal. Chem.*, **36**, 932 (1964).
[8] Christie, J. H., Lauer, G., Osteryoung, R. A. and Anson, F. C., *Anal. Chem.*, **35**, 1979 (1963).
[9] Christie, J. H., Lauer, G. and Osteryoung, R. A., *J. Electroanal. Chem.*, **7**, 60 (1964).
[10] Osteryoung, R. A. and Anson, F. C., *Anal. Chem.*, **36**, 975 (1964).
[11] Oldham, K. B. and Spanier, J., *J. Electroanal. Chem.*, **26**, 331 (1970).
[12] Oldham, K. B., *Anal. Chem.*, **44**, 196 (1972).
[13] Grenness, M. and Oldham, K. B., *Anal. Chem.*, **44**, 1121 (1972).
[14] Goto, M. and Oldham, K. B., *Anal. Chem.*, **45**, 2043 (1973).
[15] Oldham, K. B., *Anal. Chem.*, **45**, 39 (1973).
[16] Churchill, R. V., "*Operational Mathematics*", McGraw-Hill, New York–Toronto–London 1958.
[17] Doetsch, G., "*Applications of Laplace Transformation*", PWN, Warsaw 1964 (in Polish).
[18] Weber, H. F., *Wied. Ann.*, **7**, 536 (1879).
[19] Sand, H. J. S., *Phil. Mag.*, **1**, 45 (1901).
[20] Rosebrugh, T. R. and Miller, W. L., *J. Phys. Chem.*, **14**, 816 (1910).
[21] Karaoglanoff, Z., *Z. Elektrochem.*, **12**, 5 (1906).

[22] Bard, A. J., *Anal. Chem.*, **33**, 11 (1961).
[23] Bard, A. J., *Anal. Chem.*, **35**, 340 (1963).
[24] Delahay, P. and Mattax, C. C., *J. Am. Chem. Soc.*, **76**, 874 (1954).
[25] Delahay, P. and Mamantov, G., *Anal. Chem.*, **27**, 478 (1955).
[26] Kuwana, T., *Thesis*, University of Kansas, Lawrence 1959.
[27] Russell, C. D. and Peterson, J. M., *J. Electroanal. Chem.*, **5**, 467 (1963).
[28] Iwamoto, R. T., *Anal. Chem.*, **31**, 1062 (1959).
[29] Rodgers, R. S. and Meites, L., *J. Electroanal. Chem.*, **16**, 1 (1968).
[30] Olmstead, M. L. and Nicholson, R. S., *J. Phys. Chem.*, **72**, 1950 (1968).
[31] Dračka, O., *Coll. Czechoslov. Chem. Communs.*, **34**, 2627 (1969).
[32] Reinmuth, W. H., *Anal. Chem.*, **32**, 1514 (1960).
[33] Senda, M., *Rev. Polarography (Japan)*, **4**, 89 (1956).
[34] Kambara, T. and Tachi, I., *J. Phys. Chem.*, **61**, 1405 (1957).
[35] Hurwitz, H. and Gierst, L., *J. Electroanal. Chem.*, **2**, 128 (1961).
[36] Hurwitz, H., *J. Electroanal. Chem.*, **2**, 142 (1961).
[37] Hurwitz, H., *J. Electroanal. Chem.*, **2**, 328 (1961).
[38] Murray, R. W. and Reilley, C. N., *J. Electroanal. Chem.*, **3**, 64 (1962).
[39] Murray, R. W. and Reilley, C. N., *J. Electroanal. Chem.*, **3**, 182 (1962).
[40] Ishibashi, M. and Fujinaga, T., *J. Electrochem. Soc. Japan*, **24**, 375, 526 (1956).
[41] Senda, M., Kambara, T. and Takemori, Y., *J. Phys. Chem.*, **61**, 965 (1957).
[42] Ishibashi, M. and Fujinaga, T., *Anal. Chim. Acta*, **18**, 112 (1958).
[43] Fujinaga, T. and Izutsu, K., *J. Electroanal. Chem.*, **4**, 287 (1962).
[44] Fujinaga, T., in *"Progress in Polarography"*, Zuman, P. and Kolthoff, I. M., Ed., Vol. 1, Interscience, New York 1962, page 201.
[45] Kies, H. L., *J. Electroanal. Chem.*, **16**, 279 (1968).
[46] Kies, H. L., *J. Electroanal. Chem.*, **45**, 71 (1973).
[47] Barański, A. and Galus, Z., *Roczniki Chem.*, **45**, 457 (1971).
[48] Reinmuth, W. H., *Anal. Chem.*, **34**, 1446 (1962).
[49] Nicholson, R. S. and Shain, I., *Anal. Chem.*, **36**, 706 (1964).
[50] Ševčik, A., *Coll. Czechoslov. Chem. Communs.*, **13**, 349 (1948).
[51] Randles, J. E., *Trans. Faraday. Soc.*, **44**, 327 (1948).
[52] Reinmuth, W. H., *Anal. Chem.*, **33**, 1793 (1961).
[53] Matsuda, H. and Ayabe, Y., *Z. Elektrochem.*, **59**, 494 (1955).
[54] Gokhshtein, Ya. P., *Dokl. Akad. Nauk SSSR*, **126**, 598 (1959).
[55] de Vries, W. T. and Van Dalen, E., *J. Electroanal. Chem.*, **6**, 490 (1963).
[56] Delahay, P., *J. Phys. Chem.*, **54**, 630 (1950).
[57] Mueller, T. R. and Adams, R. N., *Anal. Chim. Acta*, **25**, 482 (1961).
[58] Andrieux, C. P., Nadjo, L. and Savéant, J. M., *J. Electroanal. Chem.*, **26**, 147 (1970); Savéant, J. M. and Tessier, D., *J. Electroanal. Chem.*, **61**, 251 (1975).
[59] Imbeaux, J. C. and Savéant, J. M., *J. Electroanal. Chem.*, **44**, 169 (1973); Andrieux, C. P, Savéant, J. M. and Tessier, D., *J. Electroanal. Chem.*, **63**, 429 (1975); Nadjo, L., Savéant, J. M. and Tessier, D., *J. Electroanal. Chem.*, **64**, 143 (1975).
[60] Berzins, T. and Delahay, P., *J. Am. Chem. Soc.*, **75**, 555 (1953).
[61] Carslaw, H. S. and Jaeger, J. C., *"Conduction of Heat in Solids"*, Oxford University Press, London 1947, page 372.
[62] Miller, W. L. and Gordon, A. R., *J. Phys. Chem.*, **35**, 2785 (1931).
[63] White, N. and Lawson, F., *J. Electroanal. Chem.*, **25**, 409 (1970).

[64] Ilkovič, D., *Coll. Czechoslov. Chem. Communs.*, **6**, 498 (1934).
[65] Ilkovič, D., *J. chim. phys.*, **35**, 129 (1938).
[66] De Levie, R., *J. Electroanal. Chem.*, **9**, 311 (1965).
[67] Lingane, J. J., *Ind. Eng. Chem., Anal. Ed.*, **15**, 583 (1943).
[68] Lingane, J. J. and Loveridge, B. A., *J. Am. Chem. Soc.*, **66**, 1425 (1944).
[69] Meites, L., *"Polarographic Techniques"*, Interscience Publishers, New York, London, Sydney 1965.
[70] von Karman, T., *Z. angew. Math. Mech.*, **1**, 244 (1921).
[71] Cochran, W. G., *Proc. Cambridge Phil. Soc.*, **30**, 365 (1934).
[72] Levich, V. G., *Acta Physicochim. U.R.S.S.*, **17**, 257 (1942).
[73] Levich, V. G., *"Physicochemical Hydrodynamics"*, Ed. Akad. Nauk USSR, Moscow 1952 (in Russian).
[74] Gregory, D. P. and Riddiford, A. C., *J. Chem. Soc.*, 3756 (1956).
[75] Newman, J., *J. Phys. Chem.*, **70**, 1327 (1966).
[76] Kassner, T. F., *J. Electrochem. Soc.*, **114**, 689 (1967).
[77] Siver, Yu. G. and Kabanov, B. N., *Zhurn. Fiz. Khim.*, **22**, 53 (1948).
[78] Kabanov, B. N., *Zhurn. Fiz. Khim.*, **23**, 428 (1949).
[79] Hogge, E. A. and Kraichman, M. B., *J. Am. Chem. Soc.*, **76**, 1431 (1954).
[80] Kraichman, M. B. and Hogge, E. A., *J. Phys. Chem.*, **59**, 986 (1955).
[81] Pleskov, Yu. V., *Zhurn. Fiz. Khim.*, **34**, 623 (1960).
[82] Galus, Z., Olson, C., Lee, H. Y. and Adams, R. N., *Anal. Chem.*, **34**, 164 (1962).
[83] Desiderev, G. P, and Berezina, S. I., *Dokl. Akad. Nauk SSSR*, **130**, 1270 (1960).
[84] Blurton, K. F. and Riddiford, A. C., *J. Electroanal. Chem.*, **10**, 457 (1965).
[85] Riddiford, A. C., *Adv. Electrochem. Eng.*, **4**, 47 (1966).
[86] Prater, K. B. and Adams, R. N., *Anal. Chem.*, **38**, 153 (1966).
[87] Newman, J., *J. Electrochem. Soc.*, **113**, 501, 1235 (1966).
[88] Marathe, V. and Newman, J., *J. Electrochem. Soc.*, **116**, 1704 (1969).
[89] Bruckenstein, S. and Miller, S., *J. Electrochem. Soc.*, **117**, 1044 (1970).
[90] Miller, B. and Bellavance, M. I., *J. Electrochem. Soc.*, **120**, 42 (1973).
[91] Siver, Yu. G., *Zhurn. Fiz. Khim.*, **34**, 577 (1960).
[92] Hale, J. M., *J. Electroanal. Chem.*, **6**, 187 (1963).
[93] Filinovskii, V. Yu. and Kiryanov, V. A., *Dokl. Akad. Nauk SSSR*, **156**, 1412 (1964).
[94] Fried, I. and Elving, P. J., *Anal. Chem.*, **37**, 464, 803 (1965).
[95] Girina, G. P., Filinovskii, V. Yu. and Feoktistov, L. G., *Elektrokhimiya*, **3**, 941 (1967).
[96] Matsuda, H., *J. Electroanal. Chem.*, **15**, 109 (1967).
[97] Matsuda, H. and Yamada, J., *J. Electroanal. Chem.*, **30**, 261 (1971).
[98] Yamada, J. and Matsuda, H., *J. Electroanal. Chem.*, **30**, 271 (1971).
[99] Matsuda, H., *J. Electroanal. Chem.*, **21**, 433 (1969).
[100] Matsuda, H., *J. Electroanal. Chem.*, **22**, 413 (1969).
[101] Matsuda, H., *J. Electroanal. Chem.*, **25**, 461 (1970).
[102] Matsuda, H., *J. Electroanal. Chem.*, **38**, 159 (1972).
[103] Yamada, J. and Matsuda, H., *J. Electroanal. Chem.*, **44**, 189 (1973).
[104] Levich, V. G., *"Physicochemical Hydrodynamics"*, Prentice-Hall, Inc., Englewood Cliffs, N. J., 1962, page 112.
[105] Matsuda, H., *J. Electroanal. Chem.*, **15**, 325 (1967).

References

[106] Matsuda, H., *J. Electroanal. Chem.*, **16**, 153 (1968).
[107] Fujinaga, T., Nagai, T., Okazaki, S. and Takagi, C., *Nippon Kagaku Zasshi*, **84**, 941 (1963).
[108] Fujinaga, T., Takagi, C. and Okazaki, S., *Kogyo Kagaku Zasshi*, **67**, 1798 (1964).
[109] Eckfeldt, E. L. and Shaffer, E. W., Jr., *Anal. Chem.*, **36**, 2008 (1964).
[110] Roe, D. K., *Anal. Chem.*, **36**, 2371 (1964).
[111] Blaedel, W. J. and Strohl, J. H., *Anal. Chem.*, **37**, 64 (1965).
[112] Sioda, R. E., *Electrochim. Acta*, **13**, 375 (1968).
[113] Sioda, R. E., *Electrochim. Acta*, **15**, 783 (1970).
[114] Blaedel, W. J. and Strohl, J. H., *Anal. Chem.*, **36**, 1245 (1964).
[115] Sioda, R. E., *Electrochim. Acta*, **15**, 1559 (1970).
[116] Fujinaga, T., Izutsu, K. and Okazaki, S., *Rev. Polarogr. (Kyoto)*, **14**, 164 (1967).
[117] Kihara, S., Yamamoto, T., Motojima, K. and Fujinaga, T., *Talanta*, **19**, 329 (1972).
[118] Kihara, S., Yamamoto, T., Motojima, K. and Fujinaga, T., *Talanta*, **19**, 657 (1972).
[119] Kihara, S., Yamamoto, T., Motojima, K. and Fujinaga, T., *Bunseki Kagaku*, **21**, 469 (1972).
[120] Kihara, S., Motojima, K. and Fujinaga, T., *Bunseki Kagaku*, **21**, 883 (1972).
[121] Komendarek-Stryjek, Z., *Thesis*, Warsaw University, 1970.
[122] Fujinaga, T., *Pure Appl. Chem.*, **25**, 709 (1971).
[123] Kihara, S., *J. Electroanal. Chem.*, **45**, 31 (1973).
[124] Sioda, R. E., *Elektrochim. Acta*, **16**, 1569 (1971).
[125] Sioda, R. E. and Kambara, T., *J. Electroanal. Chem.*, **38**, 51 (1972).
[126] Wroblowa, H. S., *J. Electroanal. Chem.*, **42**, 321 (1973).
[127] Lederer, E. I., *Kolloid-Z.*, **44**, 108 (1928); **46**, 169 (1928).
[128] MacGillavry, D. and Rideal, E. K., *Rec. trav. chim.*, **56**, 1013 (1937).
[129] Carslaw, H. S. and Jaeger, J. C., *"Conduction of Heat in Solids"*, Oxford University Press, London 1947, page 209.
[129a] Delahay, P., *"New Instrumental Methods in Electrochemistry"*, Fig. 3.7, Interscience, New York 1954.
[130] Lingane, J. J. and Loveridge, B. A., *J. Am. Chem. Soc.*, **72**, 438 (1950).
[131] Strehlow, H. and von Stackelberg, M., *Z. Elektrochem.*, **54**, 51 (1950).
[132] Kambara, T. and Tachi, I., *Bull. Chem. Soc. Japan*, **23**, 226 (1950).
[133] Kambara, T. and Tachi, I., *Sbornik I. Mezinar. Polarograph. Sjezdu*, Vol. I, page 126, Prirodoved, vydavatelstvi, Praha 1951.
[134] von Stackelberg, M., *Z. Elektrochem.*, **57**, 338 (1953).
[135] Koutecký, J., *Ceskoslov. cas. fys.*, **2**, 117 (1952); *Czechoslov. J. Phys.*, **2**, 50 (1953).
[136] Matsuda, H., *Bull. Chem. Soc. Japan*, **26**, 342 (1953).
[137] Kambara, T. and Tachi, I., *Bull. Chem. Soc. Japan*, **25**, 284 (1952); *Polarogr. Ber.*, **1**, 503 (1953).
[138] Koutecký, J. and von Stackelberg, M., in *"Progress in Polarography"*, Zuman, P. and Kolthoff, I. M., Eds., Interscience, New York 1962, Vol. 1.
[139] Levich, V. G., *"Physicochemical Hydrodynamics"*. Ed. Akad. Nauk. USSR, Moscow 1952, page 538.
[140] Newman, J., *J. Electroanal. Chem.*, **15**, 309 (1967).
[141] Schulman, J. H., Battey, H. B. and Jelatis, D. B., *Rev. Sci. Instr.*, **18**, 226 (1947).
[142] McKenzie, H. A., *J. Am. Chem. Soc.*, **70**, 3147 (1948).

[143] Taylor, J. K., Smith, R. E. and Cooter, J. L., *J. Research Nat. Bur. Standards*, **42**, 387 (1949).
[144] Smith, G. S., *Nature*, **163**, 290 (1949).
[145] Kambara, T., Suzuki, M. and Tachi, T., *Bull. Chem. Soc. Japan*, **23**, 219 (1950).
[146] Meites, L. and Meites, T., *J. Am. Chem. Soc.*, **72**, 4843 (1950).
[147] Gardner, H. J., *Australian J. Sci.*, **15**, 177 (1953).
[148] Lingane, J. J., *J. Am. Chem. Soc.*, **75**, 788 (1953).
[149] Vlček, A. A., *Chem. Listy*, **47**, 1440 (1953).
[150] Airey, L. and Smales, A. A. *Analyst*, **75**, 287 (1950).
[151] Hans, W. and Henne, W., *Naturwiss.*, **40**, 524 (1953).
[152] Smoler, I., *Chem. Listy*, **47**, 1667 (1953); *Coll. Czechoslov. Chem. Communs.*, **19**, 238 (1954).
[153] Smoler, I., *Chem. Zvesti*, **8**, 867 (1954).
[154] Hans, W., Henne, W. and Meurer, E., *Z. Elektrochem.*, **58**, 836 (1954).
[155] Markowitz, J. M. and Elving, P. J., *J. Am. Chem. Soc.*, **81**, 3518 (1959).
[156] Markowitz, J. M. and Elving, P. J., *Chem. Revs.*, **58**, 1047 (1958).
[157] Duffey, G. F., Rahilly, W. P. and Kidman, R. B., *J. Phys. Chem.*, **70**, 982 (1966).
[158] O'Brien, R. N. and Dieken, F. P., *J. Electroanal. Chem.*, **42**, 25 (1973).
[159] O'Brien, R. N. and Dieken F. P., *J. Electroanal. Chem.*, **42**, 37 (1973).
[160] Kirova-Eisner, E., Tchernikovski, N. and Eisner, U., *J. Electrochem. Soc.*, **120**, 361 (1973).
[161] Mairanovsky, S. G. and Neiman, M. B., *Izv. Akad. Nauk SSSR, ser. khim.*, 420 (1955).
[162] MacNevin, W. M. and Balis, E. W., *J. Am. Chem. Soc.*, **65**, 660 (1943).
[163] Smith, G. S., *Trans. Faraday Soc.*, **47**, 63 (1952).
[164] Duda, J. L. and Vrentas, J. S., *J. Phys. Chem.*, **72**, 1187 (1968).
[165] Duda, J. L. and Vrentas, J. S., *J. Phys. Chem.*, **72**, 1193 (1968).
[166] Los, J. M. and Murray, D. W., in *"Advances in Polarography"*, Longmuir, J. S., Ed., Pergamon Press, London 1960, Vol. 2, page 419.
[167] Smoler, I. and Kuta, J., *Z. physik, Chem. (Leipzig)*, Sonderheft, 58 (1958).
[168] Kuta, J. and Smoler, I., *Coll. Czechoslov. Chem. Communs.*, **26**, 224 (1961).
[169] Kuta, J. and Smoler, I., *"Progress in Polarography"*, Zuman, P. and Kolthoff, I. M., Eds., Interscience, New York 1962, Vol. 1, page 43.
[170] von Stackelberg, M. and Toome, V., *Z. Elektrochem.*, **58**, 226 (1954).
[171] Kuta, J. and Smoler, I., *Coll, Czechoslov. Chem. Communs.*, **28**, 2874 (1963).
[172] Vitek, V., *Coll Czechoslov. Chem. Communs.*, **7**, 537 (1935).
[173] Brasher, D. M. and Jones, F. R., *Trans. Faraday Soc.*, **42**, 775 (1946).
[174] Vavruch, I., *Coll. Czechoslov. Chem. Communs.*, **12**, 429 (1947).
[175] Scholander, A., *Sbornik I mezinar. polarograf. sjezdu*, Prirodoved. vydavatelstvi, Praha 1951, Vol. 1, page 260.
[176] Frankelthal, R. P. and Shain, I., *J. Am. Chem. Soc.*, **78**, 2969 (1956).
[177] Reinmuth, W. H., *J. Am. Chem. Soc.*, **79**, 6358 (1957).
[178] Mamantov, G. and Delahay, P., *J. Am. Chem. Soc.*, **76**, 5323 (1954).
[179] Koutecký, J. and Cižek J., *Coll. Czechoslov. Chem. Communs.*, **22**, 914 (1957).
[180] Gumiński, C. and Galus, Z., *Roczniki Chem.*, **44**, 1767 (1970).
[181] Carslaw, H. S. and Jaeger, J. C., *"Conduction of Heat in Solids"*, Oxford University Press, London 1947, page 280.

[182] Rius, A., Polo, S. and Llopis, J., *Anales fis. y quim. (Madrid)*, **45**, 1029 (1949).
[183] Jaeger, J. C. and Clarke, M., *Proc. Roy. Soc. Edinburgh*, **A61**, 229 (1942).
[184] Nicholson, M. M., *J. Am. Chem. Soc.*, **76**, 2539 (1954).
[185] Peters, D. G. and Lingane, J. J., *J. Electroanal. Chem.*, **2**, 1 (1961).
[186] Lingane, J. J., *J. Electroanal. Chem.*, **2**, 46 (1961).
[187] Evans, D. H. and Price, J. E., *J. Electroanal. Chem.*, **5**, 77 (1963).
[188] Dornfeld, D. I. and Evans, D. H., *J. Electroanal. Chem.*, **20**, 341 (1969).
[189] Carslaw, H. S. and Jaeger, J. C., *"Conduction of Heat in Solids"*, Oxford University Press, London 1959, page 338.
[190] Hurwitz, H. D., *J. Electroanal. Chem.*, **7**, 368 (1964).
[191] Hubbard, A. T. and Anson, F. C., in *"Electroanalytical Chemistry"*, Vol. 5, Ed. Bard, A. J., Dekker, New York 1970.
[192] Hubbard, A. T., *Crit. Reviews in Anal. Chem.*, **3**, 201 (1973).
[193] Lykhov, A. L., *"Theory of Heat Conductivity"*, Gostekhizdat, Moscow 1952 (in Russian).
[194] Chovnik, N. G. and Vashchenko, V. V., *Zhurn. Fiz. Khim.*, **37**, 538 (1963).
[195] Stromberg, A. G. and Zakharova, E. A., *Elektrokhimiya*, **1**, 1036 (1965).
[196] Stevens, W. G. and Shain, I., *J. Phys. Chem.*, **70**, 2276 (1966).
[197] Dowgird, A. and Galus, Z., *Bull. Acad. Polon. Sci., Sér. Sci. Chim.*, **18**, 255 (1970).
[198] Zakharov, M. S., Bakanov, V. I. and Pnev, V. V., *Elektrohimiya*, **3**, 820 (1967).
[199] Hubbard, A. T. and Anson, F. C., *Anal. Chem.*, **38**, 58 (1966).
[200] Carslaw, H. S. and Jaeger, J. C., *"Conduction of Heat in Solids"*, Oxford University Press, London 1959, section 3.5.
[201] de Vries, W. T. and Van Dalen, E., *J. Electroanal. Chem.*, **8**, 366 (1964).
[202] Moros, S. A., *Anal. Chem.*, **34**, 1584 (1962).
[203] Ramaley, R., Brubaker, R. L. and Enke, C. G., *Anal. Chem.*, **35**, 1088 (1963).
[204] de Vries, W. T., *J. Electroanal. Chem.*, **9**, 448 (1965).
[205] Roe, D. K. and Toni, J. E. A., *Anal. Chem.*, **37**, 1503 (1965).
[206] Nigmatullin, R. S., *Dokl. Akad. Nauk SSSR*, **151**, 1383 (1963).
[207] Bucur, R. V., Covaci, I. and Miron, C., *J. Electroanal. Chem.*, **13**, 263 (1976)
[208] de Vries, W. T. and Van Dalen, E., *J. Electroanal. Chem.*, **12**, 9 (1966).
[209] Kemula, W. and Kublik, Z., *Anal. Chim. Acta*, **19**, 104 (1958).
[210] Kemula, W., Kublik, Z. and Głodowski, S., *J. Electroanal. Chem.*, **1**, 91 (1959/60).
[211] Kemula, W., Rakowska, E. and Kublik, Z., *J. Electroanal. Chem.*, **1**, 205 (1959/60).
[212] Neeb, R., *"Inverse Polarographie und Voltammetrie"*, Akademie-Verlag, Berlin 1969.
[213] Stromberg, A. G. and Zakharova, E. A., *Zav. Lab.*, **30**, 3, 261 (1964).
[214] Shain, I. and Lewinson, J., *Anal. Chem.*, **33**, 187 (1961).
[215] Philips, S. L. and Karr, L. S., *Anal. Chem.*, **39**, 1301 (1967).
[216] Reinmuth, W. H., *Anal. Chem.*, **33**, 185 (1961).
[217] Brainina, H. Z., *"Inverse Voltamperometry of Solid Phases"*, Khimiya, Moscow 1972 (in Russian).
[218] Keller, H. E. and Reinmuth, W. H., *Anal. Chem.*, **44**, 434 (1972).
[219] Christensen, C. R. and Anson, F. C., *Anal. Chem.*, **35**, 205 (1963).
[220] Carslaw, H. S. and Jaeger, J. C., *"Conduction of Heat in Solids"*, Oxford University Press, London 1959, page 112.

[221] Galus, Z., Kemula, W. and Sacha, S., *J. Polarogr. Soc.*, **14**, 59 (1968).
[222] Crank, J., *"The Mathematics of Diffusion"*, Oxford University Press, Oxford 1956, page 91.
[223] Kemula, W. and Strojek, J., *Roczniki Chem.*, **41**, 1807 (1967).
[224] Zakharov, M. S., Pnev, V. V. and Bakhanov, V. I., *Zav. Lab.*, **36**, 643 (1970).
[225] Barański, A., *Chem. anal. (Warsaw)*, **16**, 989 (1971).
[226] Ružić, I., *J. Electroanal. Chem.*, **25**, 144 (1970).
[227] Gokhshtein, Ya. P. and Gokhshtein, A. Ya, *Dokl. Akad. Nauk SSSR*, **128**, 985 (1959).
[228] Gokhshtein, Ya. P. and Gokhshtein, A. Ya., in *"Advances in Polarography"*, Longmuir, I. S., Ed., Pergamon Press, New York 1960, Vol. 2, page 465.
[229] Poltsyn, D. S. and Shain, I., *Anal. Chem.*, **38**, 370 (1966).
[230] Gokhstein, Ya. P. and Gokhshtein, A. Ya., *Zhurn. Fiz. Khim.*, **34**, 1654 (1960).
[231] Berzins, T. and Delahay, P., *J. Am. Chem. Soc.*, **75**, 4205 (1953).
[232] Zittel, H. E. and Miller, F. J., *J. Electroanal. Chem.*, **13**, 193 (1967).
[233] Evans, D. H., *J. Electroanal. Chem.*, **6**, 419 (1963).
[234] Ružić, I. and Branica, M., *J. Electroanal. Chem.*, **22**, 422 (1969).
[235] Ružić, I. and Branica, M., *J. Electroanal. Chem.*, **22**, 243 (1969).
[236] Ružić, I., *J. Electroanal. Chem.*, **29**, 440 (1971).
[237] Ružić, I., *J. Electroanal. Chem.*, **36**, 447 (1972).
[238] Gutknecht, W. F. and Perone, S. P., *Anal. Chem.*, **42**, 906 (1970).
[239] Reilley, C. N., Everett, G. W. and Johns, R. H., *Anal. Chem.*, **27**, 483 (1955).
[240] Bos, P., *J. Electroanal. Chem.*, **33**, 379 (1971).
[241] Bos, P., *J. Electroanal. Chem.*, **34**, 475 (1972).

Chapter 6

Electrode Processes Controlled by the Rate of Charge Transfer

In the preceding chapter the problems of electrode processes controlled exclusively by the rate of transport of the reacting substance were discussed. The following boundary conditions necessary for solving the transport equations were formulated: (1) depolarizer concentration at the electrode surface was zero, an assumption that the electrode potential has an appropriate negative value, and (2) the magnitude of depolarizer concentration at the electrode surface can have any value from 0 to C_{Ox}^0, but the ratio of concentrations of the oxidized and reduced forms is related to the electrode potential by the Nernst equation.

The boundary condition for irreversible processes can be formulated in the same manner as (1), above, i.e. it is possible to create conditions in which the concentration of the depolarizer at the electrode surface equals zero. In such a case, however, a much more negative potential must be applied to the electrode than in the case of a reversible process. Nevertheless, the solutions are identical with those obtained for reversible processes.

The situation is quite different when the electrode potential is so positive that, for example, under polarographic conditions it is impossible to reach the cathodic limiting current "plateau". As in the case of a reversible process, the concentration of Ox at the electrode surface is higher than zero but in this case the relation between the concentration ratio of the oxidized to reduced forms and the potential cannot be expressed by the Nernst equation, since the concentration of the oxidized form is higher than that calculated from this equation, on account of the slow electrode reaction. On the other hand, in the anodic process the concentration of the reduced form is higher.

The deviation of the actual concentration from the value calculated from the Nernst equation is dependent on the degree of electrode process reversibility, characterized by the rate constant of this process. The more irreversible the process, the lower is its rate constant and the greater is the deviation of the process from Nernst behaviour.

6.1 Processes in Linear Diffusion Conditions

It follows from the above considerations that the rate constant of the electrode process must be introduced into the boundary condition describing the transformation of substance Ox at the electrode surface during this process.

Consider a slow reduction process:

$$\text{Ox} + ne \underset{k_{bh}}{\overset{k_{fh}}{\rightleftharpoons}} \text{Red} \tag{6.1}$$

where k_{fh} and k_{bh} are heterogeneous rate constants of the electrode process in which k_{fh} characterizes the reduction process and k_{bh} the oxidation process.

It is assumed that reduction of Ox to Red is a first-order process with one slow stage described by the constants k_{fh} and k_{bh}. Morever, it is assumed that the substance Ox approaches the electrode surface as a result of linear diffusion and the substance Red formed in the electrode process is soluble either in the solution or in the electrode phase and that it diffuses away from the electrode into the bulk of the solution.

These diffusion processes are described by the second Fick law of linear diffusion:

$$\frac{\partial C_{\text{Ox}}(x, t)}{\partial t} = D_{\text{Ox}} \frac{\partial^2 C_{\text{Ox}}(x, t)}{\partial x^2}, \tag{6.2}$$

$$\frac{\partial C_{\text{Red}}(x, t)}{\partial t} = D_{\text{Red}} \frac{\partial^2 C_{\text{Red}}(x, t)}{\partial x^2}. \tag{6.3}$$

In all the methods discussed the same initial conditions for solving equations (6.2) and (6.3) can be formulated. It is assumed that no reduced form is present in the solution before the start of the electrode process:

$$t = 0, \quad x \geqslant 0, \quad C_{\text{Ox}} = C_{\text{Ox}}^0, \quad C_{\text{Red}} = 0 \tag{6.4}$$

where C_{Ox}^0 is the initial concentration of the oxidized form.

It is further assumed that at the boundaries of the range the following conditions are fulfilled during the electrode process:

$$t > 0, \quad x \to \infty, \quad C_{\text{Ox}} \to C_{\text{Ox}}^0, \quad C_{\text{Red}} \to 0. \tag{6.5}$$

The boundary conditions describing the change in Ox and Red concentrations at the electrode surface will be formulated when the individual methods are discussed.

6.1.1 CHRONOAMPEROMETRY

It follows from the discussion in Chapter 5 that calculating the course of the process under chronoamperometric conditions gives an approximate description of the polarographic process; the chronoamperometric solution of a reversible process differs from that taking into account the convective mass transport (resulting from the increase in dropping electrode area) only by the constant appearing in the equation. It is probable that the chronoamperometric solution of irreversible processes, after a simple transformation, will also describe irreversible polarographic processes.

The first attempts to describe the theory of irreversible electrode processes under polarographic conditions were made by Eyring, Marker and Kwoh [1]. Their considerations were based on the Nernst [2] diffusion layer concept and therefore the theory is only approximate.

Similar concepts were developed by Tanaka and Tamamushi [3]. More precise treatment of this problem was presented independently by several workers: Smutek [4], Delahay [5, 6], Evans and Hush [7] and Kambara and Tachi [8]. These were mainly limited to chronoamperometric conditions, since they involved the solution of equations (6.2) and (6.3) which do not take into account the transport by convection resulting from the growth of the mercury electrode drop area, but they were adapted to polarographic conditions.

One of the boundary conditions necessary for the solution of these equations can be formulated by defining the number of moles of the Ox form which in unit time are transformed into the Red form, or in other words, by determining the rate of the reduction process. This rate is given by the equation:

$$\frac{dN_{Ox}}{dt} = k_{fh} C_{Ox}(0, t) - k_{bh} C_{Red}(0, t) \tag{6.6}$$

where N_{Ox} is the number of moles of the Ox form transformed into Red.

Since substance Ox approaches the electrode by diffusion, this rate can be described by the first Fick equation:

$$\frac{dN_{Ox}}{dt} = D_{Ox} \frac{\partial C_{Ox}(x, t)}{\partial x}. \tag{6.7}$$

Combining equations (6.6) and (6.7), the first boundary condition is obtained:

$$D_{Ox} \left[\frac{\partial C_{Ox}(x, t)}{\partial x} \right]_{x=0} = k_{fh} C_{Ox}(0, t) - k_{bh} C_{Red}(0, t). \tag{6.8}$$

The second boundary condition can be formulated as follows:

$$D_{Ox}\left[\frac{\partial C_{Ox}(x,t)}{\partial x}\right]_{x=0} = -D_{Red}\left[\frac{\partial C_{Red}(x,t)}{\partial x}\right]_{x=0}, \qquad (6.9)$$

since the Red form is formed as a result of the reduction of Ox at the electrode surface.

In order to simplify the problem, it can be assumed that the electrode process is controlled exclusively by the rate of charge transfer. In such a case, at potentials at which the cathodic process has an observable rate, the rate of the anodic process is so slow that it does not affect the overall rate of the process and the boundary condition (6.8) is reduced to the following form:

$$D_{Ox}\left[\frac{\partial C_{Ox}(x,t)}{\partial x}\right]_{x=0} = k_{fh}C_{Ox}(0,t). \qquad (6.10)$$

Applying the Laplace transformation and making use of the simple condition (6.10), it is possible to solve equation (6.2), taking into account conditions (6.4) and (6.5). This solution leads to the expression describing the concentration of the oxidized form at the electrode surface:

$$C_{Ox}(0,t) = C_{Ox}^0 \exp(l^2 t)\,\mathrm{erfc}(lt^{1/2}) \qquad (6.11)$$

where $l = k_{fh}/D_{Ox}^{1/2}$.

Combining equations (6.10) and (6.11) gives the relationship:

$$D_{Ox}\left[\frac{\partial C_{Ox}(x,t)}{\partial x}\right]_{x=0} = k_{fh}C_{Ox}^0 \exp(l^2 t)\,\mathrm{erfc}(lt^{1/2}). \qquad (6.12)$$

Since

$$i = nFDA\left[\frac{\partial C(x,t)}{\partial x}\right]_{x=0}, \qquad (6.13)$$

$$i = nFAC_{Ox}^0 k_{fh} \exp(l^2 t)\,\mathrm{erfc}(lt^{1/2}). \qquad (6.14)$$

Equation (6.14) was derived independently by Delahay and Strassner [6] and by Evans and Hush [7].

If the boundary conditions were not simplified, and relationship (6.8) used instead of (6.10), the solution of the system of equations (6.2) and (6.3) would be more complex:

$$i = nFAC_{Ox}^0 k_{fh} \exp(\bar{l}^2 t)\,\mathrm{erfc}(\bar{l} t^{1/2}) \qquad (6.15)$$

where $\bar{l} = (k_{fh}/D_{Ox}^{1/2}) + (k_{bh}/D_{Red}^{1/2})$.

Equation (6.15) was derived independently by Delahay [5], Smutek [4] and Kambara and Tachi [8]. It follows from this equation that in the case under consideration the current is determined by quantities C_{Ox}^0, A and k_{fh} which are constant for a given experiment. The rate of change of the current at constant potential applied to the electrode depends on the change in the value of function $\exp(\bar{l}^2 t)\operatorname{erfc}(\bar{l}t^{1/2})$, caused by the change of the parameter $(\bar{l}t^{1/2})$. This relationship is shown in Fig. 6.1.

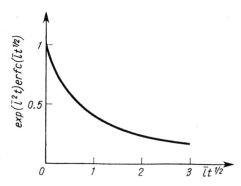

Fig. 6.1 Dependence of function $\exp(\bar{l}^2 t)\operatorname{erfc}(\bar{l}t^{1/2})$ on parameter $\bar{l}t^{1/2}$.

The values of function $\exp\lambda^2\operatorname{erfc}\lambda$ are tabulated for various λ values It follows from the relationship shown in Fig. 6.1 that the value of the function decreases as its argument increases. At $t = 0$, when the argument of the function equals zero, the value of the function equals one. When the argument increases appreciably, the value of function $\exp(\bar{l}^2 t)\operatorname{erfc}(\bar{l}t^{1/2})$ approaches zero. This results from expansion of the function into a series:

$$\exp(\bar{l}^2 t)\operatorname{erfc}(\bar{l}t^{1/2}) = \frac{1}{\pi^{1/2}\bar{l}t^{1/2}}\left\{1 - \frac{1}{2\bar{l}^2 t} + \frac{1\times 3}{(2\bar{l}^2 t)^2} - \frac{1\times 3\times 5}{(2\bar{l}^2 t)^3} + \ldots\right\}. \quad (6.16)$$

When the argument of the function is very large, all the terms of the series (6.16), except the first, may be neglected. For $\bar{l}t^{1/2} \to \infty$

$$\exp(\bar{l}^2 t)\operatorname{erfc}(\bar{l}t^{1/2}) \cong \frac{1}{\pi^{1/2}\bar{l}t^{1/2}}. \quad (6.17)$$

Relationship (6.17) is valid when the value of k_{fh} tends to infinity. In such a case the electrode process becomes reversible and equation (6.12) becomes identical with that describing the current change in a reversible potentiostatic process. Combining relationships (6.17) and (6.14):

$$i_g = \frac{nFAD_{Ox}^{1/2}C_{Ox}^0}{(\pi t)^{1/2}}. \quad (6.18)$$

In practice relationship (6.17) is valid when the constant k_{fh} is so large that the rate of the process is limited by the rate of diffusion.

Dividing equation (6.14) by equation (6.18) we obtain the relationship:

$$\frac{i}{i_g} = \pi^{1/2} l t^{1/2} \exp(l^2 t) \operatorname{erfc}(l t^{1/2}). \tag{6.19}$$

It follows from analysis of equation (6.19) and from the course of function $\exp(l^2 t)\operatorname{erfc}(lt^{1/2})$ that the current is diffusion controlled when $lt^{1/2}$ is greater than 5; then $i/i_g \simeq 1$. It has already been mentioned that such a situation exists when the constant k_{fh} is very large, or when it is relatively small but the electrolysis time is long.

The problem of determining the state of reversibility of the process can also be considered from another point of view. It was mentioned earlier that, in a reversible process, the electrode process rate is appreciable, compared with the rate of transport. Since the parameter $lt^{1/2}$, written in the developed form $k_{fh} t^{1/2} D_{Ox}^{1/2}$, determines the ratio of the electrode process rate at a given potential to the transport rate $\bar{v}$, it can be assumed that the process is reversible when:

$$\frac{k_{fh}}{\bar{v}} > 5, \tag{6.20}$$

since in such a case it is mainly limited by the rate of transport.

Inequality (6.20) can be written in the following form:

$$k_{fh} t^{1/2} > 5 D_{Ox}^{1/2}. \tag{6.21}$$

Since values of the diffusion coefficient are usually about 10^{-5} cm²/s, the criterion for potentiostatic process reversibility can be formulated as follows:

$$k_{fh} t^{1/2} > 1.6 \times 10^{-2} \text{ cm/s}. \tag{6.22}$$

Potentiostatic current–time curves have also been analysed by Johnson and Barnartt [9].

6.1.2 Stationary Electrode Voltammetry

The equation describing the stationary electrode voltammetric current of an irreversible electrode process is derived in a similar manner to that of chronoamperometry, from the equations given at the beginning of this chapter. It is not necessary to introduce the concentration of the reduced form into the boundary condition—as was done in the case of processes controlled by diffusion, provided that the assumption is made that the process is controlled exclusively by the rate of charge transfer.

The boundary condition characterizing the change of depolarizer concentration at the electrode surface can be formulated as follows:

$$t > 0, \quad x = 0, \quad D_{Ox}\left[\frac{\partial C_{Ox}(x,t)}{\partial x}\right]_{x=0} = k_{fh}C_{Ox}(0,t) \quad (6.23)$$

where

$$k_{fh} = k_s \exp\left[\frac{\alpha n_\alpha F(E-E^0)}{RT}\right], \quad (6.24)$$

k_s is the electrode process standard rate constant, and α is the transfer coefficient.

Utilizing the relationship:

$$E = E_i - Vt \quad (6.25)$$

the boundary condition can be formulated as follows:

$$D_{Ox}\left[\frac{\partial C_{Ox}(x,t)}{\partial x}\right]_{x=0} = C_{Ox}(0,t)k_i \exp(bt), \quad (6.26)$$

where

$$k_i = k_s \exp\left[-\frac{\alpha n_\alpha F(E_i - E^0)}{RT}\right], \quad (6.27)$$

$$b = \frac{\alpha n_\alpha FV}{RT}, \quad (6.28)$$

and k_i is the rate constant of the electrode reduction process at the initial potential E_i.

Proceeding as in the case of a reversible process, the integral equation first derived by Delahay [10] is obtained:

$$1 - \int_0^{bt} \frac{\chi(z)dz}{\sqrt{bt-z}} = (e^{u-bt})\chi(bt) \quad (6.29)$$

where

$$e^u = \frac{\sqrt{\pi D_{Ox} b}}{k_i} = \frac{\sqrt{\pi D_{Ox} b}}{k_s} \exp\left[\frac{\alpha n_\alpha F(E_i - E^0)}{RT}\right]. \quad (6.30)$$

For a given value of u, the solution of equation (6.29) gives the values of $\chi(bt)$ as a function of bt.

Assuming that u is greater than 7—which corresponds to an initial potential more anodic than that of the foot of the wave—the values of $\chi(bt)$

are independent of u. These values can be used for calculating the current intensity as a function of potential:

$$i = nFAC^0_{Ox}\sqrt{\pi D_{Ox}b}\,\chi(bt). \qquad (6.31)$$

The values of function $\chi(bt)$ were first calculated numerically by Delahay on the basis of equation (6.29) and later by Matsuda and Ayabe [11] and by Nicholson and Shain [12].

Function $\chi(bt)$ can also be expressed by the series [13]:

$$\chi(bt) = \frac{1}{\sqrt{\pi}}\sum_{j=1}^{\infty}(-1)^{j+1}\cdot\frac{(\sqrt{\pi})^j}{\sqrt{(j-1)!}}\exp\left[\left(-\frac{j\alpha n_\alpha F}{RT}\right)\times\right.$$
$$\left.\left(E - E^0 + \frac{RT}{\alpha n_\alpha F}\ln\frac{\sqrt{\pi D_{Ox}b}}{k_s}\right)\right]. \qquad (6.32)$$

The maximum value of function $\chi(bt)$ equals 0.280. When this value is introduced into equation (6.31), the expression for the peak current of an irreversible electrode process is obtained:

$$i_p = 0.280\,nFAC^0_{Ox}\sqrt{\pi D_{Ox}b}. \qquad (6.33)$$

At 25°C the peak current is described by the following expression:

$$i_p = 3.00\times 10^5 n(\alpha n_\alpha)^{1/2} AD^{1/2}_{Ox}V^{1/2}C^0_{Ox}. \qquad (6.34)$$

The parameters of this equation are expressed in the same units as those used for expressing the parameters of the equation describing the peak current of a reversible process. Equation (6.34) was first derived by Delahay [10] and then by Matsuda and Ayabe [11].

In both reversible and irreversible processes the observed peak currents are proportional to the square root of the rate of change of the potential applied to the electrodes. This relationship is no longer linear for electrode processes which are controlled by both the kinetics of electron transfer and diffusion. These processes will be discussed briefly in the next subsection.

6.1.2.1. *Electrode Processes Controlled by both Electron Transfer and Diffusion Rate*

The theory of such processes under stationary electrode voltammetric conditions was put forward by Matsuda and Ayabe [11]. It concerns processes for which the parameter λ, described by the equation:

$$\lambda = \frac{k_s \left(\dfrac{1}{\sqrt{D_{Ox}}}\right)^{\beta} \left(\dfrac{1}{\sqrt{D_{Red}}}\right)^{\alpha}}{a} \tag{6.35}$$

is contained in the following interval:

$$15 \geqslant \lambda \geqslant 10^{-2(1+\alpha)}. \tag{6.36}$$

In such processes the current is described by the following equation, which is similar to (6.26):

$$i = nFAC_{Ox}^0 \sqrt{D_{Ox} a}\; \chi^*(at). \tag{6.37}$$

Function $\chi^*(at)$ depends on parameter λ and on the transfer coefficients of the electrode process.

The diagrams in Fig. 6.2 represent the dependence of the $\chi^*(at)$ function on two constant α values and variable λ parameter values. These diagrams illustrate the complexity of the problem considered here.

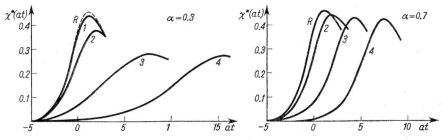

Fig. 6.2 Dependence of function $\chi^*(at)$ on at. R—reversible process, 1—$\lambda = 10$, 2—$\lambda = 1$, 3—$\lambda = 10^{-1}$, 4—$\lambda = 10^{-2}$ (from [11] by permission of the copyright holders, Verlag-Chemie).

The maximum value of function $\chi^*(at)$ depends on experimental conditions (parameter a) and on the chemistry of the process investigated. The peak current of a process which is controlled both by the rate of electron transfer and by the rate of diffusion depends on these factors and can be expressed by the general equation:

$$i_p = 0.452 \frac{n^{3/2} F^{3/2}}{R^{1/2} T^{1/2}} AD_{Ox}^{1/2} V^{1/2} C_{Ox}^0 K(\lambda, \alpha) \tag{6.38}$$

where function $K(\lambda, \alpha)$ depends on the parameter λ and on the transfer coefficient of the electrode process.

Taking into account the Randles and Ševčik equation for the peak current of a reversible process, it is possible to formulate relationship (6.32) as follows:

$$i_p = i_p^r K(\lambda, \alpha) \qquad (6.39)$$

where i_p^r is the peak current of the reversible electrode process.

The dependence of this function on λ for certain values of transfer coefficient α is shown in Fig. 6.3, which is according to Matsuda and Ayabe.

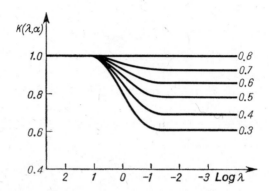

Fig. 6.3 Dependence of function $K(\lambda, \alpha)$ on $\log \lambda$. The figures stand for transition coefficient values (from [11] by permission of the copyright holders, Verlag-Chemie).

Three varied cases of electrode processes in stationary electrode voltammetry having now been discussed, the graph of the dependence of current on scan rate can be plotted. In the case of a moderately high standard rate constant for the electrode reaction, at comparatively low rates of potential change, the rate of the process may be diffusion controlled. In such a case, according to the equation of Randles and Ševčik, the dependence of peak current on the square root of polarization rate is linear. This is illustrated by straight line *1* in Fig. 6.4. A further increase in polarization rate, leading to an increase in the rate of depolarizer transport to the electrode, may render the electrode process *quasi*-reversible, provided that the rate of diffusion transport is similar to the rate of electron transfer. It follows from equation (6.38) that, in this zone (the central part of Fig. 6.4), the peak current is no longer a linear function of $V^{1/2}$. The changes of this current are schematically represented by the continuous line joining the straight lines *1* and *2*. Further increase in scan rate, leading to an increase in the rate of transport, may transform the *quasi*-reversible process into a fully irreversible one, provided that the rate of transport appreciably exceeds that of electron transfer. In such a case, according to equation (6.33), the dependence of peak current on the square root of polarization rate is again linear, but the slope of the straight line (straight line 2 in Fig. 6.4) is usually lower than that of the straight line representing the relationship i_p–$V^{1/2}$ in the case of a reversible electrode process.

The peak current equations derived have been tested by the authors, and good agreement of experimental results with theory has been found.

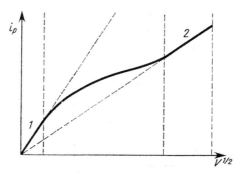

Fig. 6.4 Dependence of the peak current on the square root of the polarization rate. *1*—the region of reversibility of the process; *2*—the region of irreversibility. The middle region corresponds to *quasi*-reversibility (from [11] by permission of the copyright holders, Verlag-Chemie).

6.1.3 CHRONOPOTENTIOMETRY

In the derivation of chronoamperometric equations it was assumed that the electrode process was reversible, but this was not assumed in the derivation of the relationships describing the distribution of concentration of the electroactive substance with respect to the distance from the electrode and the time. Therefore these relationships can be utilized in the discussion of irreversible processes.

Since the formula describing $C_{Ox}(x, t)$ is valid both for reversible and irreversible processes, assuming $C_{Ox}(0, t) = 0$, an expression describing the transition time is obtained. This is identical with Sand's formula for reversible processes. Thus, on the basis of transition time measurements it is impossible to differentiate between a reversible and an irreversible process. In the case of an irreversible process, at potentials corresponding to reversible electrode processes, the rate of electrode reaction is so low that the amount of charge consumed is lower than that required for the flow of current of a defined and constant intensity. Chronopotentiometric curves of reversible (curve *1*), *quasi*-reversible (curve *2*) and irreversible (curve *3*) processes are shown schematically in Fig. 6.5.

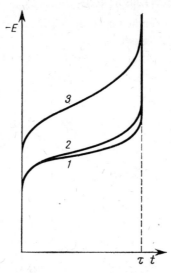

Fig. 6.5 Schematic chronopotentiometric curves: *1*—reversible process, *2*—*quasi*-reversible process, *3*—irreversible process.

The more irreversible the process, the smaller is the rate constant of the electrode reaction and, as a result, the process takes place at a considerable overvoltage. The shift of the chronopotentiometric curve in the direction of negative potentials increases with increasing current density. This can be proved, starting with relationship (6.6).

For a process which is controlled exclusively by the rate of charge transfer, this equation is reduced to the following form:

$$\frac{dN_{\text{Ox}}}{dt} = k_{fh} C_{\text{Ox}}(0, t). \tag{6.40}$$

The electrode process current is proportional to the number of moles of substance Ox reduced in unit time:

$$i = nFA \frac{dN_{\text{Ox}}}{dt}, \tag{6.41}$$

hence:

$$\frac{i_0}{nF} = k_{fh} C_{\text{Ox}}(0, t) \tag{6.42}$$

where i_0 is current density.

The left-hand side of equation (6.42) contains quantities that are constant for a given experiment, since in classical chronopotentiometry the current density is constant. In the case of an irreversible process at potentials close to the standard potential, k_{fh} is small. Therefore the process must be carried out at more negative potentials, at which k_{fh} is larger.

$C_{Ox}(0, t)$ decreases during the electrode process, and therefore, according to equation (6.42), k_{fh} should increase. This increase is observed as a shift of chronopotentiometric curves in the direction of more negative potentials. The greater the current density, the higher should be the value of k_{fh}, i.e. the more negative should be the potential at which the reduction process is observed.

Utilizing equation (6.42) it is possible to discuss the conditions at which the electrode process becomes reversible. Since:

$$C_{Ox}(0, t) = C_{Ox}^0 - \frac{2i_0 t^{1/2}}{\pi^{1/2} nFD_{Ox}^{1/2}}, \tag{6.43}$$

combining equations (6.43) and (6.42) [14]:

$$\frac{i_0}{nF} = k_{fh}\left(C_{Ox}^0 - \frac{2i_0 t^{1/2}}{\pi^{1/2} nFD_{Ox}^{1/2}}\right). \tag{6.44}$$

This equation can be formulated as follows:

$$\frac{i_0}{nF} = k_{fh}\left(\frac{2i_0 \tau^{1/2} - 2i_0 t^{1/2}}{\pi^{1/2} nFD_{Ox}^{1/2}}\right). \tag{6.45}$$

Eliminating identical quantities from both sides of the equation and multiplying both sides by $1/\tau^{1/2}$ yields the equation:

$$\frac{\pi^{1/2} D_{Ox}^{1/2}}{2k_{fh} \tau^{1/2}} = \frac{\tau^{1/2} - t^{1/2}}{\tau^{1/2}} = 1 - \frac{t^{1/2}}{\tau^{1/2}} \tag{6.46}$$

which is more conveniently written as:

$$\frac{t^{1/2}}{\tau^{1/2}} = 1 - \frac{\pi^{1/2} D_{Ox}^{1/2}}{2\tau^{1/2} k_{fh}}. \tag{6.47}$$

Equation (6.47) should be interpreted as an expression which describes the ratio of the square root of the experimental transition time to that transition time which would be observed at the same potential, if the electrode process was reversible. One of the possible values of such potentials is shown as the broken line in Fig. 6.4.

It follows from equation (6.47) that the process becomes reversible when $\pi^{1/2}D_{Ox}^{1/2}/2\tau^{1/2}k_{fh} \ll 1$. Thus reversibility can be obtained either when the value of k_{fh} is high or when the current density is so chosen that the transition time is as long as possible.

Since, in the case of chronopotentiometry, the rate of transport is given by the equation:

$$\bar{v} = \frac{\pi^{1/2}D_{Ox}^{1/2}}{2\tau^{1/2}}, \qquad (6.48)$$

equation (6.47) can be transformed into the following relationship:

$$\frac{t^{1/2}}{\tau^{1/2}} = 1 - \frac{\bar{v}}{k_{fh}}. \qquad (6.49)$$

It can be assumed that the process will be reversible, i.e. controlled exclusively by the rate of transport, when:

$$k_{fh} > 10\bar{v}. \qquad (6.50)$$

Introducing $\bar{v}$ defined by equation (6.48) into the inequality (6.50):

$$k_{fh} > 10\frac{\pi^{1/2}D_{Ox}^{1/2}}{2\tau^{1/2}}. \qquad (6.51)$$

Assuming $D_{Ox} = 10^{-5}$ cm^2/s the criterion of reversibility of chronopotentiometric processes is:

$$k_{fh}\tau^{1/2} > 2.8 \times 10^{-2} \text{ cm s}^{-1/2}. \qquad (6.52)$$

6.1.4 POLAROGRAPHY

The nature of the polarographic currents of irreversible processes can be discussed in an approximate way on the basis of the chronoamperometric equation already derived. In a similar manner it is possible to derive Ilkovič's equation approximately by solving Fick's second law, written for conditions of linear diffusion. This approach was chosen by Delahay, who multiplied the result obtained for a flat and stationary electrode by the coefficient $\sqrt{7/3}$, in order to take into account the increase of current due to the increase in area of the dropping electrode.

The exact solution for the dropping electrode was first given by Meiman [15] who considered the example of hydrogen ion discharge. A more general solution was obtained by Koutecký [16], who solved the following system of differential equations:

$$\frac{\partial C_{Ox}(x, t)}{\partial t} = D_{Ox}\frac{\partial^2 C_{Ox}(x, t)}{\partial x^2} + \frac{2x}{3t}\frac{\partial C_{Ox}(x, t)}{\partial x}, \qquad (6.53)$$

$$\frac{\partial C_{Red}(x,t)}{\partial t} = D_{Red}\frac{\partial^2 C_{Red}(x,t)}{\partial x^2} + \frac{2x}{3t}\frac{\partial C_{Red}(x,t)}{\partial x}. \qquad (6.54)$$

The initial conditions used in the solution of these equations are identical with relationship (6.4), while the boundary conditions are identical with those used in the discussion of the chronoamperometric process [(6.8) and (6.9)].

Koutecký's solution for instantaneous currents is usually expressed in the form of the $F(\chi)$ function of parameter χ. Function $F(\chi)$ determines the ratio of the current due to the slow electrode process to the current which would be observed if the rate of the electrode process was only transport controlled (i_g^*):

$$F(\chi) = \frac{i}{i_g^x} \qquad (6.55)$$

and parameter χ is described as follows:

$$\chi = \sqrt{\frac{12t}{7}\left(\frac{k_{fh}}{D_{Ox}^{1/2}} + \frac{k_{bh}}{D_{Red}^{1/2}}\right)}. \qquad (6.56)$$

Koutecký has tabulated the values of the $F(\chi)$ function for various values of the parameter.

For average currents Weber and Koutecký [17] derived a relationship similar to equation (6.55):

$$\overline{F}(\chi_1) = \frac{\overline{i}}{i_g^x} \qquad (6.55a)$$

where:

$$\chi_1 = \sqrt{\frac{12t_1}{7}\left(\frac{k_{fh}}{D_{Ox}^{1/2}} + \frac{k_{bh}}{D_{Red}^{1/2}}\right)}. \qquad (6.56a)$$

The values of function $F(\chi)$ and function $\overline{F}(\chi_1)$ are shown in Table 6.1. The function $\overline{F}(\chi_1)$ can be accurately extrapolated by means of the formula:

$$\overline{F}(\chi_1) = \frac{0.676\chi_1}{1+0.676\chi_1}. \qquad (6.57)$$

It follows from this formula and from Table 6.1 that at high values of the parameters χ or χ_1 the values of $F(\chi)$ or $\overline{F}(\chi_1)$ are equal to one, $i = i_g$; in such a case the measured current is controlled by transport alone.

In a typical irreversible electrode reduction process the parameter χ is equal to:

$$\chi = \sqrt{\frac{12t}{7}}\frac{k_{fh}}{D_{Ox}^{1/2}}. \qquad (6.56b)$$

Table 6.1 Values of functions $F(\chi)$ and $\bar{F}(\chi_1)$ (from [17] by permission of the copyright holders)

χ or χ_1	$F(\chi)$	$\bar{F}(\chi_1)$	χ or χ_1	$F(\chi)$	$\bar{F}(\chi_1)$
0.005	0.00441	0,00309	1.2	0.5552	0.4443
0.01	0.00880	0.00617	1.4	0.5970	0.4845
0.02	0.01748	0.01128	1.6	0.6326	0.5196
0.03	0.02604	0.01831	1.8	0.6623	0.5505
0.04	0.03447	0.02429	2.0	0.6879	0.5777
0.05	0.04281	0.03021	2.5	0.7391	0.6339
0.06	0.05102	0.03605	3.0	0.773	0.677
0.08	0.06712	0.04658	4.0	0.825	0.739
0.1	0.08279	0.05886	5.0	0.8577	0.781
0.2	0.1551	0.1119	6.0	0.8803	0.812
0.3	0.2189	0.1600	8.0	0.9093	0.8535
0.4	0.2749	0.2036	10.0	0.9268	0.8801
0.5	0.3245	0.2433	15.0	0.9508	0.9177
0.6	0.3688	0.2796	20.0	0.9629	0.9373
0.7	0.4086	0.3129	30	0.9752	0.9576
0.8	0.4440	0.3435	50	0.9851	0.9743
0.9	0.4761	0.3717	110	0.9932	0.9882
1.0	0.5050	0.3977	350	0.9979	0.9963

In such conditions the k_{fh} constant at the potential of the foot of the wave is small and the value of parameter χ is small; consequently the process is controlled mainly by the kinetics of charge transfer. Changes of potential in the direction of more negative values lead to increases of k_{fh} and consequently the value of $F(\chi)$ tends to one. At the potential of the wave "plateau" formation k_{fh} is so large that, as a result of high χ values, function $F(\chi)$ becomes close to one. Then the current is controlled mainly by the rate of transport.

Irreversible electrode processes taking place under polarographic conditions were also considered independently by Matsuda and Ayabe [18]. Their $\Psi(c)$ function can be connected in a simple way with Koutecký's function $F(\chi)$ [19]:

$$\Psi(c) = \frac{1}{c}\sqrt{\frac{7}{3\pi}} F\left(\sqrt{\frac{12}{7}} c\right). \tag{6.58}$$

Smith and co-workers [20] approximated Koutecký's function by means of the following relationship which is similar to (6.57):

$$F(c) = \frac{(1.030c)^{1.091}}{1 + (1.030c)^{1.091}}. \tag{6.59}$$

Oldham and Parry [21] gave another approximation:

$$F(c) = \frac{3}{2}\left[1+b-\sqrt{b^2+\frac{2b}{3}+1}\right] \quad (6.60)$$

where

$$b = \frac{9}{8}\sqrt{\frac{7}{12}}c = 0.85923c.$$

These expressions are particularly useful when computers are used. Oldham and Smith [19] compared the two approximations and came to the conclusion that the expression derived by Oldham and Parry is more exact, whereas the approximation derived by Smith and co-workers is very convenient for routine work.

6.1.5 THE ROTATING DISC METHOD

In the case of the rotating disc technique, as in the case of reversible electrode processes, the following equation has to be solved:

$$S_x \frac{dC_{Ox}}{dx} = D_{Ox}\frac{d^2C_{Ox}}{dx^2}. \quad (6.61)$$

As the problem under consideration remains unchanged, the initial and boundary conditions should be similar. The conditions used in the discussion of chronoamperometry can be used in this case, since in both methods potential is applied to the electrode and the resulting current is recorded. Thus the differences in the final results are not caused by different boundary conditions, but by the difference in the equations which describe transport.

Assuming conditions (6.4), (6.5) and (6.10) gives, as in the case of a reversible electrode process, the general solution [22]:

$$C_{Ox} = C_1\int_0^x \exp\left\{\frac{1}{D_{Ox}}\int_0^t S_x(z)dz\right\}dt + C_2 \quad (6.62)$$

where C_1 and C_2 are the integration constants.

At the electrode surface the integral expression of equation (6.62) equals zero, and therefore:

$$C_{Ox}(0) = C_2. \quad (6.63)$$

$C_{Ox}(0)$ is the concentration of the Ox form at the surface of the disc electrode.

The following relationship is also valid:

$$D_{ox}\left[\frac{\partial C_{ox}}{\partial x}\right]_{x=0} = D_{ox} C_1. \tag{6.64}$$

On the basis of relationships (6.63) and (6.64), boundary condition (6.10) can be formulated as follows:

$$D_{ox} C_1 = k_{fh} C_2. \tag{6.65}$$

Combining boundary condition (6.65) with equation (6.62) gives the expression:

$$C_{ox}^0 = C_1 \int_0^x \exp\left\{\frac{1}{D_{ox}} \int_0^t S_x(z) dz\right\} dt + C_2. \tag{6.66}$$

The integral of expression (6.66) was calculated in Chapter 5 (p. 126) and was called δ. Therefore equation (6.66) can be written as follows:

$$C_{ox}^0 = C_1 \delta + C_2 \tag{6.67}$$

or:

$$C_1 = \frac{C_{ox}^0 - C_2}{\delta}. \tag{6.67a}$$

Utilizing relationship (6.63), equation (6.67a) can be put into the following form:

$$C_1 = \frac{C_{ox}^0 - C_{ox}(0)}{\delta}. \tag{6.68}$$

The current intensity is given by equation (6.13). Therefore, by combining this equation with equations (6.64) and (6.68) an expression describing the current of a rotating disc electrode process is finally obtained:

$$i = nFAD_{ox} \frac{[C_{ox}^0 - C_{ox}(0)]}{\delta}. \tag{6.69}$$

Since

$$\delta = 1.62 D^{1/3} \nu^{1/6} \omega^{-1/2}, \tag{6.70}$$

we have:

$$i = 0.62 nFAD_{ox}^{2/3} \nu^{-1/6} \omega^{1/2} [C_{ox}^0 - C_{ox}(0)] \tag{6.71}$$

where D is the diffusion coefficient, ν the kinematic viscosity of the solution and ω the angular velocity of rotation of the electrode.

On the assumption that the process involves diffusion, and that $C_{ox}(0) = 0$, equation (6.71) can be reduced to the equation derived in the preceding chapter which describes the limiting current in the case when the disc electrode is used:

$$i_g = 0.62nFAD_{Ox}^{2/3}v^{-1/6}\omega^{1/2}C_{Ox}^0. \qquad (6.72)$$

In order to utilize equation (6.71) for the solution of the problem of irreversible electrode reactions, the concentration at the electrode surface must be related to that in the bulk of the solution. Such a relationship can be derived from relationship (6.62), assuming that the electrode reaction is of the first order, and using equations (6.65) and (6.68):

$$D_{Ox}\left[\frac{C_{Ox}^0 - C_{Ox}(0)}{\delta}\right] = k_{fh}C_{Ox}(0). \qquad (6.73)$$

Solving this equation with respect to $C_{Ox}(0)$ gives:

$$C_{Ox}(0) = \frac{D_{Ox}C_{Ox}^0}{\delta k_{fh} + D_{Ox}} \qquad (6.74)$$

and substituting $C_{Ox}(0)$ in equation (6.69):

$$i = \frac{nFAD_{Ox}C_{Ox}^0 k_{fh}}{\delta k_{fh} + D_{Ox}}. \qquad (6.75)$$

Equation (6.75) makes it possible to consider the reversibility of the electrode process at the rotating disc electrode as a function of k_{fh} and the experimental conditions (δ). When the rate constant of the process is very large, D_{Ox} in the denominator of expression (6.75) can be neglected and the equation is reduced to relationship (6.72) which describes the limiting current of a process controlled by the rate of transport alone. When the rate constant of the electrode process is very small, δk_{fh} in the denominator of equation (6.75) can be neglected and the equation is reduced to:

$$i = nFAk_{fh}C_{Ox}^0. \qquad (6.76)$$

It is convenient to consider the reversibility of an electrode process, utilizing the ratio of the measured current to that which would be observed if the process were controlled exclusively by the rate of transport. Such a ratio can be obtained by dividing equation (6.75) by equation (6.72) and utilizing relationship (6.70):

$$\frac{i}{i_g} = \frac{\delta k_{fh}}{\delta k_{fh} + D_{Ox}}. \qquad (6.77)$$

It follows from equation (6.77) that the process is reversible when the following inequality is satisfied:

$$\delta k_{fh} \gg D_{Ox}, \qquad (6.78)$$

since in such a case $i = i_g$.

Assuming $D_{Ox} = 10^{-5}$ cm^2/s the following can be written with a good approximation for the condition for reversibility of a disc electrode:

$$\delta k_{fh} > 10^{-4} \text{ cm}^2/\text{s}. \tag{6.79}$$

According to equation (6.77) the process is controlled exclusively by the rate of charge transfer when:

$$D_{Ox} \gg k_{fh} \delta. \tag{6.80}$$

Thus in the case of an irreversible process:

$$\frac{i}{i_g} = \frac{k_{fh} \delta}{D_{Ox}}. \tag{6.81}$$

If it is assumed, as previously, that $D_{Ox} = 10^{-5}$ cm^2/s, the approximate condition of irreversibility can be written:

$$\delta k_{fh} < 10^{-6} \text{ cm}^2/\text{s}. \tag{6.82}$$

On the basis of the equations derived above it is possible to show the dependence of electrode process reversibility on the velocity of disc electrode rotation. This relationship is shown schematically in Fig. 6.6. It

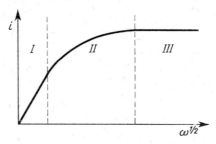

Fig. 6.6 Dependence of current (i) on the square root of rotation speed (ω) of a disc electrode. The electrode potential is in the region of wave formation, and the electrode process is not very fast. Region *I*—process controlled by the transport rate; region *III*—activation control; region *II*—process controlled simultaneously by transport and charge exchange kinetics.

could be obtained in the case of an electrode process which is reversible at a velocity of electrode rotation which is close to zero. It follows from equation (6.77) that at higher velocities of rotation the process becomes *quasi*-reversible. At very high velocities the thickness of the diffusion layer may be so low that the expression $k_{fh}\delta/D_{Ox}$ approaches zero.

Equation (6.77) also makes it possible to analyse the potential dependence of the reversibility of the electrode process under specified experimental conditions. In the case of a typically irreversible electrode process at potentials corresponding to the beginning of the wave formation the constant k_{fh} is very small and therefore the ratio i/i_g is also very small (as in the case of very high velocities of electrode rotation). When the potential of the electrode becomes more negative, in the case of a reduction process k_{fh} increases and consequently the ratio i/i_g also increases. At potentials close to the "plateau" of the wave formation the rate constant k_{fh} is so large that the ratio i/i_g approaches one. Hence the schematic dependence of i/i_g on the potential is similar to the relationship i–$\omega^{1/2}$ shown in Fig. 6.6.

6.1.6 Discussion of the Relationships Derived

In the case of a process which is controlled exclusively by the rate of charge transfer, the equations for current are very simple, since the current does not depend on the parameters determining the rate of transport. An equation common for all the methods discussed can be formulated as follows:

$$i = nFAk_{fh}C_{Ox}^0. \tag{6.83}$$

This equation is valid only in the initial sections of the recorded current–potential curves. Thus in polarography or in the rotating disc technique it concerns only the potential region related to the formation of the foot of the wave.

Since k_{fh} is related to the potential by the equation:

$$k_{fh} = k_{fh}^0 \exp\left(-\frac{\alpha n_\alpha FE}{RT}\right), \tag{6.84}$$

the initial regions of current–potential or potential–time curves are described by the following equation:

$$i = nFAk_{fh}^0 \exp\left(-\frac{\alpha n_\alpha FE}{RT}\right). \tag{6.85}$$

It follows from equation (6.85) that the relationship of log i to the potential is linear, the slope of the straight line is dependent on the transition coefficient of the electrode process.

In polarography the relationship of current to the height of the mercury reservoir (h) is frequently used as the criterion of reversibility of the electrode process. In the case of reversible processes the current increases linearly with $\sqrt{h}$. In order to determine such a relationship for irreversible processes the electrode surface area in equation (6.83) must be expressed in terms of the drop parameters m and t; then:

$$i = 0.85 nFm^{2/3}t^{2/3}k_{fh}C_{Ox}^0. \qquad (6.86)$$

Since m increases proportionally to h and h is inversely proportional to t, the conclusion can be reached that, in the case of irreversible processes, the current is independent of the height of the mercury reservoir. However, it should be borne in mind that this conclusion can be confirmed experimentally only when the potential applied to the dropping electrode is in the region at the foot of the polarographic wave.

6.2 Processes in Spherically Symmetrical Diffusion Conditions

Spherical electrodes are frequently used in studies of irreversible processes. It was shown in the preceding chapter, when discussing reversible processes, that currents or transition times observed when such electrodes are used differ appreciably from those measured for flat electrodes, and therefore it can be expected that in the case of irreversible processes certain differences will also be observed.

The electrode process investigated is defined by equation (6.1) and the diffusion equations which have to be solved must take into account the spherical diffusion:

$$\frac{\partial C_{Ox}(r,t)}{\partial t} = D_{Ox}\left[\frac{\partial^2 C_{Ox}(r,t)}{\partial r^2} + \frac{2}{r}\frac{\partial C_{Ox}(r,t)}{\partial r}\right], \qquad (6.87)$$

$$\frac{\partial C_{Red}(r,t)}{\partial t} = D_{Red}\left[\frac{\partial^2 C_{Red}(r,t)}{\partial r^2} + \frac{2}{r}\frac{\partial C_{Red}(r,t)}{\partial r}\right]. \qquad (6.88)$$

The initial conditions can be formulated as in the case of linear diffusion:

$$t = 0, \quad r \geqslant r_0, \quad C_{Ox} = C_{Ox}^0, \quad C_{Red} = 0. \qquad (6.89)$$

These conditions imply that the Red form is formed only as a result of the electrode reaction; they can be modified by assuming that the Red form is present in the solution before the start of the experiment. Then $C_{Red} = C_{Red}^0$.

When the initial condition (6.89) is assumed, the first boundary condition can be formulated as follows:

$$t > 0, \quad r \to \infty, \quad C_{Ox} \to C_{Ox}^0, \quad C_{Red} \to 0. \qquad (6.90)$$

6.2.1 Chronoamperometry

The second pair of boundary conditions can also be formulated as in the case of linear diffusion, since an identical problem is being considered, although it takes place under a different mode of diffusion:

$$t > 0, \quad r = r_0, \quad D_{Ox}\left[\frac{\partial C_{Ox}(r, t)}{\partial r}\right]_{r=r_0} = k_{fh} C_{Ox}(r_0, t) - k_{bh} C_{Red}(r_0, t) \tag{6.91}$$

$$-D_{Ox}\left[\frac{\partial C_{Ox}(r_0, t)}{\partial r}\right]_{r=r_0} = D_{Red}\left[\frac{\partial C_{Red}(r_0, t)}{\partial r}\right]_{r=r_0}. \tag{6.92}$$

This problem was solved by Shain, Martin and Ross [23] and later by Budevsky and Desimirov [24].

Shain et al. [23] obtained the following expression, using the Laplace transformation:

$$i = \frac{nFA(k_{fh}C_{Ox}^0 - k_{bh}C_{Red}^0)}{1 + \frac{r_0}{D}(k_{fh}+k_{bh})}\left\{1 + \frac{r_0}{D}(k_{fh}+k_{bh})\exp\left[\left(\frac{1}{r_0} + \frac{k_{fh}+k_{bh}}{D}\right)^2 Dt\right]\right. \times$$

$$\left. \text{erfc}\left[\left(\frac{1}{r_0} + \frac{k_{fh}+k_{bh}}{D}\right)(Dt)^{1/2}\right]\right\}. \tag{6.93}$$

In the derivation of this expression equal diffusion coefficients of the Ox and Red forms were assumed, and the common diffusion coefficient was called D ($D_{Ox} = D_{Red} = D$).

When r_0 is large, or the times of electrolysis are short, equation (6.93) is reduced to the equation which was derived for flat electrodes.

In the case of a process which is controlled exclusively by the rate of charge transfer the following expression is obtained by neglecting the terms in equation (6.93) which correspond to the reverse reaction (oxidation of Red to Ox):

$$\frac{(k_{fh}r_0 + D)i}{k_{fh}nFADC_{Ox}^0} = 1 + \frac{k_{fh}r_0}{D}\exp\left[\left(\frac{k_{fh}r_0}{D}+1\right)^2 \frac{Dt}{r_0^2}\right] \times$$

$$\text{erfc}\left[\left(\frac{k_{fh}r_0}{D}+1\right)\left(\frac{Dt}{r_0^2}\right)^{1/2}\right]. \tag{6.94}$$

When the electrode radius is very large, or the time of electrolysis is short, the problem is reduced to the irreversible chronoamperometric process under linear diffusion conditions, already discussed. Relationship (6.94) is then reduced to equation (6.83).

The relationships derived were tested by Shain *et al.*, using the reduction of iodates at pH 7.2. The correctness of the theory and its usefulness in kinetic studies of slow electrode processes were experimentally confirmed. An identical problem was solved by Budevsky and Desimirov [24] on the basis of Crank's [25] work. Their results were similar to those of Shain *et al.* This problem was also studied by Barnartt and Johnson [26, 27]. These authors and Glasser [28] considered in addition a higher order electrode process, taking place on a spherical electrode. The current–time relationship for such a reaction can be approximately expressed by the equation:

$$\frac{i}{i_{t=0}} = \gamma(1+\gamma)^{-1} + (1+\gamma)^{-1}\exp[\varrho^2(1+\gamma)^2 t]\,\text{erfc}[\varrho(1+\gamma)t^{1/2}], \quad (6.95)$$

where

$$\gamma = [r_0(\varrho_Y + \varrho_B)]^{-1},$$

$$\varrho = \varrho_Y \sqrt{D_Y} + \varrho_B \sqrt{D_B},$$

$$\varrho_Y = \frac{(y/\nu)^2 i_0^0 \exp\left[(1-\beta)/(n/\nu)\dfrac{F\eta}{RT}\right]}{(n/\nu)FC_Y^0 D_Y},$$

and

$$\varrho_B = \frac{(b/\nu)^2 i_0^0 \exp\left[-\beta/(n/\nu)\dfrac{F\eta}{RT}\right]}{(n/\nu)FC_B^0 D_B},$$

i_0^0 is density of the exchange current, ν is the stoichiometric number and the meanings of other symbols are as given earlier; the indices correspond to the general electrode reaction:

$$b\text{B}^{z+} + x\text{X} + ne \rightleftarrows y\text{Y}^{[(bz-n)/y]} + w\text{W}.$$

Equation (6.95) exactly describes the *i–t* curve in the useful time interval.

6.2.2 Stationary Electrode Voltammetry

The theory of stationary electrode voltammetric processes under conditions of spherical diffusion was put forward by De Mars and Shain [29]. They solved equation (6.87) with initial condition (6.89), and the first boundary condition was formulated as follows, assuming the full irreversibility of the process:

$$D_{Ox}\left[\frac{\partial C_{Ox}(r,t)}{\partial r}\right]_{r=r_0} = C_{Ox}(r_0, t)k_s \exp\left[-\frac{\alpha n_\alpha F}{RT}(E_i - E^0 - Vt)\right], \quad (6.96)$$

where k_s is the standard rate constant, E_i is the electrode potential at $t = 0$ and V is the polarization rate.

On the basis of equation (6.96) the current can be described by the following general formula:

$$i = nFAC_{Ox}^0 k_s \exp\left[-\frac{\alpha n_\alpha F}{RT}(E_i - E^0)\right] U(C_{Ox}, t) \exp\left[\frac{\alpha n_\alpha FVt}{RT}\right]. \quad (6.97)$$

Term $U(C_{Ox}, t)$ is a function of the concentration of substance Ox at the electrode surface; its value decreases in the course of electrolysis. Simultaneously the terminal exponential term of equation (6.97) increases and as a result a current–potential curve with the characteristic current peak is obtained. Equation (6.97) contains several terms which are related to each other in a complex way. Therefore a change of any one of them causes a complex change of current and it is impossible to obtain a current–potential curve equation on the basis of which it would be possible to determine kinetic parameters of the electrode reaction unambiguously.

In order to determine these quantities for a given system it is necessary to compare experimental curves with theoretical curves, calculated for various values of the parameters of equation (6.97). At rapid scan rates the current should approach the values observed when flat electrodes having identical surface areas are used. Changes in electrode radius result in current changes which are smaller than those observed in the case of reversible processes.

It follows from the above discussion that the relationships describing an irreversible stationary electrode voltammetric process are complex under conditions of spherical diffusion, but the kinetic parameters of the electrode process can be determined from one experimental curve. As a result of the complexity of relationship (6.97) attempts have been made to express the equation describing current in a more analytical form. Reinmuth [13] reported a solution in the form of a series convergent in the region of curve formation potentials, but this solution did not simplify the theoretical treatment of stationary electrode voltammetric curves of irreversible processes. Therefore Nicholson and Shain [12] calculated many current–potential curves theoretically and hence evaluated the differences between the currents observed in identical conditions for spherical and flat electrodes having the same surface areas.

These differences were plotted as functions of the dimensionless parameter $D^{1/2}(RT)^{1/2}/r_0(\alpha n_a FV)^{1/2}$, and as a result it was found that for values of the parameter lower than 0.1 the relationship was linear. The experiments are usually performed under such conditions that the values of this parameter do not exceed 0.1. On this basis the additional term which takes into account the spherical symmetry of diffusion was calculated. As in the case of reversible processes, the observed current was divided into two components:

$$i = i_{pl} + i_{sph} \tag{6.98}$$

where i_{pl} is the current which would be observed at a flat electrode and i_{sph} is the increase in current caused by spherical diffusion.

Equation (6.98) can be written as the full expression:

$$i = nFAC_{Ox}^0\sqrt{\pi D_{Ox}b}\chi(bt) + \frac{nFAD_{Ox}C_{Ox}^0\Phi(bt)}{r_0}. \tag{6.99}$$

The first term on the right-hand side of equation (6.99), which describes the current at a flat electrode, was discussed earlier in this chapter.

Function $\chi(bt)$ has the form of a stationary electrode voltammetric peak, since other quantities of this term are constant for the given experiment. At potentials corresponding to the beginning of the curve, the function equals zero, while for those exceeding the peak potential it is equal to one. It follows from equation (6.99) that at high r_0 values this equation is reduced to the relationship derived for processes taking place under conditions of linear diffusion.

6.2.3 Chronopotentiometry

In the case of chronopotentiometry, the discussion of irreversible processes should be based on equation (6.42), but the effect of spherical diffusion on $C_{Ox}(0, t)$ must be taken into account.

As in the case of linear diffusion, no assumptions regarding the rates of the electrode processes are made in the derivation of equations describing the distribution of the Ox form at the electrode surface during the course of electrolysis. Therefore no conclusions regarding the rate of the electrode process can be drawn from the transition time measurements. However, in the case of processes controlled by the charge transfer rate, the chronopotentiometric curve is formed at potentials more negative than those of the curve formation in a reversible process of a system characterized by an identical formal potential.

Combining equation (6.42) with that describing the distribution of the Ox form at a spherical electrode surface during chronopotentiometric electrolysis [30, 31]:

$$C_{Ox}(0, t) = C_{Ox}^0 - \frac{i_0 r_0}{nFD_{Ox}} \left[1 - \exp\left(\frac{D_{Ox} t}{r_0^2}\right) \text{erfc} \frac{(D_{Ox} t)^{1/2}}{r_0} \right], \quad (6.100)$$

the following relationship is obtained:

$$\frac{i_0}{nF} = k_{fh} \left\{ C_{Ox}^0 - \frac{i_0 r_0}{nFD_{Ox}} \left[1 - \exp\left(\frac{D_{Ox} t}{r_0^2}\right) \text{erfc} \frac{(D_{Ox} t)^{1/2}}{r_0} \right] \right\}. \quad (6.101)$$

The general remarks made in subsection 6.1.3 regarding equation (6.44) are also valid in this case.

6.3 Processes in Cylindrically Symmetrical Diffusion Conditions

Little attention has been paid to irreversible electrode processes taking place under conditions of cylindrically symmetrical diffusion.

An approximate equation for a chronopotentiometric process controlled by both the rate of transport and the rate of charge exchange was given by Johnson and Barnartt [32]. In a later work [33] these authors developed this subject, and put forward a theory of higher-order electrode processes which used equation (6.95) derived earlier in connection with spherical diffusion. They assumed that the i–t curve recorded under conditions of cylindrical diffusion, for an electrode having the radius r_0, can be approximately described by the equation for a spherical electrode having a radius of $8/3 r_0$. In the case of a chronopotentiometric process the transition time is independent of the kinetics of the electrode process. It is described by equation (5.224).

Irreversible stationary electrode voltammetric processes have not yet been treated theoretically. Generally speaking, in this case, as in the case of reversible processes, the effect of cylindrical diffusion on stationary electrode voltammetric or chronoamperometric curves should be smaller than that of spherical diffusion (assuming that radii, surface areas and other factors relating to the electrodes are identical).

References

[1] Eyring, H., Marker, L. and Kwoh, T. C., *J. Phys. Colloid Chem.*, **54**, 1453 (1949).
[2] Nernst, W., *Z. physik. Chem.*, **47**, 52 (1904).

[3] Tanaka, N. and Tamamushi, R., *Bull. Chem. Soc. Japan*, **22**, 187 (1949); Tamamushi, R. and Tanaka, N., *Bull. Chem. Soc. Japan*, **22**, 227 (1949); **23**, 110 (1950).
[4] Smutek, M., *Chem. Listy*, **45**, 241 (1951).
[5] Delahay, P., *J. Am. Chem. Soc.*, **75**, 1430 (1953).
[6] Delahay, P. and Strassner, J. E., *J. Am. Chem. Soc.*, **73**, 5218 (1951).
[7] Evans, M. G. and Hush, N. S., *J. chim. phys.*, **49**, 159 (1952).
[8] Kambara, T. and Tachi, I., *Bull. Chem. Soc. Japan*, **25**, 135 (1952).
[9] Johnson, C. A. and Barnartt, S., *J. Electrochem. Soc.*, **114**, 1256 (1967).
[10] Delahay, P., *J. Am. Chem. Soc.*, **75**, 1190 (1953).
[11] Matsuda, H. and Ayabe, Y., *Z. Elektrochem.*, **59**, 494 (1955).
[12] Nicholson, R. S. and Shain, I., *Anal. Chem.*, **36**, 706 (1964).
[13] Reinmuth, W. H., *Anal. Chem.*, **33**, 1793 (1961).
[14] Delahay, P. and Berzins, T., *J. Am. Chem. Soc.*, **75**, 2486 (1953).
[15] Meiman, N., *Zhurn. Fiz. Khim.*, **22**, 1454 (1948).
[16] Koutecký, J., *Chem. Listy*, **47**, 323 (1953); *Coll. Czechoslov. Chem. Communs.*, **18**, 597 (1953).
[17] Weber, J. and Koutecký, J., *Chem. Listy*, **49**, 562 (1955); *Coll. Czechoslov. Chem. Communs.*, **20**, 980 (1955).
[18] Matsuda, H. and Ayabe, Y., *Bull. Chem. Soc. Japan*, **28**, 422 (1955).
[19] Oldham, K. B. and Smith, D. E., *Anal. Chem.*, **40**, 1360 (1968).
[20] Smith, D. E., McCord, T. G. and Hung, H. L., *Anal. Chem.*, **39**, 1149 (1967).
[21] Oldham, K. B. and Parry, E. P., *Anal. Chem.*, **40**, 65 (1968).
[22] Levich, V. G., "*Physicochemical Hydrodynamics*", Ed. Akad. Nauk USSR, Moscow 1952, page 56 (in Russian).
[23] Shain, I., Martin, K. J. and Ross, J. W., *J. Phys. Chem.*, **65**, 259 (1961).
[24] Budevsky, E. and Desimirov, G., *Dokl. Akad. Nauk USSR*, **149**, 120 (1963).
[25] Crank, J., "*The Mathematics of Diffusion*", Oxford University Press, Oxford 1957, page 98.
[26] Johnson, C. A. and Barnartt, S., *J. Phys. Chem.*, **71**, 1637 (1967).
[27] Barnartt, S. and Johnson, C. A., *Trans. Faraday Soc.*, **65**, 1091 (1969); *J. Electroanal. Chem.*, **24**, 226 (1970).
[28] Johnson, C. A., Barnartt, S. and Glasser, C. A., *J. Electroanal. Chem.*, **28**, 1 (1970).
[29] De Mars, R. D. and Shain, I., *J. Am. Chem. Soc.*, **81**, 2654 (1959).
[30] Mamantov, G. and Delahay, P., *J. Am. Chem. Soc.*, **76**, 5323 (1954).
[31] Koutecký, J. and Čížek, J., *Coll. Czechoslov. Chem. Communs.*, **22**, 914 (1957); *Chem. Listy*, **51**, 827 (1957).
[32] Johnson, C. A and Barnartt, S., *J. Phys. Chem.*, **73**, 3374 (1969).
[33] Barnartt, S. and Johnson, C. A., *Anal. Chem.*, **43**, 2 (1971).

Chapter 7

Equations of Electroanalytical Methods. Determination of Kinetic Parameters of Electrode Processes

The description of electroanalytical techniques in previous chapters has been mainly concerned with relationships between the limiting currents recorded (or transition times in chronopotentiometry) and concentration, as well as factors determining the duration of the experiment. Consideration is now given to the full course of the curves recorded by the various electroanalytical methods. In each method, for a specific type of diffusion (linear, spherical, cylindrical) the following three processes can be theoretically distinguished and described by specific equations: (a) reversible, (b) irreversible and (c) processes controlled by both rate of diffusion and kinetics of charge transfer.

The equations for reversible and irreversible processes can be derived easily when the distributions of concentrations of the components of the redox system at the electrode surface are known. The most complicated is the *quasi*-reversible process, since in this case the shape of the curve is dependent on the kinetic parameter. When the value of this parameter decreases, the recorded curves become similar to those characteristic of irreversible processes.

Although limiting current and transition time values give no information about the kinetics of the electrode process, the knowledge of the shapes of the curves and of the potentials at which they were recorded, not only makes it possible to ascertain whether the process investigated is diffusion controlled, but can also be used for determination of the rate constant of the process, provided that it is partly irreversible. The methods and possibilities for determining kinetics of electrode processes by the four techniques under consideration will also be discussed in this chapter.

7.1 Polarography

7.1.1 Polarographic Curves for Reversible Electrode Processes

In order to obtain the equation of a reversible polarographic wave the Nernst equation should be considered:

$$E = E^0 + \frac{RT}{nF} \ln \frac{f_{Ox} C_{Ox}(0, t)}{f_{Red} C_{Red}(0, t)}. \tag{7.1}$$

This equation relates the electrode potential to the concentrations of the oxidized and reduced forms at the electrode surface; f_{Ox} and f_{Red} are the activity coefficients of these forms. Provided that these concentrations can be related to the polarographic electrolysis current the required equation of the polarographic curve can be obtained. The Ilkovič equation should be used to derive the concentration dependence of the current. If the reduction process: $Ox + ne \to Red$ is considered and it is assumed that only the Ox form is present in the solution at the commencement of the experiment, as is usually the case in practice, the following relationship is obtained:

$$\bar{i} = 605 n m^{2/3} D_{Ox}^{1/2} t_1^{1/6} [C_{Ox}^0 - C_{Ox}(0, t)] \tag{7.2}$$

where $\bar{i}$ is the mean current at any point on the polarographic wave, and $C_{Ox}(0, t)$ is the concentration of the Ox form at the electrode surface.

It follows from equation (7.2) that, when $C_{Ox}(0, t) = C_{Ox}^0$, i.e. at a potential more positive than that of the foot of the wave, the current equals zero, but when $C_{Ox}(0, t) = 0$ the current is independent of the potential and is equal to the limiting current. The change of $C_{Ox}(0, t)$ between the values 0 and C_{Ox}^0 gives rise to the change of current intensity. Putting:

$$K_{Il(Ox)} = 605 n m^{2/3} D_{Ox}^{1/2} t_1^{1/6} \tag{7.3}$$

where $K_{Il(Ox)}$ is the Ilkovič constant, we obtain equation (7.2) in the simpler form:

$$\bar{i} = K_{Il(Ox)} [C_{Ox}^0 - C_{Ox}(0, t)]. \tag{7.4}$$

The substance Red, resulting from reduction at the electrode, diffuses away from it, and the concentration of this substance at the electrode surface is proportional to the current intensity. It has a maximum value when the concentration of the substance Ox at the electrode surface equals zero. This can be expressed as:

$$\bar{i} = 605 n m^{2/3} D_{Red}^{1/2} t_1^{1/6} C_{Red}(0, t). \tag{7.5}$$

Similarly:
$$K_{\text{Il(Red)}} = 605nm^{2/3}D_{\text{Red}}^{1/2}t_1^{1/6},\tag{7.6}$$

and equation (7.5) can be simplified to:
$$\bar{i} = K_{\text{Il(Red)}}C_{\text{Red}}(0,t).\tag{7.7}$$

It follows from equations (7.3) and (7.6) that the Ilkovič constants $K_{\text{Il(Ox)}}$ and $K_{\text{Il(Red)}}$ differ only by the diffusion coefficients.

Introducing $C_{\text{Ox}}(0,t)$ and $C_{\text{Red}}(0,t)$, calculated from equations (7.4) and (7.7) into equation (7.1) gives the relationship:

$$E = E^0 + \frac{RT}{nF}\ln\frac{f_{\text{Ox}}(K_{\text{Il(Ox)}}C_{\text{Ox}}^0 - \bar{i})K_{\text{Il(Red)}}}{f_{\text{Red}}K_{\text{Il(Ox)}}\bar{i}}.\tag{7.8}$$

Since the mean limiting current is:
$$\bar{i}_g = K_{\text{Il(Ox)}}C_{\text{Ox}}^0,\tag{7.9}$$

equation (7.8) can be expressed as follows:

$$E = E^0 + \frac{RT}{nF}\ln\frac{f_{\text{Ox}}K_{\text{Il(Red)}}}{f_{\text{Red}}K_{\text{Il(Ox)}}} + \frac{RT}{nF}\ln\frac{\bar{i}_g - \bar{i}}{\bar{i}}.\tag{7.10}$$

It has already been shown that the difference between the constants of the oxidized and reduced forms is caused by the different diffusion coefficients only; therefore equation (7.10) can be expressed in the form:

$$E = E^0 + \frac{RT}{nF}\ln\frac{f_{\text{Ox}}D_{\text{Red}}^{1/2}}{f_{\text{Red}}D_{\text{Ox}}^{1/2}} + \frac{RT}{nF}\ln\frac{\bar{i}_g - \bar{i}}{\bar{i}}.\tag{7.11}$$

This equation represents the dependence of current on the electrode potential in the case of a reversible process, i.e. it describes polarographic reduction waves of a reversible process.

The equation of the polarographic wave was derived for the first time by Heyrovský and Ilkovič [1]. Since this equation gives the interval of polarographic wave potentials, the assignment of any definite characteristic potential to a depolarizer investigated by means of this equation is rather difficult. However when $\bar{i} = \frac{1}{2}\bar{i}_g$ equation (7.11) is reduced to the relationship:

$$E_{1/2} = E^0 + \frac{RT}{nF}\ln\frac{f_{\text{Ox}}D_{\text{Red}}^{1/2}}{f_{\text{Red}}D_{\text{Ox}}^{1/2}}.\tag{7.12}$$

Potential $E_{1/2}$ measured at the half-height of the recorded wave is independent of the depolarizer concentration and therefore, if the half-wave potential is known, it is frequently possible to identify the depolarizer.

An increase of the depolarizer concentration results in the shift of the deposition potential in the direction of positive potentials, but the half-wave potential remains unchanged. Introducing the half-wave potential into equation (7.11):

$$E = E_{1/2} + \frac{RT}{nF} \ln \frac{\bar{i}_g - \bar{i}}{\bar{i}}. \tag{7.13}$$

Equation (7.13) can also be written in the following form:

$$\bar{i} = \frac{\bar{i}_g}{1 + \exp \frac{nF}{RT}(E - E_{1/2})}. \tag{7.14}$$

It follows from this equation that at potentials sufficiently positive with respect to $E_{1/2}$ $\bar{i} = 0$ and at potentials sufficiently negative with respect to the half-wave potential the exponential term of equation (7.14) practically equals zero and $\bar{i} = \bar{i}_g$.

In a similar manner anodic waves of oxidation processes, such as oxidation of metals dissolved in mercury, can be considered, e.g. oxidation of zinc amalgam, or oxidation of substances dissolved in water. In this case the anodic current is described by the relationship:

$$-\bar{i} = K_{\text{II(Red)}}[C_{\text{Red}}^0 - C_{\text{Red}}(0, t)]. \tag{7.15}$$

Assuming that the Ox form is not present in the solution before the start of the experiment, but is formed during the oxidation process, the following relationship is valid:

$$\bar{i} = K_{\text{II(Ox)}} C_{\text{Ox}}(0, t). \tag{7.16}$$

Introducing concentrations $C_{\text{Ox}}(0, t)$ and $C_{\text{Red}}(0, t)$ derived from equations (7.15) and (7.16) into the Nernst equation the following relationship is obtained:

$$E = E^0 + \frac{RT}{nF} \ln \frac{f_{\text{Ox}} D_{\text{Red}}^{1/2} \bar{i}}{f_{\text{Red}} D_{\text{Ox}}^{1/2} (_{\text{anod}}\bar{i}_g - \bar{i})}, \tag{7.17}$$

or, utilizing relationship (7.12):

$$E = E_{1/2} + \frac{RT}{nF} \ln \frac{\bar{i}}{_{\text{anod}}\bar{i}_g - \bar{i}}. \tag{7.18}$$

In these equations $_{\text{anod}}\bar{i}_g$ is the anodic limiting current which is described by the equation:

$$-_{\text{anod}}\bar{i}_g = K_{\text{II(Red)}} C_{\text{Red}}^0. \tag{7.19}$$

Sometimes both forms of a reversible system are present in the solution before the start of the electrolysis. A solution containing oxalate complexes of ferrous and ferric ions, which react reversibly at a mercury electrode, is an example of such a system. Another example is a solution containing, in addition to the supporting electrolyte, a cadmium salt in low concentration, the dropping electrode being a dilute cadmium amalgam. In both examples the limiting oxidation current is observed when the applied potential is much more positive than the standard potential of the system. When the potential changes in the direction of more negative values, the current intensity decreases in the vicinity of the standard potential, but when the zero value is reached, the further polarization of the electrode to negative potentials results in the appearance of a cathodic current. The intensity of this current increases with increasing negative electrode potential. The formation of such waves is shown schematically in Fig. 7.1.

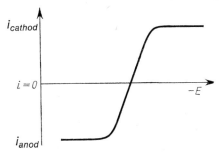

Fig. 7.1 Anodic–cathodic polarographic wave; i_{anod}—anodic current; i_{cathod}—cathodic current.

For such a case equation (7.4) and that given below are valid:

$$\bar{i} = K_{II(Red)}[C_{Red}(0, t) - C_{Red}^0]. \tag{7.20}$$

When $C_{Red}(0, t)$ equals zero (at potentials more positive with respect to the standard potential of the system), equation (7.20) is reduced to expression (7.19). Expressing $C_{Ox}(0, t)$ and $C_{Red}(0, t)$ by means of equations (7.4) and (7.20) and introducing these concentrations into the Nernst equation gives the anodic–cathodic wave equation:

$$E = E^0 + \frac{RT}{nF} \ln \frac{f_{Ox} K_{II(Red)} (K_{II(Ox)} C_{Ox}^0 - \bar{i})}{f_{Red} K_{II(Ox)} (K_{II(Red)} C_{Red}^0 + \bar{i})}, \tag{7.21}$$

or, in a shorter form:

$$E = E_{1/2} + \frac{RT}{nF} \ln \frac{\bar{i}_g - \bar{i}}{_{anod}\bar{i}_g + \bar{i}}. \tag{7.22}$$

The above equation is more general than the earlier equations describing the anodic and cathodic waves separately. This can be easily proved since, when the reduced form is not present in the solution, $_{\text{anod}}\bar{i}_g = 0$ and expression (7.22) is reduced to equation (7.13). Similarly, when only the reduced substance is present in the solution $\bar{i}_g = 0$, and equation (7.22) becomes identical with the anodic wave equation (7.18).

It follows from an analysis of equation (7.22) that the half-wave potential of the anodic–cathodic wave is determined by the potential at which:

$$\bar{i} = \frac{\bar{i}_g + {_{\text{anod}}\bar{i}_g}}{2}, \qquad (7.23)$$

since at this point the logarithmic factor of equation (7.22) equals one.

The equations discussed above indicate that in all the anodic and cathodic processes the half-wave potential is independent of the depolarizer concentration.

When several depolarizers are present in the solution, a polarographic curve with several steps is obtained, provided that the potentials of the depolarizers are sufficiently separated from one another. Well formed limiting current "plateaus" of the waves are observed when the depolarizer half-wave potentials differ by at least $200/n$ mV. If the half-wave potential differences are much smaller, one common wave is obtained and its height is proportional to the total concentration of the depolarizers.

The wave equation was derived for the first time by Heyrovský and Ilkovič. On the basis of this equation it is possible to analyse experimental polarographic curves. Considering, for the sake of simplicity, the cathodic wave equation (7.13), it is evident that the graph of the dependence of $\log(\bar{i}_g - \bar{i})/\bar{i}$ on the potential should be a straight line [2]. The slope of this straight line depends on the number of electrons n exchanged in the elementary process. The relationship is shown schematically in Fig. 7.2.

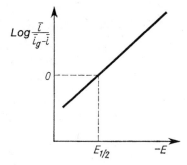

Fig. 7.2 Potential dependence of $\log(\bar{i}_g - \bar{i})/\bar{i}$ for a reversible process.

According to equation (7.13), the slope of the straight line is described by the relationship:

$$\frac{\Delta E}{\Delta \log \frac{\bar{i}_g - \bar{i}}{\bar{i}}} = \frac{2.303\,RT}{nF}. \tag{7.24}$$

Similar linear dependences are observed in the cases of anodic and anodic–cathodic waves. In the case of an anodic process the values of $\log \frac{\bar{i}}{{}_{\text{anod}}\bar{i}_g - \bar{i}}$ and in the case of an anodic–cathodic wave the values of $\log \frac{\bar{i}_g - \bar{i}}{-{}_{\text{anod}}\bar{i}_g + \bar{i}}$ are the ordinates.

Two principal conclusions can be drawn from such analyses. The first concerns the reversibility of the process. If the number of electrons exchanged in the elementary process is known, the theoretical slope of the straight line for the reversible system is easily found. Deviations from values predicted by the relationship (7.24) indicate irreversibility or *quasi*-reversibility of the electrode process. If the process is known to be reversible, it is easy to determine from the slope of the straight line in Fig. 7.2, the number of electrons exchanged in the elementary process. At 25°C the slopes for one, two and three electron processes are 59, 28.5 and 19.7 mV per logarithm unit, respectively. Even when no information about the reversibility of the process is available, but the result of analysis of the logarithmic polarographic wave is in accord with any of the above three conditions, it can be concluded with a high degree of probability that the process is reversible and the value of n can be estimated.

Plotting $\log(\bar{i}_g - \bar{i})/\bar{i}$ against the potential gives an accurate determination of the half-wave potential (this is the second conclusion). It follows from the wave equations that the half-wave potential corresponds to the potential at which the logarithmic term equals one. This method of determination of the half-wave potential is shown schematically in Fig. 7.2.

Another method for the determination of the reversibility of the process and the number of electrons, n, on the basis of the polarographic curve, is known as the Tomeš criterion. It follows from equation (7.13) that the potential measured at 1/4 of the wave height, like the half-wave potential, is independent of concentration and can be described by the equation:

$$E_{1/4} = E_{1/2} + \frac{RT}{nF} \ln 3. \tag{7.25}$$

The potential measured at 3/4 of the wave height is:

$$E_{3/4} = E_{1/2} - \frac{RT}{nF} \ln 3. \tag{7.26}$$

It follows from relationships (7.25) and (7.26) that the polarographic wave of a reversible process is symmetrical with respect to the half-wave potential, since $E_{1/4}$ and $E_{3/4}$ differ from it by the same number of mV. If equation (7.26) is subtracted from equation (7.25), it gives the relationship:

$$E_{1/4} - E_{3/4} = \frac{2RT}{nF} \ln 3, \tag{7.27}$$

and for 25°C:

$$E_{1/4} - E_{3/4} = \frac{0.056}{n} \text{ V}. \tag{7.28}$$

It follows from equation (7.28) that the difference between potentials $E_{1/4}$ and $E_{3/4}$ depends only on the number of electrons exchanged in the elementary process; for a one-electron process the difference equals 56 mV, for a two-electron process 28 mV, and so on.

The Tomeš criterion makes it possible to detect rapidly and accurately any deviation from reversibility of any process under consideration and to determine the value of n for a reversible process.

This discussion of polarographic wave equations has been based on the simple Ilkovič equation, which does not take into account the effect caused by the spherical symmetry of the diffusion. This was justified, since this approach made it possible to obtain very simple equations, giving close approximation to the experimental results. Koutecký [3] showed that when sperically symmetrical diffusion is taken into account, the wave equation is similar to the equations derived earlier in this chapter, and only the half-wave potential is described by a more complex relationship. For a polarographic reduction:

$$E = E'_{1/2} + \frac{RT}{nF} \ln \frac{\bar{i}_g - \bar{i}}{\bar{i}} + \frac{RT}{nF} \ln \frac{D_{\text{Red}}^{1/2}}{D_{\text{Ox}}^{1/2}} \tag{7.29}$$

where:

$$E'_{1/2} = E^0 - \frac{RT}{nF} 3.4 \frac{t_1^{1/6}}{m^{1/3}} (D_{\text{Ox}}^{1/2} + D_{\text{Red}}^{1/2}). \tag{7.30}$$

In this case the half-wave potential depends on the conditions of the experiment, namely on the parameters m and t of the mercury drop. The effect of the properties of the capillary on the half-wave potential is controversial [4, 5]. It should be mentioned that this effect is difficult to observe, since, as follows from equation (7.30), the effect of drop parameters on $E'_{1/2}$ is slight and possible changes of half-wave potential caused by the changes of m and t are so small that they do not exceed the experimental error.

The case of reduction of an ion to the metallic state should also be discussed, assuming that the resulting metal does not dissolve in mercury. The potential of such a system is described by the equation:

$$E = E_M^0 + \frac{RT}{nF} \ln f_{ox} C_{ox}(0, t) \tag{7.31}$$

where E_M^0 is the standard potential of the reaction $Me^{n+} + ne \rightarrow Me$.

In this case the concentration $C_{ox}(0, t)$ can be expressed by means of equation (7.4). Combining equations (7.21) and (7.4):

$$E = E_M^0 + \frac{RT}{nF} \ln \frac{f_{ox}}{K_{II(Ox)}} + \frac{RT}{nF} \ln(\bar{i}_g - \bar{i}). \tag{7.32}$$

Thus the potential corresponding to the half of the limiting current value is:

$$E_{1/2} = E_M^0 + \frac{RT}{nF} \ln \frac{f_{ox}}{K_{II(Ox)}} + \frac{RT}{nF} \ln \frac{\bar{i}_g}{2}, \tag{7.33}$$

or, taking into account relationship (7.9):

$$E_{1/2} = E_M^0 + \frac{RT}{nF} \ln f_{ox} + \frac{RT}{nF} \ln \frac{C_{ox}^0}{2}. \tag{7.34}$$

It follows from equation (7.34) that, in such a case, contrary to the cases discussed earlier, the half-wave potential is a function of the depolarizer concentration in the bulk of the solution. Tenfold increase in depolarizer concentration causes the shift of the half-wave potential in the positive direction by $2.303RT/nF$ V. Exact experimental verification of equation (7.32) is rather difficult because when a mercury electrode is used it is difficult to find a metal which is practically insoluble in the electrode and which gives ions reversibly reducible at the mercury electrode. The use of solid electrodes, for instance platinum, is unsatisfactory, since during the initial period, when the platinum is not yet fully covered by the reduced metal, the activity of the metal is variable. This is contradictory to equation (7.32) from which, in this case, the wave equation was derived.

7.1.1.1 Interpretation of Half-Wave Potentials

It follows from equation (7.12) that the half-wave potential should be very close to the standard potential. The difference between these potentials arises from differences in activity and diffusion coefficients of the oxidized and reduced forms. The activity coefficients of mercury-soluble metals are close to unity, since the concentrations of the amalgams produced during the polarographic reduction are usually very low. Also the activity coefficient of the oxidized form can be close to unity when the concentration of the supporting electrolyte is not very high. These coefficients may differ somewhat from each other, but their effect on the half-wave potential is small. This is because the decimal logarithm of the ratio of activity coefficients is multiplied by $59/n$ mV (for 25°) in the expression which determines the half-wave potential.

The effect of the difference between diffusion coefficients on the half-wave potential is also small. In moderately concentrated supporting electrolyte solutions the diffusion coefficients of the oxidized and reduced forms can differ from each other by 100 per cent but this difference is frequently much smaller, especially when both the oxidized and the reduced forms are soluble in the solution.

The effect of these differences is minimized by the fact that in the equation relating $E_{1/2}$ to E^0 the ratio of diffusion coefficients appears as the square root. Thus the half-wave potential should be approximately equal to the standard potential. This is the case when the oxidized and reduced forms are soluble in the solution. When the reduced form gives an amalgam the half-wave potential in certain cases differs appreciably from the standard potential of the ion/metal system.

Table 7.1 NORMAL POTENTIALS AND HALF-WAVE POTENTIALS OF VARIOUS DEPOLARIZERS, vs. SCE

Depolarizer	Normal potential $E_M^0(V)$	Half-wave potential $E_{1/2}(V)$
Ba^{2+}	−2.90	−1.94
Na^+	−2.961	−2.12
K^+	−3.170	−2.14
Pb^{2+}	−0.372	−0.388
Tl^+	−0.582	−0.459
Zn^{2+}	−1.008	−0.997

The standard and the half-wave potential values of a number of ions are shown in Table 7.1 from which it can be seen that in certain cases the difference between E_M^0 and $E_{1/2}$ can reach the value of one volt. In other cases, in spite of the solubility of the metal in mercury, it is small. The difference between E_M^0 and $E_{1/2}$ potentials is due mainly to the affinity of the metal for mercury.

The problem of quantitative description of this difference was worked out independently by Lingane [6] and by von Stackelberg [7].

Consider a cell in which one of the half-cells consists of a solid metal Me and its ions Me^{n+}, and the other of saturated two-phase Me amalgam which is also in equilibrium with the same solution. This cell can be expressed as:

$$Me|Me^{n+}|Me(Hg)_{sat}. \quad (7.35)$$

Call the activity of the metal in the saturated amalgam a_{sat}. The solid phase of this saturated amalgam can be formed by pure metal Me or by a solid intermetallic compound of Me with mercury. The electromotive force ΔE of such a cell can be represented by the relationship:

$$\Delta E = E^0 - E_M^0 + \frac{RT}{nF} \ln \frac{a_{Hg}}{a_{sat}} \quad (7.36)$$

where E_M^0 is the standard potential of the left-hand half of the cell system (7.35), a_{Hg} is the activity of mercury in the saturated amalgam, referred to the activity of pure mercury.

Since a reversible process is being considered:

$$-\Delta G = nF\Delta E \quad (7.37)$$

where ΔG is the thermodynamic free energy change of the reaction of the metal with mercury. When the metal does not form intermetallic compounds or solid solutions with mercury, $\Delta G = 0$ and $\Delta E = 0$.

Utilizing relationship (7.37) and assuming that the activity of mercury in the amalgam does not differ appreciably from that of pure mercury, equation (7.36) can be expressed in a different form:

$$E^0 = E_M^0 - \frac{\Delta G}{nF} + \frac{RT}{nF} \ln a_{sat}. \quad (7.38)$$

Combining equations (7.38) and (7.4) gives the relationship:

$$E_{1/2} = E_M^0 - \frac{\Delta G}{nF} + \frac{RT}{nF} \ln a_{sat} + \frac{RT}{nF} \ln \frac{f_{Ox} D_{Red}^{1/2}}{f_{Red} D_{Ox}^{1/2}}. \quad (7.39)$$

When it is possible to assume that $D_{Red}^{1/2} = D_{Ox}^{1/2}$ and $f_{Ox} = f_{Red}$ equation (7.39) is reduced to the simpler relationship:

$$E_{1/2} = E_M^0 - \frac{\Delta G}{nF} + \frac{RT}{nF} \ln a_{sat}. \quad (7.40)$$

When the metal is readily soluble in mercury and no compounds of these two metals are formed in the solid phase, then $E_{1/2}$ is close to E_M^0, since in such a case $\Delta G = 0$ and a_{sat} is close to one; zinc serves as an example of such behaviour. If the metal forms a compound with mercury, then $\Delta G < 0$ and $E_M^0 < E_{1/2}$; alkali metals shown in Table 7.1 are examples of the latter case. When the metal is very sparingly soluble in mercury but does not form any compounds with it in the solid phase, $\Delta G = 0$ and $E_M^0 > E_{1/2}$.

The work published by Lingane and von Stackelberg was discussed by Turyan [8] who pointed out minor errors in the expressions relating $E_{1/2}$ to E_M^0 given by these authors.

7.1.2 Polarographic Wave Equations of Irreversible Electrode Processes

Polarographic wave equations of irreversible electrode processes, derived by several authors, have been discussed in the preceding chapter. The most exact equation is that derived by Koutecký [9]. Delahay and Strassner [10] are the authors of another study of this problem.

Kern [11] deduced from Koutecký's work that the dependence of $\log i(i_g - i)$ on E should be linear, and that the slope of the straight line at 25°C should be equal to $0.059/\alpha n$. This result was accepted by some authors [12, 13] but other studies led to the conclusion that this dependence should have the form of a curve, and that the slopes of the asymptotes of this curve at potentials corresponding to the beginning and to the end of the polarographic wave should be in 2:1 ratio. Kivalo, Oldham and Laitinen [14] confirmed the latter conclusion experimentally, but later it was shown that their experiment was in error, as a surface active substance was present in the solution investigated.

The problem was considered again by Meites and Israel [15] and by Behr et al. [16], who also used the results of Koutecký's work as the basis for the study. Introduction of an auxiliary variable Δ made it possible to relate it to parameter χ by the following equation:

$$\Delta = \sqrt{\frac{7}{12}} \chi = k_{fh} \frac{t_1^{1/2}}{D_{\text{Ox}}^{1/2}}. \tag{7.41}$$

Equation (7.41) takes into account the reaction in one direction only and therefore it can be utilized only in the case of fully irreversible processes.

Polarography

The constant k_{fh} depends on the electrode potential:

$$k_{fh} = k_{fh}^0 \exp\left(-\frac{\alpha n_a FE}{RT}\right). \tag{7.42}$$

Combining equations (7.41) and (7.42) gives the relationship:

$$E = -\frac{0.434 RT}{\alpha n_a F} \log \frac{k_{fh}^0 t_1^{1/2}}{D_{Ox}^{1/2}} - \frac{0.434 RT}{\alpha n_a F} \log \Delta. \tag{7.43}$$

Using tables of Koutecký's function $F(\chi)$ it can be shown that, over almost all the region in which the polarographic wave is formed, $\log \Delta$ is a linear function of $\log i/(i_g - i)$. For limiting currents this relationship is described by the equation:

$$\log \Delta = -0.1300 + 0.9163 \log \frac{i}{i_g - i}, \tag{7.44}$$

which is valid for $0.1 \leqslant i/i_g \leqslant 0.94$. The average value of the error in the determination of $\log i/(i_g - i)$ is thus only ± 0.0043. This error is smaller than the experimental errors of polarographic current measurement.

Combining equations (7.43) and (7.44) gives, for 25°C, the following expression:

$$E = \frac{0.059}{\alpha n_a} \log \frac{1.349\, k_{fh}^0 t_1^{1/2}}{D_{Ox}^{1/2}} - \frac{0.0542}{\alpha n_a} \log \frac{i}{i_g - i}. \tag{7.45}$$

Thus when t_1 and α are constant and independent of potential, $\log i/(i_g - i)$ changes linearly with potential and the slope of the corresponding straight line is $-0.0542/n_a$ V.

Equation (7.45) can be formulated as follows:

$$E = E_{1/2} - \frac{0.0542}{\alpha n_a} \log \frac{i}{i_g - i} \tag{7.46}$$

where

$$E_{1/2} = \frac{0.059}{\alpha n_z} \log \frac{1.349\, k_{fh}^0 t_1^{1/2}}{D_{Ox}^{1/2}}. \tag{7.47}$$

It can easily be shown that:

$$E_{1/4} - E_{3/4} = \frac{0.0517}{\alpha n_a}. \tag{7.48}$$

It follows from equation (7.47) that, in contrast to reversible waves, the half-wave potentials of irreversible processes are drop time dependent.

According to equation (7.46) the slope of the plot of $\log i/(i_g - i)$ vs. E is smaller than in the case of reversible waves. Analysing the wave of a cathodic process, the value of αn_z can be estimated from the angular coefficient of the straight line. Accordingly to equation (7.47), this value can also be determined from the dependence of $E_{1/2}$ on the drop time. When average currents are considered, the logarithmic term $\log i/(i_g - i)$ is preceded by the coefficient $0.059/\alpha n_\alpha$ (for 25°C). This coefficient had been found earlier by Kern [11].

In the equations discussed above the effect of spherical diffusion to the dropping mercury electrode has not been considered. This can be done by taking into account the results of the studies by Koutecký and Čižek [17], and introducing the following correction to the $F(\chi)$ function:

$$\frac{i}{i^x} = F(\chi) - \varepsilon_0 H(\chi) \qquad (7.49)$$

where

$$\varepsilon_0 = 50.4 D_{Ox}^{1/2} t^{1/6} m^{1/3}. \qquad (7.50)$$

$H(\chi)$ function values have been tabulated by Koutecký and Čižek.

Meites and Israel have shown that only minor changes are caused by the introduction of the spherical effect into the equation describing the potential dependence of $\log i/(i_g - i)$. The change of ε_0 from 0 to 0.2 (the interval of ε_0 values observed in practice) results in only 1.5% change of the slope of the E vs. $\log i/(i_g - i)$ graph and a 10 mV change in the half-wave potential. Thus it appears that, in practice, the use of corrections for spherical diffusion is not always necessary, especially when the ε_0 values are much lower than 0.2.

The discussion above was limited to reduction processes, but expressions describing irreversible anodic waves can be similarly obtained.

7.1.3 Determination of Kinetic Parameters of an Electrode Process by Polarography

In the case of an irreversible polarographic process, the analysis of polarographic waves leading to the determination of the kinetic parameter αn_α and] the rate constant of the process is fairly simple, since by using equation (7.46) it is possible to determine the value of αn_z from the slope of the E vs. $\log i/(i_g - i)$ graph. From the relation obtained for $\log i/(i_g - i) = = 0$, the half-wave potential determined by equation (7.47) can be calculated. Introducing the value of αn_z, previously found, into equation (7.47), it is possible to determine the value of k_{fh}^0.

The k_{fh}^0 constant determines the rate of the cathodic reaction at a potential assumed to be the zero potential. Usually it is the standard potential of the normal hydrogen electrode. Even when the value of k_{fh}^0 is known, the determination of the standard rate constant of the electrode process may be difficult since the exact determination of the standard potential in this case may be difficult.

The determination of the rate constants of an electrode process controlled by both diffusion and charge transfer is more complex. Several procedures for this exist. That of Randles [18, 19] concerns the complex *quasi*-reversible anodic–cathodic process (it is assumed that both the Ox and the Red forms are present in the solution before start of the experiment). Apart from this exact procedure an approximate method has been proposed by Stromberg [20].

These two methods are of minor practical importance, since they involve the necessity of recording the complex anodic–cathodic waves. In practice it is easier to prepare a solution of only one form, for instance the Ox form, and record the reduction wave of this alone. Analysis of such *quasi*-reversible waves is presented in the papers of Hale and Parsons [21, 22] and in that of Sathyanarayana [23].

Consider a reversible electrode reaction of the first order:

$$\text{Ox} + ne \rightleftharpoons \text{Red}$$

and assume that the Ox form is soluble in the solution whereas the Red form is soluble either in the solution or in the electrode material (a metal soluble in mercury): then, according to Randles, the instantaneous current at the end of the drop time is described by the equation:

$$\frac{i}{i^x} = \frac{1-\exp(nf\eta)}{1+r\exp(nf\eta)} F(\chi) \qquad (7.51)$$

where i is the instantaneous current at the end of the drop life, i^x is the current which would be observed if the process were controlled by the transport rate alone, r is the ratio of the cathodic to the anodic limiting currents ($_{cat}i_g/_{anod}i_g = r$), $\eta = E - E^e$ where E is the potential at which the current i_t is measured and E^e is the electrode equilibrium potential; $f = F/RT$, and n is the number of electrons transferred in the electrode reaction rate determining step.

The Koutecký function $F(\chi)$ was described in the preceding chapter and its values were given in tables.

Consider the case when the substance Red is formed exclusively as a result of the electrode reaction and is not present in the solution before the start of the experiment. Since $_{anod}i_g = K_{11}C_{Red}^0$, and $C_{Red}^0 = 0$, the values of r

become infinitely large and equation (7.51) cannot be utilized in analysis of the kinetics of such a process.

But from the Nernst equation it follows that:

$$C_{Ox}^0 = C_{Red}^0 \exp[nf(E^e - E_f^0)], \qquad (7.52)$$

and from the Ilkovič equation written in the form:

$$_{anod}i_g = \sqrt{\frac{7}{3\pi}} \, nFAD_{Red}^{1/2} t^{-1/2} C_{Red}^0, \qquad (7.53)$$

the following relationship is obtained:

$$r \exp(nf\eta) = \left(\frac{D_{Ox}}{D_{Red}}\right)^{1/2} \exp[nf(E - E_f^0)]. \qquad (7.54)$$

Combining equations (7.51) and (7.54) gives:

$$\frac{i}{i^x} = \frac{[1 - \exp(nf\eta)]F(\chi)}{1 + \left(\dfrac{D_{Ox}}{D_{Red}}\right)^{1/2} \exp[nf(E - E_f^0)]}. \qquad (7.55)$$

If the formal potential E_f^0 is known, or if it can be calculated, it is possible to calculate $F(\chi)$ by using equation (7.55) and the experimentally determined i and i^x values.

Having calculated $F(\chi)$, the parameter χ for the potentials at which i was measured can be read from the tables of this function. The relation between the reduction rate constant and parameter χ is expressed by the equation:

$$\chi = \sqrt{\frac{12t}{7D_{Ox}}} \, k_{fh} \left\{ 1 + \left(\frac{D_{Ox}}{D_{Red}}\right)^{1/2} \exp[nf(E - E_f^0)] \right\}. \qquad (7.56)$$

The k_{fh} values calculated for various potential E values are plotted on a graph of $\log k_{fh}$ against E according to the equation:

$$k_{fh} = k_s \exp[-\alpha n_\alpha f(E - E_f^0)]. \qquad (7.57)$$

This relationship should be linear. From the slope of the straight line the transition coefficient for the cathodic process can be calculated. The k_{fh} value for $E = E_f^0$ is the required standard rate constant k_s of the electrode process.

A simple method of calculating the rate constant of the electrode process from polarographic measurement was worked out by Koryta [24]. From Koutecký's function:

$$\overline{F}(\chi_1) = \frac{i}{i^x} = \frac{0.676\chi_1}{1 + 0.676\chi_1} \qquad (7.58)$$

the following relationship is obtained:

$$\frac{\bar{i}}{\bar{i}^x - \bar{i}} = 0.676\chi_1, \qquad (7.59)$$

where:

$$\chi_1 = \sqrt{\frac{12 t_1}{7}} \; (k_{fh} D_{Ox}^{-1/2} + k_{bh} D_{Red}^{-1/2}). \qquad (7.60)$$

The rate constants k_{fh} and k_{bh} can be referred to the reversible half-wave potential, using equation (7.57); a similar relationship exists for k_{bh} and equation (7.12). Thus:

$$k_{fh} = k_s \left(\frac{D_{Ox}}{D_{Red}}\right)^{\alpha/2} \exp\left[-\alpha nf(E - E_{1/2})\right], \qquad (7.61)$$

$$k_{bh} = k_s \left(\frac{D_{Red}}{D_{Ox}}\right)^{\frac{1-\alpha}{2}} \exp\left[(1-\alpha)nf(E - E_{1/2})\right]. \qquad (7.62)$$

Call the current measured at the half-wave potential of a reversible process $\hat{i}$. Substituting k_{fh} and k_{bh} in equation (7.60) by (7.61) and (7.62) gives:

$$k_s = \frac{\hat{i}}{\hat{i}^x - \hat{i}} \times \frac{D_{Red}^{\alpha/2} D_{Ox}^{\frac{1-\alpha}{2}}}{2 \times 0.886 t_1^{1/2}}. \qquad (7.63)$$

But at $E = E_{1/2}$ the current of a reversible process is equal to half of the value of the limiting current: $i^x = \frac{1}{2} \bar{i}_g$ and therefore equation (7.63) can be expressed as follows:

$$k_s = \frac{\hat{i}}{\bar{i}_g - 2\hat{i}} \times \frac{D_{Red}^{\alpha/2} D_{Ox}^{\frac{1-\alpha}{2}}}{0.886 t_1^{1/2}}. \qquad (7.64)$$

The currents appearing in these equations are the average currents.

On the basis of Koryta's results it is possible to make a simple estimate of the kinetic parameters of an electrode process. For this purpose the relationship $\log \bar{i}/(\bar{i}_g - \bar{i})$ against E is plotted. It is already known that this relationship is linear for a reversible process. In the case of processes controlled by both the transport rate and the rate of charge transfer it is no longer linear. The curve of such a relationship for a *quasi*-reversible process is shown in Fig. 7.3.

At potentials at the foot of the polarographic wave, which are sufficiently positive, the measured current is practically equal to that observed in the case of a fully reversible process. Therefore the asymptote to the curve of the potentials of the foot of the polarographic wave cuts the potential axis at the reversible potential of the process. The transfer coefficient, α, of the cathodic process can be calculated from the slope of the asymptote

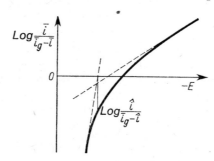

Fig. 7.3 Potential dependence of $\log \bar{i}/(\bar{i}_g - \bar{i})$ in the case of a *quasi*-reversible process (from [24] by permission of the copyright holders, Pergamon Press).

at the more negative potentials, as is shown in Fig. 7.3. In such conditions $k_{fh} \gg k_{bh}$.

The method can be used in studies of the kinetics of electrode processes, provided that their standard rate constants do not exceed 2×10^{-2} cm/s.

The problems of irreversible polarographic waves and their application to the determination of the kinetics of electrode processes have been studied by Matsuda and Ayabe [25]. They described the polarographic wave of a *quasi*-reversible process by the equation:

$$E = E_{1/2} + \frac{2.3RT}{\alpha nF} \log 0.885 t_1^{1/2} + \frac{2.3RT}{\alpha nF} \log \lambda - \frac{2.3RT}{\alpha nF} \log \frac{\bar{i}}{\bar{i}_g - \bar{i}} \quad (7.65)$$

where:

$$\lambda = k_s \frac{f_{Ox}^\beta f_{Red}^\alpha}{(D_{Ox}^\beta D_{Red}^\alpha)^{1/2}}, \quad (7.66)$$

where α and β are the transition coefficients of the anodic and the cathodic processes, respectively. Coefficient α can be determined by logarithmic analysis of the polarographic wave. At negative potentials the logarithmic relationship is linear and the coefficient αn is determined directly by the slope of the straight line:

$$-\alpha n = \frac{2.3RT}{nF} \left[\frac{\Delta \log \frac{\bar{i}}{\bar{i}_g - \bar{i}}}{\Delta E} \right]_{E \to -\infty}. \quad (7.67)$$

This method of determining α is identical to that of Koryta.

Potential $E'_{1/2}$ can be determined by extrapolation of the linear section of the logarithmic relationship to $\log \bar{i}/(\bar{i}_g - \bar{i}) = 0$. From this value and from that of the known or determined half-wave potential the parameter λ can be calculated:

$$\log \lambda = \frac{\alpha n F}{2.3 RT} (E'_{1/2} - E_{1/2}) - \log t_1^{1/2} + 0.053. \qquad (7.68)$$

The calculation of the standard rate constant from equation (7.66) is now simple.

This method for determining the kinetics of electrode processes was used by Gaur and Goswami [26] in their studies of the $Mn^{2+}/Mn(Hg)$ system.

Ružič et al. [27] have discussed the methods of determination of the kinetic parameters of electrode processes, suggested by Matsuda and Ayabe and by Koryta. These authors came to the conclusion that those methods and especially Koryta's method of reversible half-wave potential determination are not exact in the case of slow electrode reactions, and they presented a new graphical method of reversible half-wave and transition coefficient determination, which gives much better results. The advantage of this method is that the reversible half-wave potential is determined on the basis of an analysis of the whole polarographic wave. The authors tested this method by determining α and k_s for the reduction of Cu(II) and Zn(II) ions in various media and found that the method of Matsuda and Ayabe always gave low values of k_s whereas those obtained by Koryta's method were too high. The method of determining kinetic parameters of a *quasi-reversible* electrode process in the presence of complexing ligands was recently studied by Verdier et al. [28]. They discussed the theory of *quasi-reversible* reduction of complexes described by Matsuda [29]. This theory is rather complex, and so a simpler equation was suggested, which describes the dependence of the current on the potential and on the parameters of the process:

$$\frac{\bar{i}_g - \bar{i}}{\bar{i}} = \frac{\sum_{\gamma=0}^{N} \beta'_\gamma (C_x)^\gamma}{\sum_{p=0}^{N} \beta'_p (C_x)^p \exp\left[\frac{-\alpha_p n F}{RT}(E - E_p)\right]} + \exp\left[\frac{nF}{RT}(E - E_{1/2})\right]$$

(7.69)

where

$$\exp\left(\frac{\alpha_p n F E_p}{RT}\right) = 0.885 k_s \left(\frac{t_1}{D}\right)^{1/2},$$

α_p is the transfer coefficient of the MX_p complex, and β'_p is the measured stability constant of the complex. Studies of the reduction of zinc ions in the presence of thiocyanates and tartrates showed the usefulness of this simplified equation in analysis of the kinetics of *quasi*-reversible processes of complex compounds.

Oldham and Parry [30] have shown that polarographic curves of irreversible processes can be analyzed in yet another way in order to obtain kinetic parameters of the electrode reaction. They have found that the relationship:

$$\frac{0.059}{\alpha n} \log\left\{\frac{2x(3-x)}{5(1-x)}\right\} = E_{1/2} - E, \qquad (7.69a)$$

where $x = i/i_g$ (i is the current measured at potential E) and

$$E_{1/2} = E^0 + \frac{0.059}{n} \log\left(1.349 k_s \cdot \frac{t}{n}\right)$$

is in very good agreement with the relationships predicted by the theories developed by Koutecký and by Matsuda and Ayabe.

Equation (7.69a) is valid for polarographic instantaneous currents. In the case of average currents the following relationship is fulfilled:

$$\frac{0.059}{\alpha n} \log\left\{\frac{x(5.5-x)}{5(1-x)}\right\} = E_{1/2} - E. \qquad (7.69b)$$

Both relationships (7.69a) and (7.69b) are valid for 25°C.

Apart from the methods of half-wave potential determination discussed above the reversible half-wave potential of a *quasi*-reversible process can be calculated from the relationship [31]:

$$\lim_{i \to 0}\left[E - \frac{RT}{nF} \ln \frac{\bar{i}_g - \bar{i}}{\bar{i}}\right] = E_{1/2}. \qquad (7.70)$$

A chronocoulometric method has also been proposed for determination of electrode process rate constants [32]. An expression describing the current–time behaviour after a stepwise potential change has been given. After a certain time the charge is described by an asymptotic expression.

In the absence of adsorption, kinetic parameters of electrode reactions can be calculated from the slope and extrapolated intercepts with the $t^{1/2}$ axis of the relationship Q–$t^{1/2}$. This method has been improved by Lingane and Christie [33]. The analysis given by these authors is not limited to selected time intervals. In experimental studies they obtained by this method values of k_s which were in agreement with literature data for the system $Zn^{2+}/Zn(Hg)$ in 1 M $NaClO_4$.

Osteryoung and Osteryoung [34] after a rigorous consideration of the problem came to the conclusion that chronocoulometry is a better technique than chronoamperometry for determination of kinetic parameters of electrode reactions, since the determination of charge in very short times provides the data containing the largest amount of kinetic information.

Determination of kinetic parameters of electrode reactions on the basis of chronoamperometric measurements with the aid of a computer in the analysis was described by Niki *et al.* [35].

7.2 Stationary Electrode Voltammetry

7.2.1 REVERSIBLE ELECTRODE PROCESSES

The shape of stationary electrode voltammetric curves of reversible electrode processes was reported by Randles and Ševčik, but on the basis of their results it is difficult to construct theoretical curves for comparison with the experimental ones.

It is convenient, from a practical point of view, to describe the curve, not by a complex equation, but by the potential-dependence of current function $\chi(at)$, expressed in tabular form.

In the course of studies on stationary electrode voltammetry under cylindrical diffusion conditions, Nicholson [36] calculated the current function values for various potentials, and various values of the parameter $D^{1/2}/n^{1/2}V^{1/2}r_0$. Frankenthal and Shain [37] derived the potential dependence of the current function for reversible stationary electrode voltammetric processes taking place on spherical electrodes. In this case also the current function is dependent on parameter $D^{1/2}/n^{1/2}V^{1/2}r_0$. On the basis of these data theoretical stationary electrode voltammetric curves can be plotted, using the current function values for $D^{1/2}/n^{1/2}V^{1/2}r_0 = 0$, when the effect of the spherical or cylindrical symmetry of the diffusion can be neglected, due to large electrode dimensions, or as a result of fast scan rates.

Values of the $\chi(at)$ function multiplied by $\pi^{1/2}$ at various potentials, for a process carried out in linear diffusion conditions are given in Table 7.2.

Introducing these values into the general equation:

$$i = nFAC_{Ox}^0 \sqrt{\pi D_{Ox} a} \; \chi(at) \tag{7.71}$$

it is possible to plot the current–potential curve.

Table 7.2 Values of functions $\pi^{1/2}\chi(at)$ and $\Phi(at)$ of reversible electrode processes (from [40] by permission of the copyright holders, the American Chemical Society)

Potential* (mV)	$\pi^{1/2}\chi(at)$	$\Phi(at)$	Potential* (mV)	$\pi^{1/2}\chi(at)$	$\Phi(at)$
120	0.009	0.008	−5	0.400	0.548
100	0.020	0.019	−10	0.418	0.596
80	0.042	0.041	−15	0.432	0.641
60	0.084	0.087	−20	0.441	0.685
50	0.117	0.124	−25	0.445	0.725
45	0.138	0.146	−28.5	0.4463	0.7516
40	0.160	0.173	−30	0.446	0.763
35	0.185	0.208	−35	0.443	0.796
30	0.211	0.236	−40	0.438	0.826
25	0.240	0.273	−50	0.421	0.875
20	0.269	0.314	−60	0.399	0.912
15	0.298	0.357	−80	0.353	0.957
10	0.328	0.403	−100	0.312	0.980
5	0.355	0.451	−120	0.280	0.991
0	0.380	0.499	−150	0.245	0.997

* $(E - E_{1/2})n$.

The potentials in Table 7.2 are referred to half-wave potentials and therefore, by analysing this Table, simple relations between the peak and half-peak current potentials and the half-wave potentials can be found. Function $\chi(at)$ and the current–potential curve reach their maximum at a potential which is by $28.5/n$ mV more negative than the half-wave potential. This value is correct at 25°C since the values in Table 7.2 have been calculated for this temperature. The general relation between peak current potential E_p and the reversible polarographic half-wave potential has the following form:

$$E_p = E_{1/2} - (1.109 \pm 0.002)\frac{RT}{nF}. \qquad (7.72)$$

In practice the determination of the peak potential is not always accurate and therefore, for the determination of the depolarizer the half-peak potential is measured. It follows from Table 7.2 that, at 25°C, the peak current potential precedes the polarographic half-wave potential by $28/n$ mV. The general equation relating these two quantities has the following form:

$$E_{p/2} = E_{1/2} + 1.09\frac{RT}{nF}. \qquad (7.73)$$

These relationships can be used to ascertain whether the electrode process is reversible or not and to determine the number of electrons transferred in the elementary process. However, these relationships are not convenient in practice, since their use requires the knowledge of exact formal potential values. Moreover, even a very small discrepancy between actual and theoretical values of the reference electrode potentials, or the appearance of liquid junction potential, may result in erroneous conclusions. Therefore in practice it is convenient to utilize the difference of peak and half-peak current potentials. From equations (7.72) and (7.73) we obtain:

$$E_{p/2} - E_p = 2.20 \frac{RT}{nF}, \qquad (7.74)$$

and at 25°C

$$E_{p/2} - E_p = \frac{0.0565}{n} \text{ V}. \qquad (7.74a)$$

This is the simplest and probably the best criterion of the reversibility of the electrode process.

In addition to the above relationships, a logarithmic relationship, analogous with that used in polarography, has been used in analysis of stationary electrode voltammetric curves. This utilizes the current peak instead of the limiting current [38]. The method was criticized by Reinmuth [39], who found that an appreciable section of the log $(i_p - i)^2/i$–potential relationship is represented by a straight line having the slope $2.3RT/nF$.

Relationships (7.72) and (7.73) make it possible to make a rapid determination of the reversible polarographic half-wave potential. This potential can also be determined by another method: it follows from the analysis of the values of function $\chi(at)$, shown in Table 7.2, that the $E_{1/2}$ potential corresponds to 85.17% of the peak current value.

The relationships discussed above are valid only for stationary electrode voltammetric processes taking place under linear diffusion conditions, whereas in practice many stationary electrode voltammetric measurements are carried out using hanging drop mercury electrodes. In such cases, when the spherical electrodes are not large and the polarization rate is low, the deviations from linear diffusion can be so large that the above equations are no longer valid.

According to Nicholson and Shain [40], in such a case the current–potential relationship is represented by the equation:

$$i = nFAC^0_{Ox}\sqrt{\pi D_{Ox} a}\ \chi(at) + nFAD_{Ox}C^0_{Ox}\Phi(at)/r_0 \qquad (7.75)$$

Function $\chi(at)$ is identical with the current function given in Table 7.2.

It follows from comparison with equation (7.71) that the second term of the right-hand side of equation (7.75) makes a correction for spherical diffusion. As has been mentioned, this correction is small when the electrode radius is large. Its value, referred to that of the first term, decreases with increasing rate of polarization.

The values of the $\Phi(at)$ function are given in Table 7.2. The potential dependence of this function has the form of a polarographic wave. It becomes equal to one at fairly negative potentials with respect to the half-wave potential, is equal to zero at potentials much more positive than the half-wave potential, and reaches half of the maximum value at the half-wave potential.

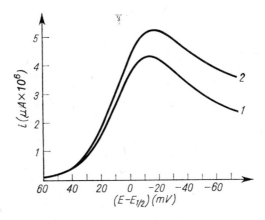

Fig. 7.4 Stationary electrode voltammetric current–potential curves of a reversible electrode process recorded by means of a flat electrode (curve *1*) and spherical electrode (curve *2*); $V = 0.2$ V/min., $r_0 = 0.05$ cm, $n = 2$, $D = 10^{-5}$ cm²/s.

The effect of spherical diffusion on the recorded curves is shown in Fig. 7.4. Curve *1* is the current–potential relationship corresponding to processes carried out under conditions of linear diffusion at a flat electrode having a surface area identical with that of a spherical electrode (curve *2*). The most important difference is the increase in peak current and in currents at other potentials. Furthermore, the reduction curve is slightly shifted in the direction of negative potentials. This shift increases with increasing value of the parameter $D^{1/2}/n^{1/2}V^{1/2}r_0$.

Both the peak and the half-peak potentials are shifted and, since the polarographic half-wave potential is independent of the experimental conditions, relationships (7.72) and (7.73) are no longer valid. Relationship

(7.74) is also no longer valid, although significant deviations from this relationship are observed only at comparatively high values of the parameter $D^{1/2}/n^{1/2}V^{1/2}r_0$. This is due to the fact that under spherical diffusion conditions the shift of peak potential in the negative direction is, as follows from relationship (7.75), larger than the shift of the half-peak potential.

It is rather difficult to describe the current–potential curve of a process carried out under conditions of spherical diffusion in such a compact form as equation (7.75), and therefore the functional relationships given by Nicholson [36] should be utilized. Since, for cylindrical diffusion, the deviations from the curves characteristic of linear diffusion are insignificant, even at high values of the parameter $D^{1/2}/n^{1/2}V^{1/2}r_0$, relationships (7.74) and (7.74a) can be utilized over a comparatively broad range of electrode radii and polarization rates.

The shapes of stationary electrode voltammetric curves are qualitatively different when the electrode process is carried out in limited diffusion zones. When the layer of the electrolysed solution is very thin and the polarization rate is slow, the peak potential is formed at the half-wave potential [41, 42]. The curve is characterized by a very rapid decrease of current to the zero value when the peak current is exceeded; this is due to the rapid disappearance of depolarizer from the bulk of the electrolysed solution.

7.2.2 Irreversible Electrode Processes

Stationary electrode voltammetric curves of irreversible processes were first described by Delahay [43] and later by Matsuda and Ayabe [44] and by Reinmuth [45]. These authors, however, did not give exact values of the current function $\chi(bt)$, with respect to the potential. The values of $\pi^{1/2}\chi(bt)$, shown in Table 7.3, were calculated by Nicholson and Shain [40].

Using the tabulated values of the $\chi(bt)$ function and equation:

$$i = FAC^0_{Ox}n\sqrt{D_{Ox}\pi b}\ \chi(bt) \tag{7.76}$$

it is possible to plot the theoretical current–potential curves of irreversible stationary electrode voltammetric processes.

It follows from Table 7.3, that the peak current is reached at -5.34 mV in the above potential scale; the peak potential is therefore described by the equation:

$$(E_p - E^0)\alpha n_z + \frac{RT}{F}\ln\frac{\sqrt{\pi D_{Ox} b}}{k_s} = -5.34\ \text{mV}. \tag{7.77}$$

Table 7.3 VALUES OF FUNCTIONS $\pi^{1/2}\chi(bt)$ AND $\Phi(bt)$ FOR IRREVERSIBLE ELECTRODE PROCESSES (FROM [40] BY PERMISSION OF THE COPYRIGHT HOLDERS, THE AMERICAN CHEMICAL SOCIETY)

Potential* (mV)	$\pi^{1/2}\chi(bt)$	$\Phi(bt)$	Potential* (mV)	$\pi^{1/2}\chi(bt)$	$\Phi(bt)$
160	0.003	0	15	0.437	0.323
140	0.008	0	10	0.462	0.396
120	0.016	0	5	0.480	0.482
110	0.024	0	0	0.492	0.600
100	0.035	0	−5	0.496	0.685
90	0.050	0	−5.34	0.4958	0.694
80	0.073	0.004	−10	0.493	0.755
70	0.104	0.010	−15	0.485	0.823
60	0.145	0.021	−20	0.472	0.895
50	0.199	0.042	−25	0.457	0.952
40	0.264	0.083	−30	0.441	0.992
35	0.300	0.115	−35	0.423	1.00
30	0.337	0.154	−40	0.406	1.00
25	0.372	0.199	−50	0.374	1.00
20	0.406	0.253	−70	0.323	1.00

* Potential scale is $(E-E^0)\alpha n_\alpha + \dfrac{RT}{F} \ln \dfrac{\sqrt{\pi D_{Ox}\, b}}{k_s}$ where $b = \alpha n_\alpha FV/RT$

This equation can be presented in a form directly describing the peak current potential:

$$E_p = E^0 - \frac{RT}{\alpha n_a F}\left[0.78 - \ln k_s + \ln \sqrt{D_{Ox} b}\,\right]. \tag{7.78}$$

It follows from this equation that the difference between the peak potential and the standard depolarizer potential increases with decreasing rate of charge transfer. E_p depends on parameter b, i.e. on the polarization rate. The dependence of E_p on $\ln V$ is linear and in the reduction process the half-wave potential becomes more negative when the polarization rate is increased.

It also follows from Table 7.3 that the current corresponding to the half-peak current is reached at the potential equal to 42.36 mV, according to the equation:

$$(E_{p/2} - E^0)\alpha n_a + \frac{RT}{F}\ln \frac{\sqrt{\pi D_{Ox} b}}{k_s} = 42.36 \text{ mV}. \tag{7.79}$$

Equations (7.78) and (7.79) can be utilized for the determination of the transfer coefficients of electrode processes, but this requires the measurement at a number of different polarization rates, of αn_a, obtained from the slope of the graph of E_p or $E_{1/2}$ vs. ln V. However, it is possible to obtain from equations (7.78) and (7.79) an expression which makes it possible to determine αn_a on the basis of only one experiment, because the difference of peak and half-peak potentials is related to parameter αn_z by the simple formula:

$$E_p - E_{p/2} = -1.857 \frac{RT}{\alpha n_\alpha F}, \qquad (7.80)$$

and at 25°C

$$E_{p/2} - E_p = \frac{0.048}{\alpha n_z} \text{ V}. \qquad (7.80\text{a})$$

A theoretical description of the curves of irreversible electrode processes under conditions of spherical diffusion was given by Reinmuth [45] and by De Mars and Shain [46], but the practical application of their work proved to be inconvenient and tedious. For this reason Nicholson and Shain [40] attempted to find a simpler solution to this problem. Having analysed many theoretical curves plotted for various r_0 and b values, they were able to express the current by an equation consisting of two terms:

$$i = nFAC_{Ox}^0 \sqrt{\pi D_{Ox} b}\; \chi(bt) + \frac{nFAD_{Ox} C_{Ox}^0 \Phi(bt)}{r_0}. \qquad (7.81)$$

The first term, identical with (7.76), describes the current observed when the experiment is carried out at a flat electrode, and the second term represents the correction for spherical diffusion. As in the case of function $\Phi(at)$ of a reversible process the shape of function $\Phi(at)$ resembles a polarographic wave.

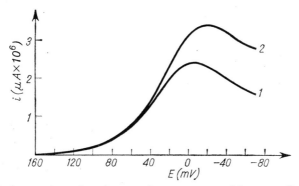

Fig. 7.5 Stationary electrode voltammetric current–potential curves of an irreversible process, recorded by means of a flat electrode (curve 1) and a spherical electrode (curve 2); $V = 0.2$ V/min; $r_0 = 0.05$ cm; $n = 2$; $\alpha n_\alpha = 0.5$; $D = 10^{-5}$ cm²/s.

The difference between the shapes of stationary electrode voltammetric curves due to spherical diffusion is shown in Fig. 7.5. It follows from equation (7.81) that the effect of spherical diffusion on the recorded curves decreases when the electrode radius and the polarization rates are increased.

Stationary electrode voltammetric curves of *quasi*-reversible processes were described by Matsuda and Ayabe [44]. The theory is very complex and has probably been used in practice only by the authors.

7.2.3 Determination of Kinetic Parameters of Electrode Processes by the Stationary Electrode Voltammetric Method

Stationary electrode voltammetry can be successfully applied to the determination of kinetics of electrode processes. The calculations are particularly simple when the process investigated is irreversible under the conditions of the experiment. In such a case it is very convenient to determine the transfer coefficients from equation (7.80) and the rate constant from the equation (7.78) describing peak current potential. It follows from the latter equation that the method of calculating the standard rate constant of an electrode process depends on knowledge of the standard potential. If this is not known, equation (7.78) can be transformed into the expression:

$$E_p = -1.14 \frac{RT}{\alpha n_z F} + \frac{RT}{\alpha n_\alpha F} \ln \frac{k_{fh}^0}{D_{Ox}^{1/2}} - \frac{RT}{2\alpha n_\alpha F} \ln \alpha n_z V, \qquad (7.82)$$

from which the rate constant k_{fh}^0 can be calculated.

Reinmuth [47] put forward an alternative method for the analysis of stationary electrode voltammetric curves, leading to the determination of kinetic parameters. His method is based on the fact that, in the initial parts of the curve, the current is independent of polarization rate and can be described by the very simple equation:

$$i = nFAC_{Ox}^0 k_{fh}. \qquad (7.83)$$

Substituting standard rate constant k_s for k_{fh} the potential can be introduced into equation (7.83) through the relationship:

$$i = nFAC_{Ox}^0 k_s \exp\left[-\frac{\alpha n_\alpha F(E-E^0)}{RT}\right]. \qquad (7.84)$$

However it should be stressed that this relationship is only valid for current values which do not exceed 10% of the peak current of the irreversible process, and that also in this method for the determination of k_s a knowledge of the standard potential E^0 is necessary.

It has already been shown that, when it is possible to record curves for the Ox and Red form, the analyses of cathodic and anodic stationary electrode voltammetric curves on the basis of relationship (7.84) give good results and make it possible to determine the formal potential, the rate constant at this potential and the transfer coefficients. This method is particularly useful in analyses of cyclic curves of irreversible processes and therefore it will be discussed in detail in the chapter dealing with cyclic methods.

In the case of *quasi*-reversible processes the determination of kinetic parameters of an electrode process is very complex. Kinetic analysis of such a process can be based on the theory of Matsuda and Ayabe [44] but it is probable that this theory has been used in practice only by the authors. From the theoretical point of view at least it is simpler to increase the polarization rate to the values at which the *quasi*-reversible process becomes irreversible; as we have mentioned in this case the interpretation of the results is no longer complex.

Production of cyclic curves of *quasi*-reversible processes is fairly simple. The standard rate constants can be determined according to the theory developed by Nicholson [48]. This method will be discussed in detail in Chapter 17.

It has already been pointed in Chapter 3 that, at high polarization rates (of the order of 100 V/s), deviations from reversibility are observed when the standard rate constant of the electrode process is smaller than 0.1 cm/s. In practice polarization rates appreciably exceeding 100 V/s have been sporadically used [49]. At polarization rates of this magnitude it is possible to investigate electrode reactions having rate constants exceeding 0.1 cm/s.

Kinetics of electrode reactions can also be studied by stationary electrode voltammetry of thin solution layers. Results of relevant theoretical studies carried out by Hubbard [50] have been used in practice [51, 52].

7.3 Chronopotentiometry

7.3.1 Reversible Electrode Processes

The description of chronopotentiometric curves of reversible electrode processes is reduced to the relationships between the electrode potential and the duration of electrolysis at constant current intensity. As in the case of polarography, the Nernst equation involving the concentrations of the oxidized and reduced forms at the electrode surface is used. Assuming that the electrode process takes place at flat electrodes in semi-infinite space and that both forms are soluble either in the solution or in the electrode material, these concentrations can be expressed, as already shown in Chapter 5, by the equations:

$$C_{Ox}(0, t) = C_{Ox}^0 - \frac{2i_0 t^{1/2}}{\pi^{1/2} nFD_{Ox}^{1/2}}, \quad (7.85)$$

$$C_{Red}(0, t) = \frac{2i_0 t^{1/2}}{\pi^{1/2} nFD_{Red}^{1/2}}. \quad (7.85a)$$

Introducing these concentrations into the Nernst equation we obtain the initial relationship:

$$E = E^0 + \frac{RT}{nF} \ln \frac{f_{Ox}}{f_{Red}} + \frac{RT}{nF} \ln \frac{C_{Ox}^0 - \dfrac{2i_0 t^{1/2}}{\pi^{1/2} nFD_{Ox}^{1/2}}}{\dfrac{2i_0 t^{1/2}}{\pi^{1/2} nFD_{Red}^{1/2}}}. \quad (7.86)$$

After appropriate transformation, simplification and utilization of the equation describing transition time equation (7.86) is reduced to the simple form [53]:

$$E = E^0 + \frac{RT}{nF} \ln \frac{f_{Ox} D_{Red}^{1/2}}{f_{Red} D_{Ox}^{1/2}} + \frac{RT}{nF} \ln \frac{\tau^{1/2} - t^{1/2}}{t^{1/2}}. \quad (7.87)$$

Since the first two terms of the right-hand side of this relationship are equal to the polarographic reversible half-wave potential:

$$E = E_{1/2} + \frac{RT}{nF} \ln \frac{\tau^{1/2} - t^{1/2}}{t^{1/2}}, \quad (7.88)$$

on the basis of this equation, a simple analysis of chronopotentiometric curves is possible. Introducing the values of the constants and substituting common logarithms for natural logarithms gives for 25°C:

$$E = E_{1/2} + \frac{0.059}{n} \log \frac{\tau^{1/2} - t^{1/2}}{t^{1/2}}. \quad (7.88a)$$

The potential dependence of $\log(\tau^{1/2} - t^{1/2})/t^{1/2}$ shown schematically in Fig. 7.6 is linear. The reciprocal of the slope of the straight line is $0.059/n$ V per logarithm unit. Hence on the basis of the measured slope it is possible to calculate the number of electrons, n, exchanged in the elementary process [54]. The half-wave potential, which is also shown in the figure, can be calculated exactly from a graph of $\log(\tau^{1/2} - t^{1/2})/t^{1/2}$ vs. E. For this purpose the potential at which $\log(\tau^{1/2} - t^{1/2})/t^{1/2} = 0$ should be determined.

In the discussion of polarographic waves it has been mentioned that the half-wave potential of a reversible electrode process is independent of the conditions of the experiment (i.e. depolarizer concentration and drop parameters). It follows from the analysis of equation (7.88) that, in the case of chronopotentiometry also, it is possible to find a similar characteristic potential. Assuming $t = \tau/4$ and substituting this value in equation (7.88) gives:

$$E_{\tau/4} = E_{1/2} \qquad (7.89)$$

where $E_{\tau/4}$ is the potential measured at $1/4$ of the transition time.

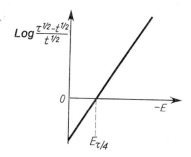

Fig. 7.6 Potential dependence of $\log(\tau^{1/2} - t^{1/2})/t^{1/2}$ for a reversible process.

It follows from equation (7.89) that the $E_{\tau/4}$ potential, like the half-wave potential, is independent of the experimental conditions, i.e. the current density used in chronopotentiometric electrolysis and the depolarizer concentration. For assessing the reversibility of a chronopotentiometric process, a criterion resembling the Tomeš criterion used in polarography could be employed, but it has not yet been used in practice.

From equation (7.88a), in the case of reversible processes studied at 25°C, the potential measured at $3/4$ of the transition time is described by the relationship:

$$E_{3\tau/4} = E_{1/2} - \frac{0.048}{n} \text{ V}. \qquad (7.90)$$

Subtracting equation (7.90) from equation (7.89) gives:

$$E_{\tau/4} - E_{3\tau/4} = \frac{0.048}{n} \text{ V}. \qquad (7.91)$$

On the basis of relationship (7.91) the reversibility of a process can be rapidly ascertained and the number of electrons exchanged in the elementary process can be readily determined.

The above relationships are valid for reversible reduction processes taking place under conditions of linear diffusion. In the case of processes taking place at spherical electrodes the relationships describing the concentrations of the Ox and Red forms should be introduced into the Nernst equation. These relationships were derived by Mamantov and Delahay [55] and by Koutecký and Čížek [56]. Due to the complexity of these relationships it is troublesome to use them in practice. The use of such complex equations is justified only when the transition times are of the order of seconds and the electrode radius is relatively small (not exceeding 0.1 cm).

The equation for thin layer chronopotentiometric reduction [57], i.e. of the process in which the substance is almost completely removed from the thin layer by electrolysis, has the following form:

$$E = E_{1/2} + \frac{RT}{nF} \ln \frac{\tau-t}{t}. \tag{7.92}$$

This equation also describes the chronopotentiometric reduction curve in the case when the current increases linearly with $t^{1/2}$ [58].

Interpretation of chronopotentiometric curves has been discussed by Reinmuth [59] who also considered a higher order reversible electrode reaction which can be expressed by the general equation:

$$m\text{Ox} + ne \rightleftharpoons p\,\text{Red}.$$

In this case the equation of the potential–time curve is as follows:

$$E = E^0 + \frac{RT}{nF} \ln \frac{\tau^{1/2} - t^{1/2}}{t^{1/2}} + \frac{RT}{nF} \ln \left(\frac{D_{\text{Red}}}{p^2}\right)^p \left(\frac{m^2}{D_{\text{Ox}}}\right)^m$$
$$+ \frac{(m-p)RT}{nF} \ln \frac{2i_0}{\pi^{1/2}nF} \tag{7.92a}$$

7.3.2 Irreversible Electrode Processes

Chronopotentiometric curves relating to irreversible electrode processes have been described by Delahay and Berzins [60]. This description is valid for processes in which one stage is slow. In such cases the current density is related to the rate of the electrode process by the equation:

$$\frac{i_0}{nF} = k_{fh}^0 C_{\text{Ox}}(0, t) \exp\left(-\frac{\alpha n_\alpha FE}{RT}\right). \tag{7.93}$$

Since no assumptions were made concerning the kinetics of electrode processes in the derivation of equation (7.85) $C_{\text{Ox}}(0, t)$ in equation (7.93) can be represented by equation (7.85). Then:

$$\frac{i_0}{nF} = k_{fh}^0 \left(C_{\text{Ox}}^0 - \frac{2i_0 t^{1/2}}{\pi^{1/2} nF D_{\text{Ox}}^{1/2}}\right) \exp\left(-\frac{\alpha n_z FE}{RT}\right). \tag{7.94}$$

Utilizing Sand's equation (7.94), after transformation:

$$E = \frac{RT}{\alpha n_\alpha F} \ln(\tau^{1/2} - t^{1/2}) - \frac{RT}{\alpha n_\alpha F} \ln \frac{\pi^{1/2} D_{Ox}^{1/2}}{2k_{fh}^0}. \quad (7.95)$$

The final time dependence of the potential, generally used in irreversible chronopotentiometric reduction, has the following form:

$$E = \frac{RT}{\alpha n_\alpha F} \ln \frac{nFC_{Ox}^0 k_{fh}^0}{i_0} + \frac{RT}{\alpha n_\alpha F} \ln \left[1 - \left(\frac{t}{\tau}\right)^{1/2}\right]. \quad (7.96)$$

This equation suggests a simple method of chronopotentiometric curve analysis, since the potential dependence of $\log[1 - (t/\tau)^{1/2}]$ is linear and the reciprocal of the slope of the straight line is equal to $2.3RT/\alpha n_\alpha F$.

Introducing the concentration $C_{Ox}(0, t)$, derived for spherical or for cylindrical diffusion conditions, into equation (7.93) gives the time dependence of electrode potential for processes carried out under these diffusion conditions.

Interpretation of chronopotentiometric curves of irreversible processes has also been discussed by Russell and Peterson [61].

Anderson and Macero [62] derived the following general equation describing the curves of *quasi*-reversible processes:

$$\left[1 - \frac{C_{Red}^0}{C_{Ox}^0 \Theta} - \frac{D_{Ox}^{1/2} \pi^{1/2}}{2k_s \tau^{1/2} \Theta}\right]\left(\frac{\Theta}{\Theta+1}\right) = \left(\frac{t}{\tau}\right)^{1/2}, \quad (7.96a)$$

where $\Theta = \exp\left[\frac{nF}{RT}-(E^0-E)\right]$ and ak_s is the heterogeneous rate constant of the electrode process. at $E = E^0$.

7.3.3 Chronopotentiometric Determination of Kinetic Parameters of Electrode Processes

The determination of the rate constants of irreversible electrode processes is particularly simple since equation (7.96) can be utilized. It follows from this equation that, at $t = 0$, the potential is described by the relationship:

$$E_{t=0} = \frac{RT}{\alpha n_\alpha F} \ln \frac{nFC_{Ox}^0 k_{fh}^0}{i_0} \quad (7.97)$$

Since $E_{t=0}$ and n are determined from the dependence of $\log[1-(t/\tau)^{1/2}]$ on E, and the values of C_{Ox}^0 and i_0 are known, it is easy to calculate the value of k_{fh}^0. From this constant the standard rate constant k_s can be calculated, provided that the standard electrode potential of the system under investigation is known.

In the case of a process controlled both by diffusion and by the kinetics of charge transfer the method of Berzins and Delahay [63] can be utilized.

The current density is expressed by the general equation:

$$i_0 = nFk_s \left\{ C_{Ox}(0,t) \exp\left[-\frac{\alpha nF(E-E^0)}{RT}\right] - C_{Red}(0,t) \exp\left[\frac{(1-\alpha)nF(E-E^0)}{RT}\right] \right\}. \quad (7.98)$$

and the concentrations $C_{Ox}(0, t)$ and $C_{Red}(0, t)$ are expressed by the following:

$$C_{Ox}(0, t) = C_{Ox}^0 \left[1 - \left(\frac{t}{\tau_{Red}}\right)^{1/2}\right], \quad (7.99)$$

$$C_{Red}(0, t) = C_{Red}^0 \left[1 + \left(\frac{t}{\tau_{Ox}}\right)^{1/2}\right], \quad (7.100)$$

where τ_{Red} and τ_{Ox} are the transition times for reduction and oxidation processes, respectively.

Combining equations (7.98), (7.99) and (7.100) gives the following relationship:

$$\frac{\exp\left[\frac{\alpha nF(E-E^0)}{RT}\right]}{nFk_s} = \frac{C_{Ox}^0\left[1-\left(\frac{t}{\tau_{Red}}\right)^{1/2}\right] - C_{Red}^0\left[1+\left(\frac{t}{\tau_{Ox}}\right)^{1/2}\right]\exp\left[\frac{nF(E-E^0)}{RT}\right]}{i_0} \quad (7.101)$$

It follows from this equation that the quantity:

$$\ln \frac{C_{Ox}^0\left[1-\left(\frac{t}{\tau_{Red}}\right)^{1/2}\right] - C_{Red}^0\left[1+\left(\frac{t}{\tau_{Ox}}\right)^{1/2}\right]\exp\left[\frac{nF(E-E^0)}{RT}\right]}{i_0}$$

should be a linear function of $(E-E^0)$, and that the slope of the straight line should be $\alpha nF/RT$. The intersection with the coordinate axis at $E-E^0 = 0$ makes it possible to determine the value of $-\ln nFk_s$.

Thus from this plot two principal parameters of electrode process kinetics can be determined: the rate constant k_s and the transition coefficient α.

The rate constant can also be determined in another way. Assuming that $C_{Ox}^0 = C_{Red}^0 = C^0$, the chronopotentiometric curve of a *quasi*-reversible process can be described by the relationship:

$$E = E^0 - \frac{i_0 RT}{nF} \left\{ \frac{\dfrac{1}{nFk_s} + \dfrac{2}{\pi^{1/2}nF}\left(\dfrac{1}{D_{Ox}^{1/2}} + \dfrac{1}{D_{Red}^{1/2}}\right)t^{1/2}}{C^0 - \dfrac{2i_0}{\pi^{1/2}nF}\left[\dfrac{\alpha}{D_{Ox}^{1/2}} - \dfrac{(1-\alpha)}{D_{Red}^{1/2}}\right]t^{1/2}} \right\}. \quad (7.102)$$

In most cases $D_{Ox} \cong D_{Red}$ and $0 < \alpha < 1$. Therefore equation (7.102) can be reduced to the following form:

$$E = E^0 - \frac{i_0 RT}{nF}\left[\frac{1}{nFC^0 k_s} + \frac{2}{\pi^{1/2}nFC^0}\left(\frac{1}{D_{Ox}^{1/2}} + \frac{1}{D_{Red}^{1/2}}\right)t^{1/2}\right]. \quad (7.103)$$

The problem of chronopotentiometric determination of rate constants has been studied by Hale and Parsons [21].

Galvanostatic single and double current pulse techniques are related to chronopotentiometry. It is suggested that readers interested in these subjects should neglect earlier work. Newer publications [64, 65], in which the whole subject is discussed in detail, are recommended.

7.4 The Rotating Disc Method

7.4.1 Reversible Electrode Processes

The curves recorded for rotating disc electrodes resemble polarographic waves. It is assumed that both forms of the reversible system are present in the solution: the Ox form in concentration C_{Ox}^0 and the Red form in concentration C_{Red}^0. In such conditions the anodic–cathodic curve of the current–potential relationship is recorded:

$$i = \frac{nFAD_{Ox}[C_{Ox}^0 - C_{Ox}(0)]}{\delta_{Ox}} \quad (7.104)$$

where δ_{Ox} is the thickness of the diffusion layer of the Ox form.

In order to describe the current–potential curve, the dependence of $C_{Ox}(0)$ on the electrode potential must be determined. For this purpose the fact that the Ox and Red fluxes are equal at the electrode surface is utilized:

$$\frac{D_{Ox}[C_{Ox}^0 - C_{Ox}(0)]}{\delta_{Ox}} = -\frac{D_{Red}[C_{Red}^0 - C_{Red}(0)]}{\delta_{Red}}. \quad (7.105)$$

For the sake of simplicity it is assumed that the diffusion coefficients and the thicknesses of the diffusion layers of the Ox and Red forms are equal: $D_{Ox} = D_{Red}$ and $\delta_{Ox} = \delta_{Red}$. Then, from equation (7.105):

$$C_{Ox}(0) = C_{Ox}^0 + C_{Red}^0 - C_{Red}(0). \tag{7.106}$$

$C_{Ox}(0)$ in equation (7.106) can be expressed by the Nernst equation, which can be written in the following form:

$$\frac{C_{Ox}(0)}{C_{Red}(0)} = \Theta \tag{7.107}$$

where:

$$\Theta = \exp\left[\frac{nF(E-E^0)}{RT}\right]. \tag{7.108}$$

Combining equations (7.106) and (7.107) we obtain:

$$C_{Ox}(0) = C_{Ox}^0 + C_{Red}^0 - \left(\frac{C_{Ox}(0)}{\Theta}\right). \tag{7.109}$$

This equation can be transformed into the following form:

$$C_{Ox}(0) = \frac{(C_{Ox}^0 + C_{Red}^0)\Theta}{(\Theta+1)}. \tag{7.110}$$

Expressing $C_{Ox}(0)$ in equation (7.104) by equation (7.110) gives the relationship:

$$i = \frac{nFAD_{Ox}C_{Ox}^0}{\delta_{Ox}} - \frac{nFAD_{Ox}}{\delta_{Ox}}\left[\frac{(C_{Ox}^0 + C_{Red}^0)\Theta}{\Theta+1}\right]. \tag{7.111}$$

Expressing the limiting cathodic current as:

$$i_g = \frac{nFAD_{Ox}C_{Ox}^0}{\delta_{Ox}} \tag{7.112}$$

and the limiting anodic current as:

$$_{anod}i_g = -\frac{nFAD_{Red}C_{Red}^0}{\delta_{Red}}, \tag{7.113}$$

and remembering that the diffusion coefficients and the thicknesses of diffusion layers have been assumed to be equal, equation (7.111) can be written in the following form:

$$i = i_g - \left[\frac{(-_{anod}i_g + i_g)\Theta}{\Theta+1}\right], \tag{7.114}$$

or, in another form:

$$\Theta = \frac{i_g - i}{i - {}_{\text{anod}}i_g}. \tag{7.115}$$

Introducing Θ [equation (7.108)] into equation (7.115) finally yields:

$$E = E^0 + \frac{RT}{nF} \ln \frac{i_g - i}{i - {}_{\text{anod}}i_g}. \tag{7.116}$$

The term E^0 appears in equation (7.116), instead of the half-wave potential, as a result of the simplifying assumptions that $f_{\text{Ox}} = f_{\text{Red}}$ and that $D_{\text{Ox}} = D_{\text{Red}}$, according to the relationship:

$$E_{1/2} = E^0 + \frac{RT}{nF} \ln \frac{D_{\text{Red}} \delta_{\text{Ox}} f_{\text{Ox}}}{D_{\text{Ox}} \delta_{\text{Red}} f_{\text{Red}}}. \tag{7.117}$$

If $E_{1/2}$ is introduced instead of E^0, an equation is obtained similar to that for the anodic–cathodic polarographic wave:

$$E = E_{1/2} + \frac{RT}{nF} \ln \frac{i_g - i}{i - {}_{\text{anod}}i_g}. \tag{7.118}$$

The half-wave potential described by equation (7.117) differs from the polarographic half-wave potential, but the difference is small, usually not exceeding a few mV (it decreases with decreasing difference between the diffusion coefficients of the Ox and Red forms).

If the corresponding limiting current is neglected, equation (7.118) can be used for the analysis of solutions in which either only the Ox, or only the Red form, is present.

The current–potential curves described by the above equations can be observed when the potential difference applied to the electrodes is increased stepwise by equal increments, or when the potential difference is applied continuously but at low rates of potential change in conjunction with a high velocity of electrode rotation.

The curves obtained for a rapidly changing potential applied to the electrodes were described by Fried and Elving [66] and by Girina, Filinovskii and Feoktistov [67].

7.4.2 Irreversible Electrode Processes

The curves of irreversible electrode processes can be treated like those of the reversible processes, but the equation relating the current to the potential (7.98) must be used instead of the Nernst equation. Concentration

$C_{Ox}(0)$ can be expressed by the relationship:

$$C_{Ox}(0) = \frac{\delta_{Ox}}{nFD_{Ox}A}(i_g - i) \quad (7.119)$$

and $C_{Red}(0)$ by equation (7.106).

Combining equations (7.98), (7.106) and (7.119) and assuming $\delta_{Ox} = \delta_{Red} = \delta$ and $D_{Ox} = D_{Red} = D$, gives the following relationship:

$$i = nFAk_s \frac{\left\{C_{Ox}^0 \exp\left[-\frac{\alpha nF}{RT}(E-E^0)\right] - C_{Red}^0 \exp\left[\frac{(1-\alpha)nF}{RT}(E-E^0)\right]\right\}}{\left\{1 + \frac{\delta k_s}{D}\left\{\exp\left[-\frac{\alpha nF}{RT}(E-E^0)\right] + \exp\left[\frac{(1-\alpha)nF}{RT}(E-E^0)\right]\right\}\right\}} \quad (7.120)$$

This equation was derived by Delahay [68] and is based on the results of Eyring, Marker and Kwoh [69].

In order to elucidate the meaning of relationship (7.120), it has been used to calculate a number of current–potential curves for various standard rate constants, k_s. These curves are shown in Fig. 7.7. It follows from this figure that the value of k_s affects the potentials of current formation but it does not affect the limiting current values.

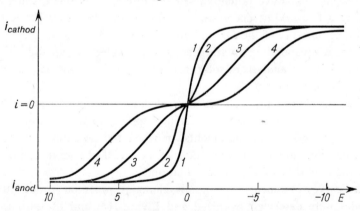

Fig. 7.7 Current–potential curves obtained by means of a rotating disc electrode. On the potential axis the unit is 2.3 RT/nF V. Curve 1—reversible process, curve 2—$k_s = 10^{-3}$ cm/s; curve 3—$k_s = 10^{-4}$ cm/s; curve 4—$k_s = 10^{-5}$ cm/s (from [68] by permission of the copyright holders, Interscience Publishers).

Equation (7.119) describes both irreversible and *quasi*-reversible processes. Assuming the process to be fully irreversible, it is possible — on the basis of relationship (7.112) — to reduce equation (7.120) to the following:

$$i = \frac{i_g}{1 + \dfrac{D}{k_s\delta}\exp\left[\dfrac{\alpha nF}{RT}(E-E^0)\right]}, \qquad (7.121)$$

or:

$$E = E^0 + \frac{RT}{\alpha nF}\ln\frac{D}{k_s\delta} + \frac{RT}{\alpha nF}\ln\frac{i_g-i}{i}. \qquad (7.122)$$

It follows from equation (7.122) that, in this case, the half-wave potential is described by the relationship:

$$E_{1/2} = E^0 + \frac{RT}{nF}\ln\frac{D}{k_s\delta}. \qquad (7.123)$$

Thus, in the case of irreversible processes the half-wave potential depends on experimental conditions and particularly on the velocity of disc electrode rotation, which determines the value of δ.

Analogously, assuming irreversibility and considering anodic processes, the equation of the irreversible anodic wave can be derived from equation (7.120).

7.4.3 Determination of Kinetic Parameters of Electrode Processes by the Rotating Disc Method

A number of methods have been suggested for the determination of kinetic parameters from measurements carried out by means of the rotating disc electrode. For this purpose the Levich equation [70] for first-order electrode reactions controlled by charge transfer rate can be used. In the case of a slow cathodic process:

$$i = \frac{nFAD_{Ox}C^0_{Ox}}{1.61 D_{Ox}^{1/3}\nu^{1/6}\omega^{-1/2} + \dfrac{D_{Ox}}{k_{fh}}}, \qquad (7.124)$$

where ν is the kinematic velocity; in simpler form:

$$i = \frac{nFAD_{Ox}C^0_{Ox}}{\delta_{Ox} + \dfrac{D_{Ox}}{k_{fh}}}. \qquad (7.124a)$$

Depending on the relation between δ_{Ox} and D_{Ox}/k_{fh}, equation (7.124) can be reduced to a relationship valid for electrode reactions controlled exclusively by either transport or the charge transfer rate.

Assuming $\delta_{Ox} = 10^{-3}$ cm and $D_{Ox} = 10^{-5}$ cm^2/s, for values of $k_{fh} > 10^{-1}$ cm/s, the Nernst and Levich equation for the limiting current of a cathodic reduction (7.112) is obtained from (7.124a).

When it is assumed that $k_{fh} < 10^{-4}$ cm/s, equation (7.124) or (7.124a) is reduced to:

$$i = nFAk_{fh}C_{Ox}^0. \tag{7.125}$$

The rate constant can be calculated from either equation (7.124) or (7.125), which are valid for a definite disc electrode potential. Having determined k_{fh} at various potentials, a graph of $\log k_{fh}$–E may be plotted, from which it is easy to determine $\log k_{fh}$ at the standard potential. The transfer coefficient can be calculated from the slope of the straight line on the $\log k_{fh}$–E graph.

The above equations are valid for cathodic reactions but they may also be employed for investigations of the kinetics of anodic processes.

This method of calculation of rate constants was used to study the kinetics of the Fe^{3+}/Fe^{2+} system [71].

The procedures developed by Randles [18] can also be used for studies of electrode kinetics. They are valid for both polarographic and rotating disc electrode measurements in cases where anodic–cathodic waves are recorded and the system studied is *quasi*-reversible. The equation used has the following form:

$$-\log nFk_{fh}C_{Ox}^0 = \log\left[\left(\frac{1}{i}-\frac{1}{i_g}\right)-\left(\frac{1}{i}+\frac{1}{_{anod}i_g}\right)\exp\left(\frac{nF\eta}{RT}\right)\right] \tag{7.126}$$

where η is the overvoltage.

A similar equation can be derived for the calculation of k_{bh}.

Another method of determining rate constants was suggested by Jahn and Vielstich [72]. From the equation:

$$i_0 = nF[k_{fh}C_{Ox}(0) - k_{bh}C_{Red}(0)] \tag{7.127}$$

where i_0 is the current density, by substituting for concentrations the expressions obtained from the relationships:

$$i_0 = \frac{nFD_{Ox}[C_{Ox}^0 - C_{Ox}(0)]}{\delta_{Ox}}, \tag{7.128}$$

$$i_0 = -\frac{nFD_{Red}[C_{Red}^0 - C_{Red}(0)]}{\delta_{Red}}, \tag{7.129}$$

the following expression is obtained:

$$i_0 = \frac{nF(k_{fh}C_{Ox}^0 - k_{bh}C_{Red}^0)}{1 + \dfrac{k_{fh}B_{Ox}}{D_{Ox}\sqrt{\omega}} + \dfrac{k_{bh}B_{Red}}{D_{Red}\sqrt{\omega}}}, \tag{7.130}$$

where $B = \delta\sqrt{\omega}$.

A more convenient form of equation (7.130) is:

$$\frac{1}{i_0} = \frac{1}{nF(k_{fh}C_{Ox}^0 - k_{bh}C_{Red}^0)}\left[1 + \left(\frac{k_{fh}B_{Ox}}{D_{Ox}} + \frac{k_{bh}B_{Red}}{D_{Red}}\right)\frac{1}{\sqrt{\omega}}\right]. \quad (7.131)$$

A graph of $1/i_0$ vs. $1/\sqrt{\omega}$ is plotted for constant overvoltage values. From the slope of the straight line and from the value of the point of its intersection with the ordinate for $\omega = \infty$, values of k_{fh} and k_{bh} are obtained. For a definite overvoltage the standard rate constants can be readily calculated from these constants. The transfer coefficients can be calculated from the relation between k_{fh} and k_{bh} and the overvoltage.

Jahn and Vielstich applied this method to the determination of kinetic parameters of the ferrocyanide–ferricyanide system on platinum electrodes.

For the cathodic wave the current–potential relationship of a *quasi-reversible* process on a disc electrode can also be expressed by the following relationship [73]:

$$\left(\frac{i_g - i}{i}\right)\frac{T_{Red}}{T_{Ox}} = \exp\left[\frac{nF(E-E^0)}{RT}\right] + \frac{T_{Red}}{T_{Ox}}\exp\left[\frac{\alpha nF(E-E^0)}{RT}\right] \quad (7.132)$$

where T_{Red} and T_{Ox} are the diffusion current constants of the oxidized and reduced species.

This equation and a treatment analogous with that suggested by Koryta for the determination of kinetic parameters by the polarographic technique was used by Kuta and Yeager [74]. Using a rotating gold disc, these authors studied the kinetics of the UO_2^{2+}/UO_2^+ system in KCl, NaClO$_4$ and Na$_2$SO$_4$ solutions. For the two first electrolytes, rate constants of the order of 10^{-2} cm/s were found.

Still another method was proposed by Malyszko [75]. Enyo and Yokoyama [76] published also a paper on the application of the rotating disc technique to the investigations of electrode kinetics.

The uneven current distribution on disc electrodes can lead to serious errors in measurements of parameters of electrode kinetics [77]. The magnitude of this error depends on the position of the reference electrode relative to the disc electrode. The cases in which the reference electrode is far removed from the disc, is situated near its centre and is situated near its edge have been considered theoretically and for each case an expression describing the ratio of the real to the observed exchange current has been derived. The error is smallest when the reference electrode is situated far from the disc, but in this case the potential drops in the solution are large, which also causes a decrease in accuracy.

In addition to the methods of determination of kinetic parameters of electrode reactions discussed in this chapter several other methods have been proposed, mainly for investigation of rapid processes. Early achievements in this field have been discussed by Delahay [78] and Vetter [79]. Later a critical review of methods of investigation of electrode kinetics was published by Damaskin [80]. Several interesting publications regarding these problems have also appeared in recent years [81–80].

References

[1] Heyrovský, J. and Ilkovič, D., *Coll. Czechoslov. Chem. Communs.*, **7**, 198 (1935).
[2] Tomeš, J., *Coll. Czechosl. Chem. Communs.*, **9**, 12 (1937).
[3] Koutecký, J., *Českoslov. cas. fys.*, **2**, 117 (1952); *Czechoslov. J. Phys.*, **2**, 50 (1953).
[4] Strehlow, H. and von Stackelberg, M., *Z. Elektrochem.*, **54**, 51 (1950).
[5] Micka, K., *Chem. Listy*, **50**, 203 (1956); *Coll. Czechoslov. Chem. Communs.*, **21**, 1246 (1956).
[6] Lingane, J. J., *J. Am. Chem. Soc.*, **61**, 2099 (1939).
[7] von Stackelberg, M., *Z. Elektrochem.*, **45**, 446 (1939).
[8] Turyan, Ya. I., *Zhurn. Fiz. Khim.*, **30**, 709 (1956).
[9] Koutecký, J., *Chem. Listy*, **47**, 323 (1953); *Coll. Czechoslov. Chem. Communs.*, **18**, 597 (1953).
[10] Delahay, P. and Strassner, J. E., *J. Am. Chem. Soc.*, **73**, 5219 (1951).
[11] Kern, D. M. H., *J. Am. Chem. Soc.*, **76**, 4234 (1954).
[12] Delahay, P., *Record of Chemical Progress*, **19**, 83 (1958).
[13] Reinmuth, W. H. and Rogers, L. B., *J. Am. Chem. Soc.*, **82**, 802 (1960).
[14] Kivalo, P., Oldham, K. B. and Laitinen, H. A., *J. Am. Chem. Soc.*, **75**, 4148 (1953).
[15] Meites, L. and Israel, Y., *J. Am. Chem. Soc.*, **83**, 4903 (1961).
[16] Behr, B., Dojlido, J. and Małyszko, J., *Roczniki Chem.*, **36**, 725 (1962).
[17] Koutecký, J. and Čížek, J., *Coll. Czechoslov. Chem. Communs.*, **21**, 836 (1956).
[18] Randles, J. E. B., *Can. J. Chem.*, **37**, 238 (1959).
[19] Randles, J. E. B., in: *"Progress in Polarography"*, Ed. Kolthoff, I. M. and Zuman, P., Interscience Publishers, New York, 1962, Vol. 1, page 123.
[20] Stromberg, A. G., *Zhurn. Fiz. Khim.*, **36**, 2714 (1962).
[21] Hale, J. M. and Parsons, R., *Coll. Czechoslov. Chem. Communs.*, **27**, 2444 (1962).
[22] Hale, J. M. and Parsons, R., *J. Electroanal. Chem.*, **8**, 247 (1964).
[23] Sathyanarayana, S., *J. Electroanal. Chem.*, **7**, 403 (1964).
[24] Koryta, J., *Electrochim. Acta*, **6**, 67 (1962).
[25] Matsuda, H. and Ayabe, Y., *Z. Elektrochem.*, **63**, 1164 (1959); Matsuda, H., *Z. Elektrochem.*, **62**, 977 (1958).
[26] Gaur, J. N. and Goswami, N. K., *Electrochim. Acta*, **12**, 1483 (1967); **12**, 1489 (1967).
[27] Ružić, I., Barić, A. and Branica, M., *J. Electroanal. Chem.*, **29**, 411 (1971).
[28] Verdier, E., Bennes, R. and Balette, B., *J. Electroanal. Chem.*, **31**, 463 (1971).
[29] Matsuda, H., *Rept. Govt. Chem. Industr. Res. Inst. Tokyo*, **61**, 315 (1966).

References

[30] Oldham, K. B. and Parry, E. P., *Anal. Chem.*, **40**, 65 (1968).
[31] Tamamushi, R. and Tanaka, N., *Z. Elektrochem.*, **39**, 117 (1963).
[32] Christie, J. H., Lauer, G. and Osteryoung, R. A., *J. Electroanal. Chem.*, **7**, 60 (1964).
[33] Lingane, P. J. and Christie, J. H., *J. Electroanal. Chem.*, **10**, 284 (1965).
[34] Osteryoung, J. and Osteryoung, R. A., *Electrochim. Acta*, **16**, 525 (1971).
[35] Niki, K., Okuda, Y., Tomonari, T., Buck, E. and Hackerman, N., *Electrochim. Acta*, **16**, 487 (1971).
[36] Nicholson, M. M., *J. Am. Chem. Soc.*, **76**, 2539 (1954).
[37] Frankenthal, R. P. and Shain, I., *J. Am. Chem. Soc.*, **78**, 2969 (1956).
[38] Muller, T. R. and Adams, R. N., *Anal. Chim. Acta*, **25**, 482 (1961).
[39] Reinmuth, W. H., *Anal. Chem.*, **33**, 1793 (1961); **34**, 1446 (1962).
[40] Nicholson, R. S. and, Shain I., *Anal. Chem.*, **36**, 706 (1964).
[41] Hubbard, A. T. and Anson, F. C., *Anal. Chem.*, **38**, 58 (1966).
[42] de Vries, W. T. and van Dalen, E., *J. Electroanal. Chem.*, **8**, 366 (1964); de Vries, W. T., *J. Electroanal. Chem.*, **9**, 448 (1965).
[43] Delahay, P., *J. Am. Chem. Soc.*, **75**, 1190 (1953).
[44] Matsuda, H. and Ayabe, Y., *Z. Elektrochem.*, **59**, 494 (1955).
[45] Reinmuth, W. H., *Anal. Chem.*, **33**, 1793 (1961).
[46] De Mars, R. D. and Shain, I., *J. Am. Chem. Soc.*, **81**, 2654 (1959).
[47] Reinmuth, W. H., *Anal. Chem.*, **32**, 1891 (1960).
[48] Nicholson, R. S., *Anal. Chem.*, **37**, 1351 (1965).
[49] Perone, S. P., *Anal. Chem.*, **38**, 1158 (1966).
[50] Hubbard, A. T., *J. Electroanal. Chem.*, **22**, 165 (1969).
[51] Cushing, J. R. and Hubbard, A. T., *J. Electroanal. Chem.*, **23**, 183 (1969).
[52] Lau, A. L. Y. and Hubbard, A. T., *J. Electroanal. Chem.*, **24**, 237 (1970).
[53] Karaoglanoff, Z., *Z. Elektrochem.*, **12**, 5 (1906).
[54] Delahay, P. and Mattax, C. C., *J. Am. Chem. Soc.*, **76**, 874 (1954).
[55] Mamantov, G. and Delahay, P., *J. Am. Chem. Soc.*, **76**, 5325 (1954).
[56] Koutecký, J. and Čižek, J., *Coll. Czechoslov. Chem. Communs.*, **22**, 914 (1957).
[57] Christensen, C. R. and Anson, F. C., *Anal. Chem.*, **35**, 205 (1963).
[58] Hurwitz, H. and Gierst, L., *J. Electroanal. Chem.*, **2**, 128 (1961); Hurwitz, H., *J. Electroanal. Chem.*, **2**, 142 (1961).
[59] Reinmuth, W. H., *Anal. Chem.*, **32**, 1514 (1960).
[60] Delahay, P. and Berzins, T., *J. Am. Chem. Soc.*, **75**, 2486 (1953).
[61] Russell, C. D. and Peterson, J. M., *J. Electroanal. Chem.*, **5**, 467 (1963).
[62] Anderson, L. B. and Macero, D. J., *Anal. Chem.*, **37**, 322 (1965).
[63] Berzins, T. and Delahay, P., *Z. Elektrochem.*, **59**, 792 (1955).
[64] Birke, D. L. and Roe, D. K., *Anal. Chem.*, **37**, 450 (1965).
[65] Birke, D. L. and Roe, D. K., *Anal. Chem.*, **37**, 455 (1965).
[66] Fried, I. and Elving, P. J., *Anal. Chem.*, **37**, 464, 803 (1965).
[67] Girina, G. P., Filinovsky, V. Yu. and Feoktistov, L. G., *Elektrokhimiya*, **3**, 941 (1967) (in Russian).
[68] Delahay, P., "*New Instrumental Methods in Electrochemistry*", Interscience Publishers, New York, 1964, Chapter 9.
[69] Eyring, H., Marker, L. and Kwoh, T. C., *J. Phys. Colloid. Chem.*, **53**, 1453 (1949).

[70] Levich, V. G., *"Physicochemical Hydrodynamics"*, Akad. Nauk USSR, Moscow, 1952, page 56 (in Russian).
[71] Galus, Z. and Adams, R. N., *J. Phys. Chem.*, **67**, 866 (1963).
[72] Jahn, D. and Vielstich, W., *J. Electrochem. Soc.*, **109**, 849 (1962).
[73] Dvořak, J., Koryta, J. and Boháčková, V. , *"Electrochemistry"*, Methuen, London, 1970, page 266.
[74] Kuta, J. and Yeager, E., *J. Electroanal. Chem.*, **31**, 119 (1971).
[75] Malyszko, J., *Chimia*, **29**, 166 (1975).
[76] Enyo, M. and Yokoyama, T., *Electrochim. Acta*, **15**, 183 (1970).
[77] Tiedemann, W. H., Newman, J. and Bennion D. N., *J. Electrochem. Soc.*, **120**, 256 (1973).
[78] Delahay, P., in *"Advances in Electrochemistry and Electrochemical Engineering"*, Vol. 1, page 233.
[79] Vetter, K. J., *"Elektrochemische Kinetik"*, Springer-Verlag, Berlin 1961.
[80] Damaskin, V. V., *"Principles of Modern Methods of Investigation of Electrochemical Reactions on Electrodes"*, Nauka, Moscow 1968 (in Russian).
[81] Pilla, A. A., *J. Electrochem. Soc.*, **117**, 467 (1970).
[82] Birke, R. L., *Anal. Chem.*, **43**, 1253 (1971).
[83] Oldham, K. B. and Osteryoung, R. A., *J. Electroanal. Chem.*, **11**, 397 (1966).
[84] van der Pol, F., Sluyters-Rehbach, M. and Sluyters, J. H., *J. Electroanal. Chem.*, **40**, 209 (1972).
[85] Doblhofer, K. and Pilla, A. A., *J. Electroanal. Chem.*, **39**, 91 (1972).
[86] Rangarajan, S. K., *J. Electroanal. Chem.*, **32**, 329 (1971).
[87] Rangarajan, S. K., *J. Electroanal. Chem.*, **41**, 459, 491 (1973).
[88] Leeuwen, H. P. and Sluyters, J. H., *J. Electroanal. Chem.*, **42**, 313 (1973).
[89] Tanaka, N., Kitani, A., Yamada, A. and Sasaki, K., *Electrochim. Acta*, **18**, 675 (1973).
[90] Agarwal, H. P. in *"Electroanalytical Chemistry"*, Vol. VII, Bard, A. J., Ed., Dekker, New York 1974.

Chapter 8

Electrode Processes Preceded by First-Order Chemical Reactions

In the earlier chapters only those electrode processes in which the depolarizer present in the solution reacts with the electrode to give a product have been considered. In such cases the intensity of the limiting current, or in chronopotentiometry the transition time, is related to the depolarizer concentration in the bulk of solution, by the equations derived by Ilkovič, Randles and Ševčik, Sand, Nernst and Levich.

However, in certain cases the measured currents are much smaller than those calculated from these equations. In such cases the substance studied can exist in solution in two forms which are in equilibrium:

$$A \underset{k_2}{\overset{k_1}{\rightleftharpoons}} Ox \qquad (8.1)$$

determined by the equilibrium constant

$$K = \frac{k_1}{k_2}. \qquad (8.2)$$

Only the substance Ox can participate in the electrode reaction in the potential range employed and, as a result, a decrease in the amount of this substance at the electrode surface pushes the equilibrium of the reaction (8.1) to the right.

The whole process can be expressed by the general scheme:

$$A \underset{k_2}{\overset{k_1}{\rightleftharpoons}} Ox + ne \rightleftharpoons Red. \qquad (8.3)$$

In some circumstances the chemical equilibrium is largely on the left-hand side of the equation, so that practically only substance A is present in the solution. However, the reduction current can be observed in this case also, provided that the magnitude of the rate constant k_1 is large. When this constant is very small, the recorded current is proportional to the concentration of the Ox form existing in equilibrium with A. In the case of polarography this situation is shown schematically in Fig. 8.1.

When k_1 is very large the loss of some of substance Ox, due to the reduction, is compensated by a shift in equilibrium according to equation (8.1), and as a result the recorded current is in agreement with the value obtained from the Ilkovič equation. Thus in this case the limiting current is recorded and the corresponding curves do not reflect the process (8.1) preceding the charge transfer (Fig. 8.1, curve *1*).

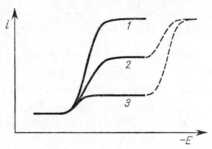

Fig. 8.1 Polarographic curves of the electrode process described by equation (8.3). Curve *1*—$k_1 = \infty$, curve *2*—finite value of k_1; curve *3*—$k_1 = 0$.

When the value of k_1 is close to zero the recorded current is proportional to the equilibrium concentration of substance Ox. If the equilibrium constant K is very large, the recorded current is also relatively large, but when both K and k_1 are very small the recorded current is also very small (curve *3*). In both cases the character of the observed current is not kinetic.

Curve *3* in Fig. 8.1 shows that substance A can be reduced at potentials more negative than the reaction (8.3) potential. The reduction wave of substance A is shown by the broken line.

The sum of both waves caused by the reduction of substances Ox and A should be practically equal to the height of the wave represented by curve *1* in the case when $k_1 \to \infty$.

The kinetic current is represented by curve *2*, since it is larger than current i_r caused by the reduction of the form Ox, but is smaller than the limiting current i_g. The difference $i_k - i_r$ is caused by the kinetic effect of conversion of A into Ox near the electrode surface and it increases with increasing rate of this conversion.

A typical example of this type of process is the electroreduction of formaldehyde in buffered solutions. In aqueous solutions formaldehyde exists in the hydrated form which is not reduced at a mercury electrode in the usual potential range. This form is, however, in equilibrium with the anhydrous form:

$$\begin{array}{c}H\\H\end{array}\!\!>\!\!C\!\!<\!\!\begin{array}{c}OH\\OH\end{array}\underset{k_2}{\overset{k_1}{\rightleftarrows}}\begin{array}{c}H\\H\end{array}\!\!>\!\!C=O + H_2O \qquad (8.4)$$

Only the anhydrous form is reduced at the electrode, and therefore, at the beginning of the process, when the equilibrium (8.4) is disturbed at the electrode surface, the formation of the anhydrous form takes place.

The equilibrium constant of reaction (8.4) is equal to about 10^{-4}; hence in the absence of the kinetic effect the current should be very small even in relatively concentrated formaldehyde solutions, provided that the current is proportional to the equilibrium concentration of anhydrous formaldehyde only.

Processes corresponding to equation (8.3) are also known. In such cases the solution contains two substances, both reducible at the electrode, in the accessible potential range, in equilibrium:

$$Ox_1 + C \underset{k_2}{\overset{k_1}{\rightleftarrows}} Ox_2. \qquad (8.5)$$

Both Ox_1 and Ox_2 react with the electrode, but Ox_2 is reduced at more positive potentials.

As in the previous case the disappearance of Ox_2 causes a shift of the equilibrium to the right. Therefore when k_1 is very large the reduction of Ox_2 alone is observed, since on the electrode surface Ox_1 is converted into Ox_2. When the rate constant is very small two currents are observed, and a polarogram consists of two waves with heights proportional to equilibrium concentrations of Ox_1 and Ox_2 (see Fig. 8.1, curve 3).

An example of such a process is the reduction of certain weak acids. Thus pyruvic acid is polarographically reduced in two waves. In solutions of pH lower than 4.5 and higher than 8 only one wave is observed, but it appears at different potentials. In the pH region 4.5–8 two polarographic waves are observed; the sum of their heights is constant and is proportional to the concentration of the acid in the solution. Brdička [1] suggested that the two waves correspond to the reduction of undissociated molecules of the acid and its anions which are in equilibrium:

$$\begin{array}{c}R-\underset{\underset{O}{\|}}{C}-\underset{\underset{O}{\|}}{C}-OH \underset{k_2}{\overset{k_1}{\rightleftarrows}} R-\underset{\underset{O}{\|}}{C}-C\!\!<\!\!\begin{array}{c}O^-\\O\end{array} + H^+\end{array} \qquad (8.6)$$

The currents of the two waves, referred to the total current, depend on the pH according to the expression:

$$\text{pH} = \text{p}K' + \log\frac{[\text{A}^-]}{[\text{HA}]} \tag{8.7}$$

where A^- and HA stand for the anion and the molecule of the weak acid which are in equilibrium:

$$\text{A}^- + \text{H}^+ \rightleftharpoons \text{HA} \tag{8.8}$$

where K' is the dissociation constant of acid HA.

If the kinetic effect were not observed the waves would be of equal size at $\text{pH} = \text{p}K'$, according to equation (8.7). In fact at this pH the first wave, corresponding to the reduction of undissociated HA molecules, is larger as a result of the existence of the kinetic effect of recombination of the acid [equation (8.8)] near the electrode surface. The heights of the waves become equal at a pH higher than $\text{p}K'$. This pH value is equal to $\text{p}\overline{K}$ which is called the apparent polarographic dissociation constant.

Another example of the occurrence of kinetic currents is the reduction of aldoses. Dextrose, mannose, and various other aldoses exist in aqueous solutions in two forms. One of them, the aldehyde form, is reduced at the mercury electrode, whereas the other, the cyclic form, is not reduced in the potential region of the reduction of the aldehyde form. The equilibrium existing in aqueous solutions between these two forms favours the cyclic form.

The nature of the reduction currents of aldoses was convincingly explained by Wiesner [2]. Earlier it had been thought [3] that the limiting currents of aldoses are proportional to the equilibrium concentration of the free aldehyde form.

Kinetic currents are often observed in the case of the reduction of metal complexes when the solution contains an equilibrium mixture of several complexes of the same metal ion with different number of ligands, provided that their reduction takes place at different potentials.

The nature of kinetic currents was first explained for polarographic conditions, but the solution was not exact. It was based on the reaction layer concept introduced by Wiesner and Brdička [4–7]. According to this concept it is assumed that a layer having the thickness μ is present at the surface of the electrode. During the electrode process in this layer additional amounts of the depolarizer are formed as a result of reduction of the Ox form, and the depolarizer reacts with the electrode. It is also assumed that the distribution of concentrations in the reaction layer is stationary. As a result of these assumptions the number of moles of the substance formed in unit time on the electrode surface can be expressed by means of the equation:

$$\left[\frac{\partial N_{Ox}(x, t)}{\partial t}\right]_{x=0} = \mu \left[\frac{\partial C_{Ox}(x, t)}{\partial t}\right]_{x=0} \qquad (8.9)$$

or

$$\left[\frac{\partial N_{Ox}(x, t)}{\partial t}\right]_{x=0} = \mu k_1 C_A(0, t) - \mu k_2 C_{Ox}(0, t). \qquad (8.10)$$

In order to find the limiting current value a potential must be applied to the electrode, sufficiently negative to ensure that the condition $C_{Ox}(0, t) = 0$, is fulfilled.

Since

$$i = nFA\frac{dN_{Ox}}{dt}, \qquad (8.11)$$

by combining equations (8.9)–(8.11) the following relationship for the mechanism (8.3) is obtained:

$$i_k = nFA\mu k_1 C_A(0, t). \qquad (8.12)$$

Equation (8.12) expresses the relation between the kinetic current i_k and the electrode surface area, the rate constant of formation of the form Ox from substance A, the concentration of substance A on the electrode surface and the thickness of the reaction layer.

In order to apply measurements of kinetic currents to such problems as kinetics of chemical reactions preceding electrode processes it is necessary to determine the thickness of the reaction layer μ and the concentration of substance A on the electrode surface.

Wiesner [8] assumed that the thickness of the reaction layer is proportional to the path traversed by depolarizer Ox during its average life time t', which is inversely proportional to constant k_2 of equations (8.1) or (8.3):

$$t' = \frac{1}{k_2}. \qquad (8.13)$$

Taking into account equations (8.2) and (8.3) gives the expression:

$$t' = \frac{K}{k_1}. \qquad (8.14)$$

The electroactive substance can reach the electrode provided that its distance from the electrode is not larger than the average path which the substance can traverse during the time t'. This average path is given by the

Einstein and Smoluchowski formula:

$$\Delta = \sqrt{2Dt'}. \tag{8.15}$$

Wiesner assumed initially that only one half of the molecules of substance Ox formed at the distance Δ from the electrode surface, would be able to reach it. Hence he obtained initially

$$\mu = 1/2\sqrt{2Dt'}, \tag{8.16}$$

and combining equations (8.16) and (8.14):

$$\mu = \sqrt{\frac{DK}{2k_1}}. \tag{8.17}$$

Comparison of exact solutions [9] with those obtained on the basis of the reaction layer concept and statistical considerations showed that:

$$\mu = \sqrt{\frac{DK}{k_1}} = \sqrt{\frac{D}{k_2}}. \tag{8.18}$$

Equation (8.18) indicates that the thickness of the reaction layer depends on the rate constants of depolarizer formation and is inversely proportional to the square root of the rate of transformation of depolarizer Ox into the inactive substance A.

Assuming $D = 10^{-5}$ cm^2/s it is found that when $k_2 = 10^2$ s^{-1}, $\mu = 3.1 \times 10^{-4}$ cm, but when $k_2 = 10^{10}$ s^{-1} the thickness of the reaction layer calculated from equation (8.18) is 3.1×10^{-8} cm, i.e. molecular size.

It has already been found that the application of equation (8.12) is possible when μ and $C_A(0, t)$ are known. Now consider briefly, $C_A(0, t)$. Assuming that the inactive substance A reaches the electrode by diffusion, it is possible to obtain the following approximate relationship by means of the Ilkovič equation:

$$i_k = K_{II}[C_A^0 - C_A(0, t)] \tag{8.19}$$

where K_{II} is the Ilkovič constant.

According to this relationship the current is proportional to the concentration of substance A on the electrode surface, that is to say, it is proportional to the concentration gradient at the electrode.

In the case when substance A reacts with the electrode the limiting current due to the reduction of A is equal to $K_{II}C_A^0$. Hence equation (8.19) can be expressed in the form:

$$C_A(0, t) = \frac{i_g - i_k}{K_{II}} \qquad (8.20)$$

where i_g is the hypothetical limiting current of substance A.

Expressing μ and $C_A(0, t)$ in equation (8.12) by means of equations (8.18) and (8.20):

$$i_k = i_g \frac{\dfrac{nFAD^{1/2}k_1}{k_2^{1/2} K_{II}}}{1 + \dfrac{nFAD^{1/2}k_1}{k_2^{1/2} K_{II}}}. \qquad (8.21)$$

Equation (8.21) indicates that when the rate constant k_1 is large the unity term in the denominator can be neglected, and the measured current becomes equal to the limiting current. On the other hand when k_1 is very small the term $nFAD^{1/2}k_1/k_2^{1/2} K_{II}$ is small as compared with unity, and in such a case equation (8.21) becomes:

$$i_k = \frac{nFAD^{1/2}k_1}{k_2^{1/2} K_{II}} i_g, \qquad (8.22)$$

or

$$i_k = \frac{nFAD^{1/2}C_A^0 k_1}{k_2^{1/2}}. \qquad (8.22a)$$

This equation describes the purely kinetic current, i.e. the current due to the kinetic change (8.1) alone, since the large value of k_2 assumed implies a large shift of the equilibrium of reaction (8.1) in the direction of form A.

A slightly different treatment of the kinetic currents has been proposed by Hanuš [9]. Instead of equation (8.12) he used the relationship:

$$i = nFAD \frac{C_A^\mu - C_A(0, t)}{\mu} \qquad (8.12a)$$

where C_A^μ is the concentration of A at the distance μ from the electrode surface.

The discussion of kinetic currents has so far been limited to chronoamperometric or polarographic conditions, depending on the type of electrode surface and the significance of the Ilkovič constant in equation (8.19). Delahay showed [10] that the method based on the reaction layer concept can be applied to the description of kinetic processes in chronopotentiometry. Application of this method to the evaluation of kinetic currents in the rotating disc technique [11] also leads to results similar to those obtained by the exact method.

Although the values of kinetic currents based on the reaction layer concept were in a fairly good agreement with experimental data, Koutecký and Brdička [12] in 1947 gave the exact description for the case of linear diffusion. It was obtained by solving the Fick equations, taking into account concentration changes resulting from the chemical reaction according to scheme (8.3). These equations for forms A and Ox are:

$$\frac{\partial C_A(x,t)}{\partial t} = D_A \frac{\partial^2 C_A(x,t)}{\partial x^2} - k_1 C_A(x,t) + k_2 C_{Ox}(x,t), \quad (8.23)$$

$$\frac{\partial C_{Ox}(x,t)}{\partial t} = D_{Ox} \frac{\partial^2 C_{Ox}(x,t)}{\partial x^2} + k_1 C_A(x,t) - k_2 C_{Ox}(x,t). \quad (8.24)$$

The initial conditions are as follows:

$$t = 0, \quad x \geq 0, \quad \frac{C_{Ox}(x,0)}{C_A(x,0)} = K, \quad (8.25)$$

$$C_{Ox}(x,0) + C_A(x,0) = C^0, \quad (8.26)$$

where C^0 is the total concentration of forms A and Ox in the bulk of the solution.

One general boundary condition reflects the electrode inactivity of substance A:

$$D_A \left[\frac{\partial C_A(x,t)}{\partial x} \right]_{x=0} = 0. \quad (8.27)$$

In addition, at the ends of the diffusion region:

$$t > 0, \quad x \to \infty, \quad C_{Ox}(x,t) + C_A(x,t) \to C^0, \quad (8.28)$$

$$\frac{C_{Ox}(x,t)}{C_A(x,t)} \to K. \quad (8.29)$$

The problem described by equation (8.3) is now considered in relation to various electrochemical methods.

8.1 Chronoamperometry

In order to solve the system of equations (8.23) and (8.24) it is necessary to formulate a second boundary condition characteristic of the chronoamperometric method. It can be expressed in a simple form, if it is assumed that the process takes place at values of applied potential which are so negative that the concentration of form Ox on the electrode surface is equal to zero. In this case:

$$t > 0, \quad x = 0, \quad C_{Ox}(0, t) = 0. \tag{8.30}$$

The system of equations (8.23) and (8.24) was solved by Koutecký and Brdička [12].

In order to facilitate the solution it is assumed that the diffusion coefficients of forms A and Ox are equal. The common coefficient is called D. It is also assumed that the equilibrium of the chemical reaction is shifted considerably in the direction of substance A. Then in the bulk of solution C_A is practically equal to C^0 and, in the expression for total current, the contribution of the current due to reduction of form Ox diffusing to the electrode from the bulk of the solution can be neglected. In such cases the equation giving the kinetic current is:

$$i_k = nFAD^{1/2}C^0K^{1/2}k_1^{1/2}\exp(Kk_1 t)\operatorname{erfc}(Kk_1 t)^{1/2} \tag{8.31}$$

where t is the time measured from the beginning of electrolysis, erfc is the complement of the error function and $(Kk_1 t)^{1/2}$ is its argument.

Equation (8.31) is similar to that describing the current of an irreversible electrode process under chronoamperometric conditions. By dividing by the equation describing the diffusion current:

$$\frac{i_k}{i_g} = \pi^{1/2}K^{1/2}k_1^{1/2}t^{1/2}\exp(Kk_1 t)\operatorname{erfc}(Kk_1 t)^{1/2}. \tag{8.32}$$

The dependence of the current ratio (i_k/i_g) on $(Kk_1 t)^{1/2}$ was discussed in Chapter 6 for the case of similar irreversible electrode processes.

When $(Kk_1 t)^{1/2}$ tends to zero, the function $\exp(Kk_1 t)\operatorname{erfc}(Kk_1 t)^{1/2}$ tends to one, and therefore equation (8.31) is reduced to

$$i_k = nFAD^{1/2}C^0K^{1/2}k_1^{1/2}. \tag{8.33}$$

This equation is equivalent to the relationship (8.22a) which was derived by the approximate method, based on the reaction layer concept. The assumptions made in the derivation of equation (8.22a) are of course also valid in the case of equation (8.33).

Equations (8.31)–(8.33) can be used for an approximate interpretation of results of polarographic studies; in this case the electrode surface A should be expressed in terms of the drop parameters and the drop time t_1 should be used instead of time t.

The problem of an electrode process preceded by a chemical reaction leading to depolarizer formation under chronoamperometric conditions has also been solved for conditions of spherical diffusion. Budevskii and

Desimirov [13] obtained a result which can be expressed in the form of the following equation:

$$i_k = nFA\left(1+\frac{1}{K}\right)C_A^0 D\left[\frac{h}{1+hr_0} + \frac{h^2 r_0}{1+hr_0}\exp(\bar{h}^2 Dt)\operatorname{erfc}(\bar{h}\sqrt{Dt})\right], \quad (8.34)$$

where

$$h = K\left(\sqrt{\frac{k_1+k_2}{D}} + \frac{1}{r_0}\right) \quad (8.35)$$

and

$$\bar{h} = h + \frac{1}{r_0}. \quad (8.36)$$

Equation (8.34) can be written in a form similar to equation (8.31):

$$i_k = nFAC^0 D\left[\frac{h}{1+hr_0} + \frac{h^2 r_0}{1+hr_0}\exp(\bar{h}^2 Dt)\operatorname{erfc}(\bar{h}\sqrt{Dt})\right] \quad (8.37)$$

where C^0 is the sum of the equilibrium concentrations of forms Ox and A.

It can be easily shown that equation (8.37) is reduced to relationship (8.31) when it is assumed that the radius of the electrode is very large. Then the equation obtained is the same as relationship (8.31), in which the sum k_1+k_2 is used instead of k_1. This equation takes into account the kinetics of formation of substance Ox and Tits conversion into substance A.

8.2 Polarography

It has already been mentioned that chronoamperometric equations can be adapted to an approximate description of results of polarographic studies. The problem of an electrode process preceded by a first-order chemical reaction, according to scheme (8.3), has been developed for polarographic conditions by Koutecký [14]. This author solved the following system of equations:

$$\frac{\partial C_A(x,t)}{\partial t} = D\frac{\partial^2 C_A(x,t)}{\partial x^2} + \frac{2x}{3t}\frac{\partial C_A(x,t)}{\partial x} - k_1 C_A(x,t) + k_2 C_{\text{Ox}}(x,t), \quad (8.38)$$

$$\frac{\partial C_{\text{Ox}}(x,t)}{\partial t} = D\frac{\partial^2 C_{\text{Ox}}(x,t)}{\partial x^2} + \frac{2x}{3t}\frac{\partial C_{\text{Ox}}(x,t)}{\partial x} + k_1 C_A(x,t) - k_2 C_{\text{Ox}}(x,t), \quad (8.39)$$

in which the initial and boundary conditions are similar to conditions (8.25)–(8.27):

$$t = 0, \quad x \geqslant 0, \quad C_A(x, 0) = C_A^0, \quad C_{Ox}(x, 0) = C_{Ox}^0; \quad (8.40)$$

$$t > 0, \quad x = 0, \quad C_{Ox}(0, t) = 0, \quad D_A\left[\frac{\partial C_A(x, t)}{\partial x}\right]_{x=0} = 0. \quad (8.41)$$

Koutecký also assumed that $K \ll 1$.

In the case of a very fast chemical reaction in the immediate neighbourhood of the electrode a stationary state exists between the reaction and diffusion, and therefore equations (8.23) and (8.24) are simplified to the following forms:

$$D\frac{d^2 C_A(x, t)}{dx^2} - k_1 C_A(x, t) + k_2 C_{Ox}(x, t) = 0, \quad (8.42)$$

$$D_{Ox}\frac{d^2 C_{Ox}(x, t)}{dx^2} + k_1 C_A(x, t) - k_2 C_{Ox}(x, t) = 0. \quad (8.43)$$

When it is possible to neglect the diffusion of substance A (e.g. in the case of processes controlled by the kinetics of the chemical reaction alone) and when $C_{Ox} \ll C_A$, the solution of equation (8.43) leads to the relationship:

$$C_{Ox}(x, t) = K C_A^0 \left[1 - \exp\left(-\frac{x}{\sqrt{\frac{DK}{k_1}}}\right)\right]. \quad (8.44)$$

The current is described by the expression:

$$i = nFAD\left(\frac{\partial C_{Ox}}{\partial x}\right)_{x=0}, \quad (8.45)$$

therefore on differentiating equation (8.44) with respect to x, assuming $x = 0$, and combining the result with equation (8.45):

$$i_k = nFDAC_A^0 K^{1/2} k_1^{1/2}. \quad (8.46)$$

This result is practically identical with equation (8.33) obtained for chronoamperometric conditions.

Čižek, Koryta and Koutecký [15, 16] and Koutecký and Koryta [17] showed that equation (8.46) can be used as the boundary condition for the solution of the diffusion equation of substance A. This method was applied by Matsuda [18] and by Gierst and Hurwitz [19–22] for the solution of the problem of double layer effects on the rate of a chemical reaction preceding the electrode process.

Mention has already been made that Koutecký [14] solved the system of equations (8.38)–(8.41) by means of the dimensionless parameters method. In the case of rapid chemical reactions, when a stationary state exists between the reaction rate and diffusion, and $K \ll 1$, the ratio of the instantaneous current to the limiting one is given by the function:

$$F(\chi) = \frac{i_k}{i_g} \qquad (8.47)$$

where parameter χ is defined by the expression:

$$\chi = \sqrt{\frac{12}{7} Kk_1 t}. \qquad (8.48)$$

The functions for the average currents are similarly described as follows:

$$\overline{F}(\chi_1) = \frac{\overline{i_k}}{\overline{i_g}}. \qquad (8.49)$$

In this case parameter χ_1 is:

$$\chi_1 = \sqrt{\frac{12}{7} Kk_1 t_1}. \qquad (8.50)$$

The values of functions $F(\chi)$ and $\overline{F}(\chi_1)$ are shown in Chapter 6 according to Koutecký and Weber [23, 24]. They are identical, irrespective of the nature of factors slowing down the electrode process, a slow chemical reaction of depolarizer formation preceding the process, or a slow exchange of electrons, and only the meaning of the parameter changes.

Function $\overline{F}(\chi_1)$ can be expressed fairly accurately by means of the following interpolative equation:

$$\frac{\overline{i_k}}{\overline{i_g}} = \frac{0.886 \sqrt{Kk_1 t_1}}{1 + 0.886 \sqrt{Kk_1 t_1}} \qquad (8.51)$$

which is usually written in the form

$$\frac{\overline{i_k}}{\overline{i_g} - \overline{i_k}} = 0.886 \sqrt{Kk_1 t_1}. \qquad (8.52)$$

The present discussions of polarographic reactions have not taken into account spherical diffusion, but only the transport due to the expansion of the dropping electrode. The effect of spherical diffusion on the polarographic limiting currents of electrode processes preceded by first-order chemical reactions, was taken into account in the theory proposed by Koutecký and Čížek [25]. They expressed the ratio of average currents in the form:

$$\frac{\bar{i}_k}{\bar{i}_g} = \bar{F}(\chi_1) - \varepsilon_{1.1} \bar{H}_c(\chi_1), \tag{8.53}$$

where $\bar{F}(\chi_1)$ is given by relationship (8.49) and parameter $\varepsilon_{1.1}$ by the equation:

$$\varepsilon_{1,1} = \sqrt{\frac{12D}{7r^2} t_1^{1/6}}, \tag{8.54}$$

in which r is the radius of the drop electrode at $t = 1$.

The values of function $H_c(\chi)$ and $\bar{H}_c(\chi_1)$ for average currents and various values of parameters χ and χ_1 are shown in Table 8.1.

Table 8.1 Values of functions $H_c(\chi)$ and $\bar{H}_c(\chi_1)$ (from [25] by permission of the copyright holders)

χ or χ_1	$H_c(\chi)$	$\bar{H}_c(\chi_1)$	χ or χ_1	$H_c(\chi)$	$\bar{H}_c(\chi_1)$
0.01	0.0069	0.0042	3.00	0.1942	0.1835
0.05	0.0327	0.0202	5.00	0.1412	0.147
0.10	0.0617	0.0388	10.00	0.0804	0.093
0.20	0.1097	0.0706	15.00	0.056	0.067
0.50	0.1977	0.1362	20.00	0.043	0.052
1.00	0.2458	0.1857	25.00	0.034	0.041
1.50	0.2446	0.2001	30.00	0.029	0.036
2.00	0.2314	0.1996	50.00	0.017	0.023
2.50	0.2136	0.1931			

It follows from equations (8.53) and (8.54) that the ratio of kinetic and limiting currents depends not only on the drop time according to equation (8.51), but also on the capillary yield m which depends on the radius r. The latter dependence was shown experimentally by Hanuš in 1951 [26].

The solutions of the polarographic problems described above are based on the assumption that the diffusion coefficients of the inactive substance A and the electroactive substance Ox are equal. Koutecký [27] showed that when these coefficients are significantly different the kinetic current is given by the equation:

$$\bar{i}_k = \bar{K}C^0 \sqrt{D} \frac{0.886\sqrt{Kk_1 t_1} M}{1 + 0.886\sqrt{Kk_1 t_1} M} \tag{8.55}$$

where D is the mean diffusion coefficient given by the equation

$$D = \frac{KD_{Ox}+D_A}{1+K}. \tag{8.56}$$

In equation (8.55) $\overline{K} = 605nm^{2/3}t_1^{1/6}$, and $C^0 = C_{Ox}+C_A$. Parameter M is given by the equation:

$$M = \frac{D}{D_A}\sqrt{\frac{D_{Ox}}{D_A}}. \tag{8.56a}$$

It follows from equation (8.56) that when K is very small it can be assumed that $D = D_A$ and M is determined by the ratio $(D_{Ox}/D_A)^{1/2}$.

Koutecký and Koryta [28] worked out the general theory of kinetic currents based on the expanding plane model of diffusion.

Ružić [29], using the generalized Brdička–Wiesner approximate method, considered several electrode reactions coupled with chemical reactions.

8.3 Stationary Electrode Voltammetry

The problem of an electrode process preceded by a first-order chemical reaction, leading to the formation of the depolarizer under conditions of stationary electrode voltammetry was solved by Savéant and Vianello [30], and was later theoretically elaborated in detail by Nicholson and Shain [31]. In order to solve this problem it is necessary to add the boundary condition to conditions (8.25)–(8.29), formulated at the beginning of this chapter. For a reversible electrode process this condition is:

$$t > 0, \quad x = 0, \quad \frac{C_{Ox}}{C_{Red}} = \Theta e^{-at}, \tag{8.57}$$

where

$$\Theta = \exp\left[\frac{nF}{RT}(E_i - E^0)\right], \tag{8.58}$$

and

$$a = \frac{nFV}{RT}. \tag{8.59}$$

The introduction of the Red form concentration into the boundary condition (8.57) requires the solution of equations (8.23), (8.24), and the Fick equation written in the form:

$$\frac{\partial C_{Red}(x,t)}{\partial t} = D_{Red}\frac{\partial^2 C_{Red}(x,t)}{\partial x^2} \tag{8.60}$$

with the initial condition
$$t = 0, \quad x \geqslant 0, \quad C_{\text{Red}} = C_{\text{Red}}^0 = 0, \tag{8.61}$$
and boundary conditions:
$$t > 0, \quad x \to \infty, \quad C_{\text{Red}} \to 0, \tag{8.62}$$

$$t > 0, \quad x = 0, \quad D_{\text{Ox}} \left[\frac{\partial C_{\text{Ox}}(x, t)}{\partial x} \right]_{x=0} = -D_{\text{Red}} \left[\frac{\partial C_{\text{Red}}(x, t)}{\partial x} \right]_{x=0}. \tag{8.63}$$

The effect of the preceding chemical reaction on the electrode process depends on several factors. The analysis of this problem can be divided into three parts according to the value of the ratio l/a, where $l = k_1 + k_2$.

When l/a is very small the kinetic effect is absent, since during the experiment the extent of conversion of A into Ox is also small. In such a case the recorded current–potential curves are identical with those recorded in the case of a diffusion process, but the current is proportional to the equilibrium concentration of the form Ox in the solution. The current function $\chi(at)$ is, in this case, given by the equation:

$$\chi(at) = \frac{1}{\sqrt{\pi}} \sum_{j=1}^{\infty} (-1)^{j+1} \sqrt{j} \left(\frac{K}{1+K} \right) \exp\left[\left(-\frac{jnF}{RT} \right)(E - E_{1/2}) \right]. \tag{8.64}$$

This equation is very similar to that of the function $\chi(at)$ of a kinetically uncomplicated reversible electrode process, derived by Reinmuth [32]. The only difference is the term $K/(1+K)$ which appears in equation (8.64). When the equilibrium of reaction (8.1) is shifted considerably to the right, i.e. when K is large, this term is practically equal to one and equation (8.64) becomes identical with that describing the diffusion process. When the equilibrium is shifted in favour of substance A the current is determined by the equilibrium concentration of substance Ox.

When the ratio l/a is large the behaviour of the function $\chi(at)$ depends primarily on the value of $\sqrt{a}/K\sqrt{l}$. When it is small the function $\chi(at)$ is given by the equation:

$$\chi(at) = \frac{1}{\sqrt{\pi}} \sum_{j=1}^{\infty} (-1)^{j+1} \sqrt{j} \exp\left[\left(-\frac{jnF}{RT} \right)\left(E - E_{1/2} - \frac{RT}{nF} \ln \frac{K}{1+K} \right) \right]. \tag{8.65}$$

This equation is identical with that of the reversible process, but in this case the potential region in which the peak current appears depends on the equilibrium constant of the chemical reaction.

When the value of $\sqrt{a}/K\sqrt{l}$ is large the equation of the function $\chi(at)$ is as follows:

$$\chi(at) = \frac{1}{\sqrt{\pi}} \sum_{j=1}^{\infty} (-1)^{j+1} \sqrt{j!} K \sqrt{1/a} \exp\left[\left(-\frac{jnF}{RT}\right) \times \left(E - E_{1/2} - \frac{RT}{nF} \ln \frac{K}{1+K} - \frac{RT}{nF} \ln \frac{\sqrt{a}}{K\sqrt{l}}\right)\right]. \quad (8.66)$$

The direct analysis of this equation is difficult. Since, however, the expression $1/\sqrt{a}$ occurs in each term it appears that the current is independent of the polarization rate. The analysis of stationary electrode voltammetric curves for intermediate values of $\sqrt{a}/K\sqrt{l}$ was carried out by Nicholson and Shain. The shape of the recorded curves depends on the value of $\sqrt{a}/K\sqrt{l}$. When it tends to zero the shape of the curves becomes similar to that of the theoretical curve of a reversible process. As the value of $\sqrt{a}/K\sqrt{l}$ increases the peak current decreases, the curves become stretched and the peak of the current becomes flat. At large values of $\sqrt{a}/K\sqrt{l}$ the curves are more like polarographic waves than the current peaks characteristic of stationary electrode voltammetry.

If the equilibrium constant of the chemical reaction is known, the rate constants of this reaction can be determined by comparing the stationary electrode voltammetric curves with the theoretical curves plotted for various values of $\sqrt{a}/K\sqrt{l}$, shown in Table 8.2. However, this procedure is very tedious, since it requires adjusting the experimental curve to the theoretical results. Therefore Nicholson and Shain proposed a different method. They plotted several theoretical curves for various values of $\sqrt{a}/K\sqrt{l}$ and used them for determining the dependence of i_k/i_p on this parameter. This dependence is given with an error of 1% (with the exception of small values of $\sqrt{a}/K\sqrt{l}$, for which, however, the error does not exceed 1.5%) by the following empirical equation:

$$\frac{i_k}{i_p} = \frac{1}{1.02 + 0.471\sqrt{a}/K\sqrt{l}}. \quad (8.67)$$

It is possible, by means of this equation, to calculate the value of $K\sqrt{l}$ rapidly, using the measured value of kinetic current i_k, the calculated value of diffusion-limited peak current i_p and the value of parameter a determined for the experimental conditions.

Table 8.2 VALUES OF CURRENT FUNCTION $\sqrt{\pi}\chi(at)$ FOR A REVERSIBLE ELECTRODE PROCESS PRECEDED BY A FIRST-ORDER CHEMICAL REACTION (FROM [31] BY PERMISSION OF THE COPYRIGHT HOLDERS, THE AMERICAN CHEMICAL SOCIETY)

Potential	Parameter $\sqrt{a}/K\sqrt{l}$						
	0.2	0.5	1.0	1.5	3.0	6.0	10.0
120	0.009	0.009	0.009	0.009	0.009	0.009	0.008
100	0.019	0.019	0.019	0.019	0.018	0.017	0.015
80	0.041	0.040	0.039	0.038	0.035	0.031	0.027
60	0.081	0.080	0.075	0.072	0.063	0.051	0.041
50	0.113	0.108	0.100	0.094	0.080	0.062	0.049
45	0.132	0.125	0.116	0.108	0.089	0.068	0.052
40	0.152	0.144	0.131	0.121	0.099	0.074	0.055
35	0.174	0.164	0.149	0.135	0.109	0.079	0.059
30	0.199	0.184	0.164	0.150	0.118	0.084	0.062
25	0.224	0.206	0.183	0.164	0.127	0.089	0.064
20	0.249	0.228	0.199	0.178	0.136	0.093	0.067
15	0.275	0.249	0.216	0.191	0.144	0.098	0.069
10	0.301	0.270	0.232	0.204	0.151	0.101	0.071
5	0.324	0.289	0.246	0.215	0.158	0.104	0.072
0	0.345	0.307	0.259	0.225	0.163	0.107	0.074
−5	0.364	0.321	0.271	0.234	0.168	0.109	0.075
−10	0.379	0.334	0.280	0.241	0.173	0.111	0.076
−15	0.391	0.344	0.288	0.247	0.176	0.113	0.077
−20	0.399	0.351	0.293	0.252	0.179	0.114	0.077
−25	0.404	0.355	0.297	0.255	0.181	0.115	0.078
−30	0.406	0.358	0.299	0.257	0.182	0.116	0.078
−35	0.405	0.358	0.300	0.258	0.183	0.116	0.079
−40	0.402	0.357	0.300	0.258	0.183	0.117	0.079
−45	0.397	0.353	0.298	0.258	0.183	0.117	0.079
−50	0.390	0.349	0.296	0.256	0.183	0.117	0.079
−60	0.373	0.338	0.289	0.252	0.181	0.116	0.079
−80	0.337	0.310	0.272	0.240	0.176	0.115	0.078
−100	0.301	0.284	0.253	0.227	0.170	0.113	0.077
−120	0.273	0.260	0.236	0.214	0.164	0.110	0.076
−140	0.250	0.240	0.222	0.203	0.158	0.108	0.075
$E_{p/2}$ mV	+29.3	+31.3	+34.4	+37.5	+42.2	+53.4	+62.2

Potentials scale is $(E - E_{1/2})n - \dfrac{RT}{nF} \ln \dfrac{K}{1+K}$.

According to equation (8.64), at small values of $\sqrt{a}/K\sqrt{l}$, the potential at which the current peak is observed does not depend on the value of this parameter. At large values of $\sqrt{a}/K\sqrt{l}$ a tenfold increase of this parameter causes a shift of the reduction peak in the positive direction by $60/n$ mV at 25°C. Since at large values of $\sqrt{a}/K\sqrt{l}$ the current peaks are flat, the exact determination of the peak potential is difficult, and it is more convenient to use half-current peak potentials which can be accurately measured.

The discussion has been limited to reversible electrode processes preceded by first-order chemical reactions. The theory of irreversible electrode processes preceded by chemical reactions in stationary electrode voltammetry has also been described by Nicholson and Shain [31]. They solved the system of equations (8.23) and (8.24) with initial conditions (8.25) and (8.26) and boundary conditions (8.27)–(8.29). In this case the second boundary condition for $x = 0$ is as follows:

$$D_{Ox}\frac{\partial C_{Ox}}{\partial x} = k_{fh}C_{Ox} = k_i C_{Ox}\exp(bt), \qquad (8.68)$$

where

$$k_i = k_s \exp\left[\left(-\frac{\alpha n_\alpha F}{RT}\right)(E_i - E^0)\right], \qquad (8.69)$$

and

$$b = \alpha n_\alpha FV/RT.$$

Equation (8.69) indicates that k_i is the rate constant of the electrode process at the initial potential E_i.

Utilizing condition (8.68) we obtain the solution, which, as in the case of the reversible process, can be expressed in various ways depending on the value of $\sqrt{b}/K\sqrt{l}$. When this value is small the shape of recorded curves is identical to that of the curves of simple irreversible processes, the potential at which the process takes place does not depend on the kinetics of the chemical reaction and the current value is proportional to the equilibrium concentration of the substance Ox. The current function $\chi(bt)$ in this case is expressed by the equation:

$$\chi(bt) = \frac{1}{\sqrt{\pi}} \sum_{j=1}^{\infty} (-1)^{j+1} \frac{(\sqrt{\pi})^j}{\sqrt{(j-1)!}} \left(\frac{K}{1+K}\right) \times$$

$$\exp\left[\left(-\frac{j\alpha n_\alpha F}{RT}\right)\left(E - E^0 + \frac{RT}{\alpha n_\alpha F}\ln\sqrt{\pi Db}/k_s\right)\right]. \qquad (8.70)$$

This equation is similar to that of function $\chi(bt)$ for kinetically uncomplicated irreversible processes, from which it differs only by the presence of the term $K/(1+K)$.

When the ratio l/b is large the shape of the curves depends on the value of $\sqrt{b}/K\sqrt{l}$. For small values of this parameter the current function is given by the expression:

$$\chi(bt) = \frac{1}{\sqrt{\pi}} \sum_{j=1}^{\infty} (-1)^{j+1} \frac{(\sqrt{\pi})^j}{\sqrt{(i-1)!}} \exp\left[\left(-\frac{j\alpha n_a F}{RT}\right) \times \right.$$
$$\left. \left(E - E^0 + \frac{RT}{\alpha n_a F} \ln \frac{\sqrt{\pi Db}}{k_s} - \frac{RT}{\alpha n_a F} \ln \frac{K}{1+K}\right)\right]. \quad (8.71)$$

This equation is similar to that of the function $\chi(bt)$ of the simple irreversible process, but it contains the additional term $(-RT/\alpha n_a F) \times \ln[K/(1+K)]$. The potential at which the current peak is recorded depends on the equilibrium constant of the chemical reaction.

When the parameter $\sqrt{b}/K\sqrt{l}$ is large the function $\chi(bt)$ is described by a different equation having the following form:

$$\chi(bt) = \frac{1}{\sqrt{\pi}} \sum_{j=1}^{\infty} (-1)^{j+1} (\sqrt{\pi})^j K\sqrt{l/b} \exp\left[\left(-\frac{j\alpha n_a F}{RT}\right) \times \right.$$
$$\left. \left(E - E^0 + \frac{RT}{\alpha n_a F} \ln \frac{\sqrt{\pi Db}}{k_s} - \frac{RT}{\alpha n_a F} \ln \frac{K}{1+K} + \frac{RT}{\alpha n_a F} \ln \frac{K\sqrt{\pi b}}{\sqrt{l}}\right)\right]. \quad (8.72)$$

In such conditions the recorded curves resemble polarographic waves, due to disappearance of the current peak. The values of the observed current and the potential range in which the wave appears do not depend on parameter b. Thus the current is fully controlled by the rate of the chemical reaction preceding the electrode process. In this case the current is described by the equation:

$$i = \frac{nFAC^0 D^{1/2} K\sqrt{l}}{1 + \exp\left[\frac{\alpha n_a F}{RT}\left(E - E^0 + \frac{RT}{\alpha n_a F} \ln \frac{\sqrt{\pi Db}}{k_s} + \frac{RT}{\alpha n_a F} \ln \frac{\sqrt{\pi b}(1+K)}{\sqrt{l}}\right)\right]}. \quad (8.73)$$

The range intermediate between the two extreme cases can be described on the basis of the values of function $\chi(bt)$ calculated by Nicholson and Shain for various electrode potential and parameter $\sqrt{b}/K\sqrt{l}$ values. These are listed in Table 8.3.

Table 8.3 VALUES OF CURRENT FUNCTION $\sqrt{\pi}\chi(bt)$ FOR AN IRREVERSIBLE ELECTRODE PROCESS PRECEDED BY A FIRST-ORDER CHEMICAL REACTION FOR VARIOUS VALUES OF PARAMETER $\sqrt{b}/K\sqrt{l}$ (FROM [31] BY PERMISSION OF THE COPYRIGHT HOLDERS, THE AMERICAN CHEMICAL SOCIETY)

Potential mV	Parameter $\sqrt{b}/K\sqrt{l}$						
	0.2	0.5	1.0	1.5	3.0	6.0	10.0
160	0.003	0.003	0.003	0.003	0.003	0.003	0.003
140	0.007	0.007	0.007	0.007	0.007	0.007	0.007
120	0.016	0.016	0.016	0.016	0.015	0.015	0.014
110	0.024	0.024	0.023	0.023	0.022	0.021	0.019
100	0.035	0.034	0.034	0.033	0.031	0.029	0.026
90	0.050	0.049	0.048	0.047	0.044	0.039	0.033
80	0.070	0.070	0.067	0.065	0.059	0.050	0.042
70	0.102	0.099	0.094	0.090	0.079	0.063	0.050
60	0.140	0.134	0.126	0.117	0.100	0.076	0.058
50	0.190	0.179	0.164	0.151	0.122	0.088	0.065
40	0.248	0.230	0.205	0.185	0.143	0.099	0.070
35	0.280	0.257	0.226	0.201	0.152	0.103	0.072
30	0.312	0.282	0.244	0.216	0.161	0.107	0.074
25	0.343	0.307	0.263	0.230	0.168	0.110	0.076
20	0.370	0.330	0.279	0.241	0.174	0.112	0.077
15	0.395	0.349	0.292	0.251	0.179	0.115	0.078
10	0.414	0.364	0.302	0.260	0.183	0.116	0.079
5	0.430	0.375	0.310	0.265	0.186	0.117	0.079
0	0.440	0.382	0.315	0.269	0.188	0.118	0.080
−5	0.444	0.385	0.318	0.271	0.189	0.119	0.080
−10	0.443	0.386	0.318	0.272	0.189	0.119	0.080
−15	0.438	0.383	0.317	0.271	0.189	0.119	0.080
−20	0.430	0.378	0.314	0.269	0.189	0.119	0.080
−25	0.419	0.371	0.310	0.267	0.188	0.119	0.080
−30	0.407	0.362	0.306	0.263	0.187	0.118	0.080
−35	0.394	0.354	0.301	0.260	0.186	0.118	0.080
−40	0.381	0.345	0.295	0.257	0.184	0.117	0.079
−50	0.355	0.327	0.283	0.248	0.180	0.116	0.079
−60	0.333	0.309	0.272	0.240	0.177	0.115	0.078
−70	0.313	0.294	0.261	0.233	0.174	0.114	0.078
$E_{p/2}$ mV	+44.2	+47.3	+51.4	+54.5	+62.2	+71.9	+82.2

Potentials scale is $(E-E^0)\alpha n_\alpha - \dfrac{RT}{F}\ln\dfrac{K}{1+K} + \dfrac{RT}{F}\ln\dfrac{\sqrt{\pi Db}}{k_s}$.

In this case, as in the case of reversible electrode processes preceded by chemical reactions, the kinetic parameters of the chemical process can be calculated from the theoretical dependence of the ratio of kinetic to diffusion currents i_k/i_p on parameter $\sqrt{b}/K\sqrt{l}$. This dependence is expressed by the following equation, similar to equation (8.67):

$$\frac{i_k}{i_p} = \frac{1}{1.02 + 0.531\,\sqrt{b}/K\sqrt{l}} \cdot \qquad (8.74)$$

It can be utilized employing a method similar to that described in the case of equation (8.67).

At low values of $\sqrt{b}/K\sqrt{l}$ the potentials of formation of the curve do not depend on this parameter, whereas at high values tenfold increase of the parameter value causes a shift of the potential peak towards more positive values.

Theoretical relationships have not yet been established for electrode processes controlled by both diffusion and charge transfer rates and simultaneously preceded by a first-order chemical reaction. This problem is difficult, because change of polarization rate causes not only a certain amount of inhibition of the electrode process by the chemical reaction, but also a shift of the *quasi*-reversible process to the region controlled by charge transfer or by transport rate alone.

8.4 Chronopotentiometry

In order to solve equations (8.23) and (8.24) it is necessary to formulate an additional boundary condition characteristic of this method.

$$\left[\frac{\partial C_{Ox}}{\partial x}\right]_{x=0} = \lambda \qquad (8.75)$$

where $\lambda = i_0/nFD_{Ox}$.

This problem was solved by Delahay and Berzins [33], who used the following substitution:

$$\Psi(x, t) = C_{Ox}(x, t) + C_A(x, t), \qquad (8.76)$$

$$\varphi(x, t) = C_A(x, t) - \frac{k_2}{k_1} C_{Ox}(x, t), \qquad (8.77)$$

and, assuming that $D_A = D_{Ox} = D$, they obtained the solution by means of Laplace transformation.

The final results lead to definition of the concentrations of A and Ox forms depending on time and on distance from the electrode, but in this case the time dependence of the Ox form concentration on the electrode surface is of primary importance:

$$C_{Ox}(0, t) = \frac{k_1}{k_1+k_2}\left\{C^0 - 2\lambda\left(\frac{Dt}{\pi}\right)^{1/2} - \lambda\frac{k_2}{k_1}\frac{D^{1/2}}{(k_1+k_2)^{1/2}}\operatorname{erf}[(k_1+k_2)^{1/2}t^{1/2}]\right\}. \quad (8.78)$$

Putting $C_{Ox}(0, t) = 0$ we obtain from this equation the expression for the transition time of the kinetic process:

$$\tau_k^{1/2} = \frac{\pi^{1/2}nFD^{1/2}C^0}{2i_0} - \frac{\pi^{1/2}}{2K(k_1+k_2)^{1/2}}\operatorname{erf}[(k_1+k_2)^{1/2}\tau_k^{1/2}]. \quad (8.79)$$

This equation is generally valid, but its use in practice is inconvenient, since the kinetic transition time τ_k also appears in the argument of the error function. Therefore it is easier to consider the cases in which the values of the argument of the error function are within certain limits. When the argument is greater than two the error function is practically equal to one, and equation (8.79) becomes simplified to

$$\tau_k^{1/2} = \frac{\pi^{1/2}nFC^0D^{1/2}}{2i_0} - \frac{\pi^{1/2}}{2K(k_1+k_2)^{1/2}}. \quad (8.80)$$

This equation is often used in practice for the calculation of kinetic parameters of chemical reactions preceding electrode processes. According to the suggestion of Gierst and Juliard [34] the transition times measured at various current densities are expressed in the form of the dependence of $i_0\tau_k^{1/2}$ on i_0.

It follows from the Sand's equation that, when the electrode process is controlled by the diffusion rate alone, the linear relationship shown by curve *1* (Fig. 8.2) parallel to the i_0 axis is observed.

It can be seen from the equation:

$$\tau_k^{1/2}i_0 = \tau^{1/2}i_0 - \frac{\pi^{1/2}i_0}{2K(k_1+k_2)^{1/2}} \quad (8.81)$$

that the products $i_0\tau_k^{1/2}$ obtained for kinetic processes are always smaller than those obtained for diffusion processes. Equation (8.81), which has been derived from equation (8.80), indicates that when the current density tends to zero $\tau_k^{1/2}i_0$ tends to $\tau^{1/2}i_0$. The product $i_0\tau_k^{1/2}$ decreases linearly with increasing current density and, according to equation (8.81) the slope of the straight line representing this relationship depends on the equilibrium constant and rate constants of the chemical process. The value of the second term of equation (8.81) decreases with increasing value of the product $K(k_1+k_2)^{1/2}$, and the straight line representing the dependence of $i_0\tau_k^{1/2}$ on i_0 approaches straight line *1* representing this dependence for diffusion processes. Straight lines *2, 3, 4* and *5* represent schematically this dependence, for increasing values of $K(k_1+k_2)^{1/2}$.

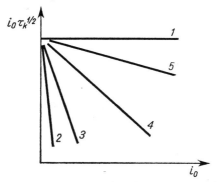

Fig. 8.2 Schematic representation of the dependence of $i_0\tau_k^{1/2}$ on i_0: *1*—diffusion process; *2, 3, 4* and *5*—the dependence for increasing values of $K(k_1+k_2)^{1/2}$ in the case of an electrode process preceded by a first-order chemical reaction.

This discussion leads to the conclusion that the determination of rate constants of chemical reactions preceding the electrode process is not possible when the value of $\pi^{1/2}i_0/2K(k_1+k_2)^{1/2}$ is small compared with $i_0\tau^{1/2}$. The exact measurement of short transition times, which are convenient in investigations of fast electrode processes, is, in practice, limited to one millisecond. Therefore if it is assumed that, in a wide range of current density it is possible to observe transition time changes exceeding 10%, we can conclude from equation (8.79) that the kinetic effect can be observed when $K(k_1+k_2)^{1/2}$ is smaller than 500 s$^{-1/2}$.

So far this discussion has been limited to the range of validity of equation (8.77) in which the argument of the error function expressed by $(k_1+k_2)^{1/2}\tau_k^{1/2}$ is greater than 2. When the value of this argument is smaller, the dependence of the error function on τ_k must be taken into account. In such cases the limiting value of $i_0\tau_k^{1/2}$ for large current densities can be determined by expanding the error function into a series:

$$\operatorname{erf} X = \frac{2}{\pi^{1/2}}\left(X - \frac{X^3}{3\times 1!} + \frac{X^5}{5\times 2!} - \frac{X^7}{7\times 3!} \cdots\right). \qquad (8.82)$$

Taking into account only the first term (which is justified because higher powers of small values of X are much smaller than X), and substituting this expression in equation (8.77) gives the relationship:

$$(i_0 \tau_k^{1/2})_{\text{for } i_0 \to \infty} = \frac{\pi^{1/2} n F D^{1/2} C^0}{2\left(1 + \dfrac{1}{K}\right)}. \tag{8.83}$$

In this case, as in the case of diffusion processes, the product $i_0 \tau_k^{1/2}$ does not depend on current density, but its values are smaller than those of $i_0 \tau^{1/2}$ in the diffusion process, and this difference increases with decreasing equilibrium constant K. When the value of K is very small, then unity in the denominator of equation (8.83) can be neglected and the following simpler form is obtained:

$$(i_0 \tau_k^{1/2})_{\text{for } i_0 \to \infty} = \frac{\pi^{1/2} n F D^{1/2} C^0 K}{2}. \tag{8.84}$$

Thus in this case the transition time is proportional to the equilibrium concentration C_{Ox}^0, which is expressed as the product KC^0.

When the equilibrium constant is very large, equation (8.83) becomes identical with the Sand equation.

In certain cases, at large current densities, when $i_0 \tau_k^{1/2}$ becomes independent of current density, it is possible to determine the equilibrium constant of the chemical reaction preceding the electrode process, by means of chronopotentiometric experiments.

Tanaka and Yamada [35] made a theoretical study of a similar case, assuming that diffusion coefficients of substances A and Ox were different. They derived equations which they used for the determination of association constants of ion pairs formed by sulphate with complex chromic and cobaltic ions.

Hale [36] considered the effect of a preceding chemical reaction on an electrode process taking place at the rotating disc electrode when electrolysis at constant current intensity was taking place.

8.5 The Rotating Disc Method

The problem represented by scheme (8.3) under rotating disc conditions was first solved by Budevskii [37], who assumed that, in the diffusion layer, the liquid is stationary. The complete solution of this problem was given by Koutecký and Levich [38], who solved the following system of convective diffusion equations, taking into account the formation and disappearance of the A and Ox forms, due to the chemical reactions preceding the electrode process:

$$S_x \frac{dC_A}{dx} = D\frac{d^2 C_A}{dx^2} - k_1 C_A + k_2 C_{Ox}, \tag{8.85}$$

$$S_x \frac{dC_{Ox}}{dx} = D\frac{d^2 C_{Ox}}{dx^2} + k_1 C_A - k_2 C_{Ox}, \tag{8.86}$$

where S_x is the component of the speed of the solution perpendicular to the disc surface.

In this case also conditions (8.25) and (8.26) are fulfilled before the start of the experiment and remain valid in the regions of solution sufficiently remote from the electrode surface. It is assumed that the electrode reaction is very fast and that the electrode potential is so negative that the concentration of the oxidized form is always equal to zero and that substance A is inactive. Conditions (8.27) and (8.30), given for chronoamperometric conditions, are also fulfilled.

The system of equations (8.85) and (8.86) can be solved by substituting:

$$\varphi = k_2 C_{Ox} - k_1 C_A, \tag{8.87}$$

and

$$\Psi = C_{Ox} + C_A. \tag{8.88}$$

Then instead of equations (8.85) and (8.86):

$$S_x \frac{d\Psi}{dx} = D\frac{d^2 \Psi}{dx^2}, \tag{8.89}$$

and

$$S_x \frac{d\varphi}{dx} = D\frac{d^2 \varphi}{dx^2} - \varphi l, \tag{8.90}$$

where

$$l = k_1 + k_2. \tag{8.91}$$

The boundary conditions should be suitably expressed in the new variables.

Equation (8.89) is identical with that for convection diffusion at the disc electrode. According to the conclusions reached in Chapter 5 its general solution is:

$$\Psi = a_1 \int_0^x \exp\left\{\int_0^t \frac{S_x(t')dt'}{D}\right\} dt + (C^0 - a_1 \delta) \tag{8.92}$$

where C^0 is the sum of concentrations of the A and Ox forms in the bulk of solution, and δ is the thickness of the diffusion layer given by:

$$\delta = 1.61 D^{1/3} \nu^{1/6} \omega^{-1/2}. \tag{8.93}$$

In the solution of equation (8.90) it was assumed that the chemical reaction is so fast that the condition:

$$S_x \frac{d\varphi}{dx} \ll \varphi l \tag{8.94}$$

is fulfilled.

In this case, instead of equation (8.90):

$$D \frac{d^2 \varphi_0}{dx^2} = \frac{l \varphi_0}{D} \tag{8.95}$$

where φ_0 is the value of φ when the convection term is not present in equation (8.90).

The solution of equation (8.95) is:

$$\varphi_0 = (C^0 - a_1 \delta) \exp \left\{ -\sqrt{\frac{l}{D}} x \right\} \tag{8.96}$$

and equation (8.92) can be expressed in the following fuller form:

$$\Psi = \frac{C^0}{\left(1 + \frac{1}{K\delta}\sqrt{\frac{D}{l}}\right) \delta} \int_0^x \exp \left\{ \int_0^t \frac{S_x(t') dt'}{D} \right\} dt + (C^0 - a_1 \delta). \tag{8.97}$$

Let us now consider the effect, on the recorded current, of a chemical reaction preceding the electrode process. The current is described by the equation:

$$i = nFDA \left(\frac{dC_{Ox}}{dx} \right)_{x=0} = nFDA \left(\frac{d\Psi}{dx} \right)_{x=0} = \frac{nFDAC^0}{\left(1 + \frac{1}{K\delta}\sqrt{\frac{D}{l}}\right) \delta}. \tag{8.98}$$

Since, in the case of a rotating disc electrode, the limiting current i_g is given by the general equation:

$$i_g = \frac{nFADC^0}{\delta}, \tag{8.99}$$

by taking into account equation (8.97) we can express equation (8.98) in the form:

$$i_k = -\frac{i_g}{1 + \frac{1}{K\delta}\sqrt{\frac{D}{l}}} \qquad (8.100)$$

Since the parameter $\sqrt{D/l}$ corresponds to the thickness of the reaction layer it can be called μ_k. Kinetic processes not governed by the equilibrium take place only in the layer of thickness μ_k, and in the case of a disc electrode μ_k is independent of the place on the disc surface.

By introducing the full description of i_g and δ into equation (8.100) the complete equation for the kinetic current is obtained:

$$i_k = \frac{DC^0}{1.61 D^{1/3}\nu^{1/6}\omega^{-1/2}\left[1 + \frac{1}{1.61K}\omega^{1/2}l^{-1/2}D^{1/6}\nu^{-1/6}\right]} \qquad (8.101)$$

In order to elucidate the significance of condition (8.95) we must calculate the correction for φ, assuming that $\varphi = \varphi_0 + \varphi_1$, where $\varphi_1 \ll \varphi_0$, and that the equation:

$$\frac{d^2\varphi_1}{dx^2} - l\varphi_1 = S_x \frac{d\varphi_0}{dx} \qquad (8.102)$$

is satisfied.

Substituting the value of S_x for the surface area of the disc:

$$S_x = -0.51 \frac{\omega^{3/2}}{\nu^{1/2}} x^2 \qquad (8.103)$$

and putting $\varphi_0 = a_3 \exp[-x(\sqrt{l/D})]$, we obtain:

$$\varphi_1 = -\frac{\omega^{3/2}}{4\nu^{1/2}l} a_3 x \exp\left(-\sqrt{\frac{l}{D}}x\right)\left(\frac{1}{2} + \frac{1}{2}\sqrt{\frac{l}{D}}x + \frac{1}{3}\sqrt{\frac{l}{D}}x^2\right). \qquad (8.104)$$

From this relationship we obtain:

$$\frac{1}{K}\left(\frac{d\Psi}{dx}\right)_{x=0} = \left(\frac{d\Psi_0}{dx}\right)_{x=0}\left[1 + \frac{1}{8}\sqrt{\frac{\omega^3}{l^3}\frac{D}{\nu}}\right], \qquad (8.105)$$

from which it follows that equation (8.101), which was derived neglecting the convection term, is valid when the following condition is fulfilled:

$$\frac{1}{8}\sqrt{\frac{\omega^3 D}{l^3 \nu}} \ll 1. \qquad (8.106)$$

Inequality (8.106) proves that, in the case of fast chemical reactions, the principal change of φ takes place in the layer μ_k which is much thinner than δ. Since the contribution of convection to the transport of the substance is small at distances $x < \delta$ the convection term in equation (8.90) can be neglected.

The problems discussed above have been limited to electrode processes taking place at rotating disc electrodes and preceded by first-order chemical reactions, in which the diffusion coefficients of A and Ox forms were equal. The solution of identical problems in which these coefficients are different has been presented by Dogonadze [39].

Vielstich and Jahn [40] applied the rotating disc technique to the investigation of the kinetics of diffusion of weak acids and expressed in a different form equation (8.101) derived by Koutecký and Levich. This makes it possible to determine the kinetic parameters of a chemical reaction from the dependence of $i_k/\omega^{1/2}$ on i_k:

$$\frac{i_k}{\omega^{1/2}} = \frac{i_g}{\omega^{1/2}} - \frac{D^{1/6}i_k}{1.61\nu^{1/6}K(k_1+k_2)^{1/2}} \quad . \tag{8.107}$$

This procedure is more convenient, since it does not require a knowledge of the limiting current value. The parameter $K(k_1+k_2)^{1/2}$ is calculated from the slope of the straight line of a plot of $i_k/\omega^{1/2}$ vs. i_k.

8.6 General Discussion

Presentation of kinetic equations in a form which makes it possible to define kinetic parameters of chemical reactions preceding the electrode process without comparing the kinetic current with the limiting current is very convenient. It is not always possible to evaluate the limiting current experimentally and an evaluation based on calculations may be erroneous. Therefore it is convenient to determine the kinetic constants of chemical reactions from the angular coefficient of the $i_0\tau_k^{1/2}-i_0$ relationship obtained from chronopotentiometric measurements. Equation (8.107) is useful for the same reason. Kinetic equations for the four methods discussed can be represented by the general relationship:

$$i_k X^{1/2} = A i_g X^{1/2} - B \frac{i_k}{K(k_1+k_2)^{1/2}} \tag{8.108}$$

where A and B are constants characteristic of the individual methods, and X is the kinetic parameter.

Table 8.4 VALUES OF A AND B IN EQUATION (8.108)

Method	A	B
Polarography	1	$\dfrac{1}{0.886}$
Stationary electrode voltammetry, reversible process	0.98	$0.462\left(\dfrac{nF}{RT}\right)^{1/2}$
Stationary electrode voltammetry, irreversible process	0.98	$0.521\left(\dfrac{\alpha n_\alpha F}{RT}\right)^{1/2}$
Chronopotentiometry	1	$\dfrac{\pi^{1/2}}{2}$
Rotating disc method	1	$\dfrac{D^{1/6}}{1.61\nu^{1/6}}$

The values of A and B are given in Table 8.4. Constant A is practically identical for all the methods discussed and is equal to one with the exception of stationary electrode voltammetry. The values of B are different for each method.

When the suitable kinetic parameter and constants A and B are introduced into equation (8.108), a specific equation is obtained. Substituting kinetic parameter t_1 for X we obtain polarographic equation (8.52) written in a different form. Similarly, the substitution of $1/V$, $1/\omega$, τ and respective constants from Table 8.4 leads to equations (8.67), (8.73) and (8.79) for stationary electrode voltammetry, rotating disc electrode and chronopotentiometry, respectively.

It is possible, as described above, to represent particular equations describing the kinetic processes for individual methods by the general relationship (8.108). It is therefore possible to suggest a general method for the determination of the kinetic constants of chemical reactions in which neither the calculation of the limiting current nor calculation of the transition time for a diffusion process is necessary.

According to equation (8.108), in order to determine the kinetic parameter of the chemical reaction $K(k_1+k_2)^{1/2}$, the relationship between $i_k X^{1/2}$ and i_k should be plotted. When parameter $K(k_1+k_2)^{1/2}$ is very large, $i_k X^{1/2}$ is independent of i_k, since the second term of the right-hand side of equation (8.108) is practically equal to zero. For finite values of $K(k_1+k_2)^{1/2}$ the relationship is linear, but the product $i_k X^{1/2}$ decreases with increasing i_k. Such a relationship is represented by straight line 2 in Fig. 8.3. The slope of this straight line depends on the equilibrium constant and the chemical reaction rate constants. $K(k_1+k_2)^{1/2}$ canbe calculated from the slope of the experimental straight line

$$\frac{\Delta(i_k X^{1/2})}{\Delta i_k} = -\frac{B}{K(k_1+k_2)^{1/2}}. \qquad (8.109)$$

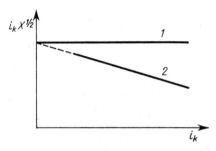

Fig. 8.3   Schematic representation of the dependence of $i_k X^{1/2}$ on i_k for an electrode process preceded by a first-order chemical reaction (curve 2). Straight line *1* shows the dependence in the case of diffusion processes.

If the equilibrium constant of the chemical reaction, K, is known, the rate constants k_1 and k_2 can be calculated from this expression. In principle the equilibrium constant of the chemical reaction can also be determined by the electrochemical methods discussed, but in practice such an approach is limited to the cases when the rate constant of the chemical reaction is small and only the Ox form, which is in equilibrium with the A form, reacts with the electrode. The duration of the experiment must be short enough to exclude the possibility of transformation of A into Ox at the electrode surface.

This method of determining rate constants from the $i_k X^{1/2}$–i_k relationship has been used for many years, mainly in chronopotentiometry.

In the case of chronopotentiometry equation (8.108) must be somewhat modified, since in this case — contrary to other methods — the current density, and not the kinetic parameter, is the independent variable. The subscripts "k" and "g" should be attached to the transition times and the current density should have the single meaning of current density given by the experimental programme.

In the case of polarography, stationary electrode voltammetry and the rotating disc technique equation (8.108) is used mainly in investigations of processes controlled by both diffusion and the kinetics of chemical reactions. When the process is controlled by the rate of the chemical reaction alone the kinetic current is independent of the kinetic parameter and in such a case the possibility of studying the $i_k X^{1/2} - i_k$ relationship is limited.

Reinmuth [41] proposed that in chronopotentiometry the experimental results should be analysed by the system $\dfrac{1}{i_0 \tau^{1/2}}$ vs. $\dfrac{1}{\tau^{1/2}}$ or $\tau^{1/2}$ vs. $\dfrac{1}{i_0}$ instead of the relationship $i_0 \tau^{1/2}$ vs. i_0 which has been used up to the present time. The first alternative is particularly interesting since it is linear over a considerably larger current density range than the classical relationship. The second alternative is equivalent to the relationship $i_0 \tau^{1/2}$ vs. i_0. Kinetic currents have been analysed and discussed in a number of papers [42–49].

8.7 Electrode Processes Preceded by Pseudo-First Order Chemical Reactions

The electrode processes discussed so far have been those preceded by first-order reactions resulting in depolarizer formation. Such processes can be represented by the general scheme (8.3). In practice, electrode processes preceded by more complex chemical reactions are also encountered. The most common processes of this type can be represented by the following schemes:

$$A + C \underset{k_2'}{\overset{k_1'}{\rightleftharpoons}} Ox + ne \rightleftharpoons Red \qquad (8.110)$$

$$A \underset{\overline{k_2}}{\overset{\overline{k_1}}{\rightleftharpoons}} Ox + C \quad Ox + ne \rightleftharpoons Red \qquad (8.111)$$

where C is an electrode-inactive substance in the range of $Ox + ne \rightleftharpoons Red$ reaction potentials.

According to equations (8.110) and (8.111) the rate of Ox formation depends on the concentration of substance C. When this concentration is comparable with that of substance A, its value in the vicinity of electrode surface changes during the experiment and the process is complex from the mathematical point of view. However, in practice it is often possible to reach such a high concentration of C that, in spite of the electrode process, even at the electrode surface, it is practically equal to that in the bulk of the solution. Reaction (8.4) is an example of such a process. Theoretically this reaction can be represented by scheme (8.111). However, since the concentration of substance C, in this case water, exceeds by several orders the concentration of formaldehyde, it can be described and investigated on the basis of the relationships derived for first-order reactions.

In such cases the rates and equilibrium constants of reactions (8.110) and (8.111) are related to K, k_1 and k_2 derived for first-order reactions in the following way:

$$k_1 = k_1' C_C^0, \quad k_2 = k_2', \quad K = K' C_C^0, \qquad (8.112)$$

where

$$K' = \frac{C_{Ox}^0}{C_C^0 C_A^0} \qquad (8.113)$$

and

$$k_1 = \bar{k}_1, \quad k_2 = \bar{k}_2 C_C^0, \quad K = \frac{\bar{K}}{C_C^0}, \qquad (8.114)$$

where

$$\bar{K} = \frac{C_{Ox}^0 C_C^0}{C_A^0}. \qquad (8.115)$$

Superscripts on the concentration symbols indicate the equilibrium state.

Combining equation (8.108) with (8.112) gives the general kinetic equation of reaction (8.110):

$$i_k X^{1/2} = A i_g X^{1/2} - \frac{B i_k}{K' C_C^0 (k_1' C_C^0 + k_2')^{1/2}}. \qquad (8.116)$$

The general kinetic equation of reaction (8.111) can be obtained by combining equation (8.114) with (8.108):

$$i_k X^{1/2} = A i_g X^{1/2} - \frac{B i_k C_C^0}{\bar{K}(\bar{k}_1 + \bar{k}_2 C_C^0)^{1/2}}. \qquad (8.117)$$

It follows from the above equations that the current of such processes depends on substance C concentration and this dependence in reaction (8.110) is different from that in reaction (8.111).

8.8 Electrode Processes Preceded by Two First-Order Chemical Reactions

In practical electrochemistry electrode processes preceded by two first-order chemical reactions are also encountered. Usually they consist of recombinations of anions with hydrogen ions, leading to the formation of dibasic acids. Such processes can also be encountered in electroanalytical investigations of complexes.

On the assumption that the chemical reactions are of the first order, the processes discussed can be represented by the general scheme:

$$A_1 \underset{k_2}{\overset{k_1}{\rightleftharpoons}} A_2 \underset{k_4}{\overset{k_3}{\rightleftharpoons}} Ox + ne \rightleftharpoons Red. \tag{8.118}$$

In the case of electrode processes preceded by recombinations of acids, scheme (8.118) has the following form:

$$A^{2-} + H^+ \underset{k_2}{\overset{k_1}{\rightleftharpoons}} HA^- + H^+ \underset{k_4}{\overset{k_3}{\rightleftharpoons}} H_2A + ne \rightleftharpoons Red. \tag{8.119}$$

H_2A is reduced at the least negative potential.

Hanuš and Brdička [50] described this type of reaction for polarographic conditions. They obtained an approximate solution using the reaction layer concept. The exact solution was given by Koutecký [51], who assumed equality of the diffusion coefficients of the undissociated acid molecules and the corresponding anions. See also paper of Tanaka et al. [70].

According to this treatment the rate constant k_3 is described by the following relationship:

$$k_3 = \left\{ \frac{1}{0.886} \frac{([H^+] + K_{a2})}{[H^+]^2} \frac{\bar{i}_k}{(\bar{i}_g - \bar{i}_k)} \right\}^2 \frac{K_{a1}}{t_1} \tag{8.120}$$

where K_{a1} and K_{a2} are the first and the second dissociation constants of acid H_2A, respectively, and i_k is the limiting current of the first wave corresponding to the reduction of H_2A. The solution based on the reaction layer concept leads to a similar result, but it gives the numerical coefficient equal to $1/0.81$ instead of $1/0.886$.

Taking into account the difference between the values of diffusion coefficients of ionized and non-ionized forms, the exact solution can be expressed as follows:

$$\frac{\bar{i}_k}{\bar{i}_g - \bar{i}_k} = 0.886 Y \sqrt{t_1} \tag{8.121}$$

where:

$$Y = \frac{k_3 [H^+]^2}{K_{a1}^{1/2} \left\{ [H^+] + K_{a2} \left(\frac{D_1}{D_2} \right) \right\}^{1/2} (K_{a2} + [H^+])^{1/2}}, \tag{8.122}$$

and D_1 and D_2 are the diffusion coefficients of A^{2-} and HA^-, respectively; it is assumed that D_2 is equal to the diffusion coefficient of H_2A.

Process (8.118) was considered by Ashley and Reilley in their elegant paper on chronoamperometric and chronopotentiometric electrode processes connected in various ways with one or two chemical reactions [52]. Although they did not derive particular solutions, such solutions can be obtained on the basis of the results of their work.

The case of a chronopotentiometric electrode process preceded by two first-order chemical reactions was also worked out by the author of this book [53], who solved the problem by means of Laplace transformation and expressed the final result in the form of an equation describing the kinetic transition time observed under the experimental conditions.

8.9 Effect of the Double Layer Structure on the Kinetics of Chemical Reactions Connected with the Electrode Processes

Homogeneous chemical reactions connected with electrode processes, such as the dissociation and recombination of acids, or formation and decomposition of ion pairs and complexes, can take place in the electric field of the double layer. For this reason the rates of these reactions are affected by the double layer effects. The influence of the double layer is particularly pronounced when the thickness of the reaction layer does not greatly exceed the diffusion part of the double layer. In such cases the chemical reaction rate can be varied by changing the structure of the double layer, e.g. by changing the supporting electrolyt econcentration or the change of potential. For this purpose the process may be carried out on thallium amalgam electrodes.

Examples of the effect of double layer structure on the kinetics of chemical reactions preceding charge transfer were quoted by Gierst [19], who showed that many abnormal features of electrode processes are due to the double layer effect. One of the examples is the reduction of nickel(II) in concentrated solutions of non-complexing electrolytes, which was formerly regarded [54] as a process leading to the formation of nickel(I). Another example is the discharge of $Cd(CN)_4^{2-}$. When the cyanide concentration exceeds 0.05 M electroreduction is preceded by the following chemical reaction [19, 55, 56]:

$$Cd(CN)_4^{2-} \underset{k_2}{\overset{k_1}{\rightleftharpoons}} Cd(CN)_3^- + CN^-. \qquad (8.123)$$

The observed kinetic limiting current can be changed by adding an electrolyte which is not specifically adsorbed onto the electrode.

The theoretical aspects of the double layer effect on kinetic currents were investigated by Gierst [22] and Hurwitz [57] but the most accurate study of these problems for chronopotentiometry was published by Matsuda [58], who considered the following two kinds of reactions preceding the electrode process:

$$\text{I} \quad A^{z_A} \underset{k_2}{\overset{k_1}{\rightleftharpoons}} Ox^{z_{Ox}} + \nu Y^{z_Y} \quad (z_A = z_{Ox} + \nu z_Y), \quad (8.124)$$

$$\text{II} \quad A^{z_A} + \nu Y^{z_Y} \underset{k_2}{\overset{k_1}{\rightleftharpoons}} Ox^{z_{Ox}} \quad (z_{Ox} = z_A + \nu z_Y). \quad (8.125)$$

Matsuda derived functions G_I and G_{II} theoretically for reactions I and II, respectively. These functions are complex functions of potential Φ_2, the charge of the ions participating in the reaction and the thickness of the reaction layer.

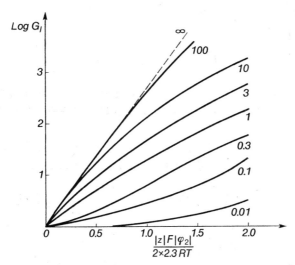

Fig. 8.4 Dependence of function G_I on potential Φ_2. The figures by the curves stand for $1/\mu\varkappa$ values (from [58] by permission of the copyright holders, the American Chemical Society).

In Fig. 8.4 log G_I is shown as a function of $\dfrac{|z|F\Phi_2}{2 \times 2.3RT}$ for various values of $\dfrac{1}{\mu\varkappa}$; it is assumed that the supporting electrolyte is of the 1-1 type, $z_A = -2$, and $z_{Ox} = -1$. This case corresponds to reaction (8.123).

Fig. 8.5 shows the dependence of G_I on $\dfrac{1}{\mu\varkappa}$ for different values of Φ_2. $\mu = \left(\dfrac{D_{Ox}}{k_2(C_Y^0)^\nu}\right)^{1/2}$, and $\dfrac{1}{\varkappa} = \left(\dfrac{RTD^0}{8\pi z^2 F^2 C^P}\right)^{1/2}$, where D^0 is the dielectric constant and C^P is the concentration of the supporting electrolyte.

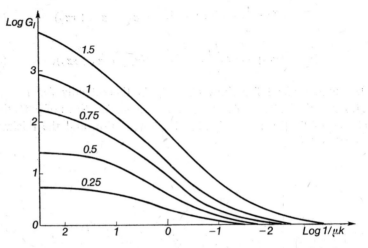

Fig. 8.5 Dependence of function G_I on $1/\mu\varkappa$ at various potentials Φ_2. The figures by the curves stand for $|z|F\Phi_2/2 \times 2.3RT$ (from [58] by permission of the copyright holders, the American Chemical Society).

The transition time for an electrode reaction preceded by chemical reaction I can be described by the following relationship:

$$i\tau_k^{1/2} = \frac{\pi^{1/2} nFD_A^{1/2} C_A^0}{2} - \frac{\pi^{1/2}(D_A/D_{Ox})^{1/2} i}{2K_d[k_2/(C_Y^0)^\nu]^{1/2}} G_I. \qquad (8.126)$$

This equation, with the exception of function G_I, is identical with that given by Delahay and Berzins [33].

The average limiting current can be described as follows:

$$\frac{\bar{i}_g - \bar{i}_k}{\bar{i}_k} = \left(\frac{D_A}{D_{Ox}}\right)^{1/2} \frac{G_I}{0.89 t_1^{1/2} K_d[k_2/(C_Y^0)^\nu]^{1/2}}. \qquad (8.127)$$

The validity of this equation was established by Delahay and Kleinerman.

In equations (8.126) and (8.127) K_d is the dissociation constant of A^{z_A} when $\dfrac{1}{\mu\varkappa} < 0.01$, i.e. when the thickness of the diffuse part of the double layer $\frac{1}{\varkappa}$ is much smaller than that of the reaction layer μ we have $G_I \approx 1$. As a result, the effect of the double layer on the kinetics of the reaction preceding the electrode process is not observed.

On the other hand, when $\frac{1}{\mu\varkappa} > 100$, i.e. when the diffusion layer is much thicker than the reaction layer, G_I increases exponentially with increasing Φ_2:

$$G_I = \exp\left[\frac{(z_{Ox}+z_A)F\Phi_2}{2RT}\right]. \qquad (8.128)$$

Then a relatively simple equation is obtained describing the kinetic current. It was used by Grabowski and Bartel [59] in their studies of the effect of ionic strength (potential Φ_2) on the kinetic current of p-dimethylaminobenzaldehyde controlled by the rate of the preceding protonation of the $-C\diagdown_H^{\nearrow O}$ (aldehyde) group.

Polarographic kinetic currents and other examples of the effect of the double layer on the kinetics of chemical reactions connected with electrode processes have been discussed by Mairanovskii [60].

The dissociating effect of the electrode field, consisting in an increase of the dissociation rate in the layer of solution neighbouring the outer Helmholtz plane, has also been considered. This effect is particularly clearly visible when the difference between the electrode potential and the electrocapillary zero potential is large, i.e. when the electrode charges are large and can become the predominant effect of the double layer on the kinetics of chemical reactions preceding the charge transfer.

This effect has been discussed for several years [61–63]. The conclusions of the discussion suggest that in the field of the layer the dissociation should be inhibited. This is contrary to the results of studies carried out by Nürnberg [64–68] who was investigating the effect of the structure of the double layer on dissociation of several organic acids, using the Faraday rectification method. He described the dissociating effect of the field of the layer and the results of his theoretical considerations are in agreement with the results of measurements. This shows that the results of kinetic studies of chemical reactions based on electrode measurements are not always reliable and the correct results can be obtained only after the introduction of corrections for the double layer effects.

8.10 Utility of the Methods for Kinetic Studies on Homogeneous Reactions

In order to evaluate the usefulness of the methods considered, in kinetic studies of chemical reactions connected with electrode processes, the rate constants of the chemical reactions should be related to those of the electrode processes by means of the following equation [69]:

$$k_{fh} = k_1 \mu \tag{8.129}$$

where μ is the thickness of the reaction layer, given by the expression:

$$\mu = \sqrt{\frac{D}{k_2}}. \tag{8.130}$$

From equations (8.129) and (8.130):

$$k_{fh} = k_1 k_2^{-1/2} D^{1/2}, \tag{8.131}$$

and by introducing the equilibrium constant of the chemical reaction $K = k_1/k_2$:

$$k_{fh} = K k_2^{1/2} D^{1/2}. \tag{8.132}$$

Since the maximum rate constant k_{fh}^m that can be determined does not exceed the maximum transport rate $\bar{V}_m$ by more than tenfold:

$$k_{fh}^m = 10\bar{V}_m, \tag{8.133}$$

combining equations (8.132) and (8.133) gives the general expression:

$$(Kk_2^{1/2})^m = \frac{10\bar{V}_m}{D^{1/2}} \tag{8.134}$$

where $(Kk_2^{1/2})^m$ is the maximum value of parameter $Kk_2^{1/2}$ which can be determined under given conditions.

Substituting the maximum values of transport rates calculated in Chapter 3 for the four methods discussed and assuming that $D = 10^{-5} \text{ cm}^2/\text{s}$, the following values of $(Kk_2^{1/2})^m$ are obtained:

polarography	$\bar{V}_m = 2.4 \times 10^{-3}$ cm/s	$(Kk_2^{1/2})^m = 7.6 \text{ s}^{-1/2}$
chronopotentiometry	$\bar{V}_m = 6 \times 10^{-2}$ cm/s	$(Kk_2^{1/2})^m = 1.9 \times 10^2 \text{ s}^{-1/2}$
stationary electrode voltammetry	$\bar{V}_m = 1.2 \times 10^{-1}$ cm/s	$(Kk_2^{1/2})^m = 3.8 \times 10^2 \text{ s}^{-1/2}$
rotating disc method	$\bar{V}_m = 1.6 \times 10^{-2}$ cm/s	$(Kk_2^{1/2})^m = 5.1 \times 10^1 \text{ s}^{-1/2}$

It follows from this discussion that, in studies of rates of chemical reactions connected with electrode processes, as in studies on electrode kinetics, chronopotentiometry and stationary electrode voltammetry are more effective than the other methods.

References

[1] Brdička, R., *Chem. Listy*, **39**, 35 (1945).
[2] Wiesner, K., *Coll. Czechoslov. Chem. Communs.*, **12**, 64 (1947).
[3] Cantor, L. M. and Peniston, Q. M. J., *J. Am. Chem. Soc.*, **62**, 2113 (1940).
[4] Wiesner, K., *Z. Elektrochem.*, **49**, 164 (1943).
[5] Brdička, R. and Wiesner, K., *Vestn. Kralov. Ceske Spolecnosti Nauk Trida Mat. Prirod.*, **18**, 1 (1943).
[6] Brdička, R. and Wiesner, K., *Naturwiss.*, **31**, 247 (1943).
[7] Brdička, R. and Wiesner, K., *Coll. Czechoslov. Chem. Communs.*, **12**, 39 (1947).
[8] Wiesner, K., *Chem. Listy*, **41**, 6 (1947).
[9] Hanuš, V., *Chem. Zvesti*, **8**, 702 (1954).
[10] Delahay, P., *"New Instrumental Methods in Electrochemistry"*, Interscience Publishers, New York 1954.
[11] Galus, Z. and Adams, R. N., *J. Electroanal. Chem.*, **4**, 248 (1962).
[12] Koutecký, J. and Brdička R., *Coll. Czechoslov. Chem. Communs.*, **12**, 337 (1947).
[13] Buđevskii, E. and Desimirov, G., *Dokl. Akad. Nauk SSSR*, **149**, 120 (1963).
[14] Koutecký, J., *Chem. Listy*, **47**, 323 (1953); *Coll. Czechoslov. Chem. Communs.*, **18**, 597 (1953).
[15] Čížek, J., Koryta, J. and Koutecký, J., *Chem. Listy*, **52**, 201 (1958); *Coll. Czechoslov. Chem. Communs.*, **24**, 663 (1959).
[16] Čížek, J., Koryta, J. and Koutecký, J., *Coll. Czechoslov. Chem. Communs.*, **24**, 3844 (1959).
[17] Koutecký, J. and Koryta, J., *Electrochim. Acta*, **4**, 318 (1961).
[18] Matsuda, H., *J. Phys. Chem.*, **64**, 336 (1960).
[19] Gierst, L., *"Cinétique d'approche et reactions d'electrode irreversibles"*, Thesis, Université Libre, Bruxelles 1958.
[20] Gierst, L. and Hurwitz, H., *Z. Elektrochem.*, **64**, 36 (1960).
[21] Hurwitz, H., *Z. Elektrochem.*, **65**, 178 (1961).
[22] Gierst, L., *"Transactions of the Symposium on Electrode Processes"*, Wiley, New York 1961, page 109.
[23] Weber, J. and Koutecký, J., *Chem. Listy*, **49**, 562 (1955).
[24] Weber, J. and Koutecký, J., *Coll. Czechoslov. Chem. Communs.*, **20**, 980 (1955).
[25] Koutecký, J. and Čížek, J., *Chem. Listy*, **50**, 196 (1956); *Coll. Czechoslov. Chem. Communs.*, **21**, 836 (1956).
[26] Hanuš, V., *Proceedings of the 1st Congress of Polarography, Prague 1951*, Vol. I, page 811.
[27] Koutecký, J., *Chem. Listy*, **47**, 1758 (1953); *Coll. Czechoslov. Chem. Communs.*, **21**, 857 (1954).
[28] Koutecký, J. and Koryta, J., *Electrochim. Acta*, **3**, 318 (1961).

[29] Ružić, I., private communication.
[30] Savéant, J. M. and Vianello, E., *Electrochim. Acta*, **8**, 905 (1963).
[31] Nicholson, R. S., and Shain, I., *Anal. Chem.*, **36**, 706 (1964).
[32] Reinmuth, W. H., *Anal. Chem.*, **33**, 1793 (1961); **34**, 1446 (1962).
[33] Delahay, P. and Berzins, T., *J. Am. Chem. Soc.*, **75**, 2486 (1953).
[34] Gierst L. and Juliard, A., *Proc. Intern. Comm. Electrochem. Thermod. and Kinet.*, 2nd Meeting, Tamburini, Milan 1950, p. 117, 279.
[35] Tanaka, N. and Yamada, A., *Z. anal. Chem.*, **224**, 117 (1967).
[36] Hale, J. M., *J. Electroanal. Chem.*, **8**, 332 (1964).
[37] Buđevskii, E., *Izv. Bulg. Akad. Nauk, ser. fiz.*, **3**, 43 (1952); **4**, 119 (1954).
[38] Koutecký, J. and Levich, V. G., *Zhurn. Fiz. Khim.*, **32**, 1565 (1958); *Dokl. Akad. Nauk SSSR*, **117**, 441 (1957).
[39] Dogonadze, R. R., *Zhurn. Fiz. Khim.*, **32**, 2437 (1958).
[40] Vielstich, W. and Jahn, D., *Z. Elektrochem.*, **64**, 43 (1960).
[41] Reinmuth, W. H., *J. Electroanal. Chem.*, **38**, 95 (1972).
[42] Brdička, R., *Z. Elektrochem.*, **64**, 16 (1960).
[43] Brdička, R., in "*Advances in Polarography*", Longmuir, I. S., Ed., Pergamon Press, Oxford 1960.
[44] Brdička, R., Hanuš, V., and Koutecký, J., in "*Progress in Polarography*", Zuman P. and Kolthoff, I. M., Eds., Interscience, New York 1962, Vol. 1, page 145.
[45] Koryta, J., in "*Progress in Polarography*", Zuman P. and Kolthoff, I. M., Eds., Interscience, New York 1962, Vol. 1, page 291.
[46] Koryta, J., *Adv. Elektrochem.*, *Electrochem. Eng.*, **6**, 289 (1967).
[47] Koryta, J., *Z. Elektrochem.*, **64**, 23 (1960).
[48] Nürnberg, H. W., and von Stackelberg, M., *J. Electroanal. Chem.*, **2**, 350 (1961).
[49] Koryta, J., *International Congress of Polarography in Kyoto*, September 1966, Plenary Lectures, Butterworths, London, 1967, page 207.
[50] Hanuš, V. and Brdička, R., *Chem. Listy*, **44**, 29 (1950); *Khimiya*, **1**, 28 (1951).
[51] Koutecký, J., *Chem. Listy*, **48**, 360 (1954); *Coll. Czechoslov. Chem. Communs.*, **19**, 1093 (1954).
[52] Ashley, J. W., Jr. and Reilley, C. N., *J. Electroanal. Chem.*, **7**, 253 (1964).
[53] Galus, Z., *Elektrokhimiya*, **4**, 553 (1968).
[54] Orlemann, E. and Sanborn, R., *J. Am. Chem. Soc.*, **78**, 4852 (1956).
[55] Gerischer, H., *Z. physik. Chem.*, Frankfurt, **2**, 79 (1954).
[56] Breiter, M., Kleinerman, M. and Delahay, P., *J. Am. Chem. Soc.*, **80**, 5111 (1958).
[57] Hurwitz, H., *Z. Elektrochem.*, **65**, 178 (1961).
[58] Matsuda, H., *J. Phys. Chem.*, **64**, 336 (1960).
[59] Grabowski, Z. R. and Bartel, E. T., *Roczniki Chem.*, **34**, 611 (1960).
[60] Mairanovskii, S. G., "*Catalytic and Kinetic Waves in Polarography*", Nauka, Moscow 1966 (in Russian).
[61] Sanfeld, A., Steinchen-Sanfeld, A., Hurwitz, H. D. and Defay, R., *J. Chem. Phys.*, **58**, 139 (1962).
[62] Hurwitz, H. D., Sanfeld, A. and Steinchen-Sanfeld, A., *Electrochim. Acta*, **9**, 929 (1964).
[63] Sanfeld, A. and Steinchen-Sanfeld, A., *Trans. Faraday Soc.*, **62**, 1907 (1966).
[64] Nürnberg, H. W. and Barker, G. C., *Naturwiss.*, **51**, 191 (1964).

[65] Nürnberg, H. W., *Fortschr. chem. Forsch.*, **8**, 24 (1967).
[66] Nürnberg, H. W., in *"Polarography 1964"*, Hills, G. J., Ed., Macmillan, London 1966, page 149.
[67] Nürnberg, H. W., Dürbeck, H. W. and Wolff, G., *Z. physik. Chem., Frankfurt,* **52**, 144 (1967).
[68] Nürnberg, H. W., *Disc. Faraday Soc.*, **39**, 136 (1965).
[69] Barker, G. C., Nürnberg, H. W. and Bolzan, J. A., in *"Basic Problems of Modern Theoretical Electrochemistry"*, Mir, Moscow 1965, page 93 (in Russian).
[70] Tanaka, N., Tamamushi, R. and Kadama, M., *Z. physik. Chem., Frankfurt,* **14**, 141 (1958).

Chapter 9

Electrode Processes Followed by First Order Chemical Reactions

---

In the previous chapter we discussed electrode processes, coupled with a chemical reaction which preceded the charge transfer. Electrode processes in which chemical reactions take place after the charge transfer are also known.

Assuming that the electrode process is a reduction of substance Ox, it can be represented as:

$$\text{Ox} + ne \rightleftharpoons \text{Red} \underset{k_2}{\overset{k_1}{\rightleftharpoons}} \text{A}. \tag{9.1}$$

The product of the electrode reduction is unstable and changes chemically to substance A at constant rate k_1.

The mechanisms encountered in electroanalysis are more complex than that represented by scheme (9.1) since the change of Red to A usually involves some other substance. However, it is usually possible to use the additional substance at a concentration so high that it does not change during the process. Hence the chemical reaction becomes a pseudo-first order process which can be described (after suitable modifications) by the expression (9.1).

One of the best known electrode reactions followed by a chemical reaction is the oxidation of ascorbic acid to dehydroascorbic acid [1, 2], which in turn rapidly becomes hydrated. The final product is inactive with respect to the electrode in the voltage region of the change of ascorbic into dehydroascorbic acid. Polarographic studies show that half-wave potentials of this oxidation process are about 200 mV more positive than the equilibrium potentials measured in static conditions.

The first attempts at a mathematical treatment of such processes taking place under polarographic conditions were made by Vavřin [2], but the correct solution of this problem is due to the work carried out by Kern [3, 4].

In order to solve the problem it is necessary to obtain a system of differential equations taking into account concentration changes caused by the chemical reaction. For (9.1) and substances Ox, Red, and A these equations are as follows:

$$\frac{\partial C_{Ox}(x,t)}{\partial t} = D_{Ox} \frac{\partial^2 C_{Ox}(x,t)}{\partial x^2}, \qquad (9.2)$$

$$\frac{\partial C_{Red}(x,t)}{\partial t} = D_{Red} \frac{\partial^2 C_{Red}(x,t)}{\partial x^2} - k_1 C_{Red}(x,t) + k_2 C_A(x,t), \quad (9.3)$$

$$\frac{\partial C_A(x,t)}{\partial t} = D_A \frac{\partial^2 C_A(x,t)}{\partial x^2} + k_1 C_{Red}(x,t) + - k_2 C_A(x,t). \quad (9.4)$$

The initial conditions may be common for the methods discussed:

$$t = 0, \quad x \geqslant 0, \quad C_{Ox} = C_{Ox}^0, \quad C_{Red} = 0, \quad C_A = 0. \qquad (9.5)$$

The boundary conditions at the ends of the region are also common:

$$t > 0, \quad x \to \infty, \quad C_{Ox} \to C_{Ox}^0, \quad C_{Red} \to 0, \quad C_A \to 0. \qquad (9.6)$$

It follows from the initial conditions assumed above that before the commencement of the electrochemical process the Ox form is present in the solution at concentration C_{Ox}^0. During the process it is possible to find in the diffusion layer, at a sufficient distance from the electrode surface, a solution at which the initial system of concentrations (9.5) remains unaltered.

9.1 Polarography

After Vavřin's attempted mathematical treatment of an electrode process followed by a chemical reaction, Kern [3] presented his interpretation of this problem based on the reaction layer concept, and then published the exact solution [4], in which he assumed that the chemical change is fast. In such a case it is possible to neglect as insignificant the third term of the right-hand side of equation (9.3) and to eliminate equation (9.4) from the solution of the problem.

Kern defined the boundary conditions on the surface of the electrode as follows:

$$t > 0, \quad x = 0, \quad \frac{C_{Ox}(0,t)}{C_{Red}(0,t)} = c \text{ (constant)}; \qquad (9.7)$$

$$D_{Ox}\left[\frac{\partial C_{Ox}(x,t)}{\partial x}\right]_{x=0} + D_{Red}\left[\frac{\partial C_{Red}(x,t)}{\partial x}\right]_{x=0} = 0. \qquad (9.8)$$

Constant c is related to the applied potential E by the relationship:

$$E = E^0 + \frac{RT}{nF} \ln c \qquad (9.9)$$

where E^0 is the standard potential of the Ox/Red system.

Kern's treatment was limited to cases in which $E_{1/2}$ is shifted by more then 0.03 V, with respect to the normal potential, i.e. to the cases in which the following chemical reaction is fast.

The solution expresses the polarographic wave by the equation:

$$E = E^0 + \frac{RT}{nF} \ln \frac{i_g - i}{i} + \frac{RT}{nF} \ln 0.65 + \frac{RT}{2nF} \ln k_1 t_1, \qquad (9.10)$$

from which it follows that the half-wave potential depends on the drop time and on the rate constant of the subsequent reaction:

$$E^k_{1/2} = E^0 + \frac{RT}{nF} \ln 0.65 + \frac{RT}{2nF} \ln k_1 t_1. \qquad (9.11)$$

It follows from equations (9.2)–(9.4) that this solution was obtained for conditions of linear diffusion and hence it could not be strictly valid when applied to the polarographic process. The strict polarographic solution, taking into account the rate of increase of the mercury drop surface was obtained by Koutecký [5, 6] for the conditions in which the equilibrium of the chemical reaction is shifted significantly in the direction of form A. Assuming that A is formed at a fast rate, it is possible to express the half-wave potential on the basis of Koutecký's solution by means of an equation very similar to that deduced by Kern [equation (9.11)]:

$$E^k_{1/2} = E^0 + \frac{RT}{nF} \ln 0.886 - \frac{RT}{2nF} \ln \frac{D_{Ox}}{D_{Red}} + \frac{RT}{2nF} \ln k_1 t_1. \qquad (9.12)$$

These relationships describe only the processes in which the charge transfer is rapid, since the Nernst equation, which is valid for reversible processes, was used in their deduction. In the case of irreversible processes the concentration of the electrode reaction product does not influence the potential and consequently the change of its concentration due to the following chemical reaction does not affect the potential of the process.

Equations (9.11) and (9.12) show that the half-wave potential is not constant, since it depends on the rate constant of the following chemical process and on the drop time. $E^k_{1/2}$ changes linearly with log t_1, and the slope of the straight line is equal to $2.3RT/2nF$. Kern obtained a linear relationship between these values in the case of the oxidation of ascorbic acid.

During polarographic reductions an increase of the drop time leads to a shift of the half-wave potential in the positive direction. During oxidation reactions the potential is shifted in the opposite direction.

Marcoux and O'Brien [7] considered a reversible electrode process with subsequent irreversible chemical reaction under chronoamperometric conditions at potentials insufficient for reaching the limiting current. An advantage of the proposed method is the determination of rate constants on the basis of measurements of the current and working curves.

9.2 Stationary Electrode Voltammetry

The strict solution for a stationary electrode voltammetric process, followed by a chemical reaction, was given in 1964 by Nicholson and Shain [8]. Earlier, in 1960, the author of this book described in his doctorate thesis a solution based on the concept of the reaction layer [9].

The solution described by Nicholson and Shain is valid for both a reversible and an irreversible chemical change.

In order to obtain solutions of equations (9.2)–(9.4) under conditions other than (9.5) and (9.6) additional boundary conditions were formulated, describing the changes of concentrations of substances Ox, Red and A, on the surface of the electrode:

$$t > 0, \quad x = 0, \quad \frac{C_{Ox}(0, t)}{C_{Red}(0, t)} = \Theta e^{-at}, \qquad (9.13)$$

$$D_{Ox}\left[\frac{\partial C_{Ox}(x, t)}{\partial x}\right]_{x=0} + D_{Red}\left[\frac{\partial C_{Red}(x, t)}{\partial x}\right]_{x=0} = 0, \qquad (9.14)$$

$$D_A\left[\frac{\partial C_A(x, t)}{\partial x}\right]_{x=0} = 0. \qquad (9.15)$$

Θ and a have been already described in Chapter 5.

According to condition (9.15) the final product A is inactive with respect to the electrode.

In the discussion of the solution for a reversible chemical reaction three cases can be considered.

In the first case the following reaction is very fast; the system is always in the equilibrium state and the only effect comparable with the diffusion process is the shift of the reduction curves in the direction of positive potentials. In this case the current function is:

$$\chi(at) = \frac{1}{\sqrt{\pi}} \sum_{j=1}^{\infty} (-1)^{j+1} \sqrt{j} \exp\left\{-\frac{jnF}{RT}\left[E - E_{1/2} - \frac{RT}{nF}\ln(1+K)\right]\right\}. \tag{9.16}$$

The only difference between this function and that of the diffusion process is the presence of expression $\frac{RT}{nF}\ln(1+K)$ in the exponential term.

In the second case the chemical reaction is very slow. When the ratio $(k_1+k_2)/a$ is small, the chemical reaction has practically no effect on the course of recorded stationary electrode voltammetric curves.

The third case is the most interesting from the kinetic point of view. In this case the ratio $(k_1+k_2)/a$ is large and the value of $K\sqrt{a/(k_1+k_2)}$ is also large. In such case the current function is:

$$\chi(at) = \frac{1}{\sqrt{\pi}} \sum_{j=1}^{\infty} (-1)^{j+1} \frac{(\sqrt{\pi})^j}{\sqrt{(j-1)!}} \exp\left\{-\frac{jnF}{RT}\left[E - E_{1/2} - \frac{RT}{nF}\ln(1+K) + \frac{RT}{nF}\ln K\sqrt{\frac{\pi a}{k_1+k_2}}\right]\right\}. \tag{9.17}$$

The current functions for various values of $K\sqrt{a/(k_1+k_2)}$ have been calculated and tabulated by Nicholson and Shain. The current function increases with increasing value of $K\sqrt{a/(k_1+k_2)}$, but these changes are not large. A change of the value of $K\sqrt{a(k_1+k_2)}$ by three orders leads to an increase in the maximum value of the function from 0.446 to 0.496. This can be observed in Fig. 9.1, which shows current functions for several values of $K\sqrt{a/(k_1+k_2)}$.

The slight changes of current connected with the changes of the kinetic parameter cannot be utilized in kinetic analysis of the secondary chemical reaction. It is possible, however, to use the equation giving the potential of the current peak for this purpose:

$$E_p^k = E_{1/2} - \frac{RT}{nF}\left[0.78 + \ln K\sqrt{\frac{a}{k_1+k_2}} - \ln(1+K)\right], \tag{9.18}$$

from which it follows that a tenfold decrease in the value of $K\sqrt{a/(k_1+k_2)}$ leads to a shift of peak potential in the positive direction by $2.3RT/nF$ V in the case of a reduction process. When the parameter is greater than unity, the relation between the potential of the peak and $\log K\sqrt{a/(k_1+k_2)}$ is

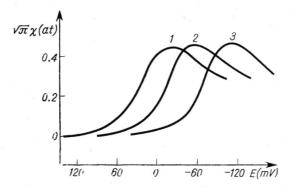

Fig. 9.1 Potential dependence of current function $\sqrt{\pi}\chi(at)$ for a reversible electrode process followed by a reversible chemical reaction. Curve 1—parameter $K\sqrt{a/(k_1+k_2)} = 0.1$, curve 2—the parameter equal to 5.0; curve 3—the parameter equal to 50 (from [8] by permission of the copyright holders, the American Chemical Society).

linear. For smaller values of the parameter this relation is no longer linear as shown in Fig. 9.2.

The theory of an electrode process followed by an irreversible chemical reaction is simpler. This has also been completely worked out by Nicholson and Shain [8] and a short treatment of these problems has been published

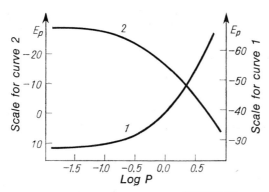

Fig. 9.2 The dependence of peak potential on log $K\sqrt{a/(k_1+k_2)}$ for reversible electrode processes. Curve 1—the following reaction is reversible; parameter $P = K\sqrt{a/(k_1+k_2)}$. Curve 2—the following reaction is irreversible; parameter $P = k_1/a$. The scale on the left hand side corresponds to curve 2 and is referred to potential $E_{1/2}$. The scale on the right hand side corresponds to curve 1 and the peak potential is referred to $E_{1/2}$ as follows: $(E_p-E_{1/2})n-[RT\ln(K+1)]/F$ (from [8] by permission of the copyright holders, the American Chemical Society).

by Reinmuth [10]. In these solutions only the rate constant k_1 appears, and the equilibrium constant of the chemical reaction is absent.

When the ratio k_1/a is small, the secondary chemical change has a small effect on the recorded curves; they have the shape of curves controlled by diffusion and are formed at potentials characteristic of diffusion processes.

When the ratio k_1/a is large the current function is:

$$\chi(at) = \frac{1}{\sqrt{\pi}} \sum_{j=1}^{\infty} (-1)^{j+1} \frac{(\sqrt{\pi})^j}{\sqrt{(j-1)!}} \exp\left[-\frac{jnF}{RT}\left(E - E_{1/2} - \frac{RT}{nF}\ln\sqrt{\frac{k_1}{\pi a}}\right)\right]. \tag{9.19}$$

It follows from the tables of function $\chi(at)$ defined by equation (9.19) that the value of the function does not change significantly with a change in the k_1/a ratio. As in the case of an electrode process followed by a reversible chemical process the increase of the ratio k_1/a by three orders of magnitude leads to an increase in the maximum value of function $\chi(at)$ by about 10%. For this reason the changes of the current function cannot be utilized for kinetic studies.

The potential of the current peak is expressed by the simple equation:

$$E_p^k = E_{1/2} - \frac{RT}{nF} 0.78 + \frac{RT}{2nF} \ln \frac{k_1}{a}, \tag{9.20}$$

or taking into account the components of a:

$$E_p^k = E_{1/2} - \frac{RT}{nF} 0.78 + \frac{RT}{2nF} \ln \frac{RT}{nF} + \frac{RT}{2nF} \ln \frac{k_1}{V}. \tag{9.21}$$

It follows from this equation that, in the case of a reduction, the potential of the peak is shifted in the positive direction with respect to the reversible half-wave potential. This shift increases with an increasing rate constant for the chemical reaction and with a decreasing polarization rate.

For large values of the ratio k_1/a a tenfold increase of this ratio shifts the potential of the peak in the positive direction by $2.3RT/2nF$ V. For small values of the ratio k_1/a the relation between E_p^k and k_1/a is no longer linear, as is shown in Fig. 9.2.

It is noteworthy that at large values of k_1/a the shape of the recorded voltammetric curve changes. They are steeper than the curves of diffusion processes involving the same number of electrons exchanged in the primary process. The increase of the slope is accompanied by an increase of peak current.

These changes of shape of the voltammetric curves in processes of this type have been demonstrated experimentally.

Evans [11] considered the effect of an irreversible first-order chemical reaction and dimerization taking place after the charge transfer on the shape of the recorded curves and their situation on the potential axis. In his theoretical considerations he assumed that the electrode reaction is *quasi*-reversible. He described by means of graphs the dependence of shift of peak potential on electrode reaction rate, subsequent reaction rate and polarization rate. He also considered the dependence of the maximum of the current function on these parameters. He has found that it is possible to use the descriptions of electrode reactions given earlier with subsequent first-order chemical reactions in which it was assumed that the process is reversible on condition that inequality $k_1 < k_s^2/100\pi D$ (where k_s is the standard rate constant of the electrode process) is fulfilled.

Similar problems have been considered independently by Nadjo and Savéant [12] in an extensive theoretical work.

A reversible electrode process with subsequent first-order chemical reaction has been theoretically considered by Laviron [13] who assumed that both depolarizer and the reaction product are adsorbed on the electrode surface.

9.3 Chronopotentiometry

The problem of a chronopotentiometric electrode process followed by a reversible chemical reaction has been elucidated by Delahay and co-workers [14]. The solution of equations (9.2)–(9.4) required the definition of the following additional boundary conditions:

$$D_{Ox}\left[\frac{\partial C_{Ox}(x, t)}{\partial x}\right]_{x=0} = \frac{i_0}{nF}, \quad (9.22)$$

$$-D_{Red}\left[\frac{\partial C_{Red}(x, t)}{\partial x}\right]_{x=0} = \frac{i_0}{nF}, \quad (9.23)$$

$$D_A\left[\frac{\partial C_A(x, t)}{\partial x}\right]_{x=0} = 0. \quad (9.24)$$

For a reversible electrode process followed by a reversible chemical reaction, Delahay *et al.* [14] obtained the following equation for a chronopotentiometric curve:

$$E = E_{1/2} + \frac{RT}{nF}\ln\frac{\tau^{1/2}-t^{1/2}}{t^{1/2}} - \frac{RT}{nF}\ln\left\{\frac{1}{1+K} + \frac{\pi^{1/2}K\mathrm{erf}[(k_1+k_2)^{1/2}t^{1/2}]}{2(1+K)(k_1+k_2)^{1/2}t^{1/2}}\right\} \quad (9.25)$$

where K is the equilibrium constant of the chemical reaction and equal to the ratio k_1/k_2.

Chronopotentiometric curves plotted by means of this equation are similar to those for the diffusion process. The transition time remains the same, and only the slope of the curve changes. The formal logarithmic analysis of such kinetic curves in the log $(\tau^{1/2}-t^{1/2})/t^{1/2}-E$ system would show a value higher than the real value of n. The reduction curves are shifted in the direction of positive potentials with increasing equilibrium constant and with increasing reaction rate constant k_1. It is convenient to discuss these shifts in terms of potential $E_{\tau/4}$. On the basis of equation (9.25) the kinetic potential $E_{\tau/4}$ can be easily expressed by means of the equation:

$$E^k_{\tau/4} = E_{1/2} - \frac{RT}{nF} \ln \left[\frac{1}{1+K} + \frac{\pi^{1/2} K \operatorname{erf}\left[(k_1+k_2)^{1/2}\left(\frac{\tau}{2}\right)^{1/2}\right]}{(1+K)(k_1+k_2)^{1/2}\tau^{1/2}} \right]. \quad (9.26)$$

When the consecutive reaction is slow it has no effect on the electrode process at high current densities provided that the transition time is very short. The electrode process is terminated before the chemical change of Red into A takes place to a significant degree. In such a case the argument of the error function in equation (9.26) is small.

By developing this function into a series:

$$\operatorname{erf} \lambda = \frac{2}{\pi^{1/2}} \left(\lambda - \frac{\lambda^3}{3 \cdot 1!} + \frac{\lambda^5}{5 \cdot 2!} - \ldots \right) \quad (9.27)$$

it can be shown that in this case $E^k_{\tau/4} = E_{1/2}$.

It can also be shown that at low current densities, when the transition time is long and the system is always at equilibrium, the potential $E_{\tau/4}$ can be expressed as follows:

$$E_{\tau/4} = E_{1/2} + \frac{RT}{nF} \ln K. \quad (9.28)$$

At low current densities the potential $E_{\tau/4}$ is shifted to positive values, proportionally to the logarithm of the equilibrium constant of the chemical reaction, but it is independent of experimental conditions.

In the intermediate region of current densities the potential $E^k_{\tau/4}$ depends on the current density according to equation (9.26). Assuming that the argument of the error function is greater than 2 and that the equilibrium of the chemical reaction is shifted considerably to the right, for the case of chronopotentiometric process followed by an irreversible chemical reaction, equation (9.26) can be expressed as [15]:

$$E^k_{\tau/4} = E_{1/2} + \frac{RT}{2nF}\ln\frac{1}{\pi} + \frac{RT}{2nF}\ln \tau k_1. \tag{9.29}$$

This equation clearly shows that, in a reduction process, the shift of potential $E_{\tau/4}$ with respect to the reversible half-wave potential in the positive direction increases with increasing transition time and reaction rate k_1. In the case of an oxidation process the potential $E^k_{\tau/4}$ is shifted in the negative direction with respect to $E_{1/2}$.

9.4 The Rotating Disc Method

The problem of an electrode process followed by a chemical reaction taking place at a rotating disc electrode was approximately solved in 1962 by Adams and also by this author [16]. For an irreversible, fast chemical change, using the concept of the reaction layer, and assuming that the electrode process is an electroreduction, the following expression for the half-wave potential was obtained:

$$E^k_{\tau/2} = E_{1/2} + \frac{RT}{nF}\ln 1.61 v^{1/6} D_{Red}^{-1/6} + \frac{RT}{2nF}\ln\frac{k_1}{\omega}. \tag{9.30}$$

It follows from equation (9.30) that the kinetic half-wave potentia- is shifted in the positive direction with respect to that measured under diffusion conditions. A tenfold increase of the k_1/ω ratio causes a potential shift of $2.3RT/2nF$ V.

The accurate solution of this problem was given more recently by two groups of workers [17, 18] independently for both reversible and irreversible chemical reactions.

Kiryanov and Filinovskii [17] obtained the following expression for he half-wave potential:

$$E^k_{1/2} = E^0 - \frac{RT}{nF}\ln \frac{1.06 + K\dfrac{[3.10 + k_2(1+K)\delta^2/D]^{1/2}}{1.65 + k_2(1+K)\delta^2/D}}{1+K} \tag{9.31}$$

where δ is the thickness of the diffusion layer.

In the case of a fast irreversible chemical reaction this equation is simplified to equation (9.30). In cases where the consecutive reaction is very slow, $E_{1/2}$ is practically equal to E^0.

The equation deduced by Tong, Kai Liang and Ruby [18] is different, but equivalent to (9.31).

Philp [19] investigated the oxidation of N,N,N',N'-tetramethyl-p-phenylenediamine and found that when $(k_1/\omega)^{1/2} > 10$ (for typical values of D and v) the equation of Tong et al. is equivalent to that of Galus and Adams.

Tong, Kai Liang and Ruby [18] have also considered the interaction of the primary electrode reaction product with another substance present in the solution. In their scheme the resulting intermediate product reacts again with the primary product of the electrode reaction.

The problems connected with a chemical reaction taking place after the charge transfer have been discussed also by Möller and Heckner [20] who considered the effect of both chemical reaction rate constant and electrode reaction rate constant on the recorded relationships.

9.5 Generalization of the Relationships Discussed

The equations deduced for a reversible electrode process followed by a chemical reaction of the first order, involving the primary product of the electrode reaction, are generally similar. This becomes obvious when the reactions relating to an irreversible consecutive reaction are compared. In such cases the polarographic equation (9.12), the chronopotentiometric equation (9.29) and the rotating disc method equation (9.30) can be represented by one general relationship, by introducing the kinetic parameter X:

$$E^k_{1/2} = E_{1/2} + B + \frac{RT}{2nF} \ln k_1 X. \qquad (9.32)$$

In order to make equation (9.21) similar to equation (9.32), it is expressed after certain transformations:

$$E^k_{1/2} = E_{1/2} + 0.33\frac{RT}{nF} + \frac{RT}{2nF}\ln\frac{RT}{nF} + \frac{RT}{2nF}\ln\frac{k_1}{V}. \qquad (9.33)$$

The values of constant B are given in Table 9.1.

Equation (9.32) makes it possible to formulate the general criterion defining this type of process. The dependence of the kinetic half-wave potential $E^k_{1/2}$ on the logarithm of the kinetic parameter X should be linear. The slope of the straight line obtained in this system is $2.3RT/2nF$. The increase of kinetic parameter causes the shift of $E^k_{1/2}$ in the positive direction in the case of reduction and in the opposite direction in the case of anodic processes.

This criterion is found to be insufficient for determination of the type of process, since the half-wave potential of irreversible processes changes similarly with changing values of X. However, in the latter case the slope of the straight line is not equal to $2.3RT/2nF$.

Table 9.1 VALUES OF B IN EQUATION (9.31)

Method	Kinetic parameter	B
Polarography	t_1	$\dfrac{RT}{nF}\ln 0.886$
Stationary electrode voltammetry	$\dfrac{1}{V}$	$0.33\dfrac{RT}{nF}+\dfrac{RT}{2nF}\ln\dfrac{RT}{nF}$
Chronopotentiometry	τ	$\dfrac{RT}{2nF}\ln\dfrac{1}{\pi}$
Rotating disc method	$\dfrac{1}{\omega}$	$\dfrac{RT}{nF}\ln 1.61 v^{1/6} D_{\text{Red}}^{-1/6}$

Equation (9.32) is valid for moderately large values of the kinetic parameter X. When this parameter is very large the consecutive reaction can reach the equilibrium state, and in this case any further increase of X would have no effect on the half-wave potential. In the case of very small values of X the half-wave potential may also be constant, since in this case the chemical reaction has no effect on the concentration of the primary product of electrode reaction. These changes of the half-wave potential,

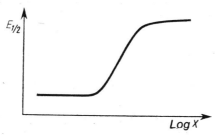

Fig. 9.3 Half-wave potential changes caused by changes of log X.

corresponding to changes of the kinetic parameter are shown schematically in Fig. 9.3.

The above discussion has been limited to the processes in which a reversible electrode process is followed by a chemical reaction.

In addition to the mechanism shown in scheme (9.1) more complex mechanisms have been proposed in which, after the charge transfer, two consecutive first-order chemical reactions of the primary product take place. Such a mechanism has been discussed in general terms by Ashley and Reilley [21]. A detailed discussion of this problem in chronopotentiometric conditions was published by Galus [22].

9.6 Examples of Electrode Processes Followed by a Chemical Reaction

In addition to the anodic oxidation of ascorbic acid mentioned above other examples of a similar mechanism of oxidation of enediols are known [23, 24].

A consecutive reaction accompanies the reduction of the complex of mercury(II) with complexone III (EDTA) in the presence of magnesium ions [25]. The chemical reaction is here the reaction of magnesium ions with complexone III which is liberated during the electrode process. Obviously its rate can be controlled by the concentration of Mg^{2+} ions. The highest determined value of the pseudo-first order rate constant was 320 s^{-1} but the condition of reversibility of the electrode process according to the inequality given by Evans [11] — $k_1 < k_s^2/100\pi D$ — is that under such conditions the standard rate constant should be greater than 0.7 cm/s. Since the standard rate constant of the examined system was lower it appears that the theory used in the calculations, assuming the reversibility of the electrode reaction, was not leading to completely correct results. Another example of an electrode reaction followed by a chemical reaction is the anodic oxidation of cadmium amalgam in a buffered solution, in the presence of EDTA [26].

Evans [11] emphasised that in these studies also it appears that the process was taking place under the conditions of incomplete reversibility of the electrode reaction, since although the standard rate constant of the electrode reaction was high, the highest determined rate constant of the pseudo-first order reaction was also very high (about 10^7 s^{-1}).

Hydrolysis of p-benzoquinoneimine (obtained by electrolytic oxidation of p-aminophenol) to p-benzoquinone has been investigated [27] by several electrochemical methods under the conditions of electrolysis of thin layers of the solution. It has been confirmed that the reaction is of first order and several rate constants have been determined for sulphuric acid solutions of various concentrations at temperatures ranging from 25 to 40°C. In future studies involving the theory deduced on the assumption of reversibility of the processes examined, the electrode reaction rate should be considered and it should be remembered that the subsequent chemical reaction considerably decreases the activity of Ox or Red forms of the redox system and hence it decreases the electrode reaction rate. Obviously this decrease increases with increasing equilibrium constant and increasing rate constant of the subsequent chemical reaction.

The work on consecutive chemical reactions in electrode processes has been reviewed by Bewick and Thirsk [28].

References

[1] Cattanec, C. and Sartori, G., *Gazz. chim. ital.*, **72**, 351 (1942).
[2] Vavřin, Z., *Coll. Czechoslov. Chem. Communs.*, **14**, 367 (1949).
[3] Kern, D. M. H., *J. Am. Chem. Soc.*, **75**, 2473 (1953).
[4] Kern, D. M. H., *J. Am. Chem. Soc.*, **76**, 1011 (1954).
[5] Koutecký, J., *Chem. Listy*, **48**, 1314 (1954).
[6] Koutecký, J., *Coll. Czechoslov. Chem. Communs.*, **20**, 116 (1955).
[7] Marcoux, L. and O'Brien, T. J. P., *J. Phys. Chem.*, **76**, 1666 (1972).
[8] Nicholson, R. S. and Shain, J., *Anal. Chem.*, **36**, 706 (1964).
[9] Galus, Z., *Doctorate Thesis*, University of Warsaw, 1960.
[10] Reinmuth, W. H., *Anal. Chem.*, **33**, 1793 (1961).
[11] Evans, D. H., *J. Phys. Chem.*, **76**, 1160 (1972).
[12] Nadjo, L. and Savéant, J. M., *J. Electroanal. Chem.*, **48**, 113 (1973).
[13] Laviron, E. J., *Electroanal. Chem.*, **35**, 333 (1972).
[14] Delahay, P., Mattax, C. C. and Berzins, T., *J. Am. Chem. Soc.*, **76**, 5319 (1954).
[15] Galus, Z., *Chem. Anal.*, Warsaw, **10**, 803 (1965).
[16] Galus, Z. and Adams, R. N., *J. Electroanal. Chem.*, **4**, 248 (1962).
[17] Kiryanov, V. A. and Filinovskii, V. D., "*Lectures on Polarography*", Kiev 1965, page 42 (in Russian).
[18] Tong, L. K. J., Kai Liang and Ruby, W. R., *J. Electroanal. Chem.*, **13**, 245 (1967).
[19] Philp, R. H. Jr., *J. Electroanal. Chem.*, **27**, 369 (1970).
[20] Möller, D. and Heckner, K. H., *J. Electroanal. Chem.*, **38**, 337 (1972).
[21] Ashley, J. W. and Reilley, C. N., *J. Electroanal. Chem.*, **7**, 253 (1964).
[22] Galus, Z., *Bull. Acad. Polon. Sci., Sér. Sci. Chim.*, **13**, 433 (1965).
[23] Šantavy, F. and Bitter, B., *Coll. Czechoslov. Chem. Communs.*, **15**, 112 (1950).
[24] Brdička, R. and Zuman, P., *Coll. Czechoslov. Chem. Communs.*, **15**, 766 (1950).

[25] Kern, D. M. H., *J. Am. Chem. Soc.*, **81**, 1563 (1959).
[26] Koryta, J. and Zabranský, Z., *Coll. Czechoslov. Chem. Communs.*, **25**, 3153 (1960).
[27] Plichon, V. and Faure, G., *J. Electroanal. Chem.*, **44**, 275 (1973).
[28] Bewick, A. and Thirsk, H., in "*Modern Aspects of Electrochemistry*", Bockris, J. O'M. and Conway, B. E., Eds., Plenum Press, New York 1969, Chapter 4.

Chapter 10

Catalytic Electrode Processes

In electrode processes discussed in the preceding chapters, either the substrate or the product were in equilibrium with another substance, the latter being electrode inactive in the potential region under consideration. In the case when the electrode reaction is preceded by a chemical reaction, the kinetic current is always lower than the limiting current calculated on the basis of diffusion equations. The chemical reaction of the electrode reaction product has negligible influence on the current intensity.

Catalytic processes considered in this section differ substantially from those discussed earlier, since the resulting current always exceeds the value of the limiting current observed for a process which is not complicated by chemical reactions.

Assuming that the electrode reaction is a cathodic reduction, these processes can be formulated by the following general equations:

$$Ox + ne \rightleftharpoons Red \qquad (10.1)$$

$$Red + Z \underset{k_2}{\overset{k_1}{\rightleftharpoons}} Ox \qquad (10.2)$$

where k_1 is the rate constant of the transformation of Red into Ox. This reaction involves substance Z, but in our discussion we will assume that it is electrode inactive, and that its concentration in the solution is so high that it suffers negligible change as a result of the electrode process.

An example of a catalytic process is the electroreduction of ferric ions in the presence of hydrogen peroxide in an acidic medium. When H_2O_2 is not present, the reduction of Fe^{3+} ions starts at the mercury oxidation potential. Introduction of hydrogen peroxide into the solution results in an increase of the recorded current, since some of the ferrous ions produced in the reduction process are reoxidized to the ferric species.

The occurrence of catalysis in polarography was discovered as long ago as 1943 by Wiesner [1] and the first attempts to describe catalytic currents in polarography were made by Brdička and Wiesner [2, 3]. Employing a method based on the concept of the reaction layer they derived an approximate equation which is valid for very fast chemical reactions participating in the catalytic process.

The exact solution requires, as in the case of electrode processes preceded by a chemical reaction, the solution of the extended Fick equation, including the terms describing the catalytic chemical transformation. For a catalytic process described by equations (10.1) and (10.2) the following system of equations is obtained:

$$\frac{\partial C_{Ox}(x,t)}{\partial t} = D_{Ox}\frac{\partial^2 C_{Ox}(x,t)}{\partial x^2} + k_1 C_{Red}(x,t) C_Z(x,t) - k_2 C_{Ox}(x,t), \tag{10.3}$$

$$\frac{\partial C_{Red}(x,t)}{\partial t} = D_{Red}\frac{\partial^2 C_{Red}(x,t)}{\partial x^2} - k_1 C_{Red}(x,t) C_Z(x,t) + k_2 C_{Ox}(x,t). \tag{10.4}$$

In these equations the concentration C_Z of the substance Z appears. If it were low it would change during the electrode process and in such a case the theoretical treatment of the process would be very difficult. Therefore, in our discussion it is assumed that the concentration of Z is so high that it remains constant during the whole electrochemical process. This concentration is designated C_Z^0.

The above assumption allows considerable simplification of the problem. For further simplification, in theoretical considerations of catalytic currents, a fast catalytic transformation is usually assumed. In this case equations (10.3) and (10.4) are reduced to

$$\frac{\partial C_{Ox}(x,t)}{\partial t} = D_{Ox}\frac{\partial^2 C_{Ox}(x,t)}{\partial x^2} + k_1 C_Z^0 C_{Red}(x,t), \tag{10.5}$$

and

$$\frac{\partial C_{Red}(x,t)}{\partial t} = D_{Red}\frac{\partial^2 C_{Red}(x,t)}{\partial x^2} - k_1 C_Z^0 C_{Red}(x,t), \tag{10.6}$$

respectively.

The initial conditions are defined by the following relationships:

$$t = 0, \quad x \geqslant 0, \quad C_{Ox} = C_{Ox}^0, \quad C_{Red} = 0. \tag{10.7}$$

Thus it is assumed that the Red form is produced only as a result of the electrode process.

At the border of the diffusion zone the following conditions are fulfilled:

$$t > 0, \quad x \to \infty, \quad C_{Ox} \to C_{Ox}^0, \quad C_{Red} \to 0. \tag{10.8}$$

The first boundary condition for $x = 0$ has the following general form for all the methods discussed:

$$t \geqslant 0, \quad x = 0, \quad D_{\text{Ox}}\left[\frac{\partial C_{\text{Ox}}(x, t)}{\partial x}\right]_{x=0} + D_{\text{Red}}\left[\frac{\partial C_{\text{Red}}(x, t)}{\partial x}\right]_{x=0} = 0. \quad (10.9)$$

This condition follows from the fact that the sum of the fluxes of the Ox and Red forms at the electrode surface is equal to zero.

10.1 Chronoamperometry

The solution of the problem of an electrode process with catalytic regeneration of the depolarizer under chronoamperometric conditions can be obtained after the formulation of one additional boundary condition describing the changes of concentration of the Ox form at the electrode surface during the electrode process. Usually it is assumed that in a reduction process a negative potential is applied to the electrode such that the concentration of the Ox form at the electrode surface equals zero. In a polarographic model this condition is fulfilled at the potential of the limiting current. Then the boundary condition takes the simple form:

$$t > 0, \quad x = 0, \quad C_{\text{Ox}} = 0. \quad (10.10)$$

The solution of the problem formulated above was presented independently by Delahay and Stiehl [4], by Pospíšil [5] and by Miller [6].

In the case of a catalytic process accompanied by a catalytic reaction in solution, the equation describing the current has the following form:

$$i_k = nFAD^{1/2}C_{\text{Ox}}^0\left\{(k_1 C_Z^0)^{1/2}\text{erf}[(k_1 C_Z^0 t)^{1/2}] + \frac{\exp(-k_1 C_Z^0 t)}{(\pi t)^{1/2}}\right\}. \quad (10.11)$$

The discussion of this result is facilitated by division of equation (10.11) by the equation describing the diffusion current for a chronoamperometric electrode process uncomplicated by chemical reactions. Then the following relationship is obtained:

$$\frac{i_k}{i_g} = \gamma^{1/2}\left[\pi^{1/2}\text{erf}(\gamma^{1/2}) + \frac{\exp(-\gamma)}{\gamma^{1/2}}\right] \quad (10.12)$$

where $\gamma = k_1 C_Z^0 t$.

Under certain conditions equation (10.12) can be presented in a simpler form. If the argument of the error function exceeds 2, the function is almost equal to 1. The second term in brackets of equation (10.12) becomes small compared with $\pi^{1/2}$ and therefore it can be neglected. Thus for $\gamma^{1/2} > 2$ equation (10.12) can be reduced to

$$\frac{i_k}{i_g} = \pi^{1/2}\gamma^{1/2}. \tag{10.13}$$

The dependence of i_k/i_g on the parameter $\gamma^{1/2}$ is shown in Fig. 10.1. This graph confirms that equation (10.13) is only valid at high $\gamma^{1/2}$ values.

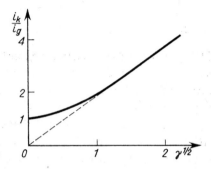

Fig. 10.1 Dependence of current ratio i_k/i_g on $\gamma^{1/2}$. The full line is obtained on the basis of equation (10.12) and the broken line on the basis of equation (10.13).

In these cases the solutions of equations (10.12) and (10.13) give identical results. In cases where arguments tend to zero, equation (10.13) leads to the erroneous conclusion that the ratio i_k/i_g should also tend to zero.

It follows from equation (10.12) that for a catalytic process the lower limit of the ratio i_k/i_g is 1.

On the basis of the equations presented above it is possible to calculate the rate constants of catalytic processes. Having measured the catalytic current, it is possible to carry out the electrode process in identical conditions, but in the absence of substance Z. In this case the catalytic reaction does not take place and it is possible to determine the limiting current value.

Given a knowledge of the ratio i_k/i_g it is possible, for example, on the basis of Fig. 10.1, to determine the corresponding value of argument γ. Taking into account the concentration of Z and the duration of the electrolysis it is then possible to determine the catalytic reaction rate constant.

The problem of an electrode process with catalytic regeneration of the depolarizer under conditions of spherical diffusion has been worked out by Delmastro and Booman [7]. Guidelli and Cozzi [8] considered a more

general case which can be described by the following chemical equations:
$$a\text{Ox} + ne \rightleftharpoons b\text{Red}$$
$$\text{Red} + z\text{Z} \xrightarrow{k_1} m\text{Ox} + \text{other products}$$
In this case the kinetic current is described by the equation:

$$i_k = \frac{nFAD^{1/2}C_{\text{Ox}}^0}{(\pi t)^{1/2}}(l+1)\left\{\frac{\pi^{1/2}lk_1^{1/2}t^{1/2}}{(l^2-1)^{3/2}}\exp\left(\frac{k_1 t}{l^2-1}\right)\left[l|\text{erf}\left(\frac{l^2 k_1 t}{l^2-1}\right)^{1/2}\right.\right.$$
$$\left.\left. - l\text{erf}\left(\frac{k_1 t}{l^2-1}\right)^{1/2}\right] + \frac{l\exp(-k_1 t)-1}{l^2-1}\right\}, \qquad (10.14)$$

where $l = \dfrac{bm}{a-bm}$. This equation is valid when $|l| > 1$.

10.2 Stationary Electrode Voltammetry

The description of catalytic currents in stationary electrode voltammetry is dependent on the degree of reversibility of the electrode reaction. For this reason it is necessary to discuss this method in two parts. The case of a reversible electrode process accompanied by regeneration of the Ox form will be discussed first.

10.2.1 Reversible Electrode Processes

The definition of the catalytic reaction current can be obtained by the solution of equations (10.5) and (10.6). In this case conditions (10.7)–(10.9) are also valid. The second boundary condition can be obtained from the Nernst equation:

$$t > 0, \quad x = 0, \quad \frac{C_{\text{Ox}}(0, t)}{C_{\text{Red}}(0, t)} = \Theta e^{-at}, \qquad (10.15)$$

where

$$a = \frac{nFV}{RT} \qquad (10.16)$$

and

$$\Theta = \exp\left[\frac{nF(E_i - E^0)}{RT}\right]. \qquad (10.17)$$

E_i is the initial potential at which polarization of the electrode starts.

This problem was solved by Savéant and Vianello [9] in 1960. Its complete treatment, although limited to irreversible chemical transformations, was given by Nicholson and Shain [10].

The effect of the depolarizer regenerated by a chemical reaction is first evident as an increase in the recorded current. Two extreme cases can be distinguished. In the first case the ratio $k_1 C_z^0/a$ is low and the current is almost exactly described by the Randles–Ševčik equation. In the second case $k_1 C_z^0/a$ values are high. The current function $\chi(at)$ is now described by the equation:

$$\chi(at) = \frac{1}{\sqrt{\pi}} \sum_{j=1}^{\infty} (-1)^{j+1} \sqrt{\frac{k_1 C_z^0}{a}} \exp\left[\frac{-jnF(E-E_{1/2})}{RT}\right] \quad (10.18)$$

and the current by the relationship:

$$i_k = \frac{nFA\sqrt{Dk_1}\, C_{Ox}^0 (C_z^0)^{1/2}}{1+\exp\left[\dfrac{nF(E-E_{1/2})}{RT}\right]}. \quad (10.19)$$

Equation (10.19) describes the full course of the potential dependence of the catalytic current. It shows that, in the case of high $k_1 C_z^0/a$ values, no current peaks are observed on the current–potential curves and the shapes of the curves resemble polarographic waves. According to equation (10.19) the current reaches half the maximum value at the polarographic half-wave potential.

At potentials which are sufficiently negative with respect to $E_{1/2}$, equation (10.19) is reduced to the relationship given by Savéant and Vianello [9]:

$$i_k = nFAC_{Ox}^0 D^{1/2} k_1^{1/2} (C_z^0)^{1/2}. \quad (10.20)$$

It follows from equation (10.20) that, in this case, the current is independent of the polarization rate, but is proportional to the square root of the depolarizer regeneration rate.

Dividing equation (10.20) by i_p, defined by the Randles–Ševčik equation, gives the relationship:

$$\frac{i_k}{i_p} = \frac{k_1^{1/2} (C_z^0)^{1/2}}{0.446 a^{1/2}} \quad (10.21)$$

from which it follows that the dependence of i_k/i_p on the square root of the polarization rate should be linear. Equation (10.21), in a similar manner to equation (10.20), is valid only in cases of high ratios of $k_1 C_z^0$ to polarization rate. In the case of decreasing polarization rates it leads to the erroneous conclusion that i_k/i_p tends to zero. Obviously, unity is the lower limit of the ratio of the catalytic current to the diffusion peak current.

Table 10.1 VALUES OF CURRENT FUNCTION $\pi^{1/2}\chi(at)$ FOR A CATALYTIC PROCESS WITH REVERSIBLE CHARGE TRANSFER (THE POTENTIALS ARE REFERRED TO THE HALF-WAVE POTENTIAL WHICH IS ARBITRARILY ASSUMED TO BE EQUAL TO ZERO) (FROM [10] BY PERMISSION OF THE COPYRIGHT HOLDERS, THE AMERICAN CHEMICAL SOCIETY)

Potential E (mV)	Parameter $\dfrac{k_1 C_Z^0}{a}$								
	0.04	0.1	0.2	0.4	0.6	1.0	1.78	3.16	10.0
120	0.009	0.010	0.010	0.011	0.012	0.013	0.015	0.019	0.030
100	0.020	0.021	0.021	0.023	0.025	0.028	0.033	0.040	0.060
80	0.042	0.043	0.045	0.049	0.052	0.059	0.069	0.086	0.139
60	0.086	0.088	0.093	0.100	0.108	0.121	0.144	0.176	0.289
50	0.120	0.123	0.129	0.140	0.150	0.170	0.201	0.249	0.409
45	0.140	0.145	0.152	0.165	0.178	0.201	0.239	0.294	0.482
40	0.163	0.168	0.177	0.193	0.207	0.234	0.279	0.345	0.567
35	0.189	0.195	0.205	0.224	0.242	0.273	0.326	0.403	0.665
30	0.216	0.224	0.236	0.258	0.278	0.315	0.378	0.467	0.773
25	0.245	0.254	0.267	0.294	0.318	0.361	0.433	0.539	0.894
20	0.275	0.285	0.301	0.331	0.359	0.409	0.493	0.614	1.022
15	0.306	0.318	0.337	0.371	0.403	0.461	0.558	0.695	1.162
10	0.336	0.349	0.370	0.410	0.447	0.512	0.623	0.782	1.310
5	0.364	0.380	0.404	0.449	0.491	0.566	0.690	0.867	1.459
0	0.391	0.408	0.436	0.487	0.534	0.617	0.756	0.955	1.614
−5	0.414	0.434	0.465	0.522	0.574	0.668	0.821	1.042	1.769
−10	0.432	0.455	0.489	0.552	0.611	0.715	0.883	1.124	1.919
−15	0.448	0.472	0.510	0.580	0.644	0.757	0.942	1.204	2.061
−20	0.459	0.485	0.527	0.604	0.673	0.796	0.996	1.278	2.197
−25	0.465	0.494	0.540	0.622	0.697	0.829	1.044	1.345	2.322
−30	0.468	0.499	0.548	0.638	0.719	0.861	1.088	1.406	2.436
−35	0.467	0.500	0.553	0.649	0.735	0.885	1.126	1.462	2.540
−40	0.463	0.499	0.556	0.658	0.749	0.907	1.159	1.510	2.633
−45	0.457	0.495	0.555	0.663	0.759	0.924	1.188	1.552	2.713
−50	0.450	0.490	0.553	0.666	0.766	0.939	1.211	1.587	2.782
−60	0.431	0.476	0.545	0.668	0.776	0.961	1.250	1.644	2.894
−80	0.390	0.442	0.522	0.662	0.782	0.984	1.295	1.715	3.034
−100	0.354	0.413	0.502	0.653	0.781	0.994	1.315	1.749	3.102
−120	0.326	0.390	0.486	0.646	0.779	0.997	1.326	1.765	3.134
−140	0.305	0.374	0.474	0.641	0.777	0.999	1.330	1.772	3.149

For intermediate values of the $k_1 C_Z^0/a$ ratio, not exceeding unity, the dependence of i_k/i_p on $V^{1/2}$ cannot be expressed by such a simple equation as (10.21). The current function $\chi(at)$ values for such a case were calculated by Nicholson and Shain [10]. The results, calculated for various potentials and for various values of the $k_1 C_Z^0/a$ ratio are presented in Table 10.1. On the basis of these data and peak current values of a reversible process it is possible to describe the dependence of i_k/i_p on $V^{1/2}$ for low $k_1 C_Z^0/a$ values. If the value of the latter ratio does not exceed 0.06, the value of i_k/i_p also depends slightly on $k_1 C_Z^0/a$.

It follows from the above discussion that the rate constant of a catalytic chemical reaction can be calculated on the basis of current measurements. This is not, at least theoretically, the only possible method of obtaining information on the kinetics of depolarizer regeneration. This information can also be deduced from the potential change corresponding to half the current maximum. In the case of low values of $k_1 C_Z^0/a$ the potential $E_{p/2}$ is constant and, as in the case of a simple process not complicated by chemical reactions, independent of $k_1 C_Z^0/a$. An increase in $k_1 C_Z^0/a$ value shifts the half-current peak potential in the negative direction. A tenfold increase in the value of $k_1 C_Z^0/a$ results in a potential shift of $60/n$ mV. At the same time, a change in the shape of the current peak is observed. It becomes broader and, for k_1/a values greater than one, it disappears. In this case the curves resemble polarographic waves.

Weber [11] derived a correction term for the calculation of deviations of current observed under linear diffusion conditions for electrolysis on spherical electrodes.

The problem of a catalytic process under stationary electrode voltammetry conditions was studied also by Kao, Chang and Chang [12], on the assumption that the equilibrium of the chemical reaction was shifted appreciably to the right-hand side of the equation.

The discussion has been concerned, so far, with a catalytic process with irreversible chemical reaction, described by equations (10.5) and (10.6). A theoretical treatment of a catalytic process with reversible electrode reaction and reversible depolarizer regeneration was presented by Rampazzo [13, 14]. She solved the more complicated systems of equations (10.3) and (10.4) with appropriately formulated initial and boundary conditions and obtained a general formula for the instantaneous current. This equation is somewhat complex. In the case of low polarization rates (as compared with the reaction rate), the general formula gives an equation similar to relationship (10.20):

$$i_k = nFAC_{Ox}^0 D^{1/2} \sqrt{k_1 C_Z^0 + k_2} \qquad (10.22)$$

In the reverse case of high polarization rates the Randles–Ševčik equation is obtained.

10.2.2 Irreversible Electrode Processes

The following boundary conditions are introduced in this case for the solution of equations (10.5) and (10.6):

$$t > 0, \quad x = 0, \quad D_{Ox}\left[\frac{\partial C_{Ox}(x,t)}{\partial x}\right] = C_{Ox}k_s \exp\left[-\frac{\alpha n \cdot F}{RT}(E_i - E^0)\right]e^{bt}. \tag{10.23}$$

This condition and conditions (10.7)–(10.9) make it possible to obtain the theoretical description of a catalytic stationary electrode voltammetric process with irreversible charge exchange.

In equation (10.23) b is defined by $b = \alpha n_z FV/RT$.

The problem for irreversible chemical reaction was worked out by Nicholson and Shain [10]. As in the case of a catalysed reversible electrode process, two extreme cases depending on the $k_1 C_Z^0/b$ ratio may be considered. In the case of low $k_1 C_Z^0/b$ values, the current is described by the Delahay equation deduced for irreversible electrode processes. When the $k_1 C_Z^0/b$ values are high the current function $\chi(bt)$ is described by the expression:

$$\chi(bt) = \frac{1}{\sqrt{\pi}} \sum_{j=1}^{\infty}(-1)^{j+1}(\sqrt{\pi})^j \sqrt{\frac{k_1 C_Z^0}{b}} \exp\left[-\frac{j\alpha n \cdot F}{RT}\left(E - E^0 + \frac{RT}{\alpha n_\alpha F}\ln\frac{\sqrt{\pi Db}}{k_s} + \frac{RT}{\alpha n_\alpha F}\ln\sqrt{\frac{k_1 C_Z^0}{\pi a}}\right)\right], \tag{10.24}$$

and the current–potential curves are defined by the equation:

$$i_k = \frac{nFAC_{Ox}^0 D^{1/2} k_1^{1/2}(C_Z^0)^{1/2}}{1 + \exp\left[\frac{\alpha n \cdot F}{RT}\left(E - E^0 + \frac{RT}{\alpha n_\alpha F}\ln\frac{\sqrt{\pi Db}}{k_s} + \frac{RT}{\alpha n_\alpha F}\ln\sqrt{\frac{k_1 C_Z^0}{\pi a}}\right)\right]}. \tag{10.25}$$

In the case of a highly negative electrode potential the exponential term in the denominator of equation (10.25) can be neglected, and the catalytic current is then described by relationship (10.20). It follows from equation (10.25) that in the case of fast chemical reaction leading to depolarizer regeneration the recorded curves resemble polarographic waves.

Such a simple description of the process is not correct when the k_1/b ratio is small. In this case the $\chi(bt)$ function table calculated by Nicholson and Shain [10] for various k_1/b ratios and various potentials should be used. On the basis of the data given in Table 10.2 and the peak current values

Table 10.2 VALUES OF CURRENT FUNCTION $\pi^{1/2}\chi(bt)$ FOR A CATALYTIC PROCESS WITH IRREVERSIBLE CHARGE EXCHANGE. THE FOLLOWING POTENTIAL SCALE IS USED:

$$(E-E^0)\alpha n_\alpha + (RT/F)\ln \frac{\sqrt{\pi Db}}{k_s}$$

(FROM [10] BY PERMISSION OF THE COPYRIGHT HOLDERS, THE AMERICAN CHEMICAL SOCIETY)

Potential E (mV)	\multicolumn{9}{c}{Parameter $\dfrac{k_1 C_Z^0}{b}$}
	0.04	0.1	0.2	0.4	0.6	1.0	1.78	3.16	10.0
160	0.004	0.004	0.004	0.004	0.004	0.004	0.004	0.004	0.004
140	0.008	0.008	0.008	0.008	0.008	0.008	0.008	0.008	0.008
120	0.016	0.016	0.016	0.016	0.016	0.016	0.016	0.016	0.016
110	0.024	0.024	0.024	0.024	0.024	0.024	0.024	0.024	0.024
100	0.035	0.035	0.035	0.035	0.035	0.035	0.035	0.036	0.036
90	0.050	0.050	0.051	0.051	0.051	0.051	0.051	0.052	0.052
80	0.072	0.073	0.073	0.073	0.074	0.074	0.075	0.075	0.076
70	0.104	0.105	0.105	0.106	0.107	0.108	0.109	0.110	0.113
60	0.145	0.147	0.147	0.148	0.150	0.152	0.155	0.157	0.162
50	0.198	0.200	0.201	0.205	0.208	0.213	0.218	0.224	0.234
40	0.264	0.267	0.271	0.278	0.283	0.291	0.302	0.313	0.334
35	0.301	0.305	0.311	0.320	0.327	0.339	0.354	0.370	0.399
30	0.334	0.344	0.349	0.362	0.372	0.388	0.409	0.430	0.471
25	0.376	0.383	0.392	0.408	0.422	0.444	0.473	0.500	0.558
20	0.410	0.418	0.431	0.453	0.470	0.498	0.534	0.574	0.650
15	0.443	0.454	0.470	0.500	0.521	0.556	0.600	0.656	0.761
10	0.469	0.483	0.503	0.538	0.567	0.612	0.673	0.740	0.879
5	0.490	0.506	0.532	0.575	0.610	0.666	0.742	0.827	1.002
0	0.504	0.524	0.555	0.608	0.651	0.719	0.813	0.920	1.145
−5	0.511	0.534	0.571	0.633	0.685	0.766	0.878	1.007	1.288
−10	0.511	0.539	0.581	0.653	0.713	0.809	0.941	1.097	1.443
−15	0.506	0.538	0.586	0.667	0.735	0.844	0.998	1.178	1.595
−20	0.497	0.532	0.585	0.676	0.753	0.875	1.049	1.258	1.755
−25	0.485	0.523	0.581	0.681	0.765	0.900	1.094	1.328	1.903
−30	0.470	0.512	0.575	0.683	0.774	0.921	1.113	1.394	2.053
−35	0.456	0.500	0.568	0.683	0.780	0.937	1.166	1.450	2.186
−40	0.440	0.487	0.559	0.681	0.783	0.951	1.195	1.502	2.317
−50	0.411	0.463	0.541	0.674	0.786	0.969	1.238	1.583	2.538
−60	0.386	0.442	0.525	0.667	0.786	0.980	1.269	1.643	2.710
−70	0.366	0.425	0.512	0.661	0.785	0.988	1.289	1.684	2.840
−80	0.348	0.409	0.501	0.655	0.783	0.992	1.303	1.714	2.936
−100	0.320	0.386	0.484	0.646	0.780	0.997	1.320	1.748	3.056
−120	0.300	0.371	0.473	0.641	0.778	0.999	1.328	1.764	3.112
−140			0.466	0.638	0.776	0.999	1.331	1.772	3.139
−160			0.461	0.636	0.776	1.000	1.332	1.776	3.152

of an irreversible process it is possible to determine, by the method described in subsection 10.2.2, the dependence of i_k/i_p on the polarization rate or, more precisely, on the $k_1 C_Z^0/b$ ratio. Again, for high (greater than one) $k_1 C_Z^0/b$ values a linear relationship is observed. At parameter values lower than 0.06 the i_k/i_p ratio is almost equal to one.

Theoretically, the rate constant k_1 can be calculated from the change of half-peak current potential, caused by the change in $k_1 C_Z^0/b$. This method of analysing the kinetics of a catalytic reaction is less important than that based on current measurements. Changes in $E_{p/2}$ potential caused by changes in $k_1 C_Z^0/b$ can be determined on the basis of the data given in Table 10.2.

A theoretical treatment of the problem of a catalytic process with irreversible chemical reaction has also been presented by Savéant and Vianello [15].

10.3 Chronopotentiometry

The problem of a catalytic process under chronopotentiometric conditions was studied by Delahay, Mattax and Berzins [16]. The solution of equations (10.5) and (10.6) required the introduction of an additional boundary condition, characteristic of chronopotentiometry:

$$t > 0, \quad x = 0, \quad D_{\text{Ox}} \left[\frac{\partial C_{\text{Ox}}(x, t)}{\partial x} \right] = \frac{i_0}{nF}. \quad (10.26)$$

This condition was utilized together with conditions (10.7)–(10.9). These authors assumed that the diffusion coefficients of the Ox and Red forms are equal and as a solution they obtained an equation describing the time dependence of the concentration of the Ox form at the electrode surface:

$$C_{\text{Ox}}(0, t) = C_{\text{Ox}}^0 - \frac{i_0 t^{1/2}}{nFD\bar{\gamma}} \operatorname{erf} \bar{\gamma}, \quad (10.27)$$

where

$$\bar{\gamma} = (k_1 C_Z^0 \tau_k)^{1/2} \quad (10.28)$$

and τ_k is the kinetic transition time.

The equation describing the transition time can be derived from equation (10.27) assuming that $C_{\text{Ox}}(0, \tau_k) = 0$. Appropriate transformations give the expression:

$$\left(\frac{\tau_k}{\tau} \right)^{1/2} = \frac{2\bar{\gamma}}{\pi^{1/2} \operatorname{erf} \bar{\gamma}} \quad (10.29)$$

where τ is the transition time in the absence of catalytic reaction, defined by Sand's equation.

The dependence of the right-hand side of this equation on $\bar{\gamma}$ is shown in Fig. 10.2. Erf $\bar{\gamma}$ tends to one as $\bar{\gamma}$ increases and the curve asymptotically approaches a straight line, having slope $2\bar{\gamma}/\pi^{1/2}$. Since the error function for arguments greater than 2 is practically equal to one, this linear approximation is valid for $(k_1 C_Z^0 \tau_k)^{1/2} > 2$.

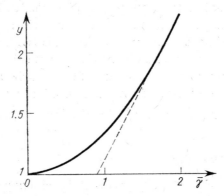

Fig. 10.2 Dependence of $y = 2\bar{\gamma}/\pi^{1/2}$ erf $\bar{\gamma}$ on $\bar{\gamma}$. Broken line—asymptotic solution valid for large values of $\bar{\gamma}$ (from [16] and [16a] by permission of the copyright holders).

When the catalytic effect is only slight, $\tau_k^{1/2}/\tau^{1/2}$ is very close to one. This is the case when the rate constant of depolarizer regeneration is small, or the time of the experiment (transition time) is very short.

On the basis of equation (10.29) constant k_1 can be calculated. For this purpose only the experimental determination of the transition time in the presence of catalytic reaction and the determination or calculation of the transition time of the diffusion controlled process are necessary. Transition time τ can be determined after removing substance Z, which is indispensable for the catalytic reaction from the solution.

10.4 Polarography

In the case of polarography, the treatment of the problem of catalytic currents requires the solution of a system of equations taking into account the convection transport resulting from the growth of the mercury drop electrode and changes in concentration due to the catalytic reaction.

Reactions (10.1) and (10.2) can be described by the following equations:

$$\frac{\partial C_{\text{Ox}}(x,t)}{\partial t} = D_{\text{Ox}} \frac{\partial^2 C_{\text{Ox}}(x,t)}{\partial x^2} + \frac{2x}{3t} \frac{\partial C_{\text{Ox}}(x,t)}{\partial x} + k_1 C_Z^0 C_{\text{Red}}(x,t),$$
(10.30)

$$\frac{\partial C_{\text{Red}}(x,t)}{\partial t} = D_{\text{Red}} \frac{\partial^2 C_{\text{Red}}(x,t)}{\partial x^2} + \frac{2x}{3t} \frac{\partial C_{\text{Red}}(x,t)}{\partial x} - k_1 C_Z^0 C_{\text{Red}}(x,t). \quad (10.31)$$

These equations were solved by Koutecký [17], who utilized the initial and boundary conditions given in section 10.1 in the discussion of catalytic current in chronoamperometry. Koutecký's solution gives the functional dependence of the ratio of the catalytic current to the limiting current on the parameter γ'. This solution can be formulated as the general relationship:

$$\frac{\bar{i}_k}{\bar{i}_g} = \Psi(\gamma'), \quad (10.32)$$

where

$$\gamma' = k_1 C_Z^0 t_1. \quad (10.33)$$

These formulae are valid for average current values. Koutecký also calculated the function $\Psi(\gamma)$ describing the dependence of the ratio of the instantaneous currents on the parameter $\gamma = k_1 C_Z^0 t$.

In Table 10.3 the values of functions $\Psi(\gamma)$ and $\Psi(\gamma')$, calculated for various parameters γ and γ', are given.

Table 10.3 VALUES OF FUNCTIONS $\Psi(\gamma)$ AND $\Psi(\gamma')$ (FROM [17] BY PERMISSION OF THE COPYRIGHT HOLDERS)

γ or γ'	$\Psi(\gamma)$	$\Psi(\gamma')$	γ or γ'	$\Psi(\gamma)$	$\Psi(\gamma')$
0	1	1	2.0	1.826	1.47
0.05	1.025	1.013	2.5	1.99	1.56
0.1	1.050	1.027	3.0	2.15	1.66
0.2	1.099	1.054	3.5	2.30	1.75
0.4	1.192	1.104	4.0	2.44	1.84
0.6	1.231	1.154	5.0	2.69	2.01
0.8	1.368	1.204	6.0	2.93	2.17
1.0	1.451	1.250	7.0	3.15	2.31
1.2	1.531	1.297	8.0	3.35	2.45
1.4	1.609	1.342	9.0	3.54	2.57
1.6	1.683	1.386	10.0	3.72	2.69
1.8	1.756	1.427			

When γ' is greater than 10, function $\Psi(\gamma')$ can be represented by the simple relationship:

$$\frac{\bar{i}_k}{\bar{i}_g} = 0.81 (k_1 C_Z^0 t_1)^{1/2} \quad (10.34)$$

It is easy to show that this equation can be utilized in analysis of catalytic currents which are at least three times greater than the magnitude of limiting current which would be observed in the absence of substance Z. The catalytic currents discussed in this subsection correspond to the potential range of the limiting current "plateau". This follows from boundary condition (10.10) which implies zero concentration of the Ox form at the electrode surface.

Henke and Hans [18] solved the problem of catalytic currents in polarography by means of the Laplace transformation. Their solution is in agreement with the calculated results of Koutecký.

In a later contribution Koutecký and Čížek [19] discussed the problem of catalytic currents, taking into account spherical diffusion.

10.5 The Rotating Disc Method

The problem of catalytic currents in relation to the rotating disc technique can be solved by means of the same conditions as those utilized in the case of chronoamperometry. The system of equations is however different. The equations describing the distribution of concentrations are as follows:

$$S_x \frac{dC_{Ox}}{dx} = D_{Ox} \frac{d^2 C_{Ox}}{dx^2} + k_1 C_{Red} C_Z^0 - k_2 C_{Ox}, \tag{10.35}$$

$$S_x \frac{dC_{Red}}{dx} = D_{Red} \frac{d^2 C_{Red}}{dx^2} - k_1 C_{Red} C_Z^0 + k_2 C_{Ox}. \tag{10.36}$$

With the aid of auxiliary functions, and assuming that the reaction is fast, Koutecký and Levich [20] derived an equation describing the current for an electrode reaction with catalytic regeneration of the depolarizer. This equation can be expressed in the form:

$$i_k = nFAC_{Ox}^0 D^{1/2} k_1^{1/2} (C_Z^0)^{1/2}. \tag{10.37}$$

It indicates that, in the case of a fast chemical reaction, the catalytic current does not depend, as in other techniques, on the mass transport determining factor, i.e. on the rate of rotation of the disc electrode. Equation (10.37) is valid when the thickness of the reaction layer is small, compared with that of the diffusion layer, $\mu \ll \delta$. The complete mathematical expression for this condition is:

$$\left(\frac{D}{k_1 C_Z^0 + k_2} \right)^{1/2} \ll 1.61 D^{1/3} v^{1/6} \omega^{-1/2}. \tag{10.38}$$

Condition (10.38) can be formulated as follows:

$$\left(\frac{\omega}{k_1 C_z^0 + k_2}\right)^{1/2} \left(\frac{D}{\nu}\right)^{1/6} \ll 1. \tag{10.38a}$$

In the practical application of the theory, it is convenient to use the ratio of the current due to the process involving catalytic reaction, to the limiting current which would be observed if the catalytic reaction rate were equal to zero. Dividing equation (10.37) by that of Nernst and Levich we obtain the equation

$$\frac{i_k}{i_g} = 1.61 D^{-1/6} \nu^{1/6} \omega^{-1/2} k_1^{1/2} (C_z^0)^{1/2}, \tag{10.39}$$

from which it follows that, in the case of a fast catalytic reaction, the ratio i_k/i_g increases linearly with increasing value of $\omega^{-1/2}$.

10.6 General Discussion

When the catalytic reaction is very fast, the current value is independent of the kinetic parameter. In such a case the catalytic current in chronoamperometry, stationary electrode voltammetry, polarography and the rotating disc technique can be described by the same equation:

$$i_k = nFAD^{1/2} C_{Ox}^0 k_1^{1/2} (C_z^0)^{1/2}. \tag{10.40}$$

As previously mentioned, in practice it is convenient to analyse the ratio of the kinetic current to the limiting current. These relationships can also be represented by a general equation using the concept of a kinetic parameter X.

$$\frac{i_k}{i_g} = B k_1^{1/2} (C_z^0)^{1/2} X^{1/2}. \tag{10.41}$$

In the case of chronopotentiometry, $\tau_k^{1/2}/\tau^{1/2}$ should be used instead of i_k/i_g.

The values of B for the methods discussed are shown in Table 10.4.

It follows from equation (10.41) that the current ratio i_k/i_g increases linearly with increasing $X^{1/2}$. This suggests the general method of analysing the catalytic currents, shown in Fig. 10.3. Such a relationship is valid only for appreciable values of parameter X, since for the values of X approaching zero it leads to the erroneous conclusion that the ratio i_k/i_g should tend to zero. Obviously, when $X \to 0$, $i_k/i_g \to 1$.

The equations derived in this chapter are valid provided that concentration polarization of substance Z is absent, as the assumption has been

Table 10.4 CONSTANT B IN EQUATION (10.41)

Method	Kinetic parameter	B
Chronoamperometry	t	$\pi^{1/2}$
Polarography	t_1	0.81
Chronopotentiometry	τ	$\dfrac{2}{\pi^{1/2}}$
Stationary electrode voltammetry	$\dfrac{1}{V}$	$\dfrac{R^{1/2}T^{1/2}}{0.466n^{1/2}F^{1/2}}$
Rotating disc method	$\dfrac{1}{\omega}$	$1.61 D^{-1/6}\nu^{1/6}$

made that the concentration of this substance at the surface of the electrode is equal to that in the bulk of the solution. This condition is fulfilled, provided that the current is much lower than the value of the limiting current which would be observed as a result of the reduction of substance Z.

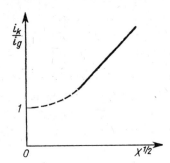

Fig. 10.3 Dependence of i_k/i_g on $X^{1/2}$.

It is known [see equation (5.169)], that the limiting current is

$$i_g = KAC_{Ox}^0 X^{-1/2}. \qquad (10.42)$$

Taking into account equation (10.41) this condition can be formulated as follows:

$$Bk_1^{1/2}(C_Z^0)^{1/2}X^{1/2}i_g \ll KAC_Z^0 X^{-1/2} \qquad (10.43)$$

Equation (10.42) gives, after transformation:

$$\frac{C_Z^0}{C_{Ox}^0} \gg Bk_1^{1/2}(C_Z^0)^{1/2}X^{1/2} \qquad (10.44)$$

This inequality must be fulfilled to maintain the validity of equations (10.40) and (10.41).

10.7 Second-Order Catalytic Reactions

Catalytic processes have been considered in which the concentration of substance Z reacting with the electrode reaction product is high enough for it to be assumed that this does not change at the electrode surface during the electrode process. It has been mentioned that the mathematical treatment of a catalytic problem involving changing concentration of substance Z is difficult. Nevertheless such a treatment of chronopotentiometry and polarography has been reported.

In the case of polarography an approximate solution was obtained by Březina [21]. It can be formulated as follows [22]:

$$\frac{i_k}{i_g} = \varphi \sqrt{1 + \frac{4C_Z^0}{\varphi C_{Ox}^0 \left(\frac{D_{Ox}}{D_Z}\right)^{1/2}} - 1}, \qquad (10.45)$$

where

$$\varphi = 0.328 k_1' t_1 C_{Ox}^0 \left(\frac{D_{Ox}}{D_Z}\right)^{1/2}. \qquad (10.46)$$

The exact solution, taking into account the growth rate of the dropping electrode, was obtained by Koutecký [23]. When the catalytic reaction is fast and the concentration of the catalyst is much lower than that of the reducible species the solution of Koutecký becomes very similar to equation (10.45).

The problem of a catalytic process taking place under chronopotentiometric conditions at a concentration of Z comparable to that of the reducible substance was solved by Fischer, Dračka and Fischerová [24]. If it is assumed that the catalytic reaction is irreversible, their solution can be expressed as follows:

$$i_0 \tau_k^{1/2} = \frac{d}{2} nFC_Z^0 \sqrt{\pi D_Z} + \frac{7}{12} nFC_{Ox}^0 D_{Ox} \sqrt{\frac{\pi}{D_Z}} - \frac{\sqrt{\pi D_Z}\, D_{Red} i_0^2}{2nFk_1 D_{Ox}^2 (C_{Ox}^0)^2}$$

$$(10.47)$$

where d is the coefficient due to the different stoichiometry of the catalytic reaction of Red with Z.

Equation (10.47) makes it possible to calculate the rate constant of the catalytic reaction k_1 from the slope of the straight line obtained from the system $i_0 \tau_k^{1/2} - i_0^2$.

10.8 Examples of Catalytic Currents

A relatively simple example of the appearance of a catalytic current is the polarographic reduction of ferric ions in the presence of hydrogen peroxide [5, 25]. Pospišil [5] determined the rate constant of the reaction of hydrogen peroxide with ferrous ions, which is 78 l. mole^{-1}s^{-1}.

Similar catalytic waves have also been observed during the reduction of several complex ions of ferric iron in the presence of hydrogen peroxide. Doskočil [26] investigated the catalytic reduction of complexes of iron with pyrocatechol, pyrogallol and ascorbic acid, and Svatek [27] studied the reduction of iron oxalate complexes. A catalytic current was also observed during the reduction of ferric iron complexes with triethanolamine in the presence of hydroxylamine. Koryta [28] has suggested a mechanism for this catalytic reaction.

Kolthoff and Parry [29] observed catalytic currents during reduction of molybdates, tungstates and vanadates in the presence of hydrogen peroxide.

Blažek and Koryta [30, 31] investigated the catalytic reduction wave of the oxalate complexes of quadrivalent titanium in the presence of hydroxylamine, and Koryta and Tenygl [32] studied the fast reaction of the oxalate complex of titanium(III) (which is formed on the electrode) with chlorates. Catalytic reduction of molybdates [33] and of tungsten(VI) to tungsten(V) [34] in the presence of chlorates was also investigated.

Catalytic currents were also observed [35] during the reduction of molybdenum(VI) and tungsten(VI), as well as of uranium(VI) in the presence of nitric acid [36].

It is noteworthy that catalytic currents are observed when one of the substances participating in the catalytic reaction is present in the form of a suspension. For example [37] during the reduction of $Ag(CN)_2^-$ in the presence of colloidal silver bromide the liberated cyanide ions react with AgBr and regenerate $Ag(CN)_2^-$.

More complex examples of catalytic currents can be observed during the reduction of iodine in the presence of iodates. In this case iodine is regenerated as a result of the reaction of the reduction product with IO_3^- [38].

An autocatalytic current is also observed in the reduction of vanadium(IV) in acidic solutions of 3 M calcium chloride at about -0.5 V. In this case small amounts of the resulting vanadium(II) react with the excess of vanadium(IV), and as a result, the current increases during the experiment when the potential is constant [39].

Examples of Catalytic Currents

Complex catalytic processes are observed [40] during the reduction of cobalt(III) complexes in the presence of hydrogen peroxide.

The examples quoted here are only a few of the studies carried out on catalytic currents. Experimental details of the determination of rate constants of catalytic reactions are given in the original publications.

References

[1] Wiesner, K., *Z. Elektrochem.*, **49**, 164 (1943).
[2] Brdička, R. and Wiesner, K., *Vestn. Kralov České Spolecnosti Nauk Trida Mat. Prirod.*, **18**, 1 (1943).
[3] Brdička, R. and Wiesner, K., *Coll. Czechoslov. Chem. Communs.*, **12**, 39 (1947).
[4] Delahay, P. and Stiehl, G. L., *J. Am. Chem. Soc.*, **74**, 3500 (1952).
[5] Pospíšil, Z., *Coll. Czechoslov. Chem. Communs.*, **18**, 337 (1953).
[6] Miller, S. L., *J. Am. Chem. Soc.*, **74**, 4130 (1952).
[7] Delmastro, J. R. and Booman, G. L., *Anal. Chem.*, **41**, 1409 (1969).
[8] Guidelli, R. and Cozzi, D., *J. Electroanal. Chem.*, **14**, 245 (1967).
[9] Savéant, J. M. and Vianello, E., in *"Advances in Polarography"*, Longmuir, J. S., Ed., Pergamon Press, New York 1960, Vol. 1, page 367.
[10] Nicholson, R. S. and Shain, I., *Anal. Chem.*, **36**, 706 (1964).
[11] Weber, J., *Coll. Czechoslov. Chem. Communs.*, **24**, 1770 (1959).
[12] Kao, H. Chang T. S. and Chang, W. B., *Sci. Sinica (Peking)*, **13**, 1411 (1964).
[13] Rampazzo, L., *Ricerca Sci.*, **36**, 998 (1966).
[14] Rampazzo, L., *J. Electroanal. Chem.*, **14**, 117 (1967).
[15] Savéant, J. M. and Vianello, E., *Electrochim. Acta*, **10**, 905 (1965).
[16] Delahay, P., Mattax, C. C. and Berzins, T., *J. Am. Chem. Soc.*, **76**, 5319 (1954).
[16a] Delahay, P., *"New Instrumental Methods in Electrochemistry"*, Interscience, New York 1954.
[17] Koutecký, J., *Chem. Listy*, **47**, 9 (1953); *Coll. Czechoslov. Chem. Communs.*, **18**, 311 (1953).
[18] Henke, K. H. and Hans, W., *Z. Elektrochem.*, **59**, 676 (1955).
[19] Koutecký, J. and Čížek, J., *Chem. Listy*, **50**, 393 (1953); *Coll. Czechoslov. Chem. Communs.*, **21**, 1063 (1956).
[20] Koutecký, J. and Levich, V. G., *Zhurn. Fiz. Khim.*, **32**, 1565 (1958).
[21] Březina, M., *Chem. Listy*, **50**, 1899 (1956); *Coll. Czechoslov. Chem. Communs.*, **22**, 339 (1957).
[22] Heyrovský, J. and Kuta, J., in *"Principles of Polarography"*, Ed. Mir, Moscow 1965, page 366 (in Russian).
[23] Koutecký, J., *Chem. Listy*, **47**, 1410 (1956); *Coll. Czechoslov. Chem. Communs.*, **22**, 160 (1957).
[24] Fischer, O., Dračka, O. and Fischerová, E., *Coll. Czechoslov. Chem. Communs.*, **26**, 1505 (1961).
[25] Kolthoff, I. M. and Parry, E. P., *J. Am. Chem. Soc.*, **73**, 3728 (1951).

[26] Doskočil, J., *1st International Congress of Polarography*, Prague 1951, Vol. 1, p. 674.
[27] Svatek, E., *1st International Congress of Polarography*, Prague 1951, Vol. 1, p. 789.
[28] Koryta, J., *Chem. Listy*, **48**, 514 (1954); *Coll. Czechoslov. Chem. Communs.*, **19** 666 (1954).
[29] Kolthoff, I. M. and Parry, E. M., *J. Am. Chem. Soc.*, **73**, 5315 (1951).
[30] Blažek, A. and Koryta, J., *Chem. Listy*, **47**, 26 (1953); *Coll. Czechoslov. Chem. Communs.*, **18**, 326 (1953).
[31] Koryta J., *Chem. Zvesti*, **8**, 723 (1954).
[32] Koryta, J. and Tenygl, J., *Chem. Listy*, **48**, 467 (1954); *Coll. Czechoslov. Chem. Communs.*, **20**, 423 (1955).
[33] Kolthoff, I. M. and Hodara, I., *J. Electroanal. Chem.*, **5**, 2 (1963).
[34] Kolthoff, I. M. and Hodara, I., *J. Electroanal. Chem.*, **4**, 369 (1962).
[35] Koryta, J., *Coll. Czechoslov. Chem. Communs.*, **20**, 667 (1955).
[36] Johnson, M. G. and Robinson, R. J., *Anal. Chem.*, **24**, 366 (1952).
[37] Bower, R. C. and Kolthoff, I. M., *J. Am. Chem. Soc.*, **81**, 1836 (1959).
[38] Desideri, P. G., *J. Electroanal. Chem.*, **9**, 218 (1965).
[39] Jędral, T. and Galus, Z., *Roczniki Chem.*, **48**, 1859 (1974).
[40] Vlček, A. A., *Coll. Czechoslov. Chem. Communs.*, **25**, 2685 (1960).

Chapter 11

Electrode Processes Preceded by Higher-Order Reactions

This chapter discusses electrode processes which are mainly limited to the dissociation of electrode inactive substances into monomers capable of exchanging a charge with the electrode. Such reactions can be formulated in the following general way:

$$B \underset{k_2}{\overset{k_1}{\rightleftharpoons}} 2Ox, \tag{11.1}$$

$$Ox + ne \rightarrow Red. \tag{11.2}$$

Only a limited number of such processes have yet been described [1, 2]. One of them is the oxidation of hyposulphites. It has been found that SO_2^- monomers originating from the reaction:

$$S_2O_4^{2-} \underset{k_2}{\overset{k_1}{\rightleftharpoons}} 2SO_2^- \tag{11.3}$$

participate in the process.

Further investigations of this problem should be fruitful, since it is to be expected that in the electrode processes of pinacols, peroxides, disulphides etc. a partial monomerization, at least, takes place under appropriately chosen conditions. Therefore the determination of monomerization criteria could be useful in the analyses of the mechanisms of electrode processes.

For the theoretical treatment of the process described by equations (11.1) and (11.2) it is necessary to solve a system of differential equations. It is assumed that the processes take place under conditions of linear diffusion:

$$\frac{\partial C_B}{\partial t} = D_B \frac{\partial^2 C_B}{\partial x^2} - \frac{k_1 C_B}{2} + \frac{k_2 C_{Ox}^2}{2}, \tag{11.4}$$

$$\frac{\partial C_{Ox}}{\partial t} = D_{Ox} \frac{\partial^2 C_{Ox}}{\partial x^2} + k_1 C_B - k_2 C_{Ox}^2, \tag{11.5}$$

$$\frac{\partial C_{Red}}{\partial t} = D_{Red} \frac{\partial^2 C_{Red}}{\partial x^2}. \tag{11.6}$$

Since the reduced form is generated in the solution as a result of the electrode reaction, the initial conditions can be formulated as follows:

$$t = 0, \quad x \geq 0, \quad C_{Ox}(x, 0) = C_{Ox}^0, \quad C_B(x, 0) = C^0, \quad C_{Red}(x, 0) = 0. \tag{11.7}$$

For $t > 0$ and $x \to \infty$ the following boundary conditions are valid:

$$C_{Ox} \to C_{Ox}^0, \quad C_B \to C^0, \quad C_{Red} \to 0. \tag{11.8}$$

The problem was first elucidated for a process carried out under polarographic conditions.

11.1 Polarography

An approximate solution of the problem was presented by Hanuš [3] who used a method based on the reaction layer concept. This solution can be formulated as follows:

$$\frac{\bar{i}_k}{i_g} \geq 0.81\alpha_1 \left(1 - \frac{\bar{i}_k}{i_g}\right)^{3/4}, \tag{11.9}$$

$$\alpha_1 = \sqrt{\frac{k_1 t_1}{(KC^0)^{1/2}}}, \tag{11.10}$$

where K is the equilibrium constant defined by the equation:

$$K = \frac{k_2}{k_1}. \tag{11.11}$$

For an exact solution of the problem it is necessary to determine the boundary conditions for $x = 0$ and $t > 0$:

$$\left[\frac{\partial C_B}{\partial x}\right]_{x=0} = 0, \tag{11.12}$$

$$\frac{C_{Ox}}{C_{Red}} = \Theta, \tag{11.13}$$

$$D_{Ox}\left[\frac{\partial C_{Ox}}{\partial x}\right]_{x=0} + D_{Red}\left[\frac{\partial C_{Red}}{\partial x}\right]_{x=0} = 0, \tag{11.14}$$

where:

$$\Theta = \exp\left[\frac{nF(E-E_0)}{RT}\right]. \tag{11.15}$$

Condition (11.12) expresses the fact that in the interval of potential under consideration the dimer does not react with the electrode.

11.1] Polarography

The solution of the system of equations (11.4)–(11.6), with added terms $2x\partial C/3t\partial x$ and with initial and boundary conditions (11.7), (11.8) and (11.12)–(11.14) was presented by Koutecký and Hanuš [4]. For instantaneous currents this solution can be represented by the relationship:

$$\frac{i}{i_g} = f(\alpha) \tag{11.16}$$

where:

$$\alpha = \sqrt{\frac{k_1 t}{(KC^0)^{1/2}}} \tag{11.17}$$

Similarly, for the average current:

$$\frac{\bar{i}}{\bar{i}_g} = f(\alpha_1). \tag{11.18}$$

α_1 has already been defined [equation (11.10)].

The values of functions $f(\alpha)$ and $f(\alpha_1)$ are given in Table 11.1.

Table 11.1 VALUES OF FUNCTIONS $f(\alpha)$ AND $f(\alpha_1)$ (FROM [4] BY PERMISSION OF THE COPYRIGHT HOLDERS)

$\sqrt{\frac{4}{7}\alpha}$	$f(\alpha)$	$\sqrt{\frac{4}{7}\alpha_1}$	$f(\alpha_1)$
0.1	0.084	0.1	0.060
0.2	0.159	0.2	0.115
0.4	0.290	0.4	0.213
0.6	0.396	0.6	0.293
0.8	0.481	0.8	0.368
1.0	0.553	1.0	0.429
1.2	0.611	1.2	0.485
1.4	0.661	1.4	0.530
1.6	0.698	1.6	0.572
1.8	0.73	1.8	0.608
		2.0	0.64
		2.5	0.70
3.0	0.83	3.0	0.75
		3.5	0.78
4.0	0.893	4.0	0.81
5.0	0.923	5.0	0.85
		6.0	0.88
7.0	0.953	7.0	0.90
		8.0	0.915
10.0	0.971	10.0	0.93
15.0	0.984	15.0	0.96

A more accurate comparison of the results of approximate calculations with those obtained by the solution of differential equations can be represented graphically. In Fig. 11.1 a graph of average currents $\bar{i}/\bar{i}_g$ against parameter α_1, calculated from equation (11.9) and from the data collected in Table 11.1, is shown.

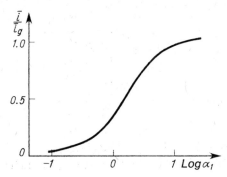

Fig. 11.1 Dependence of average current ratio $\bar{i}/\bar{i}_g$ on parameter α_1.

It follows from the above discussion that the investigation of the kinetics of monomerization processes is possible only when the values of parameter α_1 are within the interval 0.1–10, since in such a case the kinetic current is equal to 6–93% of the diffusion current which would be observed if the equilibrium of the monomerization reaction were shifted completely in the direction of monomer formation. Parameter α_1 depends on the chemistry of the reaction preceding the electrode process, i.e. on the values of k_1 and K. When the rate constant k_1 is very large, or the equilibrium constant is very small, the values of parameter α_1 are large and a kinetic analysis of the process may be impossible. In such a case the shortest possible drop time and a high concentration of substance B should be employed, but the effect of these factors on parameter α_1 is limited, since the interval of t_1 variation is narrow and the power index of C^0 is small.

The polarographic process:

$$B \underset{k_2}{\overset{k_1}{\rightleftarrows}} Ox + C$$
$$Ox + ne \rightarrow Red$$
(11.19)

with low concentration of B was described by Čížek et al. [5]. An approximate treatment of this process was presented earlier by Hanuš [3]. Čížek's description also involved diffusion to the dropping electrode, and the problem was solved by the method of dimensionless parameters.

Guidelli [6] has discussed the theory of such processes taking place in chronoamperometric conditions at flat and at spherical electrodes.

11.2 Stationary Electrode Voltammetry

The problem of an electrode reaction preceded by monomerization, leading to the formation of the depolarizer, was elucidated by Savéant and Vianello [7].

In order to solve the system of equations (11.4)–(11.6), use is made of conditions (11.12) and (11.14) and of the additional boundary condition:

$$t \geqslant 0, \quad x = 0, \quad \frac{C_{\text{Ox}}}{C_{\text{Red}}} = \exp\left[\frac{nF}{RT}(E_t - E^0)\right]. \quad (11.20)$$

E_t is the electrode potential at time t from the start of the electrolysis. It is determined by the relationship:

$$E_t = E_i - Vt$$

where E_i is the initial potential.

In discussion of the solution of this problem three principal cases can be considered, depending on the parameter described by the equation:

$$\Delta = \left(\frac{4RT}{3nF}\right)^{1/2} \left(\frac{D_{\text{Ox}}}{D_{\text{B}}}\right)^{1/2} \left(\frac{k_2}{V}\right)^{1/2} K^{-3/4}(C^0)^{-1/4}. \quad (11.21)$$

When parameter Δ is very small, the concentrations of the Ox and Red forms are also low, and the concentration of B is practically constant and equal to the initial concentration. In such a case the current is independent of the kinetics of charge transfer and of the polarization rate, and it is proportional to the 3/4 power of the initial concentration of substance B.

$$i_p = \left(\frac{4}{3}\right)^{1/2} nFD_{\text{Ox}}^{1/2} k_2^{1/2} K^{-3/4}(C^0)^{3/4} A. \quad (11.22)$$

In such a case virtually no peaks are observed, and the recorded current–potential relationship resembles polarographic waves. Thus i_p in equation (11.22) is a "plateau" current rather than a peak current. The determination of this current is sufficient for the determination of the rate constant k_2, provided that the equilibrium constant K is known. The potential corresponding to the half-height of the current "plateau" is described by the equation:

$$E_{p/2} = E_{1/2} - 0.13\frac{RT}{nF} - \frac{RT}{2nF}\ln\frac{4RT}{3nF} - \frac{RT}{2nF}\ln\frac{k_2}{V} - \frac{RT}{4nF}\ln KC^0. \quad (11.23)$$

The half "plateau" potential changes linearly with the logarithm of polarization rate, and at 25°C, a tenfold increase of V causes a shift of $30/n$ mV in the anodic direction. A tenfold increase of the initial concentration of the dimer causes a shift of $E_{p/2}$ by $15/n$ mV in the positive direction.

When the value of parameter Δ is very high the equilibrium is reached during the electrode process. In such a case the current is diffusion controlled and the current peak equation has the following form:

$$i_p = 1.087 \frac{n^{3/2} F^{3/2}}{R^{1/2} T^{1/2}} D_B^{1/2} A V^{1/2} C^0. \tag{11.24}$$

The potential corresponding to the half-current peak is defined by the equation:

$$E_{p/2} = E^0 + 0.40 \frac{RT}{nF} + \frac{RT}{2nF} \ln \frac{D_{\text{Red}}}{D_B} - \frac{RT}{2nF} \ln C^0 K. \tag{11.25}$$

$E_{p/2}$ is independent of the polarization rate, but it changes linearly with the logarithm of the initial concentration of the dimer. The equilibrium constant, K, may be calculated by determining the potential $E_{1/2}$, provided that the standard potential of the Ox/Red system and the diffusion coefficients of the Red form and the dimer are known.

According to Savéant and Vianello, the current is diffusion controlled when $\log \Delta > 1.33$, and it is a purely kinetic current when $\log \Delta < -1.36$.

11.3 Chronopotentiometry

A general mathematical solution of electrode processes preceded by a monomerization reaction was presented by Koutecký and Čížek [8]. A detailed treatment of this problem was published in 1960 by Fischer, Dračka and Fischerová [9].

The solution of this problem requires the definition of the characteristic chronopotentiometric boundary condition:

$$t > 0, \quad x = 0, \quad D_{\text{Ox}} \frac{\partial C_{\text{Ox}}}{\partial x} = -D_{\text{Red}} \frac{\partial C_{\text{Red}}}{\partial x} = \frac{i}{nFA}. \tag{11.26}$$

When the equilibrium in the solution is shifted appreciably in the direction of dimer formation ($KC_B^0 \gg 1$), and when the chemical reaction is fast ($k_1 t \gg 1$), the concentrations of the reacting substances can be described by the relationships:

$$C_{\text{Red}}(0, t) = C_{\text{Red}}^0 + \frac{2it^{1/2}}{\pi^{1/2} nF D_{\text{Red}}^{1/2} A}, \tag{11.27}$$

11.3] Chronopotentiometry

$$y^3 - 3y = \frac{3i^2}{n^2F^2k_1 DA}\sqrt{\frac{2K}{Z^3}} - 2, \qquad (11.28)$$

where:

$$y = C_{\text{Ox}}(0, t)\sqrt{\frac{2K}{Z}}, \qquad (11.29)$$

$$Z = 2C_{\text{B}}(0, t) + C_{\text{Ox}}(0, t) = 2C^0 + C_{\text{Ox}}. \qquad (11.30)$$

K is defined by equation (11.11) and $D = D_{\text{B}} = D_{\text{Ox}}$.

For transition time $t = \tau$ and $C_{\text{Ox}}(0, t) = y = 0$, the following relationship is obtained:

$$i_0\tau^{1/2} = nFC_{\text{B}}^0\sqrt{\pi D}\left(1 + \sqrt{\frac{1}{4KC_{\text{B}}^0}}\right) - \left(\frac{9K}{16nFk_1^2}\right)^{1/3}\left(\frac{\pi^3}{D}\right)^{1/6}i_0^{4/3}$$

$$= N - Pi_0^{4/3}. \qquad (11.31)$$

It follows from equation (11.31) that the relationship $i_0^{1/2} - i_0^{4/3}$ should be linear. From the slope of the straight line $P = -\Delta(i_0\tau^{1/2})/\Delta i_0^{4/3}$ the expression $k_1/\sqrt{K}$ can be determined:

$$\frac{k_1}{\sqrt{K}} = \frac{3}{4(nF)^{1/2}P^{3/2}}\left(\frac{\pi^3}{D}\right)^{1/4}.$$

When the value of D is determined on the basis of N by extrapolation to $i_0 = 0$, and the analytical concentration C^0 is substituted for the equilibrium concentration C_{B}^0, the quotient k_1/K can be described by the relationship:

$$\frac{k_1}{\sqrt{K}} = \frac{3\pi}{4}\sqrt{\frac{C^0}{NP^3}}.$$

These relationships were examined by Fischer, Dračka and Fischerová [9], using the anodic oxidation of $S_2O_4^{2-}$ ions.

11.4 The Rotating Disc Method

In the case where the electrode process, schematically described by equations (11.1) and (11.2), takes place at a rotating disc, the convective diffusion equations have the following form:

$$S_x\frac{dC_{\text{B}}}{dx} = D\frac{d^2C_{\text{B}}}{dx^2} - \frac{1}{2}k_1 C_{\text{B}} + \frac{1}{2}k_2 C_{\text{Ox}}^2, \qquad (11.32)$$

$$S_x\frac{dC_{\text{Ox}}}{dx} = D\frac{d^2C_{\text{Ox}}}{dx^2} + k_1 C_{\text{B}} - k_2 C_{\text{Ox}}^2. \qquad (11.33)$$

The initial and boundary conditions (11.7) and (11.8) are valid. The boundary conditions at the electrode surface illustrate the fact that substance B is electrode inactive and the electrode potential is so negative that the concentration of the Ox form at the electrode surface equals zero:

$$t > 0, \quad x = 0, \quad C_{Ox} = 0, \tag{11.34}$$

$$x = 0, \quad \frac{dC_B}{dx} = 0. \tag{11.35}$$

The elucidation of this problem was presented by Koutecký and Levich [10]. They assumed that the equilibrium of the chemical reaction is shifted in the direction of dimer formation. By introducing the additional function Ψ, described by the relationship:

$$\Psi = C_{Ox} + 2C_B \tag{11.36}$$

it is possible to represent the solution as follows:

$$i_k = nFAD \frac{(2C^0 - \Psi_{x=0})}{\delta}, \tag{11.37}$$

where i_k is the observed kinetic current, δ is the thickness of the diffusion layer, C^0 is the analytical concentration of substance B, and $\Psi_{x=0}$ is the value of function Ψ at the electrode surface. $\Psi_{x=0}$ is related to the rate constants of the chemical reaction and to δ and C^0 by the equation:

$$\Psi_{x=0}^{3/4} + \frac{D}{Q\delta} \Psi_{x=0} - \frac{2DC^0}{Q\delta} = 0 \tag{11.38}$$

where:

$$Q = \left(\frac{k_1}{k_2 D^2}\right)^{1/4} \frac{1}{\sqrt[4]{18}}. \tag{11.39}$$

Equation (11.38) can be solved graphically.

The relationships discussed above are valid when the diffusion coefficients of the dimer and of the monomers are equal. The case of unequal diffusion coefficients was presented by Dogonadze [11].

A further solution was produced by Hale [12] who used direct current electrolysis.

In the preceding chapters the relationships for each electrochemical method have been given in the form of one general equation, on the basis of which it was possible to discuss the principal features of the electrode process. It is not possible to derive such a general relationship in the case discussed in this chapter.

References

[1] Čermak, V., *Chem. Zvesti*, **8**, 714 (1954).
[2] Čermak, V., *Chem. Listy*, **51** 2213 (1957); *Coll. Czechoslov. Chem. Communs.*, **23**, 1871 (1958).
[3] Hanuš, V., *Chem. Zvesti*, **8**, 702 (1954).
[4] Koutecký, J. and Hanuš, V., *Chem. Listy*, **48**, 1446 (1954); *Coll. Czechoslov. Chem. Communs.*, **20**, 124 (1955).
[5] Čižek, J., Koryta, J. and Koutecký, J., *Coll. Czechoslov. Chem. Communs.*, **24**, 663 (1959).
[6] Guidelli, R., *J. Electroanal. Chem.*, **33**, 303 (1971).
[7] Savéant, J. M. and Vianello, E., *Electrochim. Acta*, **12**, 1545 (1967).
[8] Koutecký, J. and Čižek, J., *Chem. Listy*, **51**, 827 (1957); *Coll. Czechoslov. Chem. Communs.*, **22**, 914 (1957).
[9] Fischer, O., Dračka O. and Fischerová, E., *Coll. Czechoslov. Chem. Communs.*, **25**, 323 (1960).
[10] Koutecký, J. and Levich, V., *Zhurn. Fiz. Khim.*, **32**, 1565 (1958).
[11] Dogonadze, R. R., *Zhurn. Fiz. Khim.*, **32**, 2437 (1958).
[12] Hale, J. M., *J. Electroanal. Chem.*, **8**, 332 (1964).

Chapter 12

Electrode Processes Followed by Dimerization of the Primary Product

There are a large number of papers [1] describing electrode reactions accompanied by dimerization of the primary product formed during the charge transfer. Mairanovskii [2] observed dimerization of the products of the one-electron reduction of N-substituted pyridine. As a result of the electrochemical oxidation of triphenylamine and of substituted triphenylamines, taking place *via* cationic radicals and dimerization, the corresponding tetraphenylbenzidines are formed [3]. Another example is the formation of pinacols *via* dimerization of free ketyl radicals. The problem of possible dimerization during the electrode reduction of aromatic carbonyl compounds in alkaline media has been extensively studied by Nadjo and Savéant [4]. Dimerization of radicals originating during the electrode reduction of imines in acetonitrile and dimethylformamide has been investigated by Andrieux and Savéant [5]. These authors also studied the mechanism of the electrodimerization of ammonium cations in acetonitrile and in benzonitrile [6]. The existence of immonium radicals was shown by the electron paramagnetic resonance method [7]. It is possible that dimerization frequently accompanies electrochemical free radical formation.

The electrode reaction can be formulated by the general equations:

$$Ox + ne \rightleftharpoons Red, \qquad (12.1)$$

$$2Red \underset{k_2}{\overset{k_1}{\rightleftharpoons}} A. \qquad (12.2)$$

If the reduction of the Ox form takes place in the process, k_1 is the dimerization rate constant, and k_2 is the rate constant of decomposition of substance A. Substance A is the final product of the reaction and it is electrode inactive in the range within which the reaction takes place.

Assuming linear diffusion, the following system of equations must be solved:

$$\frac{\partial C_{Ox}}{\partial t} = D_{Ox} \frac{\partial^2 C_{Ox}}{\partial x^2}, \qquad (12.3)$$

$$\frac{\partial C_{Red}}{\partial t} = D_{Red} \frac{\partial^2 C_{Red}}{\partial x^2} + k_2 C_A - k_1 C_{Red}^2, \qquad (12.4)$$

$$\frac{\partial C_A}{\partial t} = D_A \frac{\partial^2 C_A}{\partial x^2} - \frac{k_2}{2} C_A + \frac{k_1 C_{Red}^2}{2}. \qquad (12.5)$$

Usually the Red and A forms are formed as a result of the electrode process, and therefore the initial conditions can be formulated as:

$$t = 0, \quad x \geqslant 0, \quad C_{Ox} = C_{Ox}^0, \quad C_{Red} = 0, \quad C_A = 0, \qquad (12.6)$$

and the boundary conditions as:

$$t > 0, \quad x \to \infty, \quad C_{Ox} \to C_{Ox}^0, \quad C_{Red} \to 0, \quad C_A \to 0. \qquad (12.7)$$

Other boundary conditions relating to the reaction at the electrode surface are specific for each method considered here.

12.1 Polarography and Chronoamperometry

The problem of electrode processes followed by dimerization of the electrode reaction products was first solved for the polarographic technique. Initially it was worked out by Hanuš [8], who utilized the method discussed earlier, based on the reaction layer concept derived by Brdička and Wiesner for the solution of kinetic problems.

The results obtained by this method are satisfactory, provided that the dimerization process is fast. The exact solution was given by Koutecký and Hanuš [9]. The equations which they solved were more complex than (12.3)–(12.5), since they contained the term $(2x/3t)(\partial C/\partial x)$ corresponding to the growth of the dropping mercury electrode. The solution was obtained by employing the assumption that the dimerization process is very fast and irreversible, i.e. that the dimerization constant is much greater than the rate constant of the reverse reaction.

For instantaneous currents the solution can be formulated in the following manner:

$$\frac{i}{i_g} = f(\gamma) \qquad (12.8)$$

where:

$$\gamma = \sqrt{\frac{D_{Red}}{D_{Ox}} k_1 C_{Ox}^0 \Theta^3 t}. \qquad (12.9)$$

A similar formula can be derived for average currents:

$$\frac{\bar{i}}{\bar{i}_g} = f(\gamma_1), \tag{12.10}$$

$$\gamma_1 = \sqrt{\frac{D_{\text{Red}}}{D_{\text{Ox}}} k_1 C_{\text{Ox}}^0 \Theta^3 t_1}. \tag{12.11}$$

The values of functions $f(\gamma)$ and $f(\gamma_1)$ are shown in Table 12.1.

Table 12.1 VALUES OF FUNCTIONS $f(\gamma)$ AND $f(\gamma_1)$. (FIGURES IN BRACKETS OBTAINED BY INTERPOLATION) (FROM [9] BY PERMISSION OF THE COPYRIGHT HOLDERS)

$\sqrt{\frac{8}{7}}\gamma$	$f(\gamma)$	$\sqrt{\frac{8}{7}}\gamma_1$	$f(\gamma_1)$	$\sqrt{\frac{8}{7}}\gamma$	$f(\gamma)$	$\sqrt{\frac{8}{7}}\gamma_1$	$f(\gamma_1)$
0.1	0.080	0.1	0.057	3.0	(0.64)	3.0	(0.52)
0.2	0.147	0.2	0.107	4.0	(0.69)	4.0	(0.61)
0.3	0.202	0.3	0.150	5.0	0.73		
0.4	0.250	0.4	0.188	6.0	0.756	6.0	(0.68)
0.5		0.5	0.222	7.0	0.779		
0.6	(0.31)	0.6	(0.25)	8.0	0.797	8.0	(0.73)
0.8	(0.38)	0.8	(0.30)	9.0	0.812		
1.0	(0.43)	1.0	(0.35)	10.0	0.825	10.0	(0.76)
1.6	(0.52)	1.6	(0.44)	15.0	0.865	15.0	(0.81)
2.0	(0.56)	2.0	(0.47)	20.0	0.887	20.0	(0.85)

Parameter Θ appearing in equations (12.9) and (12.11) is defined by the equation:

$$\Theta = \exp\left[\frac{nF(E^0 - E)}{RT}\right]. \tag{12.12}$$

The relationships describing the polarographic wave can be approximately expressed by the equation:

$$E = E^0 - \frac{RT}{nF} \ln \frac{\bar{i}^{2/3}}{\bar{i}_g - \bar{i}} \bar{i}_g^{1/3} + \frac{RT}{3nF} \ln \frac{C_{\text{Ox}}^0 k_1 t_1}{1.51}. \tag{12.13}$$

This suggests that for the determination of dimerization rate constants, the best procedure is application of the equation describing the half-wave potential. It follows from Table 12.1 that, when average currents at the half-wave potential are measured, we have:

$$\left(\sqrt{\frac{8}{7}}\gamma_1\right)_{\bar{i}=\bar{i}_g/2} = 2.7. \tag{12.14}$$

Combining equation (12.4) with equation (12.11), in which $E_{1/2}$ has been substituted for E $\left(\text{since } \bar{i} = \dfrac{1}{2}\bar{i}_g\right)$ and assuming equal diffusion coefficients of the Ox and Red forms, we obtain the relationship:

$$E^k_{1/2} = E^0 - \frac{RT}{nF}0.62 + \frac{RT}{3nF}\ln C^0_{Ox}k_1 t_1. \tag{12.15}$$

A similar relationship is obtained when equation (12.13) is considered. The expression describing the half-wave potential obtained from this equation differs from that obtained from equation (12.15) only by the constant term ($-0.36RT/nF$ instead of $-0.62RT/nF$).

It follows from equation (12.15) that, in the case of electrode processes accompanied by fast dimerization of the primary reaction product, the half-wave potential depends on several factors. An increase in the Ox form concentration and in the drop time results in shifting the half-wave potential in the positive direction when reduction takes place at the electrode. Moreover, the half-wave potential is shifted in the positive direction with respect to the reversible standard Ox/Red potential, and this shift depends on the dimerization rate constant. The faster the reaction, the greater is the shift.

When the substrate is oxidized at the electrode and it is the oxidized form which is dimerized, the half-wave potential is shifted in the negative direction.

The above criteria of dimerization after the charge transfer are valid only when the charge transfer is reversible under the experimental conditions.

An equation almost identical to (12.15) was derived by Mairanovskii [2], who utilized an approximate method based on the reaction layer concept. Dimerization of the primary electrode reaction product was further investigated by the same author [10] and by Čižek et al. [11]. Savéant et al. dealt with the problem from the aspect of diagnostic criteria in their discussion of the stationary electrode voltammetric approach to secondary dimerization.

Nelson and Feldberg [12] published a theoretical treatment of electrodimerization under chronoamperometric conditions. Utilizing working curves, they determined rate constants of the dimerization of cationic radicals. They found $k_1 = 9.4 \times 10^1$ l. mole^{-1} s^{-1} for the dimerization of 4-methyltriphenylamine and $k_1 = 6.8 \times 10^3$ l. mole^{-1} s^{-1} for the dimerization of 4-nitrotriphenylamine. Details of calculations were published in the monograph by Feldberg [13].

12.2 Stationary Electrode Voltammetry

Savéant and Vianello [14] and Nicholson [15] were the first workers to deal with the theory of stationary electrode voltammetric processes with secondary dimerization of the electrode reaction product. In 1967 the first of these authors [16] published an extensive study of reversible Ox/Red, electrode reactions in which the inequality of the diffusion coefficients of the monomeric and dimeric forms was taken into account.

The problem can be discussed in two ways, depending on the ratio of the dimerization and polarization rates.

Take first the case in which the dimerization is fast, compared with the polarization rate. In this case the peak current is described by the equation:

$$i_p = 0.527 \frac{n^{3/2} F^{3/2}}{R^{1/2} T^{1/2}} D_{Ox}^{1/2} C_{Ox}^0 V^{1/2} A \qquad (12.16)$$

and the peak potential is described by the expression:

$$E_p^k = E_{1/2} - 0.902 \frac{RT}{nF} + \frac{RT}{3nF} \ln \frac{2RT}{3nF} + \frac{RT}{3nF} \ln \frac{k_1 C_{Ox}^0}{V}. \qquad (12.17)$$

Equations (12.16) and (12.17) lead to the conclusion that investigation of the peak current cannot be used for the kinetic analysis of the chemical dimerization reaction, as the peak current depends on parameters identical with those governing the processes controlled exclusively by the diffusion rate; the numerical coefficient in equation (12.16) is, however, somewhat larger than that appearing in the Randles–Ševčik equation.

Equation (12.17) can on the other hand be utilized in kinetic studies of dimerization. It follows from this equation that the dependence of the peak potential on the dimerization rate constant and on the concentration of the oxidized form is similar to that of the half-wave potential [equation (12.15)]. Increase in the polarization rate causes a shift of the peak potential in the cathodic direction.

The half-wave potential appearing in equation (12.17) corresponds to the reversible electrode process in which the effect of the secondary dimerization on the recorded curves is very small, and hence it is identical with the half-wave potential of the Ox/Red system.

In the second case of dimerization under stationary electrode voltammetric conditions, discussed by Savéant and Vianello, the equilibrium of the dimerization is reached during the measurements. This may be observed when the rate of dimerization is very fast, compared with the polarization rate. In this case the peak current is described by the equation:

$$i_p = 0.500 \frac{n^{3/2}F^{3/2}}{R^{1/2}T^{1/2}} D_{Ox}^{1/2} A V^{1/2} C_{Ox}^0. \quad (12.18)$$

As in the previous case, the dependence of the peak current on D, A, V and C_{Ox}^0 is identical with that observed in diffusion controlled processes. The constant numerical coefficient is somewhat smaller than that of the equation describing processes in which the dimerization is not in a state of equilibrium but it is still larger than that appearing in the Randles–Ševčik equation. The peak potential no longer depends on the dimerization rate and on the polarization rate, but it depends on the concentration of the oxidized form and K:

$$E_p = E^0 - 0.70 \frac{RT}{nF} - \frac{RT}{4nF} \ln \frac{D_{Ox}}{D_A} + \frac{RT}{2nF} \ln C_{Ox}^0 K \quad (12.19)$$

where D_A is the diffusion coefficient of the dimer and K is the equilibrium constant of the dimerization reaction, determined by the ratio of k_1 to the rate constant of dissociation of the dimer into the monomers.

Equation (12.19) indicates that, in the case of a reduction, a difference between the peak potential and the reversible standard electrode potential of the Ox/Red system should be observed. This difference increases with increasing concentration of the Ox form and with increasing shift of the dimerization equilibrium in the direction of dimer formation.

The calculation of the dimerization rate constant from equation (12.19) is impossible, because of the equilibrium existing at the electrode under the experimental conditions, but this equation can be used for the determination of the dimerization equilibrium constant. The basic difficulty in this determination is the evaluation of the E^0 potential. This potential should be measured under such conditions that the dimerization does not affect the recorded curves. Theoretically this should be observed when the polarization rate is increased to values at which dimerization is negligible, on account of the very short duration of the process, but when the dimerization is very fast the corresponding polarization rates must be so high that the capacitance effects become of primary importance and, as a result, accurate measurements are no longer possible. Also the electrode reaction may be then irreversible.

Two cases of stationary electrode voltammetric reduction followed by dimerization at dimerization rates slightly or appreciably faster than the polarization rates have been considered. Theoretically a third case also exists.

This is a process in which the dimerization rate is slower than the polarization rate. This problem has not been considered, since it is practically equivalent to the diffusion controlled process. In a case where such a relationship between the chemical reaction rate and the polarization rate exists the effect of the chemical reaction on the recorded curves is negligible.

An even more extensive investigation of the dimerization problem was carried out by Andrieux et al. [17]. Electrodimerization reactions have been divided into three types:
(a) reaction of a radical with another radical (DIM 1);
(b) reaction of a radical with the substrate followed by reduction of the resulting dimer (DIM 2);
(c) reaction of an ion formed from the substrate by acceptance of two electrons with the substrate (DIM 3).

Such processes have been considered theoretically, diagnostic criteria making it possible to differentiate them from other kinetic processes have been given and procedures for determination of dimerization rate constants have been developed.

Many electrodimerization reactions are in fact electrohydrodimerizations which can be represented by the general equation:

$$2\text{Ox} + 2\text{H}^+ + 2e \rightarrow \text{DH}_2.$$

Savéant [18, 19] was dealing with such processes in his recent work and derived equations describing various mechanisms of these processes as well as diagnostic criteria.

Laviron [20] considered a number of various variants of dimerization following reversible charge transfer under the conditions of electrolysis of thin layers of solution.

Subsequent dimerization has been also considered [21] on the assumption that the substrate and the product are adsorbed on the electrode.

Laviron [22] also elaborated the case of electrode reaction followed by irreversible chemical reaction of the substrate with the primary product leading to an electrode inactive substance. He was also assuming adsorption of the substrate and the product.

Evans [23] and Nadjo and Savéant [24] considered theoretically partly reversible electrode processes followed by irreversible dimerization.

The theory of stationary electrode voltammetric electrode processes followed by DIM 1 dimerization has been elaborated also by Nicholson and co-workers [25].

The effect of spherical diffusion on recorded curves can be estimated in the case of fast chemical reactions on the basis of Olmstead and Nicholson's [26] work.

Laviron [27] considered the case where the depolarizer is strongly adsorbed at the electrode surface. Shuman [28] discussed stationary electrode voltammetric current–potential curves in cases where the electrode reactions of various orders are fast and reversible. On the basis of theoretical calculations he established the diagnostic criteria shown in Table 12.2.

Table 12.2 DIAGNOSTIC CRITERIA FOR FAST CHEMICAL REACTIONS OF VARIOUS ORDERS OF THE PRODUCT OF A REVERSIBLE ELECTRODE PROCESS (FROM [28] BY PERMISSION OF THE COPYRIGHT HOLDERS, THE AMERICAN CHEMICAL SOCIETY)

Reaction scheme	$(E_p - E_{1/2})n$ (mV)	$(E_p - E_{p/2})n$ (mV)
$Ox + ne \rightleftharpoons Red$ $3Red \rightleftharpoons A$	−17.0	35.0
$Ox + ne \rightleftharpoons Red$ $2Red \rightleftharpoons A$	−18.0	40.2
$Ox + ne \rightleftharpoons Red$ $2Red \rightleftharpoons 3A$	−52.9	70.3
$Ox + ne \rightleftharpoons Red$ $Red \rightleftharpoons 2A$	−77.6	83.9
$Ox + ne \rightleftharpoons Red$ $Red \rightleftharpoons 3A$	−191	108.6

Shuman also discussed from the point of view of cyclic stationary electrode voltammetry the mechanisms of reactions shown in Table 12.2.

12.3 Chronopotentiometry

The theory of electrode processes followed by fast dimerization has been briefly discussed by Koutecký and Čížek [29]. In this case the concentration of the Red form at the electrode surface is described by the relationship:

$$C_{\text{Red}}(0, t) = \sqrt[3]{\frac{3i^2}{2n^2 F^2 D_{\text{Red}} k_1 A^2}}. \qquad (12.20)$$

Since the concentration of the Ox form is described by an expression identical with that for the diffusion process:

$$C_{\text{Ox}}(0, t) = C_{\text{Ox}}^0 - \frac{2it^{1/2}}{\pi^{1/2} nFD_{\text{Ox}}^{1/2} A}, \qquad (12.21)$$

combining equations (12.20) and (12.21) with the Nernst equation gives an equation for stationary electrode voltammetric curves of reversible processes with fast following dimerization reaction:

$$E = E^0 + \frac{RT}{nF} \ln \frac{\tau^{1/2} - t^{1/2}}{\tau^{1/6}} + \frac{RT}{nF} \ln \frac{2D_{\text{Red}}^{1/3} k_1^{1/3} (C_{\text{Ox}}^0)^{1/3}}{3^{1/3} \pi^{1/3}}. \quad (12.22)$$

In kinetic studies of consecutive dimerization reactions it is more convenient to use the equation describing the potential at $t = \tau/4$ which can be readily derived from equation (12.22):

$$E_{\tau/4}^k = E_{1/2} + \frac{RT}{nF} \ln \frac{D_{\text{Ox}}^{1/2}}{3^{1/3} D_{\text{Red}}^{1/2}} + \frac{RT}{3nF} \ln k_1 C_{\text{Ox}}^0 \tau. \quad (12.23)$$

12.4 The Rotating Disc Method

The half-wave potential of a reversible electrode process followed by dimerization, carried out at a rotating disc electrode, was first described by Galus and Adams [30]. For cathodic processes the equation has the following form:

$$E_{1/2}^k = E^0 - \frac{RT}{nF} \ln(2.62)^{2/3} v^{-1/6} + \frac{RT}{3nF} \ln \frac{k_1 C_{\text{Ox}}^0}{\omega}. \quad (12.24)$$

A somewhat different equation is given by Pleskov and Filinovskii [31]:

$$E_{1/2}^k = E^0 + \frac{RT}{3nF} \ln \frac{0.863 v^{1/3}}{D^{1/3}} + \frac{RT}{3nF} \ln \frac{k_1 C_{\text{Ox}}^0}{\omega}. \quad (12.25)$$

This equation is in accord with the relationship derived by Bonnaterre and Cauquis [32] for irreversible chemical dimerization reactions.

In this case the wave is described by the equation [31]:

$$E = E_{1/2}^k + \frac{RT}{nF} \ln \frac{i_g - i}{i_g^{1/3} i^{2/3}}. \quad (12.26)$$

When the dimerization is very fast the electrode process is also fast, and the equilibrium is shifted appreciably in the direction of dimer formation. According to the general considerations presented by these authors, the current–potential curve in such a case is described by the relationship:

$$E = E^0 + \frac{RT}{2nF} \ln \frac{2\delta K}{FD} + \frac{RT}{2nF} \ln \frac{(i_g - i)^2}{i}, \quad (12.27)$$

and the half-wave potential is:

$$E_{1/2} = E^0 + \frac{RT}{2nF} \ln C_{\text{Ox}}^0 K, \quad (12.28)$$

where K is the equilibrium constant of the dimerization. Thus the half-wave potential is independent of the velocity of electrode rotation.

When the chemical equilibrium is not markedly shifted in the direction of dimer formation, the equation of the curve is more complex. These more complex consecutive dimerization processes, involving comparatively slow electron transfer were considered by Bonnaterre and Cauquis [32, 33]. The general wave equation proposed by these authors has the following form:

$$\frac{i_g - i}{i} = \frac{k_{bh}}{k_{fh}} L + \frac{D}{\delta k_{fh}}. \qquad (12.29)$$

Parameter L is related to concentrations at the electrode surface. In the case of an irreversible dimerization it has the following simple form:

$$L = \left(\frac{2}{3} \frac{\delta^2 k_1 C_{Ox}^0 i}{D i_g}\right)^{-1/3}. \qquad (12.30)$$

The results obtained by these authors were used [34] in studies of the electroreduction of SO_2 in dimethylsulphoxide at a platinum electrode. In this reaction the capture of one electron is followed by the dimerization:

$$2SO_2^- \rightleftharpoons S_2O_4^{2-}.$$

Dimerization of the primary product of an electrode reaction taking place on a disc electrode as well as more complex DIM 2 and DIM 3 reactions have been studied by Nadjo and Savéant [18]. As in other methods the half-wave potential differs from the potential of the normal Ox/Red system. The difference increases with increasing rate of the subsequent dimerization reaction and with decreasing speed of rotation of the disc electrode. It is also influenced by the increase of concentration of the Ox form.

The theory of electrode processes followed by polymerization at the rotating disc electrode has also been dealt with [11].

12.5 Discussion of General Relationships

The equations derived for the four methods considered can be represented by the general relationship:

$$E_{1/2}^k = E_{1/2} + B + \frac{RT}{3nF} \ln k_1 C_{Ox}^0 X \qquad (12.31)$$

where X is the kinetic parameter and B is a constant; the values of this constant for these methods are shown in Table 12.3.

Table 12.3 VALUES OF CONSTANT B IN EQUATION (12.31)

Method	Kinetic parameter	B
Polarography	t_1	$-\dfrac{RT}{nF} 0.62$
Stationary electrode voltammetry	$\dfrac{1}{V}$	$\dfrac{RT}{nF} 0.21 + \dfrac{RT}{3nF} \ln \dfrac{2RT}{3nF}$
Chronopotentiometry	τ	$-\dfrac{RT}{3nF} \ln 3$
Rotating disc method	$\dfrac{1}{\omega}$	$-\dfrac{RT}{nF} \ln (2.62)^{2/3} v^{-1/6}$

In the derivation of equation (12.31) it has been assumed that the diffusion coefficients of the Ox and Red forms are equal. On the basis of this equation it is possible to make a general consideration of the characteristic features of electrode processes followed by dimerization. In the case of electroreduction the kinetic half-wave potential ($E_{\tau/4}^k$ in chronopotentiometry) is shifted in the positive direction when the kinetic parameter increases. The dependence of $E_{1/2}^k$ on the common logarithm of X is linear, and the slope of the straight line is $2.3\ RT/3nF$. This relationship is similar to the changes observed in the case of reversible electrode processes followed by first-order chemical reactions, but in the latter case the slope is equal to $2.3RT/2nF$. The differences between the two coefficients are not large and, especially in the case of multi-electron electrode processes, they might prove to be insufficient to allow differentiation between charge transfer followed by a first-order chemical reaction and that followed by dimerization. In such cases the determination of the dependence of kinetic half-wave potential on Ox concentration can help in solving the problem. This relationship is observed only when dimerization takes place. A tenfold increase in concentration of the Ox form at 25°C causes a shift of $E_{1/2}^k$ by $20/n$ mV in the positive direction.

Calculation of rate constants of dimerization reactions requires a knowledge of the reversible half-wave potential of the Red/Ox system. This is a serious problem and frequently, in the case of the fast chemical reactions, it is difficult to solve. Theoretically an appreciable decrease in the kinetic parameter X should shorten the process to such an extent that the consecutive reaction does not affect the concentration of the Red form. It is possible to determine $E_{1/2}$ directly from the theoretical curve, provided that, under the modified conditions, the electrode process is still reversible.

The conclusions reached in the discussion above are valid, provided that the charge exchange between the Ox and Red forms is fast and that the system is reversible under the conditions of the experiment.

The relationships discussed are those of reduction processes but they can be readily adapted to anodic oxidation, followed by dimerization of the Ox forms, by reversing the signs of the right hand side terms of the equations (obviously with the exception of E^0 or $E_{1/2}$).

References

[1] Elving, P. J., *Pure and Appl. Chem.*, **15**, 297 (1967).
[2] Mairanovskii, S. G., *Dokl. Akad. Nauk. SSSR*, **110**, 593 (1956).
[3] Seo, E. T., Nelson, R. F., Fritsch, J. M., Marcoux, L. S., Leedy, D. W. and Adams, R. N., *J. Am. Chem. Soc.*, **88**, 3498 (1966).
[4] Nadjo, L. and Savéant, J. M., *J. Electroanal. Chem.*, **33**, 419 (1971).
[5] Andrieux, C. P. and Savéant, J. M., *J. Electroanal. Chem.*, **33**, 453 (1971).
[6] Andrieux, C. P. and Savéant, J. M., *J. Electroanal. Chem.*, **26**, 223 (1970).
[7] Andrieux, C. P. and Savéant, J. M., *J. Electroanal. Chem.*, **28**, 446 (1970).
[8] Hanuš, V., *Chem. Zvesti*, **8**, 702 (1954).
[9] Koutecký, J. and Hanuš, V., *Chem. Listy*, **48**, 1446 (1954); *Coll. Czechoslov. Chem. Communs.*, **20**, 124 (1955).
[10] Mairanovskii, S. G., *Izv. Akad. Nauk SSSR, Otd. Khim. Nauk*, 2140 (1961).
[11] Čížek, J., Koryta, J. and Koutecký, J., *Coll. Czechoslov. Chem. Communs.*, **24**, 663 (1959).
[12] Nelson, R. F. and Feldberg, S. W., *J. Phys. Chem.*, **73**, 2623 (1969).
[13] Feldberg, S. W., in *"Electroanalytical Chemistry"*, Vol. 3, Bard, A. J., Ed., Dekker, New York 1969.
[14] Savéant, J. M. and Vianello, E., *Compt. rend.*, **256**, 2597 (1963); **259**, 4017 (1964).
[15] Nicholson, R. S., *Anal. Chem.*, **37**, 667 (1965).
[16] Savéant, J. M. and Vianello, E., *Electrochim. Acta*, **12**, 1545 (1967).
[17] Andrieux, C. P., Nadjo, L. and Savéant, J. M., *J. Electroanal. Chem.*, **26**, 147 (1970).
[18] Nadjo, L. and Savéant, J. M., *J. Electroanal. Chem.*, **44**, 327 (1973).
[19] Andrieux, C. P. and Savéant, J. M., *Bull. soc. chim. France*, 3281 (1972).
[20] Laviron, E., *J. Electroanal. Chem.*, **39**, 1 (1972).
[21] Laviron, E., *Electrochim. Acta*, **16**, 409 (1971).
[22] Laviron, E., *J. Electroanal. Chem.*, **34**, 463 (1972).
[23] Evans, D. H., *J. Phys. Chem.*, **76**, 1160 (1972).
[24] Nadjo, L. and Savéant, J. M., *J. Electroanal. Chem.*, **48**, 113 (1973).
[25] Olmstead, M. L., Hamilton, R. G. and Nicholson, R. S., *Anal. Chem.*, **41**, 260 (1969).
[26] Olmstead, M. L. and Nicholson, R. S., *Anal. Chem.*, **41**, 862 (1969).
[27] Laviron, E., *Electrochim. Acta*, **16**, 409 (1971).
[28] Shuman, M. S., *Anal. Chem.*, **42**, 521 (1970).

[29] Koutecký, J. and Čížek, J., *Chem. Listy*, **51**, 827 (1957); *Coll. Czechoslov. Chem. Communs.*, **22**, 914 (1957).
[30] Galus, Z. and Adams, R. N., *J. Electroanal. Chem.*, **4**, 248 (1922).
[31] Pleskov, Yu. V. and Filinovskii, V. Yu. *"Rotating Disc Electrode"*, Nauka, Moscow 1972, page 209 (in Russian).
[32] Bonnaterre, R. and Cauquis G., *J. Electroanal. Chem.*, **32**, 199 (1971).
[33] Bonnaterre, R. and Cauquis, G., *J. Electroanal. Chem.* **32**, 215 (1971).
[34] Gray, D. G. and Harrison, J. A., *J. Electroanal. Chem.*, **24**, 187 (1970).

Chapter 13

Electrode Processes Followed by Disproportionation Reactions

Electroreduction followed by disproportionation can be formulated by the following scheme:

$$Ox + ne \rightleftharpoons Red_1, \tag{13.1}$$

$$2Red_1 \xrightarrow{k_1} Ox + Red_2. \tag{13.2}$$

The electrode reduction product Red_1 is unstable and the Ox form is partly regenerated according to equation (13.2).

Electroreduction of sexivalent uranium in acidic media is an example of such a process. This has been studied in detail:

$$UO_2^{2+} + e \rightleftharpoons UO_2^+, \tag{13.3}$$

$$2UO_2^+ + H^+ \xrightarrow{k_1} UO_2^{2+} + UOOH^+. \tag{13.4}$$

The rate constants of disproportionation (13.4) have been determined by various methods and in the presence of various reagents, which complex either U(V) or the disproportionation products. The correct mechanism of the U(VI) reduction was shown for the first time by Herasymenko [1].

Disproportionation is also frequently observed in generation of free radicals at an electrode [2, 3].

13.1 Polarography

The description of currents observed in processes (13.1) and (13.2), carried out under polarographic conditions, was first published by Orleman and Kern [4, 5], who used the reaction layer concept [5]. An exact treatment of this problem was given by Koutecký and Koryta [6]. Their solution can be represented by the formula:

$$\frac{\bar{i}_k}{\bar{i}_g} = 1 + \sum_{i=1}^{\infty} L_i \xi_1^i \tag{13.5}$$

where:

$$\xi_1 = 2C_{Ox}^0 k_1 t_1. \tag{13.6}$$

A useful solution was obtained by these authors for the first seven values of parameter L_i. The dependence of $\bar{i}_k/\bar{i}_g$ on ξ_1 described by equation (13.5) is shown in Fig. 13.1.

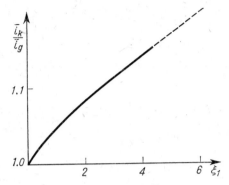

Fig. 13.1 Dependence of average currents ratio $\bar{i}_k/\bar{i}_g$ on parameter ξ_1 (from [6] by permision of the copyright holders).

Koryta and Koutecký [7] utilized the disproportionation of uranium(V) as a model for checking their theory. This reaction has also been investigated polarographically by other authors [8–10].

The theory of polarographic processes followed by disproportionation of any stoichiometry was described in 1970 by Kastening [11].

Pence and Booman [12] worked out the theory of electrode processes followed by disproportionation of the product taking place under chronoamperometric conditions on spherical electrodes. This was later improved by Delmastro and Booman [13]. A theoretical treatment of chronoamperometric disproportionation has been published by Feldberg [14].

13.2 Stationary Electrode Voltammetry

Although stationary electrode voltammetry has been used for a relatively long time in studies on uranium [15], the exact theoretical treatment of reactions (13.1) and (13.2) was not published until 1968, by Mastragostino, Nadjo and Savéant [16]. The result of this work can be expressed by the following relationship:

$$\chi(\lambda_d) = 0.446 \frac{i_k}{i_p} \qquad (13.7)$$

where 0.446 is the value of the current function for the peak of the diffusion process, and i_k is the peak current of the kinetic process. The function is connected with parameter λ_d by means of the relationship shown in Fig. 13.2. Parameter λ_d is defined by the equation:

$$\lambda_d = \frac{k_1 C_{Ox}^0 RT}{VnF}. \tag{13.8}$$

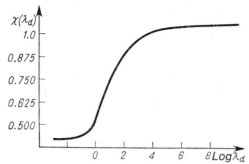

Fig. 13.2 Dependence of $\chi(\lambda_d)$ on parameter λ_d (from [16] by permission of the copyright holders, Pergamon Press).

The current peak potential for very large values of λ_d and for a reversible cathodic reduction, is described by the following relationship:

$$E_p = E^0 - \frac{0.071}{n} - \frac{0.020}{n}\log n + \frac{0.020}{n}\log\frac{k_1 C_{Ox}^0}{V} \tag{13.9}$$

which shows that this potential depends on C_{Ox}^0 and V, but the shape of the peak remains unchanged. The difference between the peak potential and the half-peak potential is:

$$E_{p/2} - E_p = \frac{0.041}{n} V. \tag{13.10}$$

Equations (13.9) and (13.10) are valid for 25°C. As in the case of electrode processes controlled by diffusion alone the peak current is proportional to $V^{1/2}$. Mastragostino and Savéant [17] utilized these relationships in their studies on the rate of disproportionation of uranium(V). Olmstead and Nicholson [18] further developed the theory of Savéant and co-workers by taking into account the spherical diffusion. As an illustration a number of values of current function peaks corresponding to the parameter $\varrho = \frac{D}{r_0^2 C_{Ox}^0 k_1}$ are quoted in Table 13.1 [18].

When λ_d is very large the effect of spherical diffusion on the current function can be expressed by the following empirical equation:

$$\sqrt{\pi}\,\chi_s(y) = \sqrt{\pi}\,\chi_{pl}(y) + \Phi\Theta(y) \tag{13.11}$$

Table 13.1 CURRENT FUNCTION PEAK FOR A SPHERICAL ELECTRODE (FROM [18] BY PERMISSION OF THE COPYRIGHT HOLDERS, THE AMERICAN CHEMICAL SOCIETY)

λ_d	$\varrho \times 10^4$			
	0.0	6.25	25.0	100.0
0.005	0.447	0.449	0.450	0.453
0.025	0.450	0.453	0.456	0.462
0.100	0.461	0.467	0.473	0.485
0.225	0.478	0.487	0.497	0.515
0.675	0.531	0.547	0.564	0.598
1.500	0.594	0.620	0.646	0.699
6.00	0.724			
24.00	0.832			

Table 13.2 CURRENT FUNCTION FOR A FLAT ELECTRODE AND THE FUNCTION REPRESENTING THE SPHERICAL DIFFUSION EFFECT (FROM [18] BY PERMISSION OF THE COPYRIGHT HOLDERS, THE AMERICAN CHEMICAL SOCIETY)

Potential* (mV)	$\sqrt{\pi}\chi_{pl}(y)$	$\Theta(y)$
120	0.004	0.0
100	0.010	
90	0.018	
80	0.032	
70	0.056	
60	0.100	0.004
50	0.170	0.014
40	0.280	0.042
30	0.440	0.110
25	0.536	0.172
20	0.636	0.254
15	0.740	0.368
10	0.834	0.504
5	0.918	0.672
0	0.982	0.848
−5	1.026	1.028
−10	1.048	1.212
−15	1.052	1.376
−20	1.040	1.528
−25	1.016	1.650
−30	0.986	1.756
−35	0.952	1.836
−40	0.912	1.902
−50	0.840	1.988
−60	0.774	2.034

* The potential is given in the following scale:
$$(E-E^0)n + \frac{RT}{3F}\ln\left(\frac{6\pi nFVk_1 C_{Ox}^0}{RT}\right) \text{ at } 25°C.$$

where $\chi_s(y)$ and $\chi_{pl}(y)$ are current functions for spherical and planar diffusion, respectively. $\Phi = \sqrt{D/r_0}\sqrt{\dfrac{nFV}{RT}}$, and $\Theta(y)$ is an empirical correlative function having the values shown in Table 13.2.

The problems connected with electrode reactions accompanied by disproportionation of the primary products have also been discussed by Nadjo and Savéant [19].

13.3 Chronopotentiometry

Chronopotentiometry was used in studies of uranium(V) disproportionation by Iwamoto [20], and an exact treatment of disproportionation under chronopotentiometric conditions was given by Fischer and Dračka [21]. These authors derived the following relationship for a one-electron electrode process followed by irreversible disproportionation:

$$i_0 \tau^{1/2} = \pi^{1/2} F D_{Ox}^{1/2} C_{Ox}^0 - \left(\dfrac{3}{16}\right)^{1/3} \pi^{1/2} F^{1/3} D_{Ox}^{-1/2} D_{Red}^{2/3} k_1^{-1/3} i_0^{2/3}. \quad (13.12)$$

The disproportionation rate constant can be calculated from this equation.

Equation (13.12) shows that the dependence of $i_0 \tau^{1/2}$ on $i_0^{2/3}$ should be linear. The experimentally determined slope of the straight line is directly connected to k_1, but in order to find the exact value of this constant it is necessary to know the values of the diffusion coefficients of the Ox and Red forms. The authors of this theory and many other investigators utilized it in their studies of uranium(V) disproportionation.

13.4 The Rotating Disc Method

The expression obtained by Ulstrup [22] for reduction on a rotating disc electrode, followed by disproportionation of the product, can be transformed into the following equation:

$$\dfrac{i_k}{\omega^{1/2}} = \dfrac{2nFAD^{2/3}C_{Ox}^0}{1.61\nu^{1/6}} - \dfrac{3^{1/3} n^{1/3} F^{1/3} A^{1/3} D^{1/3}}{2^{1/3} \times 1.61 k_1^{1/3} \nu^{1/6}} i_k^{2/3}. \quad (13.13)$$

Assuming that the thickness of the diffusion layer $\delta = 10^{-3}$ cm, and that the diffusion coefficient is equal to 10^{-5} cm^2/s the rotating disc method can be used for analysis of processes having a disproportionation rate constant between 10^6 l. mole^{-1}s^{-1} < k_1 < 10^8 l. mole^{-1}s^{-1}. The lower limit of the rate constant results from the assumption that the chemical reaction is fast.

This was investigated also by Filinovskii and Potapov [23] who obtained results similar to those reported by Ulstrup, by means of a more accurate method.

In 1971 Holub [24] published a general theory of electrode reactions followed by disproportionation, taking place on rotating disc electrodes. Although the solution of Ulstrup was valid for i_k/i_g values close to 2, the solution obtained by Holub made it possible to determine the rate constants over the whole range of i_k/i_g values between 1 and 2. When the accuracy of the approximate analytical solution was insufficient the problem was solved numerically.

In the case of fast reactions when $i_k/i_g > 1.56$ the Holub solution is in good agreement with that proposed by Ulstrup. Bonnaterre and Cauquis [25] also briefly investigated the problem of electrode reactions followed by disproportionation. These authors proposed a scheme for solution which is more general than those worked out by Holub and Ulstrup.

13.5 General Discussion

Equations (13.12) and (13.13), which were expressed in a similar form, can be rewritten as the following general relationship:

$$iX^{1/2} = B - Gi^{2/3}. \tag{13.14}$$

In this equation B and G are constants and the concept of the kinetic parameter X is employed.

This equation can also be regarded as the general form of the solutions obtained for polarography and for stationary electrode voltammetry, as has been done in the case of other kinds of kinetic currents. Thus relationship (13.14) makes it possible to determine disproportionation rate constants on the basis of analysis of the results of the $iX^{1/2}-i^{2/3}$ system. It also leads to the generalization that when the current tends towards small values, $iX^{1/2}$ tends towards B. Thus when the processes are carried out over a relatively long period (i.e. when the transition time is long or when the electrode rotation speed is low) the limiting currents or the transition time in chronopotentiometry are twice as high as the values expected for diffusion processes (13.1). This conclusion is, however, not entirely correct when spherical electrodes are used.

References

[1] Herasymenko, P., *Trans. Faraday Soc.*, **24**, 272 (1928).
[2] Senne, J. K. and Marple, L. W., *Anal. Chem.*, **42**, 1147 (1970).
[3] Kastening, B. and Vavřička, S., *Ber. Bunsenges. physik. Chem.*, **72**, 27 (1968).
[4] Kern, D. M. H., and Orleman, E. F., *J. Am. Chem. Soc.*, **71**, 2102 (1949).
[5] Orleman, E. F. and Kern, D. M. H., *J. Am. Chem. Soc.*, **75**, 3058 (1953).
[6] Koutecký, J. and Koryta, J., *Chem. Listy*, **48**, 996 (1954); *Coll. Czechoslov. Chem. Communs.*, **19**, 845 (1954).
[7] Koryta, J. and Koutecký, J., *Coll. Czechoslov. Chem. Communs.*, **20**, 423 (1955).
[8] Kraus, K. A., Nelson, F. and Nelson, G. L., *J. Am. Chem. Soc.*, **71**, 2510 (1949).
[9] Duke, F. R. and Pinkerton, R. C., *J. Am. Chem. Soc.*, **73**, 2361 (1951).
[10] McEwen, D. J. and de Vries, T., *Can. J. Chem.*, **35**, 1225 (1957).
[11] Kastening, B., *J. Electroanal. Chem.*, **24**, 417 (1970).
[12] Pence, D. T. and Booman, G. L., *Anal. Chem.*, **38**, 1112 (1966); Booman, G. L. and Pence, D. T., *Anal. Chem.*, **37**, 1366 (1965).
[13] Pence, D. T., Delmastro, J. R. and Booman, G. L., *Anal. Chem.*, **41**, 737 (1969).
[14] Feldberg, S., *J. Phys. Chem.*, **73**, 1238 (1969).
[15] Kemula, W., Rakowska, E. and Kublik, Z., *Coll. Czechoslov. Chem. Communs.*, **25**, 3105 (1960).
[16] Mastragostino, M., Nadjo, L. and Savéant, J. M., *Electrochim. Acta*, **13**, 721 (1968).
[17] Mastragostino, M. and Savéant, J. M., *Electrochim. Acta*, **13**, 751 (1968).
[18] Olmstead, M. L. and Nicholson, H. S., *Anal. Chem.*, **41**, 862 (1969).
[19] Nadjo, L. and Savéant, J. M., *J. Electroanal. Chem.*, **33**, 419 (1971).
[20] Iwamoto, R. T., *J. Phys. Chem.*, **65**, 303 (1959).
[21] Fischer, O. and Dračka, O., *Coll. Czechoslov. Chem. Communs.*, **24**, 3046 (1959).
[22] Ulstrup, J., *Electrochim. Acta*, **13**, 1717 (1968).
[23] Pleskov, Yu. V. and Filinovskii, V. Yu., "*Rotating Disc Electrode*", Nauka, Moscow 1972 (in Russian).
[24] Holub, K., *J. Electroanal. Chem.*, **30**, 71 (1971).
[25] Bonnaterre, R. and Cauquis, G., *J. Electroanal. Chem.*, **31**, App. 15–18 (1971).

Chapter 14

Electroanalytical Investigations of Complexes

Electroanalytical methods, and especially polarography, have played an important part in investigations of the composition and stability of complexes and of the mechanisms of complex formation. There are many papers dealing with the theoretical and experimental aspects of these reactions. Complexes are, therefore, discussed as a separate chapter in this book. Polarographic methods will be discussed in detail, since they have been more extensively developed than those of stationary electrode voltammetry or chronopotentiometry.

Some of the relationships derived for electrode reactions involving complexes studied under polarographic conditions can be readily adapted for stationary electrode voltammetric or chronopotentiometric conditions. The rotating disc technique seems to be the least suitable for investigations of complexes, since the use of platinum as the electrode material limits the application of this method to the region of positive potential. The electrode processes involving complex ions take place mainly at potentials that can be studied with mercury electrodes.

14.1 Investigations of Reversibly Reducible Complexes

Consider two reversible electrode reactions: the reaction of the hydrated Me^{n+} ion:

$$Me^{n+} + ne \rightleftharpoons Me, \tag{14.1}$$

and that of a complex ion of the same metal:

$$MeL_p^{n+} + ne \rightleftharpoons Me + pL. \tag{14.2}$$

For the sake of simplicity it is assumed that the ligand in the MeL_p complex is an inert molecule, such as the molecule of ammonia. The complex is formed in the reaction:

$$Me^{n+} + pL \rightleftharpoons MeL_p^{n+} \tag{14.3}$$

which very quickly reaches a state of equilibrium. The equilibrium constant of this reaction is given by:

$$\beta_p = \frac{[\text{MeL}_p^{n+}]}{[\text{Me}^{n+}][\text{L}]^p} \qquad (14.4)$$

where β_p is the stability constant of the MeL_p^{n+} complex.

It has been shown experimentally that the half-wave potential of reaction (14.2) is always more negative than that of reaction (14.1). The shift of the complex ion reduction potential in the negative direction increases with increasing stability of the complex. This is due to the fact that, in the case of the complex ion discharge, the concentration of free Me^+ ions is so low that in the region of the potential of reaction (14.1) the process hardly occurs.

It is quite easy to describe the shift of complex ion reduction potential with respect to that of the reduction of simple ions when only one complex ion MeL_p^{n+} is present in the solution and when its stability is high. The Nernst equation for the reaction (14.1) is as follows:

$$E = E^0 + \frac{RT}{nF} \ln \frac{[\text{Me}^{n+}]_0}{[\text{Me}]_0} \qquad (14.5)$$

which is valid here on account of electrode reversibility of reaction (14.1); the $[\text{Me}_p^{n+}]_0$ concentration should be expressed by equation (14.4), giving:

$$E = E^0 + \frac{RT}{nF} \ln \frac{[\text{MeL}_p^{n+}]_0}{[\text{Me}]_0 [\text{L}]_0^p \beta_p} . \qquad (14.6)$$

The subscripts "0" appearing with concentration symbols in equations (14.5) and (14.6) indicate that concentrations at the electrode surface are being considered.

The experiment is usually carried out using a large excess of ligand L, compared with the concentration of the metal ions. Therefore it can be assumed that the concentration of L at the electrode surface does not change during the electrode process. For this reason the subscript "0" for L in equation (14.6) can be omitted.

The concentration of MeL_p^{n+} ions at the electrode surface in a polarographic process can be obtained from the Ilkovič equation:

$$\bar{i} = K_{\text{Il}}([\text{MeL}_p^{n+}]^0 - [\text{MeL}_p^{n+}]_0), \qquad (14.7)$$

where $[\text{MeL}_p^{n+}]^0$ is the initial equilibrium concentration of the complex, $\bar{i}$ is the average current and K_{Il} is the Ilkovič constant.

Since the average limiting current is reached at zero concentration of the complex on the electrode surface, equation (14.7) can be written in the following form:

$$[MeL_p^{n+}]_0 = \frac{\bar{i}_g - \bar{i}}{K_{11}}. \tag{14.8}$$

The concentration of metal in the amalgam is expressed by the relationship:

$$i = K'_{11}[Me]_0. \tag{14.9}$$

Substituting in equation (14.6) the concentrations $[MeL_p^{n+}]_0$ and $[Me]_0$ given by relationships (14.8) and (14.9):

$$E^k = E^0 + \frac{RT}{nF}\ln\frac{K'_{11}}{K_{11}} + \frac{RT}{nF}\ln\frac{\bar{i}_g - \bar{i}}{\bar{i}} - \frac{RT}{nF}\ln\beta_p[L]^p. \tag{14.10}$$

Since the diffusion coefficients of metal in mercury D_{Me} and of the complex ion $D_{MeL_p^{n+}}$ can be substituted for the Ilkovič constants, equation (14.10) can be formulated as follows:

$$E^k = E^0 + \frac{RT}{2nF}\ln\frac{D_{Me}}{D_{MeL_p^{n+}}} + \frac{RT}{nF}\ln\frac{\bar{i}_g - \bar{i}}{\bar{i}} - \frac{RT}{nF}\ln\beta_p[L]^p. \tag{14.11}$$

As was shown in Chapter 7, for the discharge of simple M^{n+} ions, the dependence of current on potential has the following form:

$$E = E^0 + \frac{RT}{2nF}\ln\frac{D_{Me}}{D_{Me^{n+}}} + \frac{RT}{nF}\ln\frac{\bar{i}_g - \bar{i}}{\bar{i}}, \tag{14.12}$$

where $D_{Me^{n+}}$ is the diffusion coefficient of the simple metal ion.

Taking the difference of half-wave potentials determined on the basis of equations (14.11) and (14.12):

$$E_{1/2}^k - E_{1/2} = \frac{RT}{2nF}\ln\frac{D_{Me^{n+}}}{D_{MeL_p^{n+}}} - \frac{RT}{nF}\ln\beta_p[L]^p. \tag{14.13}$$

It follows from equation (14.13) that the shift of half-wave potential of the reduction process in the negative direction increases with increasing stability of the complex β_p. $E_{1/2}^k$ is also shifted in the negative direction when the ligand concentration in the solution increases. The dependence of $E_{1/2}^k$ on log [L] is linear, and from the slope of this straight line it is possible to determine p, the number of ligand molecules complexing the Me^{n+} ion, provided that n is known. Using this treatment Lingane[1] showed that in the case of the reduction of lead in strongly alkaline solutions, $n = 2$ and $p = 3$.

The equations discussed above were obtained in the preliminary form by Heyrovský and Ilkovič [2], and later by von Stackelberg and Freyhold [3] and by Lingane [1].

When a soluble complex ion having a smaller charge is formed in the electroreduction process:

$$\text{MeL}_p^{n+} + me \rightleftharpoons \text{MeL}_q^{(n-m)+} + (p-q)\text{L}, \qquad (14.14)$$

the difference between the half-wave potentials of the complex ion ($E_{1/2}^k$) and that of the simple ion is described by the relationship:

$$E_{1/2}^k - E_{1/2} = \frac{RT}{2nF}\ln\frac{D_{\text{MeL}_q^{(n-m)+}}}{D_{\text{MeL}_p^{n+}}} + \frac{RT}{nF}\ln\frac{\beta_q}{\beta_p} - \frac{(p-q)RT}{nF}\ln[\text{L}]. \qquad (14.15)$$

In this case the shift of the half-wave potential is proportional to the ratio of the stability constants of the MeL_p^{n+} and $\text{MeL}_q^{(n-m)+}$ complexes. The dependence of $E_{1/2}^k$ on log [L] is again linear, but the slope of the straight line is related to the difference $(p-q)$. These relationships have been derived on the assumption that the activity coefficients are constant and independent of the value of [L] and that the MeL_p^{n+} concentration is small, compared with the concentration of the ligand.

The above reasoning is valid when only one complex is present in the solution. When several complexes having similar stabilities are present, the dependence of $E_{1/2}^k$ on log [L] is no longer linear. In this case the number of ligands, p, complexing the ion Me^{n+} can be calculated from the limiting slope of the function at high ligand concentrations. Such a relationship is schematically represented by curve *1* in Fig. 14.1.

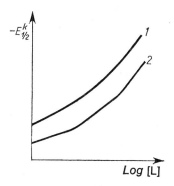

Fig. 14.1 Dependence of $E_{1/2}^k$ on log[L] for a system in which complexes having the composition 1:1, 1:2 and 1:4 are consecutively formed; the stabilities of the complexes considerably differ from each other.

Sometimes the dependence of $E_{1/2}^k$ on log [L] has linear intervals (curve 2 in Fig. 14.1). This is observed when several complexes having different p values and appreciably different stabilities, are present in the solution. From the slopes of the straight line sections it is possible to determine the composition of the complexes at certain values of ligand concentration and to calculate the formation constants.

The behaviour described above has been observed, for example in the cases of the Fe(III)–Fe(II) system in oxalate solutions [2, 3], Zn(II) in ammonia [4], and the cadmium–pyrazole system [5]. Such cases are rather rare, since usually the stability constants of MeL^{n+}, MeL_2^{n+},..., MeL_p^{n+} are similar. The last complex is formed after a number of intermediate stages:

$$Me^{n+} + L \rightleftharpoons MeL^{n+}$$
$$MeL^{n+} + L \rightleftharpoons MeL_2^{n+} \quad (14.16)$$
$$\dots\dots\dots\dots\dots\dots\dots$$
$$MeL_{p-1}^{n+} + L \rightleftharpoons MeL_p^{n+} \quad (14.17)$$

and the Me^{n+}, MeL^{n+}, ..., MeL_p^{n+} ions are in equilibrium with one other.

The authors of early polarographic studies of these phenomena were not aware of the existence of such equilibria, and therefore their results are now of minor value. A new approach to the investigation of complexes was presented in 1951 by DeFord and Hume [6]. The starting point of this work was the expression describing electrode potential, in which the activity coefficients were taken into account:

$$E = E_a^0 + \frac{RT}{nF} \ln \frac{[Me^{n+}]_0 f_{Me^{n+}}}{[Me]_0 f_{Me}} \quad (14.18)$$

where E_a^0 is the standard potential of the electrode consisting of Me^{n+} ions and the Me amalgam. The terms $f_{Me^{n+}}$ and f_{Me} are the activity coefficients of the reduced ion and of the metal in the amalgam, respectively.

The concentration of each MeL_j^{n+} complex in the bulk of solution is given by the expression:

$$[MeL_j^{n+}]^0 = \frac{\beta_j [Me^{n+}] f_{Me^{n+}} [L]^j}{f_{MeL_j^{n+}}}. \quad (14.19)$$

The concentration of complex MeL_j^{n+} at the electrode surface is described by a relationship similar to equation (14.19), in which the concentration symbols are marked with the subscript "0", with the exception of [L], since it has been assumed, as previously, that it is high and constant in the system under investigation. Moreover, it is assumed that the activity coefficients have the same values at the electrode surface as in the bulk of solution.

Summing the surface concentrations of complex ions appearing in the solution from $j = 1$ to $j = N$ (where N is the highest co-ordination number with respect to L):

$$\sum_{0}^{N} [MeL_j^{n+}]_0 = [Me^{n+}]_0 f_{Me^{n+}} \sum_{0}^{N} \frac{\beta_j [L]_i^j}{f_{MeL_j^{n+}}}. \tag{14.20}$$

Introducing $[Me^{n+}]_0 f_{Me^{n+}}$ into equation (14.18) from equation (14.20) gives the relationship:

$$E = E_a^0 + \frac{RT}{nF} \ln \frac{\sum_{0}^{N} [MeL_j^{n+}]_0}{[Me]_0 f_{Me} \sum_{0}^{N} \frac{\beta_j [L]^j}{f_{MeL_j^{n+}}}}. \tag{14.21}$$

On the basis of the Ilkovič equation the reduction current due to MeL_j^{n+} can be expressed as:

$$\bar{i}_j = {}_jK_{II}([MeL_j^{n+}]^0 - [MeL_j^{n+}]_0) \tag{14.22}$$

and the total current as:

$$\bar{i} = \sum_{0}^{N} {}_jK_{II}([MeL_j^{n+}]^0 - [MeL_j^{n+}]_0). \tag{14.23}$$

Introducing the experimental Ilkovič constant $_{exp}K_{II}$ into equation (14.23) gives:

$$\bar{i} = {}_{exp}K_{II} \sum_{0}^{N} ([MeL_j^{n+}]^0 - [MeL_j^{n+}]_0). \tag{14.23a}$$

The concentration of metal in the amalgam is given by equation (14.9). Combining this equation with (14.21) and (14.23a) gives the relationship:

$$E = E_a^0 + \frac{RT}{nF} \ln \frac{K_{II}}{{}_{exp}K_{II}f_{Me}} \left(\frac{\bar{i}_g - \bar{i}}{\bar{i}}\right) + \frac{RT}{nF} \ln \sum_{0}^{N} \frac{f_{MeL_j^{n+}}}{[L]^j \beta_j} \tag{14.24}$$

where

$$\bar{i}_g = {}_{exp}K_{II} \sum_{0}^{N} [MeL_j^{n+}]. \tag{14.25}$$

From equation (14.24) an expression for the half-wave potential is readily obtained:

$$E_{1/2}^k = E_a^0 + \frac{RT}{nF} \ln \frac{K_{II}}{_{exp}K_{II}f_{Me}} + \frac{RT}{nF} \ln \sum_0^N \frac{f_{MeL_j^{n+}}}{[L]^j \beta_j}. \quad (14.26)$$

The half-wave potential of a complex ion reduction is shifted with respect to that of the reduction of the simple Me^{n+} ion. This shift is given by the expression:

$$E_{1/2} - E_{1/2}^k = -\frac{RT}{nF} \ln \frac{K_{II}}{_{exp}K_{II}f_{Me^{n+}}} \sum_0^N \frac{f_{MeL_j^{n+}}}{[L]^j \beta_j}. \quad (14.27)$$

Equation (14.27) can be rewritten in the following form [7]:

$$F_0([L]) = \sum_0^N \frac{\beta_j [L]^j f_L^j f_{Me^{n+}}}{f_{MeL^{n+}}} =$$

$$\text{antilog} \left\{ 0.4343 \frac{nF}{RT} [E_{1/2} - E_{1/2}^k] + \log \frac{K_{II}}{_{exp}K_{II}} \right\}. \quad (14.28)$$

The right-hand side of equation (14.28) is a function of the concentration $F_0([L])$ of ligand L and it can be determined experimentally. The function can also be described by the expression:

$$F_0([L]) = \beta_0 + \frac{\beta_1 [L] f_L f_{Me^{n+}}}{f_{MeL^{n+}}} + \frac{\beta_2 [L]^2 f_L^2 f_{Me^{n+}}}{f_{MeL_2^{n+}}} + \frac{\beta_3 [L]^3 f_L^3 f_{Me^{n+}}}{f_{MeL_3^{n+}}} + \cdots$$

$$(14.29)$$

where β_0 is the formation constant of the complex having the coordination number 0 and is equal to one. The terms $\beta_1, \beta_2, \beta_3, \ldots$ are the consecutive overall stability constants.

In order to determine the stability constants of the complex a new function $F_1([L])$ is introduced, assuming, for the sake of simplicity, that $f_L f_{Me^{n+}}/f_{MeL_j^{n+}} = 1$ and $f_L = 1$. Then:

$$F_1([L]) = \frac{\{F_0([L])-1\}}{[L]} = \beta_1 + \beta_2 [L] + \beta_3 [L]^2 + \cdots. \quad (14.30)$$

Function $F_1([L])$ can also be determined experimentally for various ligand concentrations. It follows from formula (14.30) that the curve of $F_1([L])$ against [L] gives the value of β_1 on the function axis, and that the limiting slope of this curve at $[L] \to 0$ is β_2.

Then the next function $F_2([L])$ given by the equation

$$F_2([L]) = \frac{\{F_1([L])-\beta_1\}}{[L]} = \beta_2+\beta_3[L]+\beta_4[L]^2 + \ldots \quad (14.31)$$

may be introduced. It can also be calculated readily from the known values of function $F_1([L])$.

From the initial slope of $F_2([L])$ against [L] and the value of the function at [L] = 0 the values of β_3 and β_2, respectively, can be calculated.

Continuing this treatment, function $F_j([L])$ is obtained, which is linearly dependent on [L]. The straight line is parallel to the [L] axis and from its intersection with the function $F_j([L])$ axis it is possible to determine the stability constant of the complex characterized by the highest coordination number with respect to the ligand L.

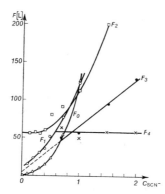

Fig. 14.2 Dependence of $F[L]$ on [L] for a system in which three complexes exist in equilibrium (from [8] by permission of the copyright holders).

Fig. 14.2 represents schematically the dependence of function $F([L])$ on [L] for a system in which three complexes are in equilibrium. When the concentration of L reaches an appreciable value, only the complex MeL_N having the highest coordination number with respect to L exists in the solution. In this case equation (14.28) is reduced to the following form:

$$\frac{0.4343nF}{RT}[E_{1/2}-E^k_{1/2}]+\log\frac{K_{11}}{{}_{exp}K_{11}} = \log\beta_N+N\log[L]. \quad (14.32)$$

Equation (14.32) is practically identical to (14.13).

DeFord and Hume used this method in their studies of complex cadmium [8] and zinc [9] thiocyanates. Later it was used in studies of other systems. When a charged ligand is used, the same equations are applicable, but the charges on the complexes are altered appropriately.

Kivalo and Rastas [10, 11] showed that successive stability constants of complexes can be calculated by the least squares method. The values of constants calculated by this method are more accurate than those determined graphically. Schaap and McMasters [12] developed the method of DeFord and Hume and adapted it to investigations of two-ligand systems, in which the following complex formation reaction takes place:

$$Me^{n+} + iL + jY \rightleftharpoons MeL_iY_j^{n+}. \qquad (14.33)$$

Methods of calculation of stability constants have also been developed by Karmalkar [13] but theyhave not found a wide application since they require complex calculations.

The methods of determining stability constants from the shift of reversible half-wave potentials discussed above were analysed by Klatt and Rouseff [14]. These authors came to the conclusion that when one of the complexes formed is much more stable (a hundred times or more) than the others, the $E_{1/2}$ vs. log [L] relationship is linear, and the slope of the straight line is determined by the co-ordination number of this complex. Equation (14.13) is not sensitive to stepwise equilibria of the complexes and even very accurate experiments do not show any distinct deviation of the semilogarithmic graph from linearity unless the condition $\beta_j[L]^j \gg 1$ is not fulfilled. The $E_{1/2}$ vs. log[L] relationship with linear sections, discussed earlier in this chapter, is very difficult to obtain. According to these authors, curves having such shapes were obtained only as a result of considering an insufficient number of experimental points.

Small errors in measurements of the half-wave potential shifts eliminate the deviations of the $\Delta E_{1/2}$ vs. log [L] relationship from linearity but do not change the slope of the curves significantly. Therefore when the $\Delta E_{1/2}$ vs. log [L] relationship is linear and when it gives an integral co-ordination number, it can be regarded as a sufficient criterion of the existence of only one stable complex. Analysis of experimental data by a least squares fit to the DeFord and Hume $F([L])$ function is the preferred procedure according to Klatt and Rouseff. The maximum precision of the calculated equilibrium constants was found to depend only upon the ratio of successive formation constants. Momoki and co-workers [15–17] have published a series of papers on statistical methods of calculation of stability constants of complexes on the basis of polarographic data and DeFord and Hume's method. Filipović and co-workers [18] presented an improved method which makes it possible to decrease the errors resulting from the preparation of solutions, measurements and treatment of results and which gives stability constants having the maximum statistical certainty. These improvements have been obtained as a result of a more exact method of preparation of the solutions directly in the three-electrode system cell and thanks to complete computerization of the interpretation of polarograms as well as calculation of stability constants from the $F([L])$ function by the least squares method. The accuracy of determination of the complex formation constants by the polarographic method has been discussed by Meites [19].

The above reasoning was based on the assumption that the ligand concentrations at the electrode surface and in the bulk of the solution are equal. The case of the reduction of complexes in the absence of an excess of the complexing agent was considered by Butler and Kaye [20]. Ringbom and Eriksson [21] showed how stability constants of complexes can be determined when the ligand is not present in excess in the solution.

The theory of the polarography of reversibly reducible complexes was further developed by Macovschi [22]. It follows from his work that the DeFord and Hume equation is the extreme case of a more general solution. This author discussed systems in which the slope of the polarographic wave is also modified.

An extensive work on the theoretical aspects of equilibria in solutions containing many acceptors and complexing agents was published by Jacq [23]. Guidelli and Cozzi [24] considered the problem of homogeneous equilibria and their effect on polarographic currents. They derived equations which are in good agreement with the relationships previously derived. Special attention was paid by the authors to the polarographic currents of co-ordination compounds.

In the majority of studies of complexes by polarography it is assumed that only the central ion is active at the electrode while the ligands are inactive. However sometimes the ligands are polarographically active, either in the free state, or when they are bound to the metal ion. Obviously, in the latter case the electrode process takes place at a different potential. Certain treatments [25, 26] make it possible to calculate the stabilities and the compositions of complexes from the differences between these potentials.

14.2 Investigations of Irreversibly Reducible Complexes

The discussion presented in section 14.1 was concerned with reversible electrode reactions and therefore the Nernst equation was used in the derivations of the relationships. However there are complexes which are not reduced at accessible potentials or which are reduced irreversibly.

In order to determine formation constants of such complexes, for instance those of $\overline{Me^{n+}}$ ion complexes with ligand L, indicator ions [21, 27] are introduced into the system under investigation. They are reversibly reducible Me^{n+} ions which form complexes of known formation constants $\beta_1, \beta_2, ..., \beta_p$ with ligand L. Moreover, it is assumed that if $\overline{Me^{n+}}$ is reduced, the reduction takes place at a more negative potential than the reduction of the indicator ion. In this case the following relationship is valid:

$$\Delta E = -\frac{RT}{nF}\ln(1+\beta_1[L]_0+\beta_2[L]_0^2+ ... +\beta_p[L]_0^p) \qquad (14.34)$$

where ΔE is the difference between the potentials in the presence and in the absence of ligand, measured at a defined current intensity.

Knowledge of the complex formation constants makes it possible to calculate the concentration of [L] at the electrode surface. The potential shifts are shown schematically in Fig. 14.3: curve *1* is the polarographic wave

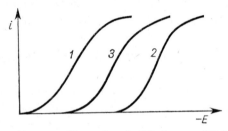

Fig. 14.3   Polarographic waves illustrating the Ringbom and Eriksson method [21, 27].

of Me^{n+} in the absence of ligand, curve *2* is the wave of the ion bonded to ligand L. When irreversibly reducible $\overline{Me^{n+}}$ ions are introduced into the solution (containing ligand L, Me^{n+} ions and complexes of the two components), a new equilibrium is established. Since a part of the ligand combines with $\overline{Me^{n+}}$ ions the reduction wave of the complex Me^{n+} ion is shifted in the positive direction. This is illustrated by curve *3*.

The calculation of $\overline{Me^{n+}}$ complex formation constants is comparatively simple. For this purpose the average number of ligands bonded to one Me^{n+} ion at the electrode surface must be calculated for a definite ligand concentration $[L]_0$:

$$_0\bar{n}_{Me^{n+}} = \frac{\beta_1[L]_0+2\beta_2[L]_0^2+ ... +p\beta_p[L]_0^p}{1+\beta_1[L]_0+\beta_2[L]_0^2+ ... +\beta_p[L]_0^p}. \qquad (14.35)$$

$_0\bar{n}_{Me^{n+}}$ depends on $[L]_0$ but not on the Me^{n+} concentration, provided that only mononuclear complexes are formed.

The average number of ligands L bonded to one $\overline{\text{Me}^{n+}}$ ion can be calculated from the equation

$$_0\bar{n}_{\overline{\text{Me}^{n+}}} = \frac{C_\text{L} - [\text{L}]_0 - {_0\bar{n}}_{\text{Me}^{n+}} \dfrac{\bar{i}_g - \bar{i}}{\bar{i}_g} C_{\text{Me}^{n+}}}{C_{\overline{\text{Me}^{n+}}}} \qquad (14.36)$$

where C_L and $C_{\overline{\text{Me}^{n+}}}$ are the total concentrations of L and $\overline{\text{Me}^{n+}}$, respectively.

The required formation constants of the irreversibly reducible complexes of $\overline{\text{Me}^{n+}}$ with ligand L can be determined by the Bjerrum method, since the formation curve, i.e. the relationship $_0\bar{n}_{\overline{\text{Me}^{n+}}} = f(-\log[\text{L}]_0)$ is now known. The accuracy of this method increases with increasing stability of the indicator ion complexes. It should be stressed that the determination of formation constants is possible only when neither the central ion nor the ligand is polarographically active. Fast equilibration in the solutions has been assumed in the above discussion. This means that, when the metal ions existing in equilibrium with the complexes are reduced at the electrode, the complexes decompose rapidly and the equilibrium upset by the electrode process is regained. When the time of reaching the equilibrium is short compared with the drop time, the current is determined by diffusion of the complex alone. In such a case only one wave is recorded. It is shifted in the negative direction and the value of the shift depends, as already mentioned, on the formation constants of the complexes and the ligand concentration. When the central ion is reduced irreversibly, the reduction of the complex may take place at less negative potentials. An example of such behaviour is the reduction of Ni(II) ions. In non-complexing media the ion is reduced at about -1.0 V. In concentrated $CaCl_2$ solution Ni(II) exists as a chloride complex and it is reduced at -0.5 V [28]. There is no relationship between the potential difference and the formation constant of the chloride Ni(II) complex.

When the time of reaching the equilibrium is long, compared with the drop time, or when both times are comparable, two waves are observed on the polarogram; the more positive wave corresponds to the reduction of the free ion and the second corresponds to the reduction of the complex, provided that the reduction is reversible.

In cases where the decomposition and formation rates of the complex are not fast, and the time of reaching the equilibrium of complexing reaction is longer than the drop time, the height of the first wave is proportional to the concentration of the free ions of the metal in the solution. In such a case the complex formation constant can easily be determined [29, 30].

The total concentration of the ligand is given by the equation:

$$C_L = p[\text{MeL}_p^{n+}] + [\text{L}]. \tag{14.37}$$

The analytical concentration of metal ions can be similarly described:

$$C_{\text{Me}^{n+}} = [\text{Me}^{n+}] + [\text{MeL}_p^{n+}]. \tag{14.38}$$

If it is assumed that the diffusion coefficients of free and complex ions are equal, and the symbol i_g is assigned to the limiting current in the absence of L, and symbol i'_g to the wave height after the introduction of the complexing agent:

$$i_g = K_{\text{II}} C_{\text{Me}^{n+}} = K_{\text{II}} \{[\text{Me}^{n+}] + [\text{MeL}_p^{n+}]\} \tag{14.39}$$

and

$$i'_g = K_{\text{II}} [\text{Me}^{n+}]. \tag{14.40}$$

Thus i'_g is proportional to the equilibrium concentration of Me^{n+}.

The complex formation constant given by equation (14.4) can be written in the following form

$$\beta_p = \frac{(i_g - i'_g)}{i'_g \left[C_L - p C_{\text{Me}^{n+}} \left(1 - \frac{i'_g}{i_g}\right) \right]^p}. \tag{14.41}$$

When the time required to reach equilibrium of the complex formation is comparable to the drop time, the first wave has a kinetic character. The current i'_g is no longer proportional to the equilibrium concentration of Me^{n+}. It is now larger due to partial decomposition of the complex at the electrode surface. This is observed in the case of the polarographic investigation of complexes of heavy metals with nitrilotriacetic acid [31, 32]. Equation (14.41) cannot be used in this case.

Good results are sometimes obtained in such cases when streaming mercury electrodes are used. Mercury flows rapidly from such electrodes and as a result the depolarizer transport rate is fast. When nitrilotriacetic acid complexes were investigated by means of these electrodes the kinetic character of the process disappeared. Thus results obtained with this electrode could be utilized for the calculation of complex formation constants from a somewhat modified form of equation (14.41).

In the case of very stable complexes, characterized by the very long time required to attain equilibrium either of complex formation or decomposition, polarographic investigation of displacement reactions can be employed for the determination of formation constants. This method was developed by Schwarzenbach et al. [33, 34] and by Bril and Krumholz [35].

Let it be assumed that the complex $\overline{\text{MeL}}^{n+}$ is irreversibly reduced at the dropping electrode or that it is not reduced in the accessible region of potential. In order to determine the formation constant of the complex a reversibly reducible Me^{n+} ion of another metal which forms a complex of similar stability with ligand L must be found. The Me^{n+} ion reduction wave should be observed at a potential more positive than the $\overline{\text{Me}}^{n+}$ reduction potential and it should be not visible when the $\overline{\text{Me}}^{n+}$ and L concentrations are equivalent.

In order to determine the formation constant, polarograms of Me^{n+} in the presence and in the absence of ligand L are recorded. Then a definite amount of $\overline{\text{Me}}^{n+}$ ion which forms complexes with ligand L is introduced into the solution. This causes partial decomposition of the earlier formed MeL^{n+} complex. Finally the equilibrium state is reached:

$$\overline{\text{Me}}^{n+} + \text{MeL}^{n+} \rightleftharpoons \overline{\text{MeL}}^{n+} + \text{Me}^{n+}. \tag{14.42}$$

As a result of this procedure Me^{n+} ions are liberated and two waves corresponding to the Me^{n+} and MeL^{n+} reduction are observed; the heights of these waves are proportional to concentration. The equilibrium constant of reaction (14.42) can be calculated from the relationship:

$$K = \frac{\beta_{\overline{\text{MeL}}^{n+}}}{\beta_{\text{MeL}^{n+}}} = \frac{[\text{Me}^{n+}]^2}{(C_{\text{Me}^{n+}} - [\text{Me}^{n+}])(C_{\overline{\text{Me}}^{n+}} - [\overline{\text{Me}}^{n+}])}$$

$$= \frac{(i'_g)^2 C_{\text{Me}^{n+}}}{(i_g - i'_g)(C_{\overline{\text{Me}}^{n+}} i_g - C_{\text{Me}^{n+}} i'_g)} \tag{14.43}$$

where i_g is the limiting current of the Me^{n+} in the absence of the complexing agent, and i'_g is the limiting current of Me^{n+} liberated from the complex after the addition of $\overline{\text{Me}}^{n+}$.

Since the $\beta_{\text{MeL}^{n+}}$ constant is known, the required formation constant $\beta_{\overline{\text{MeL}}^{n+}}$ can be calculated after the determination of K. This method can be used when the equilibria existing in the solution are not disturbed during the drop life.

Schwarzenbach and Sanders [36] used this method in investigations of EDTA–vanadium complexes in which Cu(II) was the indicator ion. In the determinations of formation constants of complexes of many metals with complexones, ions of manganese, zinc, cadmium, mercury and iron were used [37–41].

In the above and also in the discussion of Fig. 14.3 it has been assumed, for simplicity, that the charges on ions Me and $\overline{\text{Me}}$ are identical. This is obviously not a condition limiting the usefulness of these methods.

In certain cases complex formation constants can be calculated from differences between the diffusion coefficients of the simple and complex ions [42, 43]. These differences are sometimes appreciable, e.g. in the case of simple and complex thallium(I) ions.

In the case of simultaneous diffusion to the electrode of simple and complex ions (the ratio of metal to ligand in the latter being 1:1) the average diffusion coefficient $\bar{D}$ is given by the equation:

$$\bar{D} = \frac{D_{Me^{n+}} + D_{MeL^{n+}}\beta_{MeL^{n+}}[L]}{1 + \beta_{MeL^{n+}}[L]} \tag{14.44}$$

where $D_{Me^{n+}}$ and $D_{MeL^{n+}}$ are the diffusion coefficients of the simple and complex ions, respectively, $\beta_{MeL^{n+}}$ is the formation constant of complex MeL^{n+}.

Diffusion coefficients of simple and complex ions can be determined from the values of the limiting current, in the absence and in the presence of an excess of ligand. This method can be used only in the cases where the concentrations of Me^{n+} and MeL^{n+} are comparable and the equilibrium of the complex-forming reaction is rapidly established.

Yet another method of determining formation constants has been worked out by Crow [44]. It is based on measurements of changes of the limiting current, corresponding to changes of the average coordination number. Biernat [45, 46] determined the formation constants of irreversibly reducible complexes from changes of half-wave potential corresponding to changes of ligand concentration.

A simple procedure was suggested by Bond [47]. It is known that the half-wave potential of an irreversible reduction can be described by the equation:

$$E_{1/2}^{irr} = E_{1/2} + \frac{RT}{\alpha nF} \ln \frac{0.886 k_s t_1^{1/2}}{D^{1/2}}. \tag{14.45}$$

Therefore when the change of potential $E_{1/2}^{irr}$ is caused by complexing alone, and the change of ligand concentration does not affect the values of k_s, α, t_1 and D, the following relationship is approximately valid:

$$E_{1/2}^{irr} = E_{1/2} + \text{const.} \tag{14.46}$$

Under these conditions the measured values of $\Delta E_{1/2}^{irr}$ and $\Delta E_{1/2}$ are equal. This reasoning should also be correct in the case of *quasi*-reversible processes. Thus the changes of $\Delta E_{1/2}^{irr}$ with changing ligand concentration can be interpreted by the methods described earlier.

A similar reasoning is also correct in the case of changes of peak potentials recorded by a.c. polarographic methods. Bond [47] analysed the *quasi*-reversible Zn^{2+}–F^- system, using classical and a.c. polarographic methods. He obtained β_1 values which were in good agreement with those given in the literature. It is noteworthy that Bond and Hefter [48] considered effect, on polarographically determined formation constants, of ligands adsorbed on the electrode surface (however see also [70]).

Branica *et al.* suggested an original method based on the differences between the electrode reaction rates of various complexes [49–51]. Changes in the degree of complexing of ions were evaluated from the changes of the reduction peak heights corresponding to changes of ligand (acetylacetone) concentration. The peaks were recorded by a.c. and square-wave polarography.

14.3 The Mechanism of Reduction of Complex Compounds

In section 14.2 it has been mentioned that, when the time taken to reach the equilibria of complex formation reactions is comparable with the drop time, kinetic currents are observed. When the concentration of the free metal at the electrode surface decreases with respect to the equilibrium concentration, the electrode process is preceded by a chemical complex decomposition reaction. In such kinetic studies of complexes all the four electroanalytical methods can be successfully applied, although for a long time polarography was one of the few techniques used in analysis of the kinetics of complex formation.

The theory of kinetic electrode processes was discussed in Chapter 8. It was largely limited to first-order processes, but its equations can be successfully adapted to complex formation reactions and for this reason these problems will not be discussed here. They have been discussed in detail by Crow [52].

We have found that electrochemical analytical methods make it possible to determine the composition and the stability of complexes formed in solution, particularly when the processes are reversible; in certain cases they also make it possible to determine the rate constants of the reactions of formation and decomposition of the complexes.

Little attention has so far been paid to the analysis of the current–potential curves of irreversibly reducible complexes. However this problem is of importance, since from the analyses of such curves it is possible to determine the composition of complexes during reduction, the reduction rate constants and the transfer coefficients of those processes.

The above problems can also be solved by other methods which are not discussed in this book. Further information on this subject is given in a monograph by Vetter [53]. The problem of the analysis of waves corresponding to the irreversible reduction of complexes is of particular importance because many complexes are irreversibly reduced. The problem of determining the composition of complexes that are reduced directly at the electrode is discussed by Matsuda and Ayabe [54], Koryta [55], Tanaka and Tamamushi [56], Subrahmanya [57], Strombergh and Ivantsova [58] and Verdier et al. [59].

Koryta [55] assumed that an excess of ligand L is present in the solution and that the equilibrium between the complexes MeL^{n+}, MeL_2^{n+}, ..., MeL_p^{n+} is rapidly established. If it is assumed that complex MeL_i^{n+} is irreversibly reduced, the average current is described by the equation:

$$\bar{i} = nFAk_{fh}[MeL_i^{n+}]_0 \tag{14.47}$$

where $[MeL_i^{n+}]_0$ is the surface concentration of complex MeL_i^{n+} and $\bar{i}$ is the irreversible reduction current.

$[MeL_i^{n+}]_0$ can be found from the equilibrium constants of the complex formation. Assuming that:

$$\frac{[MeL_p^{n+}]}{[Me^{n+}]} \gg \frac{[MeL_{p-1}^{n+}]}{[Me^{n+}]} \gg \ldots \gg \frac{[MeL^{n+}]}{[Me^{n+}]} \gg 1$$

which can also be written as:

$$K_p K_{p-1} \ldots K_1 [L]^p \gg K_{p-1} K_{p-2} \ldots K_2 [L]^{p-1} \gg \ldots \gg K_1 [L] \gg 1$$

by introducing the successive formation constants, this concentration can be expressed by the relationship:

$$[MeL_i^{n+}]_0 = \frac{[MeL_p^{n+}]_0}{K_p K_{p-1} \ldots K_{i+1} [L]^{p-i}}. \tag{14.48}$$

The subscripts "0" indicate concentrations at the electrode surface. Since it has been assumed that an excess of the ligand is present in the solution, this subscript does not appear in the ligand L concentration.

Introducing $[MeL_i^{n+}]_0$ described by relationship (14.48) into equation (14.47) gives the relationship:

$$\bar{i} = nFAk_{fh} \frac{[MeL_p^{n+}]_0}{K_p K_{p-1} \ldots K_{i+1} [L]^{p-i}}. \tag{14.49}$$

14.3] The Mechanism of Reduction of Complex Compounds

Since $[MeL_p^{n+}]_0$ is proportional to $(\bar{i}_g - \bar{i})$, on the basis of the Ilkovič equation, equation (14.49) can be reduced to the following form:

$$\frac{\bar{i}}{\bar{i}_g - \bar{i}} = 0.886 k_{fh} t_1^{1/2} \frac{1}{D^{1/2}(K_p K_{p-1} \ldots K_{i+1})[L]^{p-i}}. \quad (14.50)$$

In the derivation of this relationship it has been assumed that the diffusion coefficients of Me^{n+} and the complex ions were equal. This average coefficient was called D. It follows from equation (14.50) that the composition of the ion directly transformed at the electrode can be determined from the analysis of the $\bar{i}/(\bar{i}_g - \bar{i})$–ligand concentration relationship. Obviously, these investigations must be carried out at constant potential, since the ratio $\bar{i}/(\bar{i}_g - \bar{i})$ depends on the potential via the k_{fh} constant.

It is very convenient to analyse the dependence of $\log \bar{i}/(\bar{i}_g - \bar{i})$ on $\log [L]$. This relationship should be linear, and the slope of the straight line should be equal to:

$$-\frac{\Delta \log \dfrac{\bar{i}}{\bar{i}_g - \bar{i}}}{\Delta \log [L]} = p - i. \quad (14.51)$$

Guidelli and Cozzi [60] obtained equation (14.50) by a rigorous process of reasoning.

The half-wave potential of the processes discussed can easily be determined by substituting the standard rate constant for k_{fh} in equation (14.50):

$$k_{fh} = k_s \exp\left[-\frac{\alpha n F(E - E^0)}{RT}\right] \quad (14.52)$$

and assuming $\bar{i} = \tfrac{1}{2} \bar{i}_g$. The following relationship is then obtained:

$$E_{1/2} = E^0 + \frac{RT}{\alpha n F} \ln 0.886 k_s \frac{t_1^{1/2}}{D^{1/2}} - \frac{RT}{\alpha n F} \ln(K_p K_{p-1} \ldots K_{i+1})$$

$$- \frac{RT}{\alpha n F} \ln [L]^{p-i}. \quad (14.53)$$

It follows from equation (14.53) that when α does not change with ligand concentration, the half-wave potential of the recorded wave changes linearly with the logarithm of ligand concentration. From the slope of the straight line the difference $(p-i)$ can be obtained, provided that α is known. This analysis gives correct results when the drop time is constant during the experiment.

It follows from the above relationships that when the composition of the complex present in the solution is known, the composition of the complex directly involved in the electrode reaction can be determined.

Matsuda and Ayabe [61] presented a general theory of the polarographic reduction of complexes which is valid for irreversible, reversible and *quasi*-reversible processes. They assumed that only one MeL_p^{n+} complex is present in the solution and that it is the complex with the lower coordination number which is reduced. In such cases the following relationship is valid:

$$\ln \frac{\bar{i}}{\bar{i}_g - \bar{i}} = -\ln\left\{1.13(k\sqrt{t_1})^{-1} \exp\left[\frac{\alpha nF}{RT}(E - E_{1/2})\right] + \exp\left[\frac{nF}{RT}(E - E_{1/2})\right]\right\}. \quad (14.54)$$

In this case the reversible half-wave potential $E_{1/2}^*$ is described by the relationship:

$$E_{1/2} = E^0 - \frac{RT}{2nF} \ln \frac{D_{MeL_p^{n+}}}{D_{Me}} - \frac{RT}{nF} \ln[L]^p \quad (14.55)$$

and k is the parameter dependent on the rate constant of the electrode process.

If the reversible half-wave potential is determined for various L concentrations (by the method of Koryta [62] discussed in Chapter 7, or by the method described by Matsuda and Ayabe [61]) it is possible to determine the number of ligands in the complex present in the solution. The method of determination follows directly from equation (14.55). The dependence of $E_{1/2}$ on the logarithm of [L] should be linear, and the slope of the straight line is related to p.

A similar theory was used by Matsuda and Ayabe [61] in their studies on hydroxo zinc complexes. In a later paper [63] they considered a more complicated case in which several complexes of differing coordination number with respect to ligand L are present in the solution.

In recent years complexes have been the subject of considerable interest as a result of the general development of the ligand field theory and a better understanding of the electronic structure of the complexes.

Rates and mechanisms of electrode processes depend on the energy of the lowest unoccupied or singly filled orbital of the reduced ion. During reduction, electrons are introduced into these orbitals [64, 65]. When the affinity of the reduced ion or molecule for the electron is very pronounced a simple transition of the electron to the vacant orbital takes place; in the opposite case this transition is possible only after the rearrangement of the electronic structure of the complex [66]. The energy of this rearrangement constitutes a considerable, and sometimes the main, part of the activation energy of the electrode process.

Vlček [67] elucidated the relation between the polarographic and the spectroscopic behaviour of various cobalt(III) complexes. The details of these studies and the general discussion of the relation between the structure of complexes and their behaviour at the electrode are given in the review articles [68, 69] by this author.

The problems connected with the mechanisms of electrode reduction of complexes have been presented in an elegant way by Crow [52] in his monograph.

References

[1] Lingane, J. J., *Chem. Revs.*, **29**, 1 (1941).
[2] Heyrovský, J. and Ilkovič, D., *Coll. Czechoslov. Chem. Communs.*, **7**, 198 (1935).
[3] von Stackelberg, M. and von Freyhold, H., *Z. Elektrochem.*, **46**, 120 (1940).
[4] Cernătescu, R., Popescu, P., Crăciun, A., Bostan, M. and Iorga, N., *Acad. rep. populare Romine, Studii si Cercetari Stiint. Chim. Fil. Iasi*, **2**, 1 (1958).
[5] Andrews, A. C. and Romary, J. K., *Inorg. Chem.*, **2**, 1060 (1963).
[6] DeFord, D. D. and Hume, D. N., *J. Am. Chem. Soc.*, **73**, 5321 (1951).
[7] Irving, H. N. M. H., in *"Advances in Polarography"*, Longmuir, I. S., Ed., Pergamon Press, Oxford 1960, Vol. 1, page 42.
[8] Hume, D. N., DeFord, D. D. and Cave, G. C. B., *J. Am. Chem. Soc.*, **73**, 5323 (1951).
[9] Frank R. E. and Hume, D. N., *J. Am. Chem. Soc.*, **75**, 1736 (1953).
[10] Kivalo, P. and Rastas, J., *Suomen Kemistilehti*, **B30**, 128 (1957).
[11] Rastas, J. and Kivalo, P., *Suomen Kemistilehti*, **B30**, 143 (1957).
[12] Schaap, W. B. and McMasters, D. L., *J. Am. Chem. Soc.*, **83**, 4699 (1961).
[13] Karmalkar, P. K., *Z. phys. Chem.*, **218**, 189 (1961).
[14] Klatt, L. N. and Rouseff, R. L., *Anal. Chem.*, **42**, 1234 (1970).
[15] Momoki, K., Sato, H. and Ogawa, H., *Anal. Chem.*, **39**, 1072 (1967).
[16] Momoki, K., Ogawa, H. and Sato. H., *Anal. Chem.*, **41**, 1826 (1969).
[17] Momoki, K. and Ogawa, H., *Anal. Chem.* **43**, 1664 (1971).
[18] Piljac, I., Grabarić, B. and Filipović, I., *J. Electroanal. Chem.*, **42**, 433 (1973).
[19] Meites, L., *"Polarographic Techniques"*, 2nd Ed., Interscience, New York 1965.
[20] Butler, C. G. and Kaye, R. C., *J. Electroanal. Chem.*, **8**, 463 (1964).
[21] Ringbom, A. and Eriksson, L., *Acta Chem. Scand.*, **7**, 1105 (1953).
[22] Macovschi, M. E., *J. Electroanal. Chem.*, **16**, 457 (1968); **18**, 47 (1968); **19**, 219 (1968); **20**, 393 (1969).
[23] Jacq, J., *J. Electroanal. Chem.*, **9**, 8 (1965).
[24] Guidelli, R. and Cozzi, D., *J. Phys. Chem.*, **71**, 302 (1967).

[25] Najdeker, E., Thesis, Warsaw University, 1968.
[26] Casassas, E. and Eek, L., *J. chim. phys.*, **64**, 971 (1967).
[27] Eriksson, L., *Acta Chem. Scand.*, **7**, 1146 (1953).
[28] Galus, Z., "*Electrode Reactions of Nickel (II) on Mercury Electrodes*", PWN, Warsaw 1966 (in Polish); Reynolds, G. F., Shalgosky, H. I. and Webber, T. J., *Anal. Chim. Acta*, **8**, 564 (1953); Florence, T., *Australian J. Chem.*, **19**, 1343 (1965).
[29] Koryta, J., *Chem. Technik*, **7**, 464 (1955).
[30] Koryta, J., *Acta Chim. Acad. Sci. Hung.* **9**, 363 (1956).
[31] Koryta, J. and Kössler, I., *Chem. Listy*, **44**, 128 (1950).
[32] Koryta, J. and Kössler, I., *Coll. Czechoslov. Chem. Communs.*, **15**, 241 (1950).
[33] Schwarzenbach, G. and Ackermann, H., *Helv. Chim. Acta*, **35**, 485 (1952).
[34] Schwarzenbach, G., Gut, R. and Anderegg, G., *Helv. Chim. Acta*, **37**, 937 (1954).
[35] Bril, K. and Krumholz, P., *J. Phys. Chem.*, **57**, 874 (1953).
[36] Schwarzenbach, G., and Sanders, J., *Helv. Chim. Acta*, **36**, 1089 (1953).
[37] Holleck, L. and Liebold, G., *Naturwiss.*, **22**, 582 (1957).
[38] Kern, D. M. H., *J. Am. Chem. Soc.*, **81**, 1563 (1959).
[39] Spedding, F. H., Powell, J. E. and Wheelwright, E. J., *J. Am. Chem. Soc.*, **78**, 34 (1956).
[40] Schwarzenbach, G. and Gut, R., *Helv. Chim. Acta*, **39**, 1589 (1956).
[41] Wheelwright, E. J., Spedding, F. H. and Schwarzenbach, G., *J. Am. Chem. Soc.*, **75**, 4196 (1953).
[42] Kačena, V. and Matoušek, L., *Chem. Listy*, **46**, 525 (1952); *Coll. Czechoslov. Chem. Communs.*, **18**, 294 (1953).
[43] Zábranský, Z., *Coll. Czechoslov. Chem. Communs.*, **24**, 3075 (1959).
[44] Crow, D. R., *J. Electroanal. Chem.*, **16**, 137 (1968).
[45] Biernat, J., "*Theory of Structure of Complex Compounds*", Pergamon Press–WNT, Warsaw 1964, page 627.
[46] Biernat, J. and Baranowska-Zralko, M., *Electrochim. Acta*, **17**, 1867, 1877 (1972).
[47] Bond, A. M., *J. Electroanal. Chem*, **28**, 433 (1970).
[48] Bond, A. M. and Hefter, G., *J. Electroanal. Chem.*, **31**, 477 (1971).
[49] Jeftić, Lj. and Branica, M., *Croat. Chim. Acta*, **35**, 203 (1963).
[50] Jeftić, Lj. and Branica, M., *Croat. Chim. Acta*, **35**, 211 (1963).
[51] Petek, M., Jeftić, Lj. and Branica, M., "*Polarography 1964*", Hills, G. J., Ed., Macmillan, London 1965, page 491.
[52] Crow, D. R., "*Polarography of Metal Complexes*", Academic Press, London and New York, 1969.
[53] Vetter, K. J., "*Electrochemische Kinetik*", Springer-Verlag, Berlin 1961.
[54] Matsuda, H. and Ayabe, Y., *Bull. Chem. Soc. Japan*, **29**, 134 (1956).
[55] Koryta, J., *Chem. Listy*, **51**, 1544 (1957); *Coll. Czechoslov. Chem. Communs.*, **24**, 3057 (1959).
[56] Tanaka, N. and Tamamushi, H., *Bull. Chem. Soc. Japan*, **22**, 187 (1949).
[57] Subrahmanya, R. S., in "*Advances in Polarography*", Longmuir, I. S., Ed., Pergamon Press, Oxford 1960, Vol. 2, page 674.
[58] Strombergh, A. G. and Ivantsova, M. K., *Dokl. Akad. Nauk SSSR*, **100**, 303 (1955).
[59] Verdier, E., Bennes, R. and Balette, B., *J. Electroanal. Chem.*, **31**, 463 (1971).
[60] Guidelli, R. and Cozzi, D., *J. Phys. Chem.*, **71**, 3027 (1967).

[61] Matsuda, H. and Ayabe, Y., *Z. Elektrochem.*, **63**, 1164 (1959).
[62] Koryta, J., *Electrochim. Acta*, **6**, 67 (1962).
[63] Matsuda, H. and Ayabe, Y., *Z. Elektrochem.*, **66**, 469 (1962).
[64] Vlček, A. A., *Coll. Czechoslov. Chem. Communs.*, **20**, 894 (1955).
[65] Vlček, A. A., *Nature*, **177**, 1043 (1956).
[66] Vlček, A. A., *Coll. Czechoslov. Chem. Communs.*, **22**, 948 (1957).
[67] Vlček, A. A., *Disc. Faraday Soc.*, **26**, 164 (1958).
[68] Vlček, A. A., *Progr. Org. Chem.*, **5**, 216 (1963).
[69] Vlček, A. A., in *"Progress in Polarography"*, Zuman, P. and Kolthoff, I. M., Eds., Interscience, New York, London, 1962, page 269.
[70] Sluyters-Rehbach, M. and Sluyters, J. H., *J. Electroanal. Chem.*, **39**, 339 (1972).

Chapter 15

ECE Processes

In previous chapters electrode processes accompanied by chemical reactions have been discussed which are either preceded or followed by the charge transfer. By calling the chemical stage C and the charge transfer E it is possible to express these electrode processes by means of symbols EC or CE. Electrode processes in which the charge transfer is preceded by two chemical reactions of depolarizer formation have been briefly discussed. Such processes are called CCE.

Reactions taking place according to an ECC mechanism have also been mentioned. In this case the initial electrode reaction product is the substrate of two consecutive chemical reactions and the process consists of two chemical stages and one electrochemical stage.

Three-stage mechanisms, consisting of two electrochemical stages with one chemical stage between them, are also known, and can be called ECE. Testa and Reinmuth [1] showed that reduction of o-nitrophenol takes place according to such a mechanism.

An electrode process taking place according to an ECE mechanism can be expressed by the following general scheme:

$$\text{Ox} + n_1 e \rightleftharpoons \text{Red} \underset{k_2}{\overset{k_1}{\rightleftharpoons}} \text{Ox}' + n_2 e \rightleftharpoons \text{Red}'. \qquad (15.1)$$

Depolarizer Ox, present in the solution, accepts n_1 electrons from the electrode and as a result, the product of the first stage of the reaction Ox' is formed. This reaction usually involves another substance, but its concentration often exceeds that of Red, and therefore it can be assumed that the chemical reactions is of the pseudo-first order type. Substance Ox' is further reduced at the electrode from which it accepts n_2 electrons.

Electrode reactions of the ECE type can be studied by any of the methods discussed in this book, but it appears that polarography is the least suitable for this purpose, since variation of the time scale is difficult and limited in scope.

15.1 Chronoamperometry and Chronocoulometry

Investigation of the dependence of current on time of electrolysis carried out at a constant potential is of great value in elucidations of the mechanism of ECE reactions and in determinations of Red → Ox' reaction rates. The theory of such chronoamperometric processes was worked out by Alberts and Shain [2] and a detailed discussion of these problems was published by Adams [3]. When the chemical reaction in process (15.1) is very slow and the time of the experiment is short, the whole process is limited to the exchange of n_1 electrons. In this case the current is expressed by the well known relationship:

$$\frac{i_g}{(k_1 \to 0)} = \frac{n_1 FAD^{1/2} C^0_{Ox}}{\pi^{1/2} t^{1/2}}. \tag{15.2}$$

When the chemical reaction is very fast and the time of the experiment is long, substance Ox can be completely reduced to substance Red. In this case the current is expressed by a relationship similar to (15.2), but it is proportional to $n_1 + n_2$:

$$\frac{i_g}{(k_1 \to \infty)} = \frac{(n_1 + n_2) FAD^{1/2} C^0_{Ox}}{\pi^{1/2} t^{1/2}}. \tag{15.3}$$

For the intermediate conditions, in which the electrode process Ox' ⇌ Red' only partly takes place Alberts and Shain expressed the ratio of the experimental value of $it^{1/2}$ to that $it^{1/2}$ resulting from equation (15.3) when the chemical reaction is infinitely fast, by the following equation:

$$\frac{(it^{1/2})_{\text{expt}}}{(it^{1/2})_\infty} = 1 - \frac{n_2}{(n_1 + n_2)} e^{-k_1 t}. \tag{15.4}$$

In many systems $n_1 = n_2$. For such cases the following expression is obtained from equation (15.4):

$$\frac{(it^{1/2})_{\text{expt}}}{(it^{1/2})_\infty} = 1 - 0.5 e^{-k_1 t}. \tag{15.5}$$

By ascribing numerical values to parameter $k_1 t$ it is possible to calculate the working curve of dependence of the ratio $(it^{1/2})_{\text{expt}}/(it^{1/2})_\infty$ on $\log k_1 t$. Such a curve is shown in Fig. 15.1.

In order to determine the rate constant of the chemical reaction the current is measured at a potential corresponding to very rapid reduction of Ox and Ox' on the electrode surface. Then $(it^{1/2})_{\text{expt}}$ is obtained. Determination of $(it^{1/2})_\infty$ can be difficult. When the diffusion coefficient and $n_1 + n_2$

are known, this can be calculated from equation (15.3). Sometimes by changing the pH the process can be stopped after the first stage. Then the value of $it^{1/2}$ determined for the first stage is usually 2, 3 or 4 times lower

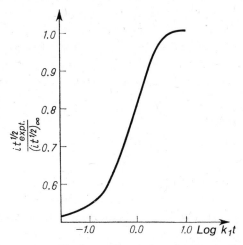

Fig. 15.1 Working curve used in the treatment of results of potentiostatic measurements of ECE reactions.

than $(it^{1/2})_\infty$. When the ratio $n_1/(n_1+n_2)$ is known, by assuming that the diffusion coefficient is not greatly affected by the pH change $(it^{1/2})_\infty$ can be easily calculated.

The value of $k_1 t$ for various times can be obtained from the curve of $(it^{1/2})_{\text{expt}}/(it^{1/2})_\infty$ vs. $\log k_1 t$. By plotting the measured $k_1 t$ values against time a straight line having the slope k_1 is obtained. This procedure was first described by Hawley [4].

Feldberg [5] pointed out that deviations from scheme (15.1) are possible when Ox' is more readily reduced than Ox, since in this case the normal potential of the Ox'/Red' system is more positive than that of the Ox/Red system and the chemical reaction:

$$\text{Red} + \text{Ox}' \rightleftharpoons \text{Ox} + \text{Red}' \quad (15.6)$$

can take place.

In this discussion it has been assumed that the rates of process (15.6) are equal to zero. In fact, when the Ox/Red and Ox'/Red' systems are reversible the rates of this process should be appreciable. The final electrochemical effect of this process is the same as in the case of the direct electrode reduction of Ox', but in such a case the process has the ECC mechanism.

After the first stage of the charge transfer leading to the formation of Red the first chemical reaction Red → Ox' takes place and is followed by the second reaction according to scheme (15.6). Obviously this scheme represents the simplest case, in which the stoichiometric coefficients of the reaction are equal to one, but this does not affect the essence of the problem.

Hawley and Feldberg [6] showed that, when an additional chemical reaction is included in the electrode process, i.e. when the process is, in fact, of the ECC type, the working curves for ECE (Fig. 15.1) are similar to those for ECC, provided that $k_1 t$ is small. For large $k_1 t$ values these curves are different. The deviations from ECE have been discussed in detail by Adams, Hawley and Feldberg [7] who have also described the methods of experimental verification of such deviations.

According to Adams [3] we can express the whole process as a sequence of reactions:

$$Ox + n_1 e \rightleftharpoons Red \quad \text{at potential } E_1, \tag{15.7}$$

$$Red + Z \xrightarrow{k_1} Ox'. \tag{15.8}$$

Then Ox' can react in two ways:

$$Ox' + n_2 e \rightleftharpoons Red' \quad \text{at potential } E_2, \tag{15.9}$$

$$Ox' + Red \rightleftharpoons Red' + Ox. \tag{15.10}$$

When potential E_2, at which process (15.9) can take place, is much more negative (by about 200 mV) than E_1, the potential of reaction (15.7), the ECE mechanism can be studied after the application of potential E_2 to the electrode, since at potential E_1 the simple diffusion process consuming n_1 electrons per reacting ion or molecule takes place. Obviously the chemical reaction (15.8) also takes place but it does not affect the current observed in the chronoamperometric experiment.

When E_2 is only slightly more negative than E_1, and when E_2 is equal to, or more positive than, E_1, the chemical reaction (15.10) always competes with the electrochemical reaction, and its effect, which depends on the equilibrium and rate constants of the process, should be taken into account in calculations.

Feldberg [5] calculated potentiostatic curves for various kinds of ECE processes including the case in which the electrode process involves dimerization.

Nelson and Feldberg [8] also considered the case in which the chemical reaction was a second order dimerization of Red and obtained working curves for the calculation of the rate constants of the chemical reaction.

Their theory was applied to the determination of the rate constants of dimerization of substituted triphenylamine cation radicals. In the development of this theory the Feldberg method [5] was used.

Another example of an ECE reaction was considered by Feldberg and Jeftić [9]. In this case Red was the substrate of a first order chemical reaction, but Red also reacted with Ox.

In all the discussions in this section it has been assumed that the chemical reaction is irreversible.

Herman and Blount [10] considered ECE processes carried out under chronocoulometric conditions on flat electrodes and on spherical electrodes. In the case of irreversible chemical reactions and a flat electrode:

$$(Q/t^{1/2})/(Q/t^{1/2})_\infty = 1 - \frac{n_2}{2(n_1+n_2)t^{1/2}} \int_0^t \frac{e^{-k_1 t}}{t^{1/2}} dt. \qquad (15.11)$$

This equation can also be written in the following, more useful, form:

$$(Q/t^{1/2})/(Q/t^{1/2})_\infty = 1 - \left\{ \frac{n_2 \pi^{1/2}}{2(n_1+n_2)(k_1 t)^{1/2}} \right\} \text{erf}(k_1 t)^{1/2}. \qquad (15.12)$$

When such a process takes place on a spherical electrode the equation is more complex. Herman and Blount also considered an ECE process, with a reversible chemical reaction, taking place on a flat electrode.

15.2 Polarography

The mechanism of an ECE process taking place under polarographic conditions was worked out a long time ago by Kastening and Holleck [11, 12]. The diagnostic criteria for ECE mechanism under polarographic conditions were discussed by Savéant [13] on the basis of the work of Alberts and Shain [2] and Nicholson [14]. The subject of the discussion was the case in which the product of the chemical reaction is reduced at more negative potentials and, as a result, two waves are observed. The criteria involve the relation between the limiting current of the second wave and the height of the mercury reservoir h. There are three forms of this relation, depending on the value of $k_1 t_1$.

When $k_1 t_1 < 0.20$ the limiting current is proportional to $h^{-1/2}$. The normal linear relation between i_g and $h^{1/2}$, characteristic of diffusion currents, is observed when $k_1 t_1 > 6$. In the intermediate region of $k_1 t_1$ values between 1.0 and 1.8 the limiting current does not depend on the height of the mercury reservoir. The dependence of i_g on $h^{1/2}$, typical for this kind of current is observed only when the rate of the chemical reaction taking place between the two charge transfer stages is slow. Assuming that $t_1 = 4$ s this relation can be observed when the rate constant is smaller than 0.05 s^{-1}. According to Guidelli [15], when $k_1 t_1 \geqslant 4$ the following relationship is approximately valid:

$$\frac{i_g - i}{i} = \frac{n_2}{n_1}\left(1 - \frac{0.573}{\sqrt{k_1 t_1}}\right) \tag{15.13}$$

where i is the current due to the reduction of Ox, and i_g that due to total reduction of Ox and Ox'.

For lower values of $k_1 t_1$ the relations between this dimensionless parameter and current given in the original papers [11, 14] must be used.

In the ECE mechanisms in polarography discussed above the type of the limiting current has been taken into account.

The problem of describing not only the limiting currents, but also the relation between the ECE mechanism and the shape of polarographic curves was investigated by Sobel and Smith [16], who obtained a solution for reversible electrode processes. The irreversible processes are discussed by Sobel in a Ph.D. thesis [17].

Sobel and Smith [16] have taken into account the growth of the mercury drop electrode, but not spherical diffusion.

They obtained the following complex general equation:

$$i = FAC_{Ox}^0 D_{Ox}^{1/2} k^{1/2}(1+K)\left(\frac{7}{3\pi t_1}\right)^{1/2}\left[(n_1 - n_2 K^{-1})\left(\frac{V}{M}\right) + \right.$$

$$\left. (n_1 Y + n_2 V K^{-1})(R - BV)\left(\frac{12 t_1}{7}\right)^{1/2} \chi^{-1} F(\chi)\right] \tag{15.14}$$

where:

$$\chi = \left(\frac{12 t_1}{7}\right)^{1/2} MBY,$$

and $F(\chi)$ is the Koutecký function of parameter χ,

$$k = k_1 + k_2, \quad K = \frac{k_2}{k_1}, \quad M = (1+K)k^{1/2}(1+e^{j_1}),$$

$$B = \left[\frac{\left(\frac{1}{1+K} + e^{j_2}\right)(1+e^{j_1})}{e^{j_1}(1+e^{j_2})} - \frac{1}{1+K}\right]\frac{1}{(1+e^{j_1})},$$

$$Y = \frac{C}{1+C} = CV, \quad C = \frac{e^{j_1}(1+e^{j_2})}{K(1+e^{j_1})},$$

$$R = \frac{\left(\frac{1}{1+K}\right)+e^{j_2}}{e^{j_1}(1+e^{j_2})}, \quad j_1 = \frac{n_1 F}{RT}(E-E^1_{1/2}), \quad j_2 = \frac{n_2 F}{RT}(E-E^2_{1/2}),$$

$E^1_{1/2}$ is the reversible half-wave potential of the first process and $E^2_{1/2}$ that of the second.

Several curves have been calculated by means of a computer, assuming that $n_1 = n_2$, diffusion coefficients are equal, and the chemical reaction is irreversible.

It should be stressed that, as a result of simplifications, equation (15.14) is valid only for large kt_1 values. By comparing the ratio of limiting currents obtained on the basis of this equation with that obtained on the basis of the theory in which the simplifying assumptions were not made [11, 14], Sobel and Smith showed that equation (15.14) gives good results when $kt_1 \geqslant 4$. These problems were further developed by Smith and co-workers in a recent paper [30].

15.3 Stationary Electrode Voltammetry

The theory of ECE processes carried out under stationary electrode voltammetric conditions was first put forward by Nicholson and Shain [18], who discussed the case in which the irreversible chemical process is of the first or pseudo-first order. Four cases may be considered corresponding to the reversibility of the charge transfer stages.

(1) When both charge transfer stages are reversible and the product of the chemical reaction is less readily reduced than Ox, two peaks are observed provided that the potentials of standard systems Ox/Red and Ox'/Red' are sufficiently different. The current of the first peak corresponds to the reduction of the Ox form only. This peak is in accord with the theory of electrode processes followed by an irreversible chemical reaction, which was discussed in Chapter 9. When k_1/a is small this peak is identical with that of a kinetically uncomplicated reversible electrode process. When k_1/a increases the current peak is shifted in the positive direction (in the case of a reduction). The effect of k_1/a on the height of the second peak is very large. When this ratio is very small the second peak is not observed. As k_1/a increases, the second peak appears but it is considerably stretched out along the potential axis. When k_1/a reaches very high values the shape and height of the second peak become identical with those predicted by the theory of two-stage electrode diffusion processes.

Nicholson and Shain obtained the relation between the height of the second peak and k_1/a in a tabular form. They also discussed the effects of the numbers of electrons involved in the first and the second stages of charge transfer and of the difference between the potentials of the normal Ox/Red and Ox'/Red' system, on the recorded curves.

(2) When the second electron transfer stage is irreversible the recorded curves are similar to those described for the case in which both stages are reversible. As a result of irreversibility the second peak is small and diffuse. It has the properties of the current peaks of irreversible processes described in Chapter 7, but they may be slightly modified as a result of the chemical reaction Red $\xrightarrow{k_1}$ Ox'.

For example, when the polarization rate increases, the peak potential is shifted in the negative direction, as a result of the irreversibility of the process. At the same time k_1/a decreases with increasing polarization rate, and as a result the potential shift becomes larger. The total value of the shift depends on k_1/a.

At low k_1/a values the second peak is so small and diffuse that it can be overlooked, especially in view of the fact that it is formed on the decreasing current slope of the first peak. At large k_1/a values the shape of the second peak resembles that of the peak due to irreversible processes.

Both peaks are observed when the difference between the potentials of the normal Ox/Red and Ox'/Red' systems is large, but the potential of the second system must be more negative in the case of a reduction. As the potential difference, ΔE^0, decreases, the peaks begin to overlap and at $\Delta E^0 = 0$ only one peak is formed. Its shape is very similar to that observed for zero potential difference between standard potentials in the case of reversibility of both electrode stages. When the potential of the Ox'/Red' system is at least 180 mV more positive than the E^0 value of the Ox/Red system the shape of the peak is independent of the degree of reversibility of the second stage.

(3) When the first electrode stage is irreversible but the second is reversible the peak is described by the theory of simple irreversible processes and the chemical reaction has no effect on its shape and size. The second peak is very similar to that observed in the case of the reversibility of both electrode stages.

(4) When both electrode stages are irreversible the chemical reaction does not affect the size of the first peak, whereas the second one depends on the k_1/a ratio. Both peaks are very stretched out along the potential axis.

As has been mentioned in all these cases, two peaks can be observed when the standard potential of the Ox'/Red' system is much more negative than that of the first system. The effect of the polarization rate on the sizes of the peaks at various degrees of reversibility is shown in Fig. 15.2.

The above relationships are valid when the peaks are clearly separated. When $\Delta E^o = 0$ the relation between the cathodic current function and the polarization rate in all the four cases is similar, and the observed effect of the polarization rate is so small that it is difficult to see any difference between the four cases. The changes of the corresponding current functions are shown in Fig. 15.3.

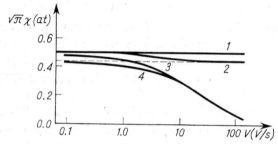

Fig. 15.2 Effect of polarization rate (V) on current function in ECE processes. $\Delta E^o = -180$ mV. Curves *1* and *2* describe the first current peak: *1*—first process irreversible, second reversible or irreversible; *2*—first process reversible, second reversible or irreversible. Curves *3* and *4* describe the second current peak: *3*—first process irreversible or reversible, second irreversible; *4*—first process reversible or irreversible, second reversible. The numbers of electrons exchanged in the two elementary processes are equal (from [18] by permission of the copyright holders, the American Chemical Society).

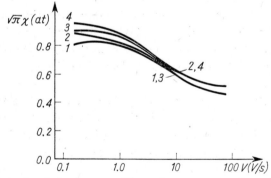

Fig. 15.3 Effect of polarization rate (V) on the current function of ECE processes $\Delta E^o = 0$. Curve *1*—both processes reversible; curve *2*—first process irreversible, second reversible; curve *3*—first process reversible, second irreversible; curve *4*—both processes irreversible. The numbers of electrons exchanged in the two elementary processes are equal (from [18] by permission of the copyright holders, the American Chemical Society).

When the standard potential of the Ox'/Red' system is much more positive than that of the first system the current functions of the reduction process in all the four cases discussed depend on the polarization rate. This dependence is similar to that shown in Fig. 15.3, and therefore it is very difficult to say, on the basis of the cathodic current, whether the first or the second stage is reversible. In such cases cyclic polarization should be employed. The theory of anodic currents observed in an anodic cycle following a cathodic reduction in the case of ECE mechanisms, was published by Nicholson and Shain [18], who later used this theory in their studies of the mechanism of the electroreduction of nitrosophenol [19]. They also described [19] methods for determining the rate constants of chemical reactions based on this theory.

The theory of ECE processes taking place in stationary electrode voltammetry was also dealt with by Savéant [13], who considered various kinds of such processes, including the case of a second order chemical reaction separating two stages of charge transfer. These problems were also investigated by Mastragostino et al. [20].

Nadjo and Savéant [21] investigated the criteria of the differences between disproportionation reactions and ECE processes, using the example of the electroreduction of aromatic carbonyl compounds in alkaline media.

15.4 Chronopotentiometry

Chronopotentiometric processes of the ECE type have been investigated by Testa and Reinmuth [22]. As in the case of stationary electrode voltammetry, where one or two current peaks can be observed, depending on the difference between standard potentials of Ox/Red and Ox'/Red' systems, in the case of chronopotentiometry one or two transition times can be observed, depending upon this difference between the standard potentials. In the case of two transition times the description of the first one is easy. The behaviour of the system at the second transition time is described by the following equation:

$$\left(\frac{n_1+n_2}{n_2}\right)\tau_1^{1/2} - \frac{n_1}{n_2}(\tau_1+\tau_2)^{1/2} - 0.5\left(\frac{\pi}{k_1}\right)^{1/2} \mathrm{erf}[k_1(\tau_1+\tau_2)]^{1/2}$$
$$+ 0.5 \int_{\tau_1}^{\tau_1+\tau_2} \frac{\exp(-k_1\lambda/2)I_0(k_1\lambda/2)}{(\tau_1+\tau_2-\lambda)^{1/2}} d\lambda = 0 \qquad (15.15)$$

where I_0 is the hyperbolic Bessel function of zero order.

It follows from equation (15.15) that, when the rate constant is very high, the system behaves like a kinetically uncomplicated two-stage process, in which n_1 and n_2 electrons are exchanged during the first and the second stages, respectively. When k_1 is close to zero the system behaves like a single-stage process.

The practical application of equation (15.15) is very difficult, since it contains an integral, but when $(k_1 \tau_1)^{1/2} > 8$ this equation can be expressed in the following simpler form:

$$(\tau_1+\tau_2)^{1/2} = (\tau_1+\tau_2)_\infty^{1/2} - \frac{n_2}{n_1}(\pi k_1)^{-1/2} \arctan\left(\frac{\tau_1}{\tau_2}\right)^{1/2}. \quad (15.16)$$

In this simplification it is assumed that the argument of the error function is greater than 2 and, therefore, that this function is practically equal to one; $(\tau_1+\tau_2)_\infty$ is the sum of the transition times which would be observed at infinitely large rate constants of the chemical reaction.

Testa and Reinmuth also obtained the solution for the case when Ox' is reduced in the same region of potentials as that at which the reduction of Ox takes place. In this case only one transition time is observed, irrespective of the value of the rate constant of the chemical reaction. When the rate constant k_1 is infinitely large, the recorded transition time corresponds to the electrode reaction in which n_1+n_2 electrons are exchanged. When k_1 is infinitely small, n_1 electrons are exchanged in the process. In the intermediate cases the following equation, derived by Testa and Reinmuth, can be used:

$$\frac{(i\tau_\infty^{1/2})}{i\tau^{1/2}} = 1+\varrho\left(\frac{\pi}{4k_1}\right)^{1/2} \operatorname{erf}(k_1\tau)^{1/2} + \frac{\varrho^2 \exp[k_1\tau/(\varrho^2-1)]}{(k_1\tau)^{1/2}(1-\varrho^2)^{1/2}} \times$$

$$\{\Phi[\varrho(k_1\tau)^{1/2}/(1-\varrho^2)^{1/2}]+\Phi[(k_1\tau)^{1/2}/(1-\varrho^2)^{1/2}]\} \quad (15.17)$$

where $\varrho = n_2/(n_1+n_2)$, τ_∞ is the transition time which would be observed at $k_1 = \infty$, and:

$$\Phi(x) = \int_0^x \exp(u^2)\,du. \quad (15.18)$$

It follows from equation (15.17) that the dependence of $i\tau^{1/2}$ on i should be similar to that observed in the case of an electrode process preceded by a first order chemical reaction.

At low current densities the second term of the right-hand side of equation (15.17) can be neglected, and therefore, assuming that the argument of the error function is greater than 2:

$$i\tau^{1/2} = (i\tau^{1/2})_\infty - \varrho i \left(\frac{\pi}{4k_1}\right)^{1/2}. \qquad (15.19)$$

In this case the dependence of $i\tau^{1/2}$ on i is linear, and the rate constant of the chemical reaction can be readily determined from the slope of the straight line.

At large and small current densities the product $i\tau^{1/2}$ should be independent of current density. When $i \to 0$ the first product $i\tau^{1/2}$ corresponds to $(i\tau^{1/2})_\infty$, and when $i \to \infty$ it corresponds to the exchange of only n_1 electrons. When the ratio of the products determined in this way is an integer it proves that the reaction being studied takes place according to ECE, and not according to CE or CCE mechanisms.

These relationships were used by Testa and Reinmuth [1] in their studies of the reduction of o-nitrophenol.

Theoretical studies of the chronopotentiometry of ECE processes have also been carried out by Tanaka and Yamada [23] and by Herman and Bard [24].

In addition to the description of the transition times Tanaka and Yamada gave a description of chronopotentiometric curves recorded for ECE processes.

15.5 The Rotating Disc Method

The theory of ECE processes taking place on rotating disc electrodes was first worked out by Malachesky et al. [25]. Their approximate treatment leads to inaccurate results in the case of fast chemical reactions and it was therefore modified by Karp [26].

A more accurate theory was published by Filinovskii [26] who obtained the following general relationship:

$$i_g = 0.94 i_g^{k_1=0} \left\{ 1 + \frac{n_2}{n_1} \left[1 - \frac{\sqrt{1 + \frac{\delta^2 k_1}{1.9D}}}{1 + \frac{\delta^2 k_1}{D}} \right] \right\} \qquad (15.20)$$

for the case when Ox' is reduced at potentials more positive that the Ox reduction potential.

In equation (15.20) $i_g^{k_1=0}$ is the limiting current of the reduction of the Ox to the Red form which is not affected by the chemical reaction and δ is the thickness of the diffusion layer.

When the transformation of Red into Ox' is fast $k_1 \gg \dfrac{D}{\delta^2}$ and equation (15.20) becomes:

$$\frac{i_g}{i_g^{k_1=0}} \approx 0.94\left[\left(1+\frac{n_2}{n_1}\right) - \frac{n_2}{n_1}\left(\frac{D}{1.9\gamma^2 k_1}\right)^{1/2}\omega^{1/2} + \ldots\right] \quad (15.21)$$

where $\gamma = 1.61 D^{1/3} \nu^{1/6}$.

It follows from equation (15.21) that, for small values of ω, the dependence of $i_g/i_g^{k_1=0}$ on $\omega^{1/2}$ should be linear.

Extrapolation of the straight line to $\omega = 0$ makes it possible to determine the ratio n_2/n_1, while the rate of the chemical reaction can be calculated from the slope.

The case of an ECE reaction in which the two charge transfer stages are separated by a dimerization reaction was considered by Marcoux et al. [28]. In theoretical investigations the method developed by Feldberg [5] was used. As in the case of studies described in reference [8] the subjects of the investigations were cation radicals of triphenylamine derivatives, which are formed during electro-oxidation at the rotating disc electrode in acetonitrile solutions.

Finally Jacq's work [29] on general equations of the polarization curves of a "square scheme" in convective diffusion conditions should be mentioned.

References

[1] Testa, A. C. and Reinmuth, W. H., *J. Am. Chem. Soc.*, **83**, 784 (1961).
[2] Alberts, G. S. and Shain, I., *Anal. Chem.*, **36**, 1859 (1963).
[3] Adams, R. N., "*Electrochemistry at Solid Electrodes*", Dekker, New York 1969.
[4] Hawley, M. D., Thesis, University of Kansas, Lawrence 1965.
[5] Feldberg, S. W., in "*Electroanalytical Chemistry*", Bard, A. J., Ed., Dekker, New York 1966, page 373.
[6] Hawley, M. D. and Feldberg, S. W., *J. Phys. Chem.*, **70**, 3459 (1966).
[7] Adams, R. N., Hawley, M. D. and Feldberg, S. W., *J. Phys. Chem.*, **71**, 851 (1967).
[8] Nelson, R. F. and Feldberg, S. W., *J. Phys. Chem.*, **73**, 2623 (1969).
[9] Feldberg, S. W., Jeftić, Lj., *J. Phys. Chem.*, **76**, 2439 (1972).
[10] Herman, H. B. and Blount, N. N., *J. Phys. Chem.*, **73**, 1406 (1969).
[11] Kastening, B. and Holleck, L., *Z. Elektrochem.*, **63**, 166 (1959).
[12] Kastening, B., *Anal. Chem.*, **41**, 1142 (1969).
[13] Savéant, J. M., *Electrochim. Acta*, **12**, 753 (1967).
[14] Nicholson, R. S., Wilson, J. M. and Olmstead, M. L., *Anal. Chem.*, **38**, 542 (1966).

[15] Guidelli, R., in *"Electroanalytical Chemistry"*, Vol. 5, Bard, A. J., Ed., Dekker, New York 1971.
[16] Sobel, H. R. and Smith, D. E., *J. Electroanal. Chem.*, **26**, 271 (1970).
[17] Sobel, H. R., Ph. D. Dissertation, Northwestern University, Evanston, Illinois, 1969.
[18] Nicholson, R. S. and Shain, I., *Anal. Chem.*, **37**, 178 (1965).
[19] Nicholson, R. S. and Shain, I., *Anal. Chem.*, **37**, 190 (1965).
[20] Mastragostino, M., Nadjo, L. and Savéant, J. M., *Electrochim. Acta*, **13**, 721 (1968).
[21] Nadjo, L. and Savéant, J. M., **33**, 419 (1971).
[22] Testa, A. C. and Reinmuth, W. H., *Anal. Chem.*, **33**, 1320 (1961).
[23] Tanaka, N. and Yamada, A., *Rev. Polarography Japan*, **14**, 234 (1967).
[24] Herman, H. B. and Bard, A. J., *J. Phys. Chem.*, **70**, 396 (1966).
[25] Malachesky, P. A., Marcoux, L. S. and Adams, R. N., *J. Phys. Chem.*, **70**, 4068 (1966).
[26] Karp, S., *J. Phys. Chem.*, **72**, 1082 (1968).
[27] Filinovskii, V. Yu., *Elektrokhimiya*, **5**, 635 (1969).
[28] Marcoux, L. S., Adams, R. N. and Feldberg, S. W., *J. Phys. Chem.*, **73**, 2611 (1969).
[29] Jacq, J., *J. Electroanal. Chem.*, **29**, 149 (1971).
[30] Ružić, I., Sobel, H. R. and Smith, D. E., *J. Electroanal. Chem.*, **65**, 21 (1975).

Chapter 16

Influence of Adsorption on Electrode Processes

In previous chapters various kinds of diffusion and kinetic electrode processes have been discussed which had the common property that the distribution of the electrolysed substance before the start of the experiment was homogeneous throughout the diffusion field. Hence, disregarding the effect of the Φ_2 potential it can be said that the concentration of the depolarizer at the electrode surface was equal to that in the bulk of solution.

However, there are substances which tend to accumulate in the solution layer adjoining the electrode. These are called surface-active substances or substances specifically adsorbable on the electrode. Usually, as a result of adsorption on the electrode, a monolayer of the adsorbed molecules is formed. When the concentration of the adsorbable substance in the bulk of solution is low the coating of the electrode may, of course, be incomplete. As a result of the monolayer formation the adsorbed molecules or ions are situated in the inner Helmholtz plane of the double layer. The depolarizer itself, or its oxidation or reduction products can be surface active. Also the ions of the supporting electrolyte and molecules or ions of any electrode inactive substances present in the solution can be adsorbed. For this reason the discussion constituting this chapter will be divided into two parts. In the first part electrode processes of specifically adsorbable substances will be discussed and in the second part, electrode processes of non-adsorbable depolarizers in the presence of adsorbed electrode inactive substances.

16.1 Electrode Processes of Specifically Adsorbable Substances

It is preferable to commence this discussion of adsorption currents with chronoamperometry and chronocoulometry, although polarography was the first method used in the studies of depolarizer adsorption. Polarographic adsorption waves were reported by Brdička and Knobloch [1, 2] soon after 1940 for the cases of the reduction of lactoflavine and methylene blue.

16.1.1 Chronoamperometry and Chronocoulometry

Both these methods have been used in recent years in studies of the adsorption of depolarizers and their electrode reaction products, but chronocoulometry has been much more widely applied [3–5] and therefore this discussion will be mainly concerned with this method.

In the case of potential-step chronocoulometry the potential is at first maintained at the value corresponding to a definite degree of adsorption on the electrode. In the case of an electrode reduction of the depolarizer this potential should be so positive that the electroreduction is negligible. After the time in which the adsorption equilibrium is reached at this potential E_i the potential is rapidly and stepwise changed to the value at which the reduction rate is limited by the rate of transport of the depolarizer to the electrode. As a result the depolarizer concentration at the electrode surface is equal to zero. In these studies the electrode should have unchangeable surface, and the solution should not be stirred. Then charge Q flowing in time t is described by the relationship

$$Q = \frac{2nFD_{Ox}^{1/2}C_{Ox}^0 t^{1/2}}{\pi^{1/2}} + Q_{dl} + nF\Gamma_{ox} \tag{16.1}$$

where Γ_{ox} is the number of moles of the depolarizer adsorbed on one square cm of the electrode surface, and Q_{dl} is the charge required for charging the double layer to the stepwise applied potential. If Q is plotted against $t^{1/2}$, $Q_{dl}+nF\Gamma_{ox}$ is obtained from the intercept on the charge axis. This can be resolved into its components by means of an auxiliary experiment with a solution having an identical concentration of the supporting electrolyte but not containing the depolarizer. In the auxiliary experiment the potential is changed in the same way as in the main experiment. From the auxiliary experiment the value of Q_{dl} is obtained and, as a result, the required value of Γ_{ox} can be determined.

Thus chronocoulometry can be used in studies of depolarizer adsorption, but the accuracy of the determination of Γ_{ox} is not very high on account of the introduction of the Q_{dl} value from the auxiliary experiment.

It should be stressed that, in order to obtain fairly satisfactory results, it is necessary to carry out the charge measurements over very short times, such that the first term of equation (16.1) is not very large compared with the last. Usually the electrolysis is carried out over a few milliseconds, and for this reason, the time of change of the potential from the initial value E_i to that of the electrode reaction must be much shorter.

Considerable progress in the studies of the adsorption of electroactive substances was brought about by the introduction [6] of the chronocoulometric method based on a double potential step. The principles of this method will be discussed in the next chapter.

The theoretical basis of the application of double potential step chronocoulometry has been worked out by Christie et al. [7]. These authors assumed that before the start of the electrode process only the Ox form is present in the solution. The Ox/Red system has the standard potential E^0. The initial potential E_i is sufficiently negative with respect to E^0, to ensure that at this potential the current in the circuit is negligible. The sudden potential change must be large enough to cause a rapid decrease to zero of the depolarizer concentration on the electrode surface. After time t' the potential returns to the value E_i and the concentration of the Red form generated at the electrode decreases to zero. The transport of Ox and Red from the bulk of the solution is controlled by diffusion alone. It is also assumed that each adsorbed substance reacts immediately after the change of the potential at which the non-adsorbed substance reacts, and that double

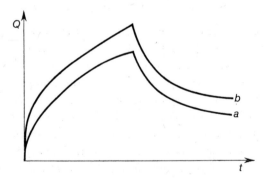

Fig. 16.1 Time dependence of charge obtained by the chronocoulometric method. Curve a—in the absence of specific adsorption of depolarizer; curve b—in the presence of specific adsorption.

layer charging takes place immediately. During the whole experiment the total charge flowing in the circuit is the measured parameter. The rate of change of charge is shown schematically in Fig. 16.1.

16.1] Electrode Processes of Specifically Adsorbable Substances

Consider the case in which the Ox form is adsorbed, while the Red form is not specifically adsorbable on the electrode. The charge flowing during time intervals $t \leqslant t'$ is described by relationship (16.1) and is the sum of the contributions of reductions of the substances diffusing to the electrode and adsorbed onto it, as well as the contribution of the double layer charging process.

When the potential returns to the value $E_i(t > t')$ the charge is described by the relationship

$$Q(t > t') = \frac{2nFD_{Ox}^{1/2}C_{Ox}^0}{\pi^{1/2}}[t^{1/2} - (t-t')^{1/2}]$$

$$+ nF\Gamma_{Ox}\left[\frac{2}{\pi}\sin^{-1}\left(\frac{t'}{t}\right)^{1/2}\right]. \quad (16.2)$$

From equations (16.1) and (16.2) we obtain the expression for charge $Q_r = Q - Q(t > t')$:

$$Q_r = \frac{2nFD_{Ox}^{1/2}C_{Ox}^0}{\pi^{1/2}}[(t-t')^{1/2} + (t')^{1/2} - t^{1/2}]$$

$$+ nF\Gamma_{Ox}\left[1 - \frac{2}{\pi}\sin^{-1}\left(\frac{t'}{t}\right)^{1/2}\right] + Q_{dl}. \quad (16.3)$$

Dividing both sides of this expression by the Cottrell equation:

$$Q_c = \frac{2nFD_{Ox}^{1/2}C_{Ox}^0(t')^{1/2}}{\pi^{1/2}} \quad (16.4)$$

gives the following relationship which can be used for the experimental determination of Γ_{Ox}:

$$\frac{Q_r}{Q_c} = \frac{Q_r - Q_{dl}}{Q_c} = \left[\left(\frac{t}{t'} - 1\right)^{1/2} + 1 - \left(\frac{t}{t'}\right)^{1/2}\right]$$

$$+ \frac{nF\Gamma_{Ox}}{Q_c}\left[1 - \frac{2}{\pi}\sin^{-1}\left(\frac{t'}{t}\right)^{1/2}\right]. \quad (16.5)$$

When both Ox and Red are adsorbable on the electrode surface, but Red is more weakly adsorbed, equation (16.3) becomes:

$$Q_r = \frac{2nFD_{Ox}^{1/2}C_{Ox}^0}{\pi^{1/2}}[(t-t')^{1/2} + (t')^{1/2} - t^{1/2}] +$$

$$nF(\Gamma_{Ox} - \Gamma_{Red})\left[1 - \frac{2}{\pi}\sin^{-1}\left(\frac{t'}{t}\right)^{1/2}\right] + nF\Gamma_{Red} + Q_{dl} \quad (16.6)$$

whereas equation (16.2) remains valid.

The authors [7] describe methods for employing these equations in the determination of Γ_{Ox} and Γ_{Red}. Details will be found in the original publication.

The methods described above have been used [8–13] by various investigators for the quantitative description of anion-induced adsorption of metal cations on mercury.

Chronoamperometry has been used for the same purpose [14], but its precision is lower than that of double potential step chronocoulometry.

Guidelli [15] described theoretically the chronoamperometric curves recorded at both flat and spherical electrodes, and assumed that the depolarizer and the electrode reaction product are adsorbed according to a linear isotherm.

The general theory of reversible processes limited by diffusion with adsorption of depolarizer taking place under potentiostatic conditions has been developed by Reinmuth and Balasubramanian [16]. Earlier these authors [17] carried out exact studies on adsorption governed by the Langmuir isotherm coupled with a reversible electrode reaction taking place on a flat electrode.

16.1.2 Polarography

The characterization of adsorption currents depends on the rate of the electrode process of the adsorbable depolarizer.

16.1.2.1 *Reversible Processes*

In the case where the substrate is strongly adsorbed on the electrode surface from dilute solutions only one wave is observed. The size of this wave is proportional to the depolarizer concentration up to a certain value, at which a new wave appears at a potential more positive than that of the first wave (when a reduction process is considered). Further increase of depolarizer concentration has no effect on the height of the first wave, but the height of the second wave increases linearly with increasing concentration. These changes are shown in Fig. 16.2.

The theory of the behaviour of the adsorption waves has been given by Brdička [18]. The interpretation of the adsorption waves of α-hydroxyphenazine by Müller [19, 20] is incorrect.

When the reduction of adsorbed depolarizer is reversible the wave is formed at potentials more negative than those corresponding to the reduction of unadsorbed molecules, or ions of depolarizer, since in the latter case additional energy is required to overcome the adsorption forces.

In Fig. 16.2 (curve c) the second wave is the adsorption wave, since it is formed at a more negative potential. The independence of this wave from the depolarizer concentration is due to the saturation of the electrode

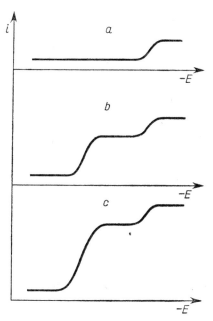

Fig. 16.2 Polarographic waves of reduction of a depolarizer which is specifically adsorbable. The depolarizer concentration increases from curve a to curve c.

surface with adsorbed molecules; although their number in the solution may increase, vacant sites no longer exist for them on the electrode surface.

The relation between the concentration of adsorbed molecules and their concentration in the bulk of solution is expressed by the Langmuir isotherm:

$$\Gamma = \frac{\beta_{ox}\Gamma_{\infty}[Ox]_0}{1+\beta_{ox}[Ox]_0} \tag{16.7}$$

where Γ_{∞} is the maximum value of Γ, which is usually defined as the number of moles of adsorbed substance per unit surface area, β_{ox} is the equilibrium constant of adsorption, and $[Ox]_0$ is the equilibrium volume concentration of the oxidized form at the electrode surface.

In the case where the equilibrium is rapidly established the charge consumed during the reduction of adsorbed depolarizer is:

$$q = nFA_t \Gamma = nFA_t \frac{\beta_{ox}\Gamma_\infty[Ox]_0}{1+\beta_{ox}[Ox]_0} \qquad (16.8)$$

where A_t is the electrode surface at the time t.

Differentiating charge q with respect to time gives the following expression for the instantaneous current:

$$i_t = \frac{dq}{dt} = nF \frac{\beta_{ox}\Gamma_\infty[Ox]_0}{1+\beta_{ox}[Ox]_0} \frac{2}{3} \times 0.85 m^{2/3} t^{-1/3}. \qquad (16.9)$$

Thus the average current is:

$$\bar{i} = \frac{1}{t_1} \int_0^{t_1} i_t \, dt = nF \frac{\Gamma_\infty \beta_{ox}[Ox]_0 \, 0.85 m^{2/3} t_1^{-1/3}}{1+\beta_{ox}[Ox]_0}. \qquad (16.10)$$

In the case of saturation of the electrode surface the unity term in the denominator of equation (16.10) can be neglected and the following simpler relationship for the average limiting current of adsorbed molecules is then obtained:

$$\bar{i}_a = nF\Gamma_\infty 0.85 m^{2/3} t_1^{-1/3}. \qquad (16.11)$$

The value of Γ_∞ can be found on this basis. The limiting adsorption current does not depend on depolarizer concentration. This is in agreement with the behaviour of the waves of surface active compounds described earlier.

The equation of the wave of an adsorbed substance is derived from the Nernst equation, since reversible processes are being considered:

$$E = E^0 + \frac{RT}{nF} \ln \frac{[Ox]_0}{[Red]_0}. \qquad (16.12)$$

Concentration $[Ox]_0$ can be determined from the equation describing the average adsorption current:

$$\bar{i} = K_{11}([Ox]-[Ox]_0) - nF \frac{2}{3} \times 0.85 m^{2/3} t^{-1/3} \frac{\beta_{ox}\Gamma_\infty[Ox]_0}{1+\beta_{ox}[Ox]_0} \qquad (16.13)$$

and the concentration $[Red]_0$ from the following simple relationship:

$$\bar{i} = K_{11}[Red]_0, \qquad (16.14)$$

since it is assumed that the reduced form is not adsorbed on the electrode.

16.1] Electrode Processes of Specifically Adsorbable Substances

Assuming that the diffusion coefficients of the two forms are equal the final relationship from equations (16.11)–(16.14) is obtained:

$$E = E^0 + \frac{RT}{nF} \times \tag{16.15}$$

$$\times \ln \left\{ \frac{\beta_{\text{Ox}}(\bar{i}_g - \bar{i}_a - \bar{i}) - K_{\text{II}} + \sqrt{[\beta_{\text{Ox}}(\bar{i}_g - \bar{i}_a - \bar{i}) + K_{\text{II}}]^2 + 4K_{\text{II}}\beta_{\text{Ox}}(\bar{i}_g - \bar{i})}}{2\beta_{\text{Ox}}\bar{i}} \right\}.$$

When the total limiting current $\bar{i}_g$ is smaller than the adsorption current $\bar{i}_a$ only one polarographic wave is observed at potentials more negative than E^0. This is the reduction wave of the adsorbed substance. When $\bar{i}_g > \bar{i}_a$ the second wave appears in the region of the standard potential, E^0. It corresponds to the reduction of the unadsorbed form.

If, in the case of an electroreduction, the depolarizer is not specifically adsorbed on the electrode surface, while the reduction product is adsorbed, the adsorption wave appears at a potential more positive than that of the diffusion wave, since the energy liberated by the adsorption causes a decrease in the amount of energy required for the reduction of the Ox to the Red form.

As before, at very small depolarizer concentrations only one polarographic wave is observed, which increases up to a certain depolarizer concentration. At higher concentrations this wave is independent of concentration, but the second wave, which increases with increasing concentration, appears at a more negative potential. The concentration dependence of these waves is shown schematically in Fig. 16.3.

It should be mentioned that the phenomena described above are observed when the adsorption energy of the product is large.

In this case the average adsorption current is:

$$\bar{i} = nF \frac{d\bar{A}}{dt} \frac{\beta_{\text{Red}} \Gamma_\infty [\text{Red}]_0}{1 + \beta_{\text{Red}}[\text{Red}]_0} . \tag{16.16}$$

Its limiting value is described by equation (16.11) in which Γ_∞ represents the number of moles of the Red form per unit surface area of the saturated electrode.

Equations of these polarographic waves can be readily derived from the Nernst equation (16.12). The concentration of the Ox form is obtained from the Ilkovič equation:

$$\bar{i} = \bar{i}_g - K_{\text{II}}[\text{Ox}]_0 . \tag{16.17}$$

The concentration $[\text{Red}]_0$ is given by the following relationship:

$$\bar{i} = K_{\text{I}1}[\text{Red}]_0 + nF\frac{2}{3} \times 0.85 m^{2/3} t_1^{-1/3} \frac{\beta_{\text{Red}} \Gamma_\infty [\text{Red}]_0}{1 + \beta_{\text{Red}}[\text{Red}]_0}, \qquad (16.18)$$

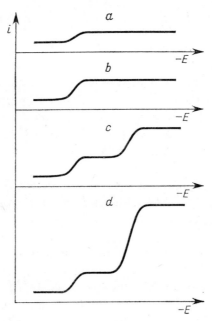

Fig. 16.3 Polarographic waves of reduction in which the product is specifically adsorbed on the electrode. The depolarizer concentration increases from curve a to curve d.

and combination of equations (16.11), (16.12), (16.17) and (16.18) gives the following relation between the electrode potential and the polarographic current:

$$E = E^0 + \frac{RT}{nF} \ln \frac{2\beta_{\text{Red}}(\bar{i}_g - \bar{i})}{\beta_{\text{Red}}(\bar{i} - \bar{i}_a) - K_{\text{I}1} + \sqrt{[K_{\text{I}1} - \beta_{\text{Red}}(\bar{i} - \bar{i}_a)]^2 + 4 K_{\text{I}1} \beta_{\text{Red}} \bar{i}}} \qquad (16.19)$$

where E^0 is close to the half-wave potential of the non-adsorption wave.

For certain conditions equation (16.19) can be reduced to a simpler form. Thus when:

$$[K_{\text{I}1} - \beta_{\text{Red}}(\bar{i} - \bar{i}_a)]^2 \gg 4 K_{\text{I}1} \beta_{\text{Red}} \bar{i} \qquad (16.20)$$

equation (16.19) becomes:

$$E = E^0 + \frac{RT}{nF} \ln \frac{(\bar{i}_g - \bar{i})K_{\text{II}} - \beta_{\text{Red}}(\bar{i} - \bar{i}_g)}{K_{\text{II}}\bar{i}}. \tag{16.21}$$

When the depolarizer concentration is such that $\bar{i}_a = \bar{i}_g$, then $\bar{i} = \frac{1}{2}\bar{i}_a$ at the half-wave potential of the adsorption wave.

Then the following relationship is obtained from equation (16.21):

$$E^a_{1/2} - E^0 = \frac{RT}{nF} \ln \left(1 + \frac{\beta_{\text{Red}}\bar{i}_a}{2K_{\text{II}}}\right) \tag{16.22}$$

where $E^a_{1/2}$ is the half-wave potential of the adsorption wave.

Since in the case under consideration the product is strongly adsorbed the unity term in equation (16.22) can be neglected. Putting $E^a_{1/2} - E^0 = \Delta E$ the following is obtained from this equation:

$$\Delta E = \frac{RT}{nF} \ln \frac{\beta_{\text{Red}}\bar{i}_a}{2K_{\text{II}}}. \tag{16.23}$$

It follows from equation (16.23) that the difference between the half-wave potentials of the diffusion and the adsorption waves is related through β_{Red} to the adsorption energy. Since this energy decreases with increasing temperature, ΔE also decreases with increasing temperature.

From equation (16.11) it follows that the height of the adsorption wave is proportional to the height of the mercury reservoir corrected for the back pressure, h_1, since:

$$\bar{i}_a = \text{const}\, m^{2/3} t_1^{-1/3} = \text{const}(k'h_1)^{2/3}(k''h_1^{-1})^{-1/3} = \overline{\text{const}\, h_1}. \tag{16.24}$$

This is an important criterion since it makes possible the ready distinction between adsorption currents, diffusion currents and kinetic currents. The second criterion is the concentration independence of the height of adsorption waves and the decrease of these waves with increasing temperature.

The adsorption of the depolarizer does not always cause the appearance of two waves. Sometimes only one wave is observed, but its shape is different from that of normal polarographic waves, since a large maximum is observed on the limiting current. Such maxima can be observed in the case of the so-called surface kinetic waves.

Now assume that substance M, which will adsorb onto the electrode, is present in the solution, and that in the adsorbed state it reacts with hydrogen ions and gives MH^+. The equilibrium of this reaction is heavily biased in the direction of M and only the protonated form MH^+ reacts with the electrode.

Assuming that the adsorption rate is determined by the transport rate, the solution is buffered, and that the adsorption is potential dependent according to the Frumkin equation [21, 22], then

$$\beta = \beta_0 \exp[-a(E-E^m)^2], \qquad (16.25)$$

$$\log \frac{i_k}{i_g - i_k} = Z - 0.43a(E-E^m)^2, \qquad (16.26)$$

where $a = \dfrac{(\overline{C}-\overline{C'})}{RT\Gamma_\infty}$ ($\overline{C}$ and $\overline{C'}$ being the integral capacities of the double layer in the absence of the surface active agent and at complete saturation of the electrode, respectively), β is the adsorption equilibrium constant and β_0 its maximum value at the maximum adsorption potential E^m, and Γ_∞ is the number of moles of adsorbed substance per cm^2 of saturated electrode surface area.

In equation (16.26) Z is given by the relationship:

$$Z = \log \frac{nFA\bar{k}_1 \beta_0 \Gamma_\infty}{i_g} \qquad (16.27)$$

where k_1 is the protonation rate constant of the adsorbable substance.

Expressing the surface area of the dropping mercury electrode in terms of drop parameters and using the Ilkovič equation, the following expression for the average value of $\bar{Z}$ is obtained:

$$\bar{Z} = \log \frac{0.85 \bar{k}_1 \beta_0 \Gamma_\infty t_1^{1/2}}{D^{1/2}}. \qquad (16.28)$$

From equation (16.26) it follows that the limiting kinetic current i_k of the surface kinetic wave decreases with increasing electrode potential.

Examples of substances reacting with the electrode according to this mechanism and equation (16.26) are given by Mairanovskii [22]. Mairanovskii et al. [23] derived the following expression for the average current of a bulk-surface wave in the case where the preceding reaction takes place at a comparable rate on the surface of the electrode and in the bulk reaction layer:

$$\frac{\bar{i}_k}{\bar{i}_g - \bar{i}_k} \frac{1}{t_1^{1/2}} = \frac{0.46(\bar{k}_1 \beta \Gamma_\infty + k_1 D^{1/2} K^{1/2}) t_1^{1/2}}{\beta \Gamma_\infty + 0.60 D^{1/2} t_1^{1/2}} \qquad (16.29)$$

where K is the equilibrium constant of protonation and $\bar{k}_1$ is the protonation rate constant in the reaction layer. If follows from this equation that when $\left(\dfrac{\bar{i}_k}{i_g - \bar{i}_k}\right)\left(\dfrac{1}{t_1^{1/2}}\right)$ depends on the drop time the adsorbed electrode-inactive form of the depolarizer participates in the electrode process. However when this time dependence is not observed the possibility of the adsorption of this form of depolarizer cannot be excluded.

A detailed discussion of the consequences of equation (16.29) can be found in the original publications [22, 23].

16.1.2.2 *Irreversible Processes*

In the case of irreversible electrode processes two waves can be observed only when the substrate or the product is strongly adsorbed on the electrode surface, but in this case the adsorption of the product does not always facilitate the reduction process. In certain cases the adsorbed product hinders the reduction at the saturated electrode surface [24, 25].

Kemula and Cisak [26] investigated the irreversible reduction of dibromocyclohexanes and came to the conclusion that in the case of irreversible processes the adsorption of the substrate can facilitate reduction by decreasing the activation energy of the electrode process.

Guidelli [27] made a theoretical investigation of the effect of the adsorption of the substrates and the products of electrode processes which obeyed the Langmuir isotherm, on the charge transfer in polarography. He considered processes having various standard rate constants and with adsorption equilibrium constants of various magnitudes for both Red and Ox forms. The results of this work make it possible to predict when, in the case of irreversible processes, the additional adsorption wave will be formed.

16.1.3 STATIONARY ELECTRODE VOLTAMMETRY

Stationary electrode voltammetry was first applied to studies of adsorption by Mirri and Favero [28] in 1958. Later this method was used by Kemula, Kublik and Axt [29] and by Hartley and Wilson [30]. These qualitative studies showed that stationary electrode voltammetry is a very valuable method for investigations of adsorption processes and that theoretical studies, leading to better understanding of experimental results, were needed. Such theoretical studies were carried out by Wopschall and Shain [31], who considered reversible electrode processes, assuming rapid adsorption and desorption and the Langmuir type of adsorption. It is difficult to express the solution obtained by these workers in an analytical form, but it can be used for predicting the dependence of adsorption currents on various experimental conditions.

There are four principal kinds of adsorption currents corresponding to the following conditions:

(1) the depolarizer is weakly adsorbed;
(2) the product is weakly adsorbed;
(3) the depolarizer is strongly adsorbed;
(4) the product is strongly adsorbed.

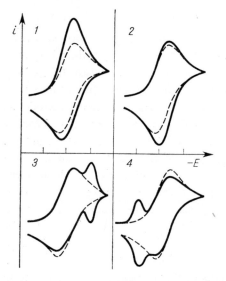

Fig. 16.4 Cyclic stationary electrode voltammetric curves in the case of adsorption of the substrate or the product. Curve *1*—weak adsorption of the substrate; curve *2*—weak adsorption of the product; curve *3*—strong adsorption of the substrate; curve *4*—strong adsorption of the product. Broken lines show the curves of the diffusion process (from [31] by permission of the copyright holders, the American Chemical Society).

The curves typical for these cases according to Wopschall and Shain are shown in Fig. 16.4. They are compared with the cyclic curves of simple diffusion-controlled processes.

It is desirable to discuss briefly the properties characteristic of the above adsorption currents.

(1) When the depolarizer is weakly adsorbed the cathodic current also increases during the return half-cycle of the anodic current, but this

increase is larger (Fig. 16.4, curve *1*) and can lead to serious errors in analyses based on this method, while certain changes of the current peak shape can lead to an erroneous interpretation of the reduction mechanism. Obviously these errors occur when the recorded currents, which are very similar to current peaks of diffusion controlled processes, are regarded as diffusion currents. Both kinds of current can be recognized by plotting the curves at different polarization rates. The ratio of adsorption to diffusion currents increases with increasing polarization rate, since under such conditions a considerable fraction of the reducible ions or molecules is adsorbed on the electrode and does not need to be supplied by diffusion. The relation between this current ratio and the polarization rate is shown in Fig. 16.5.

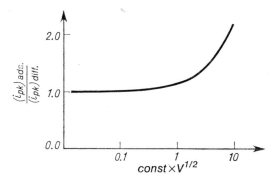

Fig. 16.5 Dependence of the ratio of adsorption to diffusion currents on the polarization rate in the case of weak adsorption of the depolarizer (from [31] by permission of the copyright holders, the American Chemical Society).

In this case the effect of adsorption on the recorded current peaks is the most pronounced at low depolarizer concentrations.

(2) The weak adsorption of the product has a slight effect on cathodic curves (Fig. 16.4, curve *2*). When the polarization rate increases the current peak is shifted in the positive direction and the cathodic current peak decreases slightly.

In cyclic stationary electrode voltammetry the anodic current increases with increasing rate of polarization as a result of retention of the product on the electrode surface.

(3) In the case of strong adsorption of the substrate two current peaks are recorded (Fig. 16.4, curve *3*). The current peak observed at less negative potentials is connected with reduction of diffusing depolarizer, and the second peak, observed at more negative values, corresponds to the reduction of the adsorbed depolarizer. The potential difference between these peaks

increases with increasing energy of adsorption of the product. Sometimes the adsorption peaks are very sharp, and their width at the half-height can be equal to only a few millivolts. After reaching the maximum the adsorption current decreases rapidly to zero.

The ratio of adsorption to diffusion currents increases with increasing rate of polarization. The maximum value of the adsorption current function increases and that of the diffusion function decreases with increasing V. When the scan rate is very large the maximum value of the diffusion current function can be small, compared with the maximum of the adsorption current function. Both the diffusion and the adsorption peaks are shifted slightly in the negative direction when the scan rate is increased.

The ratio of adsorption to diffusion controlled current decreases with increasing depolarizer concentration and the adsorption peak is shifted in the positive direction.

(4) The case of a strong adsorption of the product is similar to that of strong adsorption of the substrate, but in this case the adsorption current peak is observed at a potential more positive than that of the diffusion current peak (Fig. 16.4, curve *4*). When either the depolarizer, or the product, is strongly adsorbed, it is possible to separate the properties of the adsorption controlled peak from those of the diffusion controlled one. Therefore the results obtained in case (3) above can be used for an interpretation of the currents in the case of a strong adsorption of the product.

These currents can be very readily distinguished from other kinds of current, when the substrate or the product of the electrode process is strongly adsorbed, since in such cases adsorption peaks are formed. When the adsorption is weak the changes of peak height, due to changes of depolarizer concentration, or of polarization rate, should be carefully observed.

Wopschall and Shain [32] studied the validity of the above theory by examining the adsorption currents of methylene blue. They also considered theoretically [33] the reduction of an adsorbable depolarizer followed by an irreversible chemical first order reaction, and they used the results of these studies in an investigation of the mechanism of reduction of azobenzene followed by the benzidine rearrangement.

Hulbert and Shain [34] made a theoretical investigation of stationary electrode voltammetric processes in which the adsorption of depolarizer does not reach the equilibrium state.

Feldberg [35] considered in detail the effect of adsorption on the shape of cyclic stationary electrode voltammetric curves. Unlike Wopschall and Shain he did not limit his considerations to adsorption governed by the Langmuir isotherm but by means of examples he discussed systems which can be described by the Frumkin isotherm. Frumkin considered redox systems which are adsorbed as well as those which are not adsorbed on the electrode. He assumed that the adsorbed and the non-adsorbed forms are at equilibrium (e.g. $C_{Ox} \stackrel{B_{Ox}}{\rightleftharpoons} \Gamma_{Ox}$) which depends on equilibrium constant B_{Ox}. He also considered different electron transport rates in redox systems. The results of these considerations are illustrated by curves obtained on the assumption that the interaction constant in the Frumkin isotherm varies from 3.0 to -3.0. Frumkin also considered the kinetics of adsorption in the absence of Faradaic reaction.

Adsorption of depolarizer was also investigated by Laviron [36] using the stationary electrode voltammetric method. This author also considered a reversible electrode process in which the substrate and the product are strongly adsorbed and the electrode reaction product is the substrate of a first-order chemical reaction.

The results of these theoretical studies can be applied to the same reaction scheme in thin-layer voltammetry [37].

The problem of depolarizer adsorption in stationary electrode voltammetry was also considered by Nesterov and Korovin [38].

16.1.4 Chronopotentiometry

Adsorption processes in chronopotentiometry have been investigated by several authors. The first study was published by Lorenz [39] and this was followed by that of Reinmuth [40]. Anson [41], Laitinen [42], and Bard [43] also looked into these processes. The results of these studies have been systematically reviewed by Tatwawadi and Bard [44] in their paper on the chronopotentiometric investigation of the adsorption of riboflavine on a mercury electrode; their calculations were criticized by Lingane [45]. The studies carried out by all these authors related to four kinds of adsorption processes, depending on the mechanism of continued reduction of compounds in the layer adjoining the electrode. These processes are discussed below.

(1) The adsorbed layer is electrolysed at the beginning of the process. At first the current is consumed by reduction of adsorbed molecules or ions, and then the depolarizer diffusing to the electrode is reduced. When

its concentration at the electrode decreases to zero the transition time is described by the following relationship:

$$i_0 \tau = \frac{n^2 F^2 \pi D (C_{Ox}^0)^2}{4 i_0} + nF\Gamma. \qquad (16.30)$$

This equation indicates that the dependence of $i_0 \tau$ on $(C_{Ox}^0)^2/i_0$ should be linear. By extrapolation to $i_0 = 0$, $nF\Gamma$ is obtained, giving the concentration of the adsorbed substance on the surface. Equation (16.30) was derived by Lorenz [39] and this adsorption model was suggested by Laitinen [42].

(2) The adsorbed layer is electrolysed at the end of the process.

At the beginning of electrolysis the substance diffusing to the electrode from the bulk of the solution is reduced. When its concentration on the electrode surface decreases to zero, time τ_1 is reached and the reduction of the adsorbed layer begins. This is accompanied by electrolysis of the depolarizer diffusing to the electrode from the bulk of the solution.

Lorenz [39] developed a similar model, assuming that the current applied in the electrolysis of the diffusing substance, during the electrolysis of the adsorbed layer, changes linearly with $t^{-1/2}$ and he obtained the following equation:

$$(i_0 \tau)^{1/2} = \frac{nF(\pi D)^{1/2} C_{Ox}^0}{2 i_0^{1/2}} + (nF\Gamma)^{1/2}. \qquad (16.31)$$

The dependence of $(i_0 \tau)^{1/2}$ on $C_{Ox}^0/i_0^{1/2}$ is linear and the extrapolation to $C_{Ox}^0/i_0^{1/2} = 0$ gives $(nF\Gamma)^{1/2}$.

Reinmuth [40] obtained a more accurate equation:

$$\frac{nF\pi\Gamma}{i_0} = \tau \arccos \frac{\tau_1 - \tau_2}{\tau} - 2(\tau_1 \tau_2)^{1/2} \qquad (16.32)$$

where τ_1 is the first transition time described by the Sand equation, and $\tau = \tau_1 + \tau_2$.

Another form of this equation was derived by Anson [41].

(3) Electrolysis of the adsorbed substance takes place simultaneously with that of the substance diffusing to the electrode.

Lorenz [39] used a simple model for this case in which the ratio of the currents consumed in the electrolysis of the diffusing and the adsorbed substances is constant. He obtained the following relationship:

$$\frac{i_0 \tau}{C_{Ox}^0} = \frac{nF(\pi D)^{1/2} \tau^{1/2}}{2} + \frac{nF\Gamma}{C_{Ox}^0}. \qquad (16.33)$$

Γ can be readily determined from the linear relation between $i_0 \tau/C_{Ox}^0$ and $\tau^{1/2}$.

(4) Electrolysis of the adsorbed substance takes place simultaneously with that of the substance diffusing to the electrode and the system obeys a linear isotherm.

The equation describing this case was derived by Lorenz [39]. After integration it can be expressed in the following form:

$$\Gamma = \frac{2C_{Ox}^0(D_{Ox}\tau/\pi)^{1/2} - nF(C_{Ox}^0)^2 D_{Ox}/i_0}{1 - \exp(a^2)\operatorname{erfc} a} \qquad (16.34)$$

where $a = C_{Ox}^0(D_{Ox}\tau)^{1/2}/\Gamma$.

When $C_{Ox}^0(D_{Ox}\tau)^{1/2} \gg \Gamma$ the term $\exp(a^2)\operatorname{erfc}(a)$ approaches zero and equation (16.34) yields the following relationship:

$$C_{Ox}^0 \tau^{1/2} = \frac{nF(\pi D_{Ox})^{1/2}(C_{Ox}^0)^2}{2i_0} + \frac{\pi^{1/2}\Gamma}{2D_{Ox}^{1/2}}. \qquad (16.35)$$

Γ can be readily determined from the graph of $C_{Ox}^0 \tau^{1/2}$ vs. $(C_{Ox}^0)^2/i_0$.

Other models can be obtained using other isotherms, but difficulties in solving the Fick equation then appear. All the models using isotherms are based on the assumption that both adsorption and desorption are fast and that adsorption equilibrium is maintained during the electrolysis.

The value of Γ determined usually depends on the kind of model selected. In the method which allows Γ to be determined irrespective of the model, i_0 should be so large that the contribution of diffusion to the recorded transition times can be neglected. Therefore $i_0\tau$ becomes practically constant.

According to Lingane [46] chronopotentiometry is not a convenient method in studies of adsorption, since it is difficult to obtain accurate transition times. The method is also tedious. Lingane therefore recommends chronocoulometry as a superior method for investigation of these phenomena.

16.2 Electrode Processes in the Presence of Adsorbed Electrode-Inactive Substances

Surface active substances can either decrease or increase the rates of the electrode processes of other substances. This is often observed when the adsorbed substances are the inorganic anions—chloride, bromide, iodide or thiocyanate. The accelerating effect of halides on the kinetics of hydrogen ion discharge has been known for many years [47], and lately, the catalytic effect of adsorbed halide ions on the discharge of various cations has been described.

Adsorption of neutral organic molecules usually decreases the rate of the electrode process in the region of the adsorption potential. These effects have been discussed in review articles [48–50].

The effect of adsorption on electrode processes is more complex in polarography than in other methods, since the adsorption equilibrium cannot be reached at the changing surface of growing mercury drops. This is especially the case at slow rates of adsorption.

The effect of diffusion on adsorption equilibrium at very fast adsorption rates was investigated by Delahay and Trachtenberg [51] for systems obeying the Henry isotherm and by Delahay and Fike [52] for those obeying the Langmuir isotherm. The solution of the latter problem was criticized by Reinmuth [53] who considered in detail the diffusion to a plane with Langmuir adsorption. Delahay and Fike gave also a formal solution for a shifting plane but they did not carry out detailed calculations for this case. The result obtained by Delahay and Trachtenberg may be given by the following relationship:

$$\Gamma = \Gamma_r - \Gamma_r \exp\frac{Dt}{\beta^2} \operatorname{erfc}\frac{(Dt)^{1/2}}{\beta} \tag{16.36}$$

where Γ_r is the equilibrium surface concentration at the given concentration in the bulk of solution, and β is the isotherm constant.

The results obtained by Delahay and Fike are qualitatively similar to those predicted by equation (16.36).

An important result of these studies is the conclusion that adsorption equilibria are established over a long period. For the values of D (4×10^{-6} cm^2/s) and β (4×10^{-3}) cm usually encountered, the ratio Γ/Γ_r reaches the value 0.99 over a time greater than 10^4 s. Obviously this long period is shortened as a result of convection, and therefore, according to Parsons [54], the time dependence of the surface concentration of adsorbed substance is more correctly expressed by the following equation:

$$\Gamma = \Gamma_r \left[1 - \exp\left(-\frac{Dt}{\beta\delta}\right)\right] \tag{16.37}$$

where δ is the thickness of the diffusion layer.

According to this equation the equilibrium is reached more rapidly. When D and β have the values previously quoted and $\delta = 0.05$ cm, the value of $\Gamma/\Gamma_r = 0.99$ is reached during a period of 230 s. Obviously the value of δ has been taken quite arbitrarily as the diffusion layer thickness in the presence of natural convection.

16.2] Influence of Adsorbed Electrode-Inactive Substances 415

The above discussion was limited to flat stationary electrodes. Delahay and Trachtenberg [55] considered the time dependence of Γ/Γ_r in the case of mercury dropping electrodes. Figure 16.6 shows that, in this case, the equilibrium is established much more rapidly. The concentrations are expressed in the form of the ratio of concentration of the adsorbable substance in the bulk of solution to parameter $1/\beta$ of the Langmuir isotherm. The results shown in

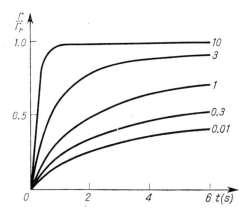

Fig. 16.6 Time dependence of Γ/Γ_r in the case of adsorption on a growing mercury drop for various concentrations of the adsorbable compound, expressed in the form of $C^0\beta$. The curves are calculated for $D = 5 \times 10^{-6}$ cm^2/s and $\Gamma_m = 5 \times 10^{-10}$ mole/cm^2 (from [52] by permission of the copyright holders, the American Chemical Society).

Fig. 16.6 indicate that the equilibrium is only reached during the drop time, when the concentration in the bulk of solution is greater than $10/\beta$.

The effect of the diffusion rate on the time in which the equilibrium described by the Frumkin isotherm is established was considered by Rampazzo [56].

The kinetics of adsorption in cases when the diffusion rate is much faster has also been examined [57]. This subject is not considered here since the available results of polarographic studies indicate that adsorption kinetics is not the limiting factor in the establishment of adsorption equilibrium.

A theory giving the complete and general description of the dependence of the kinetics of electrode processes on the presence of a layer adsorbed on the electrode surface and the character of the adsorption, is still lacking.

The experimental results obtained by various electrochemical methods have been interpreted on the assumption that the dependence of the standard rate constant on the degree of electrode coating is described by the equation:

$$k_{s,\Theta} = k_{s,0}(1-\Theta)+k_{s,1}\Theta \qquad (16.38)$$

where $k_{s,\Theta}$, $k_{s,0}$ and $k_{s,1}$ are the standard rate constants for the surface coverage Θ, 0 and 1, respectively.

This equation implies that the electrode reaction rate is different on different parts of the surface and that the rates on the completely coated and uncoated parts are proportional to the depolarizer concentration at the electrode surface.

When $k_{s,1} < k_{s,0}$ the rate of the electrode process is slowed down by the presence of the compound adsorbed on the electrode surface. When $k_{s,1} > k_{s,0}$ the adsorbed compound accelerates the rate of the electrode process.

Adsorption of neutral organic substances generally inhibits the electrode process. According to Frumkin [50] the mechanism of this inhibition changes as the degree of electrode coverage increases from zero to one hundred per cent. In this case the dependence of $k_{s,\Theta}$ on Θ is probably not linear over the whole range of Θ values.

The change of the inhibiting mechanism may be connected with the dependence of the transition coefficient on Θ.

The inhibition of electrode processes has been explained in various ways, such as:

(a) the "blocking effect", consisting of a decrease in the electrode surface area on which the process can take place;

(b) the additional activation energy which is required by the depolarizer in finding a site in the adsorbed layer [58] where it can react with the electrode;

(c) the electrostatic effect consisting in the change of the effective potential difference between the electrode and the closest approach plane in the presence of adsorbed molecules [59];

(d) the effect on the chemical reaction leading to the formation of a depolarizer reacting directly with the electrode surface.

The first quantitative description of the effect of surface active substances on electrode processes taking place on mercury dropping electrodes was produced by Weber, Koutecký and Koryta [60].

Equation (16.38) has been used for interpreting the effect of the degree of electrode coating on the kinetics of electrode processes.

Aramata and Delahay [61] investigated the effect of n-amyl alcohol on the kinetics of zinc(II) discharge and found that the relation between the exchange current and Θ was linear only at low values of Θ. A similar non-linear relationship was observed by Biegler and Laitinen [62] in the case of the effect of n-butanol, thymol and leucoriboflavin on the kinetics of Cd^{2+} and $HPbO_2^-$ discharges.

Niki and Hackerman [63] found that the slowing down effect of aliphatic alcohols on the $V^{3+} + e \rightleftharpoons V^{2+}$ electrode reaction is much stronger than that predicted by the relationship:

$$k_{s,\Theta} = k_{s,0}(1-\Theta) \qquad (16.39)$$

which is obtained from equation (16.38) by assuming that the rate of the process on the covered surface is much slower than that of the process on a free surface.

Sathyanarayana [64] investigated the possibility of obtaining a method which would allow the analysis of the polarographic waves of irreversible processes in the presence of adsorbed neutral molecules, without making assumptions regarding the mechanism of inhibition of the polarographic process. He found that the effect of butanol on the rates of reduction of Cu^{2+}, Cd^{2+} and Zn^{2+} on mercury electrodes can be expressed by the relationship:

$$k_{s,\Theta} = k_{s,0}(1-\Theta)^b \qquad (16.40)$$

where b (greater than one) is related, according to him to the repulsion effect on the electrode surface.

A similar relationship was derived by Niki and Hackerman [65] but in their equation the meaning of constant b is different.

Parsons [66] has published a more general description of the effect of specific adsorption on the rates of electrode processes where he explained also the significance of parameter b. Recently the problem of electrode reaction inhibition was discussed by Lipkowski and Galus [67] in the light of some of the more important concepts and experimental work.

References

[1] Brdička, R. and Knobloch, E., *Z. Elektrochem.*, **47**, 721 (1941).
[2] Brdička, R., *Z. Elektrochem.*, **48**, 278 (1942).
[3] Anson, F. C., *Anal. Chem.*, **36**, 932 (1964).
[4] Christie, J. H., Lauer, G. and Osteryoung, R. A., *J. Electroanal. Chem.*, 7, 60(1964).
[5] Osteryoung, R. A. and Anson, F. C., *Anal. Chem.*, **36**, 975 (1964).
[6] Anson, F. C., *Anal. Chem.*, **38**, 54 (1966).
[7] Christie, J. H., Osteryoung, R. A. and Anson, F. C., *J. Electroanal. Chem.*, **13**, 236 (1967).
[8] Murray, R. W. and Gross, D. J., *Anal. Chem.*, **38**, 392 (1966).

[9] Anson, F. C., Christie, J. H. and Osteryoung, R. A., *J. Electroanal. Chem.*, **13**, 343 (1967).
[10] O'Dom, G. W. and Murray, R. W., *J. Electroanal. Chem.*, **16**, 327 (1968).
[11] Barclay, D. J. and Anson, F. C., *J. Electrochem. Soc.*, **116**, 438 (1969).
[12] Kowalski, Z. and Anson, F. C., *J. Electrochem. Soc.*, **116**, 1208 (1969).
[13] Barclay, D. J. and Anson, F. C., *J. Electroanal. Chem.*, **28**, 71 (1970).
[14] Caselli, M. and Papoff, P., *J. Electroanal. Chem.*, **23**, 41 (1969).
[15] Guidelli, R., *J. Electroanal. Chem.*, **18**, 5 (1968).
[16] Reinmuth, W. H. and Balasubramanian, K., *J. Electroanal. Chem.*, **38**, 271 (1972).
[17] Reinmuth, W. H. and Balasubramanian, K., *J. Electroanal. Chem.*, **38**, 79 (1972).
[18] Brdička, R., *Coll. Czechoslov. Chem. Communs.*, **12**, 522 (1947).
[19] Müller, O. H., *J. Biol. Chem.*, **145**, 425 (1942).
[20] Müller, O. H., *J. Electrochem. Soc.*, **87**, 441 (1942).
[21] Mairanovskii, S. G., *Electrochim. Acta*, **9**, 803 (1964).
[22] Mairanovskii, S. G., "*Catalytic and Kinetic Waves in Polarography*", Nauka, Moscow 1966, page 167 (in Russian).
[23] Mairanovskii, S. G., Gul'tyai, V. P. and Lishcheta, L. I., *Elektrokhimiya*, **2**, 693 (1966).
[24] Schmidt, R. W. and Reilley, C. N., *J. Am. Chem. Soc.*, **80**, 2087 **(1**958).
[25] Laviron, E., *Bull. soc. chim. France*, 418 (1962); Laviron, E. and Degrand, C., *Bull. soc. chim. France*, 865 (1964); 2194 (1966); Degrand, C. and Laviron, E., *Bull. soc. chim. France*, 4603 (1969).
[26] Kemula, W. and Cisak, A., *Roczniki Chem.*, **31**, 337 (1957).
[27] Guidelli, R., *J. Phys. Chem.*, **74**, 95 (1970).
[28] Mirri, A. M. and Favero, P., *Ricerca sci.*, **28**, 2307 (1958).
[29] Kemula, W., Kublik, Z. and Axt, A., *Roczniki Chem.*, **35**, 1009 (1961).
[30] Hartley, A. M. and Wilson, G. S., *Anal. Chem.*, **38**, 681 (1966).
[31] Wopschall, R. H. and Shain, I., *Anal. Chem.*, **39**, 1514 (1967).
[32] Wopschall, R. H. and Shain, I., *Anal. Chem.*, **39**, 1527 (1967).
[33] Wopschall, R. H. and Shain, I., *Anal. Chem.*, **39**, 1535 (1967).
[34] Hulbert, M. H. and Shain, I., *Anal. Chem.*, **42**, 162 (1970).
[35] Feldberg, S. W., in "*Electrochemistry. Calculation, Simulation, Instrumentation*", Matson, J.S., Mark, H. B., Jr. and MacDonald, H. C., Eds., Dekker, New York 1972, Vol. 2, page 185.
[36] Laviron, E., *Bull. soc. chim. France*, 3717 (1967); 2256 (1968).
[37] Laviron, E., *J. Electroanal. Chem.*, **35**, 333 (1972).
[38] Nesterov, B. P. and Korovin, N. V., *Elektrokhimiya*, **5**, 1405 (1969).
[39] Lorenz, W., *Z. Elektrochem.*, **59**, 730 (1955).
[40] Reinmuth, W. H., *Anal. Chem.*, **33**, 322 (1961).
[41] Anson, F. C., *Anal. Chem.*, **33**, 1123 (1961).
[42] Laitinen, H. A., *Anal. Chem.*, **33**, 1458 (1961).
[43] Bard, A. J., *Anal. Chem.*, **35**, 340 (1963).
[44] Tatwawadi, S. V. and Bard, A. J., *Anal. Chem.*, **36**, 2 (1964).
[45] Lingane, J. J., *Anal. Chem.*, **39**, 541 (1967).
[46] Lingane, J. J., *Anal. Chem.*, **39**, 485 (1967).
[47] Iofa, Z. A., Kabanov, B., Kuczinsky, E. and Tchistyakov, F., *Acta Physicochimica U.R.S.S.*, **10**, 317 (1939).

[48] Nürnberg, H. W. and von Stackelberg, M., *J. Electroanal. Chem.*, **4**, 26 (1962).
[49] Reilley, C. N. and Stumm, W., in *"Progress in Polarography"*, Zuman P., Ed., Interscience, New York 1962, Vol. 1, page 81.
[50] Frumkin, A. N., *Electrochim. Acta*, **9**, 465 (1964).
[51] Delahay, P. and Trachtenberg, J., *J. Am. Chem. Soc.*, **79**, 2355 (1957).
[52] Delahay, P. and Fike, C. T., *J. Am. Chem. Soc.*, **80**, 2628 (1958).
[53] Reinmuth, W. H., *J. Phys. Chem.*, **65**, 473 (1961).
[54] Parsons, R., *Adv. Electrochem. Electrochem. Eng.*, **1**, 27 (1961).
[55] Delahay, P. and Trachtenberg, J., *J. Am. Chem. Soc.*, **80**, 2094 (1958).
[56] Rampazzo, L., *Electrochim. Acta*, **14**, 733 (1969).
[57] Parsons, R., *Adv. Electrochem. Electrochem. Eng.*, **1**, 20 (1961).
[58] Gierst, L., Bermane, D. and Corbusier, P., *Ricerca sci.*, **29**, Suppl. Contributi di Polarografia, 1959, p. 75.
[59] Kuta, J. and Weber, J., *Electrochim. Acta*, **9**, 541 (1964).
[60] Weber, J., Koutecký, J. and Koryta, J., *Z. Elektrochem.*, **63**, 583 (1959).
[61] Aramata, A. and Delahay, P., *J. Phys. Chem.*, **68**, 880 (1964).
[62] Biegler, T. and Laitinen, H. A., *J. Electrochem. Soc.*, **113**, 852 (1966).
[63] Niki, K. K. and Hackerman, N., *J. Phys. Chem.*, **73**, 1023 (1969).
[64] Sathyanarayana, S., *J. Electroanal. Chem.*, **10**, 119 (1965).
[65] Niki, K. K. and Hackerman, N., *J. Electroanal. Chem.*, **32**, 257 (1971).
[66] Parsons, R., *J. Electroanal. Chem.*, **21**, 35 (1969).
[67] Lipkowski, J. and Galus, Z., *J. Electroanal. Chem.*, **61**, 11 (1975).

Chapter **17**

Cyclic Methods. Diffusion Processes

In the preceding chapters processes in which the indicator electrode was polarized in one direction only have been discussed. The potential changed from the initial value in either the positive or the negative direction, but the direction of polarization was not changed during the experiment.

It is possible, however, at least in some of the techniques discussed, to change the direction of the potential applied to the indicator electrode, or the direction of current flow. In such a case the electrode process of the primary electrode product can be studied after changing the direction of polarization.

If it is assumed that an oxidant is present in the solution, and that it is reduced during cathodic polarization in the vicinity of the electrode to the Red form:

$$Ox + ne \rightleftharpoons Red \quad \text{(primary process)}, \quad (17.1)$$

the reverse process is observed when the polarization direction is reversed:

$$Red - ne \rightleftharpoons Ox \quad \text{(secondary process)}, \quad (17.2)$$

Obviously, the experiment can be continued when the cycle is completed and when in the secondary process the electrode reaches the initial potential the cathodic polarization, at which reaction (17.1) takes place at the electrode, can be started again. The experiment can be continued according to this scheme, and as a result the electrode can be polarized in a cyclic manner.

It is necessary to consider what advantages can be obtained from uni- or multicyclic electrode polarization, and whether the advantages justify the modifications which have to be introduced into the apparatus, even though such modifications are relatively slight.

It is easy to show the usefulness of cyclic methods. They make it possible to carry out electrode studies on substances which are difficult to obtain on a macro scale. Amalgams, and especially amalgams of metals having relatively negative standard redox potentials, are examples of such substances. Preparation of unstable amalgams for further investigation, for instance by polarography, is difficult and tedious. On the other hand, amalgams of metals can be easily prepared by reduction of their ions at a mercury microelectrode:

$$Me^{n+} + ne \rightarrow Me(Hg). \quad (17.3)$$

The amalgam formed in the primary half-cycle can be studied after changing the direction of polarization, since the metal is then oxidized out of the amalgam:

$$Me(Hg) - ne \rightarrow Me^{n+}. \quad (17.4)$$

In electrochemical studies of organic compounds also the investigation of the behaviour of the primary product in the secondary reaction often leads to conclusions concerning the mechanism of the primary electrode reaction.

Moreover, cyclic methods make it possible to investigate the kinetics of chemical reactions taking place after the charge transfer in the primary electrode process. Assuming electroreduction as the primary process and a first-order secondary chemical reaction we have:

$$Ox + ne \rightarrow Red \xrightarrow{k_1} A \quad \text{(primary process)} \quad (17.5)$$

where A is electrode inactive in the region of Ox/Red reaction potentials.

In the secondary process the oxidation of that part of the Red form which was not transformed into A during the primary process is studied. Obviously, when the rate of transformation of Red into A is fast, and the duration of the primary experiment is long, the surface concentration of the Red form can remain close to zero. Thus conclusions regarding the rate of the Red $\xrightarrow{k_1}$ A reaction can be drawn from the measured current or transition time values of the secondary process.

Generally, in such processes electrodes with constant surfaces should be used. In practice it is sufficient to employ electrodes with surface which practically do not change during a single cycle. Therefore, dropping electrodes with drop times of several seconds can be used, but in such a case the duration of the experiment should not exceed several hundredths of a second. If the cycle time was longer than this, for example, twice as long as the drop time, then after the primary process, the drop would break off, taking with it the electrode reaction products. Therefore no interesting information could be obtained from the investigation of the second half of the cycle.

Cyclic methods have been developed mainly for use with stationary electrode voltammetry and chronopotentiometry. A variant of chronoamperometry also belongs to this group. In this technique the potential changes stepwise, initially to the value at which the electrode reaction of the depolarizer present in the solution takes place. It then changes again stepwise, to the value at which the primary electrode reaction product reacts in another electrode reaction.

In this chapter the principles of the rotating ring-disc technique will also be discussed. In this case the electrode potential does not change in the cyclic manner, but due to the electrode rotation the reaction product formed at the inner electrode during the measurement is rapidly transferred into the field of the ring electrode having a potential different from that of the inner electrode.

17.1 Cyclic Chronoamperometry and Double Step Chronocoulometry

In cyclic chronoamperometry the electrolysis potential of the substance being studied is applied to the electrode, and then it is changed stepwise to the value at which the product formed in the first part of the cycle reacts with the electrode. This potential can be changed back again to the initial value. The potential changes are schematically shown in Fig. 17.1.

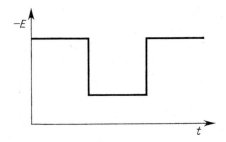

Fig. 17.1 Time dependence of the potential in cyclic chronoamperometry.

The application of this method to diffusion controlled processes is not of great importance, compared with stationary electrode voltammetry or cyclic chronopotentiometry. It is potentially more important in kinetic studies of the chemical reactions of primary electrode products. In this case the potential can be changed as shown in Fig. 17.1. Linear potential changes in the positive direction [1] can also be used after an initial potential jump to the value at which the reduction process limited by diffusion alone takes place.

Kimmerle and Chevalet [2] gave a theoretical description of the currents for the case when reactions (17.1) and (17.2) are controlled by transport

or electron transfer kinetics. They considered all the four possible cases:
(1) reactions (17.1) and (17.2) are transport controlled;
(2) (17.1) kinetic controlled, (17.2) transport controlled;
(3) (17.1) transport controlled, (17.2) kinetic controlled;
(4) both reactions controlled by the kinetics of electron exchange.

Obviously, in the first case the current of reaction (17.1) is described by the Cottrell equation:

$$i_f = \frac{nFAD_{Ox}^{1/2} C_{Ox}^0}{(\pi t)^{1/2}}, \qquad (17.6)$$

provided that the potential applied to the electrode is sufficiently negative.

The current of the reverse process (17.2), flowing when the potential is changed to a positive value, is given by the formula:

$$i_r = \frac{nFAD_{Ox}^{1/2} C_{Ox}^0}{(\pi t')^{1/2}} \left[1 - \frac{\Delta}{(1+\Delta^2)^{1/2}}\right] \qquad (17.7)$$

where $t' = (t-\tau)$ and $\Delta = (t'/\tau)^{1/2}$; τ is the time from the start of the electrolysis to the moment at which the potential initiating the reverse reaction (17.2) is changed.

In other cases it is impossible to describe the current of reaction (17.2) by such a simple analytical equation as (17.7). For instance, case (2), according to the information given in Chapter 6, gives the following expression for the current of reaction (17.1):

$$i_f = nFAD_{Ox}^{1/2} C_{Ox}^0 \frac{\chi}{t^{1/2}} \exp \chi^2 \operatorname{erfc} \chi \qquad (17.8)$$

and for $t > \tau$

$$i_r = -\frac{nFAD_{Ox}^{1/2} C_{Ox}^0}{(t')^{1/2}} K(\chi, \Delta) \qquad (17.9)$$

where for $t < \tau$ $\chi = \frac{k_{fh} t^{1/2}}{D_{Ox}^{1/2}}$, and for $t > \tau$ $\chi = \frac{k_{fh} \tau^{1/2}}{D_{Ox}^{1/2}}$. $K(\chi, \Delta)$ is an algebraic function described in detail by the authors.

The original paper should be consulted for further information. Some of the relationships derived were tested experimentally by the authors [3]. They studied the kinetics of the UO_2^{2+}/UO_2^+ system in carbonate solutions, using measurements involving charge recording.

During recent years chronocoulometry with double potential step [4–6] has developed very rapidly. In this method the potential is changed according to the function shown in Fig. 17.1 and the charge is measured during the whole process. After the first potential change, when the electrode process starts, the time-dependence of the charge can be formulated as follows:

$$Q = \frac{2nFAD^{1/2}C^0 t^{1/2}}{\pi^{1/2}} + Q_{dl}. \qquad (17.10)$$

Equation (17.10) is valid when the depolarizer is not adsorbed at the electrode surface, and the potential applied is so positive or so negative (in anodic and cathodic processes, respectively) that the electrode process is diffusion controlled and the depolarizer concentration at the electrode surface is equal to zero. Q_{dl} is the charge connected with charging or discharging the double layer when the electrode potential is changed stepwise from the initial value to that at which the electrode process takes place.

When the potential returns stepwise to the initial value, it is given by the equation:

$$Q(t > \tau) = \frac{2nFD^{1/2}C^0}{\pi^{1/2}} [\sqrt{t} - \sqrt{t-\tau}] \qquad (17.11)$$

where τ is the time in which the potential was changed back to the initial value.

According to Anson [4], the relation between $Q(t > \tau)$ and $[\sqrt{t} - \sqrt{t-\tau}]$ should be linear, and the straight line should cut the origin of coordinates (when Ox and Red are not specifically adsorbed on the electrode surface).

It should be stressed that the Q_{dl} term is not present in equation (17.11), because the electrode potential has regained its initial value. The layer has been charged and discharged, and in both processes the same charge Q_{dl} was involved.

The charge which has built up from the moment τ contains the constant component due to charging of the double layer. This charge, understood as the difference $Q_r = Q(\tau) - Q(t-\tau)$ for $t > \tau$ can be easily derived from equations (17.10) and (17.11).

$$Q_r = \frac{2nFD^{1/2}C^0}{\pi^{1/2}} [\sqrt{t-\tau} + \sqrt{\tau} - \sqrt{t}] + Q_{dl}. \qquad (17.12)$$

It follows from equation (17.12) that the charge Q_r should be linearly dependent on parameter $\sqrt{t-\tau} + \sqrt{\tau} - \sqrt{t}$; in this case Q_{dl} can be obtained from the graph.

Chronocoulometry has great importance in electrode kinetic studies. Kimmerle and Chevalet [2, 3], considering the currents of the primary and secondary processes (17.1) and (17.2) also described the charges flowing during the electrolysis and related them to the rate constants of the electrode processes. All the four cases discussed in connection with the chronoamperometric results were considered.

Dropping mercury electrodes can also be used in such experiments, the potential being applied in a definite drop growth time. The initial, very short (of the order of 1 msec), period of time is excluded from the measurements in order to eliminate the effect of the capacitance current. Then the charge is measured over an accurately determined time, of the order of several hundredths of a second. As a rule, when the dropping electrode is polarized near the end of its life, the change of the electrode surface area during the charge measurement can be neglected.

In cases where the substrate or the product, or both, are adsorbed at the electrode, the equations describing the change of charge become even more complex. The method discussed has been widely used in studies of adsorption processes (see Chapter 16).

The potential changes used in this method are very fast. When the charge on the electrode is changed rapidly, the structure of the diffusion layer is also changed and the double layer may not attain the equilibrium structure. This is observed particularly when dilute electrolytes are studied. Initially a layer is formed which reaches equilibrium during a time of the order of milliseconds.

It is essential to know the conditions under which the double layer reaches the equilibrium during the times of pulse application, since in the corrections for the double layer structure (the effect of potential Φ_2) it is assumed, when potential Φ_2 is calculated, that the layer is in equilibrium. The problem of the double layer relaxation time after the introduction of the charge during a very short period of time has been discussed by Anson [7], Buck [8] and Feldberg [9].

A modification of cyclic chronoamperometry is the polarographic method of Kalousek [10, 11] developed in the period 1940–1950. This technique has been included rather arbitrarily in this group of methods since it is also similar to alternating current polarography. The difference is the use of a larger voltage amplitude and lower current frequency.

Two basic varieties of this method exist. In the first a rectangular voltage having an amplitude of 20–50 mV and a frequency of 5 Hz is superimposed on a linearly increasing potential, as in the classical polarographic method. This mode of changing the indicator electrode potential is shown in Fig. 17.2. The electrolysis current is measured only during the half-cycles related to more positive potential values and therefore it is recorded on the polarograms as a distinct minimum in the vicinity of the half-wave potential of the reversible system. The formation of this minimum commences at potentials at which the reduction process starts in the half-cycles of the more negative potentials. At still more negative potentials the minimum disappears, since the electrode potential is so negative that even during the anodic half-cycle the oxidation rate is very slow. When the alternating voltage oscillations take place in the potential region of limiting current, the measured current corresponds to the limiting cathodic current.

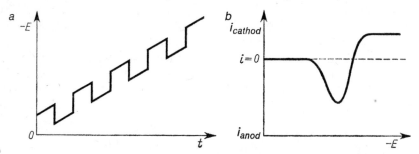

Fig. 17.2 a—Time dependence of the potential in Kalousek polarography (first variant); b—a typical curve recorded by this method.

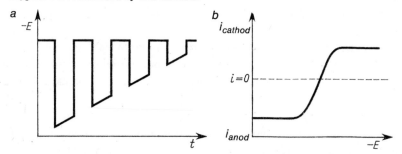

Fig. 17.3 a—Time dependence of the potential in Kalousek polarography (second variant); b—a typical curve recorded by this method.

In the second variant of the Kalousek method the potential changes from a definite value in the limiting current region of potential to a linearly increasing potential. This is shown schematically in Fig. 17.3. The recorded curves resemble classical polarograms of a redox system, when both Ox and Red forms are present in the solution. The anodic current is observed due to oxidation of Red formed during the time when the electrode had a negative potential.

On the basis of the work of Kambara [12] and Barker [13] it is possible to obtain an equation describing the current observed in the Kalousek method at a flat electrode:

$$i = \frac{1}{2}\left[nFA\vec{C}_{\text{Red}}\left(\frac{D_{\text{Red}}}{\pi t}\right)^{1/2} - nFAD_{\text{Red}}\sum_{m=1}^{j}(-1)^{m+1}\frac{\vec{C}_{\text{Red}} - \overleftarrow{C}_{\text{Red}}}{[\pi D_{\text{Red}}(t-m\tau)]^{1/2}}\right]$$
(17.13)

where $\vec{C}_{\text{Red}}$ is the concentration of the reduced form at the electrode surface during the cathodic cycle, $\overleftarrow{C}_{\text{Red}}$ is the concentration of the reduced form on the electrode surface during the anodic cycle of the rectangular wave, τ is the half-cycle time, t is the time measured from the start of the polarization, m is the number of half-cycles from the start of the polarization, and j is an integer equal to or smaller than t/τ.

The current observed in the Kalousek polarographic method is the algebraic sum of the currents observed in normal polarography [the first term of the right-hand side of equation (17.13)] and the alternating current due to the application of the rectangular alternating current.

Equation (17.13) describes the current recorded in both above variants of the Kalousek method. In the case of the second variant, in which the voltage is changed as shown in Fig. 17.3 the following relationship derived from equation (17.13) for a reversible process is valid:

$$E - E_{1/2} = \frac{RT}{nF}\ln\frac{i_g - i}{i - i_{\text{anod}}}$$
(17.14)

where i_g is the limiting diffusion current, i is the instantaneous Kalousek current, i_{anod} is the maximum anodic current, and E is the potential, which changes in the direction of the constant negative potential value during polarization.

The theoretical description of the average current in the Kalousek method has also been published by Koutecký [14]. In recent years these problems were studied by Ružić [15].

The Kalousek method was used initially in studies of mechanisms of electrode processes. It also proved to be useful in investigations of the reversibility of electrode processes [16, 17], but it can in addition be useful in analytical chemistry [18]. The sensitivity of the method in the analysis of reversible systems is several times greater than that of classical polarography.

17.2 Cyclic Stationary Electrode Voltammetry

Kemula and Kublik [19, 20] developed the cyclic stationary electrode voltammetric method in 1958 and applied it to studies of the mechanisms of electrode processes of organic compounds. During the following years the method has been widely used in the investigation of mechanisms of the reduction of organic compounds at mercury electrodes, mechanisms of oxidation on platinum, carbon and carbon paste electrodes, and the mechanisms of inorganic processes. It was also used in determinations of differential double layer capacitance, studies of the kinetics of chemical reactions following primary electrode processes, and studies of the kinetics of electrode reactions.

In cyclic stationary electrode voltammetry the indicator electrode is polarized by a linearly changing potential. The potential changes are shown schematically in Fig. 17.4.

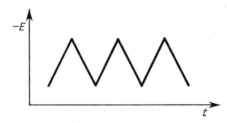

Fig. 17.4 Time dependence of the electrode potential in cyclic stationary electrode voltammetry.

It is necessary to consider the basic relationships obtained by various workers before any discussion of the advantages offered by this technique in the above fields.

The theory of cyclic stationary electrode voltammetric curves of reversible processes was put forward by Matsuda [21] in 1957. The problem was also considered by Gokhshtein [22]. In addition Nicholson and Shain [23] made a detailed study of the subject in an extensive work on various aspects of stationary electrode voltammetry. They found that, in the case of reversible processes, the consecutive curves have identical shapes when the potential point of the reversal in the direction of polarization of the primary reduction process is at least $35/n$ mV more negative than the cathodic peak potential. Such a curve is shown in Fig. 17.5. These curves are obtained when the reduction product is soluble in the solution. If the hanging mercury drop is used as the electrode and if the ions are reduced to free metal in the primary cycle, in cases where the solubility of the metal in mercury is high the diffusion field for the reduced metal is small. Under these circumstances the concentration of metal in the drop is increased in relation to that in the solution. Therefore the peak in the anodic cycle is much larger than that in the cathodic cycle. When the electrode reaction product is soluble in the solution the anodic peak current is equal to the cathodic peak current irrespective of the potential E_z at which the direction of polarization is reversed. The method of determination of the anodic peak current is shown

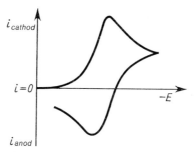

Fig. 17.5 Cyclic stationary electrode voltammetric curve obtained by means of a flat electrode in a large diffusion space.

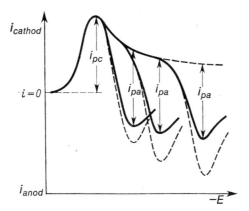

Fig. 17.6 Cyclic stationary electrode voltammetric curves. Broken lines show oxidation currents in the case of dissolution of the amalgam generated in the cathodic half-cycle in a small drop of mercury.

in Fig. 17.6. This form of the relationship is obtained by using a polarograph in which the reversal of the direction of polarization does not automatically change the direction in which the chart paper moves.

The broken line in Fig. 17.6 shows the anodic currents observed when the electrode process is carried out at a hanging mercury drop and then the primary reaction product is a metal readily soluble in mercury. In this case the anodic current increases with the difference of the potential at which the direction of polarization changes, from that of the peak potential. The ratio of the anodic to the cathodic currents for a given potential (E_z), at which the direction of polarization changes, increases with decreasing electrode radius and with decreasing polarization rate.

The position of the anodic peak with respect to the potential axis is a function of the potential E_z when $(E_z - E_{1/2})n$ is small. This follows from the fact that, at potentials close to the peak in the cathodic cycle, the concentration of the electrode reaction product is not exactly equal to that of the depolarizer in the bulk of the solution. When, under such conditions, the direction of polarization is reversed, the anodic current peak is shifted in the positive direction. This shift decreases with increasingly negative values of potential E_z. These changes are summarized in Table 17.1.

Table 17.1 DEPENDENCE OF ANODIC PEAK POTENTIAL ON THE POLARIZATION REVERSAL POTENTIAL (IN mV) (FROM [23] BY PERMISSION OF THE COPYRIGHT HOLDERS, THE AMERICAN CHEMICAL SOCIETY)

$(E_{1/2} - E_z)n$	$(E_{pa} - E_{1/2})n$	$(E_{1/2} - E_z)n$	$(E_{pa} - E_{1/2})n$
65	34.4	100	32.0
70	33.7	150	30.7
75	33.3	200	29.8
80	32.9	300	29.3

In the case of a reversible electrode process the difference between the anodic and the cathodic peak potentials is not large. It is already known from the discussion in the preceding chapters that the reversible half-wave potential differs from the peak potential by $1.11 RT/nF$ V. Expressing the anodic and the cathodic peak potentials E_{pa} and E_{pc} by the expressions:

$$E_{pc} = E_{1/2} - 1.11 \frac{RT}{nF}, \qquad (17.15)$$

$$E_{pa} = E_{1/2} + 1.11 \frac{RT}{nF}, \qquad (17.16)$$

and subtracting, the following is obtained:

$$E_{pa} - E_{pc} = 2.22 \frac{RT}{nF}. \qquad (17.17)$$

This is a good criterion of electrode process reversibility.

At 25°C equation (17.17) has the following form:

$$E_{pa} - E_{pc} = \frac{0.057}{n} \text{ V}. \qquad (17.17a)$$

It follows from this equation that the difference between the peak potentials decreases with an increasing number of electrons n, exchanged in the primary reduction process. At 25°C it is 57, 29 and 19 mV for one-, two- and three-electron processes, respectively. In the case of irreversible processes the difference is much greater.

The transition from unicyclic to multicyclic polarization leads to a gradual equalization of the concentration gradients during subsequent cycles at the same potentials, which in turn results in a gradual change of the shapes of the anodic and the cathodic curves. After about 50 cycles further changes in shape are slow and, in principle, a stationary state is reached.

Nicholson and Shain [23] have discussed the case of linear diffusion. In order to adapt their calculations to the results obtained with spherical electrodes [24], a modified Reinmuth [25] correction was used. For the reduction process:

$$i = i_{pl} + \frac{nFAD_{Ox}C^0_{Ox}\Phi(at)}{r_0} \qquad (17.18)$$

where:

$$\Phi(at) = \frac{1 - S_z(at)}{1 + \Theta S_z(at)}. \qquad (17.19)$$

$S_z(t) = e^{-at}$ for values of t shorter than the time of the polarization reversal, and $S_z(t) = e^{at - 2az}$ when t exceeds this time; Θ is given by the expression:

$$\Theta = \exp\left[\frac{nF}{RT}(E_i - E^0)\right],$$

$a = nFV/RT$ and i_{pl} is the current which would be observed if flat electrodes were used.

The diffusion field area has not so far been limited in these discussions. However, when the reduction of ions of mercury-soluble metals is carried out on a small hanging mercury drop, the above relationships between the anodic and cathodic currents are no longer valid. On account of the small dimensions of the drop the concentration of metal rapidly increases and therefore the ratio of the anodic to the cathodic currents is greater than one. Gumiński and Galus [26] suggested for this case the following empirical equation, relating the anodic current to the cathodic current and to the experimental conditions:

$$-i_{\text{anod}} = i_{\text{cathod}} \left\{ 1 + 3.2 \left[\frac{D_{\text{Red}}(E_{pc} - E_z)}{V r_0^2} \right]^{1/2} \right\}. \tag{17.20}$$

It follows from this equation that the ratio of the anodic to the cathodic current increases with increasing diffusion coefficient of the metal in mercury and increasing difference between the potential of the peak current and that at which the direction of polarization is reversed. The numerical value of the ratio of the currents decreases with increasing polarization rate and with increasing radius of the hanging mercury electrode.

Gumiński and Galus tested equation (17.20) for various experimental conditions [changing V, r_0 and $(E_{pc} - E_z)$]. It was found possible, on the basis of equation (17.20), to establish conditions under which the equation is reduced to the simple relationship $i_{\text{anod}} = i_{\text{cathod}}$. Assuming that $D_{\text{Red}} = 10^{-5}$ cm^2/s, $(E_{pc} - E_z) = 0.3$ V and $r_0 = 0.043$ cm, then $i_{\text{anod}} = i_{\text{cathod}}$ when the polarization rate exceeds 16 V/s.

Beyerlein and Nicholson [27] considered the effect of amalgam formation in the hanging mercury drop, during the cathodic cycle, on the ratio of the anodic to the cathodic current. In this treatment only the effect of the spherical surface was taken into account and the effect of the finite diffusion field for the reduced metal was neglected. Theoretical calculations showed that the ratio of the anodic to the cathodic currents is always greater than unity and that the difference between the peak potentials of the anodic and cathodic processes also changes under certain conditions.

The application of cyclic stationary electrode voltammetry to studies of the kinetics of chemical reactions following the primary electrode processes, will be discussed in the next chapter. In this chapter the discussion of this technique will be limited to the studies of the kinetics of charge exchange in electrode processes.

Equation (17.17), which describes the difference between the anodic and the cathodic peak current potentials, is valid only for reversible processes. When the process is controlled simultaneously by diffusion and by charge transfer, the difference between the anodic and the cathodic peak current potentials depends on the degree of reversibility of the process, i.e. on the relationship between the rate of the electrode process and of the polarization. If the process is slow, and the polarization is fast, the process is irreversible and the difference between the anodic and the cathodic peak potentials is large. In the opposite case the process is reversible, and the potential difference is $57/n$ mV.

The problem of determination of kinetic parameters on the basis of cyclic stationary electrode voltammetric curves was considered by Nicholson [28] in 1965. Assuming that the Ox and the Red forms are soluble, either in the solution or in the electrode material, and that both forms are transported by diffusion alone, he obtained a solution which cannot be expressed in an analytical form. The required standard rate constant of the electrode process is related to the function Ψ as follows:

$$\Psi = \left(\frac{D_{Ox}}{D_{Red}}\right)^{\alpha/2} \frac{k_s(RT)^{1/2}}{(\pi nFVD_{Ox})^{1/2}}. \qquad (17.21)$$

Function Ψ is in turn related to the difference between the peak potentials; this relationship is shown in Table 17.2.

Table 17.2 DEPENDENCE OF $E_{pa}-E_{pc}$ IN CYCLIC STATIONARY ELECTRODE VOLTAMMETRY ON THE STANDARD RATE CONSTANT OF AN ELECTRODE PROCESS (FROM [28] BY PERMISSION OF THE COPYRIGHT HOLDERS, THE AMERICAN CHEMICAL SOCIETY)

Ψ	$(E_{pa}-E_{pc})n$ (mV)	Ψ	$(E_{pa}-E_{pc})n$ (mV)
20	61	1	84
7	63	0.75	92
6	64	0.5	105
5	65	0.35	121
4	66	0.25	141
3	68	0.1	212
2	72		

The values of $(E_{pa}-E_{pc})$ are almost independent of α in the region $0.3 < \alpha < 0.7$. This dependence is particularly small when Ψ is large. For instance when $\Psi = 0.5$ the change of $(E_{pa}-E_{pc})$ with α changing from 0.3 to 0.7 is 5%, but for $\Psi = 0.1$ the same change in the transfer coefficient leads to a 20 per cent change of $(E_{pa}-E_{pc})$. In the latter case the deviation from reversibility is appreciable.

The values of Ψ given in Table 17.2 were calculated for $\alpha = 0.5$. These data combined with equation (17.21) relate the difference between the anodic and the cathodic peak potentials to the standard rate constants of the electrode reaction and to the rate of polarization. This makes it possible to choose the experimental conditions (mainly V) in such a way that the difference $(E_{pa}-E_{pc})n$ leads to the Ψ values given in Table 17.2. Theoretically, this makes it possible to study the kinetics of electrode processes of very fast reactions. Very fast polarization rates should be used in such a case, in order to produce observable deviations from reversibility. This, in turn, is related to appreciable Faradaic currents; the ohmic potential drop iR may cause large errors, even when the resistance is comparatively small. The ohmic potential drops can be minimized by using a Luggins capillary or concentrated supporting electrolytes. It is not advisable to decrease the current by decreasing the depolarizer concentration, since this would increase the ratio of capacitance current to the measured current.

Nicholson [28] used the method described for the determination of the standard rate constant of the $Cd^{2+}/Cd(Hg)$ system in 1.0 M Na_2SO_4. The value determined, 0.24 cm/s, is in quite good agreement with values obtained by other workers. According to Nicholson, the application of this method makes it possible to determine standard rate constants up to 5 cm/s.

17.3 Chronopotentiometry with Change of Direction of Current

The theoretical principles of this technique were worked out in 1953 by Berzins and Delahay [29]. If the Ox form alone is present in the solution, and a reduction process takes place at the electrode under the influence of direct current, when the direction of current is reversed before the cathodic transition time is reached, the dependence of the Red form on the distance from the electrode and on the duration of electrolysis is described by the expression:

$$C_{Red}(x, t') = 2\Omega \left[\frac{D_{Red}(t_{Red}+t')^{1/2}}{\pi} \right] \exp\left[-\frac{x^2}{4D_{Red}(t_{Red}+t')} \right]$$
$$-\Omega x \operatorname{erfc}\left[\frac{x}{[4D_{Red}(t_{Red}+t')]^{1/2}} \right]$$
$$-2(\Omega+\lambda') \left[\frac{D_{Red}t'}{\pi} \right]^{1/2} \exp\left[-\frac{x^2}{4D_{Red}t'} \right] + (\Omega+\lambda')x \operatorname{erfc}\left[\frac{x}{2(D_{Red}t')^{1/2}} \right].$$
$$(17.22)$$

In this equation, which was derived by assuming linear diffusion, Ω and λ' are given by:

$$\Omega = \frac{i_0}{nFD_{Red}}, \qquad (17.23)$$

and

$$\lambda' = \frac{i'_0}{nFD_{Red}}, \qquad (17.24)$$

17.3] Chronopotentiometry with Change of Direction of Current

t' is the time that elapses from the moment of reversal of the direction of current: $t' = t - t_{\text{Red}}$; i_0 and i_0' are current densities in the cathodic and anodic processes, respectively, and t_{Red} is the duration of the reduction process.

Equation (17.22) makes it possible to obtain an equation describing the transition time for $C_{\text{Red}}(0, t)$:

$$\tau_{\text{Ox}} = \frac{\Omega^2}{(\Omega + \lambda')^2 - \Omega^2} t_{\text{Red}}. \tag{17.25}$$

When $\Omega = \lambda'$, as is frequently observed, the following simple relationship is obtained:

$$\tau_{\text{Ox}} = \tfrac{1}{3} t_{\text{Red}}. \tag{17.26}$$

Obviously, when oxidation is the primary process and reduction takes place when the direction of current is reversed, the transition time of the reduction process is equal to 1/3 of that of the oxidation.

Equation (17.26) was tested by Berzins and Delahay using the reduction of thallium, cadmium and zinc ions at large mercury electrodes (under linear diffusion conditions). The average value of $\tau_{\text{Red}}/\tau_{\text{Ox}}$ obtained from sixteen measurements was 3.02, in good agreement with the theoretically predicted value.

The expression describing the concentration of the oxidized form $C_{\text{Ox}}(x, t)$ can be derived in a similar manner to equation (17.22). Putting in this expression and making $x = 0$ in equation (17.22) the expressions describing $C_{\text{Ox}}(0, t)$ and $C_{\text{Red}}(0, t)$ are obtained. Introducing these expressions into the Nernst equation, the equation of the chronopotentiometric curve after the reversal of the current direction is obtained:

$$E = E^0 + \frac{RT}{nF} \ln \frac{f_{\text{Ox}}}{f_{\text{Red}}}$$

$$+ \frac{RT}{nF} \ln \frac{C_{\text{Ox}}^0 - 2\lambda [D_{\text{Ox}}(t_{\text{Red}} + t')/\pi]^{1/2} + 4\lambda (D_{\text{Ox}} t'/\pi)^{1/2}}{2\Omega [D_{\text{Red}}(t_{\text{Red}} + t')/\pi]^{1/2} - 4\Omega (D_{\text{Red}} t'/\pi)^{1/2}} \tag{17.27}$$

where $\lambda = i_0/nFD_{\text{Ox}}$, and f_{Ox} and f_{Red} are the activity coefficients of the Ox and Red forms, respectively.

If it is assumed that the reversal of the current direction takes place at the moment when the cathodic transition time is reached, equation (17.27) can be written as follows:

$$E = E_{1/2} + \frac{RT}{nF} \ln \frac{\tau^{1/2} - [(\tau + t')^{1/2} - 2t'^{1/2}]}{(\tau + t')^{1/2} - 2t'^{1/2}} \tag{17.28}$$

where the half-wave potential is defined by the expression:

$$E_{1/2} = E^0 + \frac{RT}{nF} \ln \frac{f_{\text{Ox}} D_{\text{Red}}^{1/2}}{f_{\text{Red}} D_{\text{Ox}}^{1/2}}. \tag{17.29}$$

Using equation (17.28), the time in which the potential reaches the half-wave potential value can be calculated. This is done by making the logarithmic term of equation (17.28) equal to zero; then the following relationship is obtained:

$$t'_{E=E_{1/2}} = 0.222 \tau_{\text{Ox}}. \tag{17.30}$$

Since for a reduction process $E_{\tau/4} = E_{1/2}$, then:

$$E_{\tau/4} = E_{0.222\tau_{\text{Ox}}}. \tag{17.31}$$

This relationship can be used for checking that the investigated process is reversible under the experimental conditions. However it should be borne in mind that relationship (17.31) is valid when the process is carried out under the conditions of semi-infinite linear diffusion.

When the process is irreversible, utilizing the relationship:

$$\frac{i'_0}{nF} = k_{bh}^0 C_{\text{Red}}(0, t') \exp\left[\frac{(1-\alpha)nFE}{RT}\right] \tag{17.32}$$

and the expression describing $C_{\text{Red}}(0, t')$, which can be derived from equation (17.22), gives:

$$E = \frac{RT}{(1-\alpha)nF} \ln \frac{\pi^{1/2} D_{\text{Red}}^{1/2}}{2k_{bh}^0} - \frac{RT}{(1-\alpha)nF} \ln [(\tau+t')^{1/2} - 2t'^{1/2}]. \tag{17.33}$$

It follows from equation (17.33) that the potential dependence of the common logarithm of the expression $[(\tau+t')^{1/2} - 2t'^{1/2}]$ should be linear, and the slope of the straight line should be equal to $2.3RT/(1-\alpha)nF$.

Thus it is possible to determine the anodic transfer coefficient of an irreversible oxidation process. The rate constant of the anodic process k_{bh}^0 can be calculated from the value of the potential at $t = 0$, according to equation (17.33).

So far, the discussion of chronopotentiometry has been limited to primary cathodic reduction processes, the product formed being oxidized when the direction of current was reversed. Analogous equations can be derived for primary electro-oxidation followed by electroreduction.

Equation (17.26) is valid for the case of semi-infinite linear diffusion. For processes taking place under conditions of semi-infinite spherical diffusion, Murray and Reilley [30] derived the following equation, which relates the anodic transition time to the primary cathodic electroreduction time:

$$1+\exp[D_{Red}(\tau_{Ox}+t_{Red})/r_0^2]\mathrm{erfc}[D_{Red}^{1/2}(\tau_{Ox}+t_{Red})^{1/2}/r_0]$$
$$= 2\exp[D_{Red}\tau_{Ox}/r_0^2]\mathrm{erfc}[D_{Red}^{1/2}\tau_{Ox}^{1/2}/r_0] \qquad (17.34)$$

where r_0 is the radius of the spherical electrode.

The same authors have also given an equation describing the relation between reduction time and oxidation time under conditions of cylindrical diffusion:

$$\frac{\tau_{Ox}+t_{Red}}{4\tau_{Ox}} = \frac{\left[1 - \frac{D_{Red}^{1/2}\pi^{1/2}\tau_{Ox}^{1/2}}{4r_0} + \frac{D_{Red}\tau_{Ox}}{4r_0^2} - \cdots\right]^2}{\left[1 - \frac{D_{Red}^{1/2}\pi^{1/2}(\tau_{Ox}+t_{Red})^{1/2}}{4r_0} + \frac{D_{Red}(\tau_{Ox}+t_{Red})}{4r_0^2} - \cdots\right]^2}. \qquad (17.35)$$

Gumiński and Galus [26] have shown that, when ions of mercury-soluble metals are reduced at small, hanging mercury drop electrodes, and when the reduction times are longer, the following equation is valid:

$$\frac{\tau_{Ox}}{t_{Red}} = 1 - \frac{r_0^2}{15 D_{Red} t_{Red}}. \qquad (17.36)$$

In the cases of sufficiently long reduction times, or very small hanging mercury drop electrode radii, equation (17.36) is reduced to the following:

$$\frac{\tau_{Ox}}{t_{Red}} = 1. \qquad (17.37)$$

The ranges of D, r_0 and t_{Red} values over which equation (7.37) is valid can be calculated from equation (17.36).

Equation (17.36) was tested by studies of the reduction of zinc, cadmium and lead ions on small hanging mercury drop electrodes, followed by oxidation of the resulting amalgams after reversal of the current. The experimental results were in good agreement with those calculated from the equations, when the reduction times exceeded 25 seconds.

In this discussion it has been assumed that the current does not change in a programmed manner, apart from reversal of direction. Murray and Reilley [30] have worked out a general method for calculating the relation between the time of the primary process and that of the process taking place after the reversal of the current direction, in the case where the current changes in a programmed way during the experiment. This problem has also been considered by Testa and Reinmuth [31].

A new method of investigation of electrode reactions kinetics has been proposed [32]. In this method, after the initial galvanostatic current pulse a coulostatic pulse having the opposite sign is applied. This method was checked experimentally [33] in kinetic studies on the $Cd^{2+}/Cd(Hg)$ system in solutions containing chlorides and perchlorates. The results obtained for $1M$ $NaClO_4$ ($k_s = 0.38 \pm 0.02$ cm/s and $\beta = 0.67 \pm 0.04$) are in agreement with those obtained by other relaxation methods.

17.4 Cyclic Chronopotentiometry

The chronopotentiometric technique with reversal of current direction described above is a unicyclic method, since in the investigation of the Ox form, the process is limited to the reduction of this form at the electrode in the cathodic cycle and to the oxidation of the resulting Red form. However it is possible to reverse the direction of the current after the transition time is reached and to repeat both reduction and the oxidation processes. This technique, suggested by Herman and Bard [34], has been called cyclic chronopotentiometry. It has wide possibilities, like cyclic stationary electrode voltammetry, in studies of mechanisms of electrode processes.

The theoretical description of this technique in the case of reversible electrode processes without kinetic complications, was worked out on the basis of general relationships derived by Murray and Reilley [30].

For the $Ox + ne \rightleftharpoons Red$ reaction Herman and Bard derived the following equations:

$$C_{Ox}^0 = \frac{2i_0^{Red}}{nF(\pi D_{Ox})^{1/2}} [(\tau_1 + \tau_2 + \ldots \tau_n)^{1/2} - R(\tau_2 + \ldots \tau_n)^{1/2} + \ldots R(\tau_n)^{1/2}],$$

(17.38)

$$C_{Red}^0 = \frac{-2i_0^{Red}}{nF(\pi D_{Red})^{1/2}} [(\tau_1 + \tau_2 + \ldots \tau_n)^{1/2} - R(\tau_2 + \ldots \tau_n)^{1/2} + \ldots R(\tau_n)^{1/2}].$$

(17.39)

Equation (17.38) describes the first, third and further odd transition times, whereas (17.39) describes the even ones. $R = (i_0^{Red} + i_0^{Ox})/i_0^{Red}$, where i_0^{Ox} and i_0^{Red} are the current densities of the oxidation and reduction processes, respectively.

Using relationships $\tau_1^{1/2} = \pi^{1/2} D_{Ox}^{1/2} nFC_{Ox}/2i_0^{Red}$ and $\tau_n = a_n \tau_1$, where a_n is the nth transition time referred to the time τ_1 (for instance in chronopotentiometry with current reversal $a_n = 1/3$), from equations (17.38) and (17.39) the following relationships are obtained:

$n = 3, 5, 7 \ldots$
$$1 = (a_1+a_2+a_3+ \ldots a_n)^{1/2} - R(a_2+a_3+ \ldots a_n)^{1/2} + \ldots R(a_n)^{1/2}, \quad (17.40)$$
$n = 2, 4, 6 \ldots$
$$\frac{D_{\text{Red}}^{1/2} C_{\text{Red}}^0}{D_{\text{Ox}}^{1/2} C_{\text{Ox}}^0} = (a_1+a_2+ \ldots a_n)^{1/2} - R(a_2+ \ldots a_n)^{1/2} + \ldots R(a_n)^{1/2}, \quad (17.41)$$

where $a_1 = 1$, and the values of $a_2, a_3 \ldots a_n$ can be obtained by solving equations (17.40) and (17.41). Herman and Bard [34] solved this system of equations, using a computer. Some of the a_n values calculated for various R and $C_{\text{Red}}^0/C_{\text{Ox}}^0$ values, assuming that $D_{\text{Red}} = D_{\text{Ox}}$, are shown in Table 17.3.

Table 17.3 RELATIVE TRANSITION TIMES IN CYCLIC CHRONOPOTENTIOMETRY IN VARIOUS CONDITIONS (FROM [34] BY PERMISSION OF THE COPYRIGHT HOLDERS, THE AMERICAN CHEMICAL SOCIETY)

	a_n		
	Ox and Red soluble		Ox and Red insoluble
n	$i_0^{\text{Red}} = i_0^{\text{Ox}}$ $C_{\text{Ox}}^0 = 0$	$i_0^{\text{Ox}} = i_0^{\text{Red}}$ $C_{\text{Ox}}^0 = C_{\text{Red}}^0$	$i_0^{\text{Red}} = i_0^{\text{Ox}}$
1	1.000	1.000	1.000
2	0.333	1.778	1.000
3	0.588	1.713	1.174
4	0.355	1.740	1.174
5	0.546	1.724	1.263
6	0.366	1.735	1.263
7	0.525	1.727	1.319
8	0.373	1.733	1.319
9	0.513	1.728	1.359
10	0.378	1.732	1.359
11	0.504	1.728	1.390
12	0.382	1.732	1.390

When the Red form is insoluble and is deposited on the electrode, the values of a_n corresponding to even values of n can be calculated on the basis of the equation for $n = 2, 4, 6, \ldots$:

$$a_n = \frac{i_0^{\text{Red}}}{i_0^{\text{Ox}}} a_{n-1}, \quad (17.42)$$

whereas a_n corresponding to odd values of n can be calculated from equation (17.40). Some of the calculated a_n values are also shown in Table 17.3.

The theoretical results have been tested by the authors, using the reduction of cadmium ions on mercury, and also ferric ions in 0.2 M potassium oxalate and silver ions in 0.2 M KNO_3 on platinum. In all these cases the experimental results were in a good agreement with theory.

17.5 The Rotating Ring-Disc Electrode Method

This technique, although it is not cyclic, is discussed in this chapter because, like cyclic methods, it provides experimental data which permit the estimation of the stability of the products of primary electrode processes.

The electrode shown in Fig. 2.8 was introduced by Frumkin and Nekrasov [35], and the first theoretical studies of currents observed at such electrodes were carried out by Ivanov and Levich [36]. The electrode region can be divided into three parts.

The inner part, having the radius r_1, is occupied by the disc electrode. The circular plane, having the radius r_2, contains the surface of the disc and that of the surrounding ring of the insulator. The circular plane is surrounded by a ring of a conductor which constitutes the outer electrode. The radius of the circle which contains the disc, the ring and the insulating plane is equal to r_3. A linearly increasing potential is usually applied to the inner disc electrode, while the potential of the ring electrode is held constant at such a value that the product of the electrode reaction taking place at the disc could be oxidized or reduced at the ring electrode, depending on whether reduction or oxidation takes place at the disc.

Exact solutions of the various kinetic problems of this method have been given by Albery [37], who based this work on the Laplace transformation and on the Airy function.

When the electrode process consists of a rapid electron transfer, the major interest is in the definition of the efficiency of the ring electrode. This collection efficiency, N, can be written as $N = i_p/i_D$, where i_p is the current in the ring electrode circuit, and i_D is the current in the disc electrode circuit. Thus it is possible to determine N by calculating the ring electrode current for a given current in the disc electrode circuit.

Albery and Bruckenstein [38] described N as follows:

$$N = 1 - F\left(\frac{\alpha}{\beta}\right) + \beta^{2/3}[1 - F(\alpha)] - (1+\alpha+\beta)^{2/3}\left\{1 - F\left[\left(\frac{\alpha}{\beta}\right)(1+\alpha+\beta)\right]\right\} \tag{17.43}$$

where $\alpha = (r_2/r_1)^3 - 1$, $\beta = (r_3/r_1)^3 - (r_2/r_1)^3$ and

$$F(\Delta) = \frac{3^{1/2}}{2\pi} \int_0^\Delta \frac{d\lambda}{\lambda^{2/3}(1+\lambda)}. \tag{17.44}$$

The integer in equation (17.44) can be expressed by the equation:

$$F(\Delta) \cong \frac{3^{1/2}}{4\pi} \ln \frac{(1+\Delta^{1/3})^3}{1+\Delta} + \frac{3}{2\pi} \arctan\left(\frac{2\Delta^{1/3}-1}{3^{1/2}}\right) + \frac{1}{4} \quad (17.45)$$

and when $\Delta \to 0$, $F(\Delta) \to 0$, and when $\Delta \to \infty$, $F(\Delta) \to 1$.

The values of N, calculated for typical values of the r_2/r_1 and r_3/r_2 ratios, are given in Table 17.4 (Albery and Bruckenstein [38]). The values of N for the common electroanalytical electrodes can be taken from this table.

Table 17.4 VALUES OF N FOR $\dfrac{r_2}{r_1}$ AND $\dfrac{r_3}{r_2}$ USED IN PRACTICAL WORK (FROM [38] BY PERMISSION OF THE COPYRIGHT HOLDERS)

$\dfrac{r_3}{r_2}$	$\dfrac{r_2}{r_1}$								
	1.02	1.03	1.04	1.05	1.06	1.07	1.08	1.09	1.10
1.02	0.1013	1.0976	0.0947	0.0922	0.0902	0.0884	0.0869	0.0855	0.0843
1.03	0.1293	0.1250	0.1215	0.1186	0.1162	0.1140	0.1121	0.1104	0.1089
1.04	0.1529	0.1483	0.1444	0.1412	0.1385	0.1360	0.1339	0.1320	0.1302
1.05	0.1737	0.1687	0.1647	0.1612	0.1582	0.1556	0.1533	0.1512	0.1493
1.06	0.1923	0.1872	0.1829	0.1793	0.1761	0.1733	0.1708	0.1686	0.1665
1.07	0.2092	0.2039	0.1996	0.1958	0.1925	0.1896	0.1869	0.1846	0.1824
1.08	0.2247	0.2194	0.2149	0.2110	0.2076	0.2046	0.2019	0.1994	0.1972
1.09	0.2392	0.2338	0.2292	0.2252	0.2217	0.2186	0.2158	0.2133	0.2110
1.10	0.2526	0.2472	0.2426	0.2385	0.2350	0.2318	0.2289	0.2263	0.2240
1.12	0.2772	0.2717	0.2670	0.2629	0.2593	0.2560	0.2530	0.2503	0.2479
1.14	0.2992	0.2938	0.2890	0.2849	0.2812	0.2778	0.2748	0.2720	0.2695
1.16	0.3192	0.3138	0.3090	0.3048	0.3011	0.2977	0.2947	0.2919	0.2893
1.18	0.3375	0.3321	0.3274	0.3232	0.3194	0.3161	0.3130	0.3101	0.3075
1.20	0.3544	0.3490	0.3443	0.3402	0.3364	0.3330	0.3290	0.3271	0.3245
1.22	0.3701	0.3648	0.3601	0.3560	0.3523	0.3489	0.3458	0.3429	0.3403
1.24	0.3848	0.3795	0.3749	0.3708	0.3671	0.3637	0.3606	0.3577	0.3551
1.26	0.3985	0.3810	0.3887	0.3847	0.3810	0.3776	0.3745	0.3717	0.3691
1.28	0.4115	0.4063	0.4018	0.3977	0.3941	0.3907	0.3877	0.3849	0.3822
1.30	0.4237	0.4186	0.4141	0.4101	0.4065	0.4032	0.4001	0.3973	0.3947
1.32	0.4353	0.4302	0.4258	0.4218	0.4183	0.4150	0.4119	0.4092	0.4066
1.34	0.4463	0.4413	0.4369	0.4330	0.4294	0.4262	0.4232	0.4204	0.4178
1.36	0.4567	0.4518	0.4475	0.4436	0.4401	0.4369	0.4339	0.4311	0.4286
1.38	0.4667	0.4619	0.4576	0.4538	0.4503	0.4471	0.4441	0.4414	0.4389
1.40	0.4762	0.4715	0.4673	0.4635	0.4600	0.4568	0.4539	0.4512	0.4487

Bruckenstein and Feldman [39] showed that the approximate Ivanov and Levich equation can be written in the form:

$$N = \frac{0.8}{3}\left[\frac{1}{2}\ln\frac{y^3+x^3}{(y+x)^3}+3^{1/2}\arctan\left(\frac{2x-y}{3^{1/2}y}\right)+0.9069\right] \quad (17.46)$$

where $y^3 = 1-\frac{3}{4}+(\alpha 1)^{-1}$, $x^3 = \beta(\alpha+1)^{-1}$.

These relationships were tested by the authors, using the following reactions: reduction of Cu^{2+} to Cu^+ and oxidation of Cu^+ to Cu^{2+} (10^{-4} M $CuCl_2$ in 0.5 M KCl) at ring-disc electrodes, oxidation of Br^- at the disc and reduction of Br_2 at the ring electrodes in 1 M H_2SO_4, oxidation of silver deposited on a gold disc and reduction of silver ions at a ring electrode.

Six electrodes having various r_1, r_2 and r_3 values were used. The experimental results were in good agreement with theory, whereas the values of N calculated from the Ivanov and Levich equation were higher, in some cases by as much as 25%, than the experimental results. The greatest discrepancies were observed when the distance between the ring and the disc was small and the ring was narrow.

Matsuda [40] also considered in a general way the effectiveness of collection N for a system of two electrodes situated close to one another and placed in a current of liquid. He derived a general equation and applied it to description of collection efficiency of a ring electrode surrounding a disc. This equation leads to the relationship given by Albery and Bruckenstein.

The processes taking place at a ring electrode have also been described by the numerical simulation method introduced by Feldberg [41, 42]. By means of this method Bard and Prater [43] obtained results which were in agreement with those of earlier work [38] on N. The problem of comparison of solutions regarding N obtained by the analytical method and by the simulation method has been discussed by Albery and Drury [44].

Prater and co-workers [45] studied the distribution of current density on the ring electrode. They showed theoretically by the simulation method that this distribution is not homogeneous even in the case when the current density on the disc electrode is homogeneous, which is observed in the case of limiting current. Newman [46] reached a similar conclusion.

Tokuda and Matsuda [47] considered the description of current–potential curves on the ring electrode for sizes of electrodes which can be readily used in practice. In their considerations they have taken into account the possibility of variable rate of electrode reaction in addition to variable rate of convective diffusion.

In the case of a reversible reduction leading to a soluble product the equation describing the current flowing in the ring electrode circuit has the following form:

$$i_p = 2\pi \int_{r_2}^{r_3} i_r r\,dr = 0.62\pi nF(r_3^3 - r_2^3)^{2/3} v^{1/6} \omega^{1/2} D_{Ox}^{2/3} C_{Ox}^0 \Phi^r(Z_0; L, \Theta) \quad (11.47)$$

where $\Phi^r(Z_0; L, \Theta) = \dfrac{1}{1+e^\Theta} - \dfrac{l^2 N}{[(1+Z_0^3)^3 - 1]^{2/3}}$; $\Theta = \dfrac{nF}{RT}(E - E_{1/2}^r)$;

$$l = \frac{r_1}{r_2}; \quad Z_0 = \left(\frac{r_3 - r_2}{r_2}\right)^{1/3}.$$

In the case of an irreversible electrode process the exact solution cannot be represented in the form of an analytical relationship, but Tokuda and Matsuda gave a general approximate relationship which is sufficiently accurate for the description of currents flowing in the case of electrodes having the sizes suitable for practical application.

Miller and Visco [48] have developed a variant of the ring-disc electrode in which the ring is divided into two half-rings. Thanks to the possibility of maintaining different potentials on the two half-rings such electrodes are particularly useful in kinetic studies.

Ring-disc electrodes have been used in studies on electrode and chemical kinetics. The latter applications will be discussed in the next chapter.

In the studies on electrode kinetics ring-disc electrodes are particularly useful in the investigation of mechanism of electrode processes taking place in several stages [49, 50]. For example, this method has been used in the studies on electroreduction of copper(II). It has been shown [50] that copper(I) is formed on a disc electrode in 1 M Na_2SO_4 when the disc electrode potential is more positive than the $E_{1/2}$ potential of the reduction wave on the disc. In such a case it has been found by maintaining the potential of ring electrode at a value sufficient for oxidation of copper(I) that a current flows in the ring electrode circuit as a result of oxidation of copper(I) which is formed as a transitional product on the disc, from which it penetrates to the ring electrode region. The curve of dependence of this current on disc electrode potential was bell-shaped.

Studies on anodic oxidation of copper in 1 M NaOH on a divided ring electrode have been described by Miller [50].

Kiss and Farkas [51, 52] carried out a series of studies on ionization of metals by means of a rotating electrode with a ring. The purpose of these studies was derivation of general relationships and the experimental work consisted in the investigation of anodic oxidation of copper.

In the studies on anodic oxidation of indium in $HClO_4$ solutions [48] it has been found that indium(I) is formed. Electroreduction of oxygen in NaOH solutions on a platinum disc electrode, in which hydrogen peroxide is formed as an intermediate product, was also studied [53].

Theoretical problems connected with ring-disc electrodes have been discussed by Albery and Hitchman [54] and by Pleskov and Filinovskii [55] in their monographs.

References

[1] Lundquist, J. T., Jr. and Nicholson, R. S., *J. Electroanal. Chem.*, **16**, 445 (1968).
[2] Kimmerle, F. M. and Chevalet, J., *J. Electroanal. Chem.*, **21**, 237 (1969).
[3] Chevalet, J. and Kimmerle, F. M., *J. Electroanal. Chem.*, **25**, 275 (1970).
[4] Anson, F. C., *Anal. Chem.*, **38**, 54 (1966).
[5] Anson, F. C. and Payne, D. A., *J. Electroanal. Chem.*, **13**, 35 (1967).
[6] Christie, J. H., Osteryoung, R. A. and Anson, F. C., *J. Electroanal. Chem.*, **13**, 236 (1967).
[7] Anson, F. C., *J. Phys. Chem.*, **72**, 727 (1968).
[8] Buck, R. P., *J. Electroanal. Chem.*, **23**, 219 (1969).
[9] Feldberg, S. W., *J. Phys. Chem.*, **74**, 87 (1970).
[10] Kalousek, M., *Chem. Listy*, **40**, 149 (1946); *Coll. Czechoslov. Chem. Communs.*, **13**, 105 (1948).
[11] Kalousek, M. and Rálek, M., *Chem. Listy*, **48**, 808 (1954); *Coll. Czechoslov. Chem. Communs.*, **19**, 1099 (1954).
[12] Kambara, T., *Bull. Chem. Soc. Japan*, **27**, 529 (1954).
[13] Barker, G. C., "*Polarographic Theory*", Part 1, AERE C/R 1553, Harwell 1957.
[14] Koutecký J., *Coll. Czechoslov. Chem. Communs.*, **21**, 433 (1956).
[15] Ružić, I., *J. Electroanal. Chem.*, **39**, 111 (1972).
[16] Mašek, M., *Coll. Czechoslov. Chem. Communs.*, **26**, 195 (1959).
[17] Vlček, A. A., "*Progress in Polarography*", Zuman, P., Ed. Interscience, New York 1962, Vol. 1, page 269.
[18] Kinard, W. F., Philp, R. H. and Propst, R. C., *Anal. Chem.*, **39**, 1556 (1967).
[19] Kemula, W. and Kublik, Z., *Roczniki Chem.*, **32**, 941 (1958).
[20] Kemula, W. and Kublik, Z., *Bull. Acad. Polon. Sci.*, cl. III, **6**, 653 (1958).
[21] Matsuda, H., *Z. Elektrochem.*, **61**, 489 (1957).
[22] Gokhshtein, Ya. P., *Dokl. Akad. Nauk SSSR*, **126**, 598 (1959).
[23] Nicholson, R. S. and Shain, I., *Anal. Chem.*, **36**, 706 (1964).
[24] Olmstead, M. L. and Nicholson, R. S., *Anal. Chem.*, **38**, 150 (1966).
[25] Reinmuth, W. H., *J. Am. Chem. Soc.*, **79**, 6358 (1957).
[26] Gumiński, C. and Galus, Z., *Roczniki Chem.*, **43**, 2147 (1969).
[27] Beyerlein, F. H. and Nicholson, R. S., *Anal. Chem.*, **44**, 1647 (1972).
[28] Nicholson, R. S., *Anal. Chem.*, **37**, 1351 (1965).
[29] Berzins, T. and Delahay, P., *J. Am. Chem. Soc.*, **75**, 4205 (1953).

[30] Murray, R. W. and Reilley, C. N., *J. Electroanal. Chem.*, **3**, 182 (1962).
[31] Testa, A. C. and Reinmuth, W. H., *Anal. Chem.*, **33**, 1324 (1961).
[32] van Leeuwen, H. P. and Sluyters, J. H., *J. Electroanal. Chem.*, **39**, 25 (1972).
[33] van Leeuwen, H. P. and Sluyters, J. H., *J. Electroanal. Chem.*, **39**, 233 (1972).
[34] Herman, H. B. and Bard, A. J., *Anal. Chem.*, **35**, 1121 (1963).
[35] Frumkin, A. N. and Nekrasov, L. N., *Dokl. Akad. Nauk SSSR*, **126**, 115 (1959).
[36] Ivanov, Yu. B. and Levich, V. G., *Dokl. Akad. Nauk SSSR*, **126**, 1029 (1959).
[37] Albery, W. J., *Trans. Faraday Soc.*, **62**, 1915 (1966).
[38] Albery, W. J. and Bruckenstein, S., *Trans. Faraday Soc.*, **62**, 1920 (1966).
[39] Bruckenstein, S. and Feldman, G. A., *J. Electroanal. Chem.*, **9**, 395 (1965).
[40] Matsuda, H., *J. Electroanal. Chem.*, **16**, 153 (1968).
[41] Feldberg, S. W. and Auerbach, C., *Anal. Chem.*, **36**, 505 (1964).
[42] Feldberg, S. W., in *"Electroanalytical Chemistry"* Vol. 3, Bard, A. J., Ed., Dekker, New York 1969.
[43] Bard, A. J. and Prater, K. B., *J. Electrochem. Soc.*, **117**, 207 (1970).
[44] Albery, W. J. and Drury, J. S., *J. Chem. Soc., Faraday Trans. I*, **68**, 456 (1972).
[45] Neubert, G., Gorman, E., Van Reet, R. and P.ater, K. B., *J. Electrochem. Soc.*, **119**, 677 (1972).
[46] Newman, J., *J. Electrochem. Soc.*, **119**, 212 (1972).
[47] Tokuda, K. and Matsuda, H., *J. Electroanal. Chem.*, **44**, 199 (1973).
[48] Miller, B. and Visco, R. E., *J. Electrochem. Soc.*, **115**, 251 (1968).
[49] Nekrasov, L. N. and Berezina, N. P., *Dokl. Akad. Nauk SSSR*, **142**, 855 (1962).
[50] Miller, B., *J. Electrochem. Soc.*, **116**, 1675 (1969).
[51] Kiss, L. and Farkas, J., *Acta Chim. Acad. Sci. Hung.*, **64**, 241 (1970); **65**, 7 (1970); **65**, 141 (1970); **66**, 33 (1970); **66**, 395 (1970); **69**, 167 (1971).
[52] Kiss, L., Farkas, J. and Körösi, A., *Acta Chim. Acad. Sci. Hung.*, **67**, 179 (1971); **68**, 359 (1971).
[53] Nekrasov, L. N. and Myuller, L., *Dokl. Akad. Nauk SSSR*, **149**, 1107 (1963); Muyller, L. and Nekrasov, L. N., *Dokl. Akad. Nauk SSSR*, **157**, 416 (1964).
[54] Albery, H. J. and Hitchman, M. L., *"Ring-Disc Electrodes"*, Clarendon Press, Oxford 1971.
[55] Pleskov, Yu. V. and Filinovskii, V. Yu., *"Rotating Disc Electrode"*, Nauka, Moscow 1972 (in Russian).

Chapter 18

Cyclic Methods. Kinetic Processes

In the previous chapter we discussed the application of cyclic methods to diffusion limited processes. However, it has been mentioned that the cyclic methods are suitable for investigation of chemical reactions following the primary charge transfer. When the oxidized substance Ox is present in the solution and when the product of its reduction is the substrate of a rapid chemical reaction, the current at the transition time, measured in the second half of the cycle, during the oxidation of the reduction product, is smaller than the corresponding quantity measured in the absence of chemical reactions.

The first mathematical treatments of such processes were put forward in 1960, for current reversal chronopotentiometry. Later, similar treatments of cyclic stationary electrode voltammetry and chronoamperometry were published.

18.1 Cyclic Chronoamperometry and Double Potential Step Chronocoulometry

The case of an electrode process followed by an irreversible chemical reaction was considered by Schwarz and Shain [1]. After the potential jump from the initial value to that at which the reduction:

$$Ox + ne \rightarrow Red \tag{18.1}$$

takes place rapidly and the product is transformed into substance A which is inactive in the range of applied potentials in use:

$$Red \underset{k_2}{\overset{k_1}{\rightleftharpoons}} A \tag{18.2}$$

the potential then rapidly returns to the initial value at which the form Red is oxidized:

$$Red - ne \rightarrow Ox. \tag{18.3}$$

The solution of this, obtained by means of the Laplace transformation, can be expressed in the following form:

$$\frac{i_k}{i_g} = \Phi\left[k_1 t', \frac{(t-t')}{t'}\right] - \sqrt{\frac{\left(\frac{t-t'}{t'}\right)}{1+\frac{(t-t')}{t'}}} \tag{18.4}$$

where t' is the reduction time after which the potential returns to the initial value, i_g is the limiting cathodic current, and i_k is the kinetic anodic current.

Since equation (18.4) contains function Φ, it is convenient to determine k_1 from working curves plotted on the basis of this equation. These curves, representing the dependence of i_k/i_g on $k_1 t'$, are plotted for various values of the $(t-t')/t'$ ratio. Many such curves have been calculated by Schwarz and Shain [1] who also checked the theoretical relationship, using the reduction of azobenzene to hydrazobenzene, which in strongly acid media is followed by the benzidine rearrangement. According to these authors, in suitable conditions the rate constants of this reaction can be determined up to 5×10^4 s^{-1}, but as a result of convection it is difficult to measure rate constants lower than 3.5×10^{-2} s^{-1}.

The method proposed by Schwarz and Shain was modified by Lundquist and Nicholson [2]. After the initial rapid potential change, where reaction (18.1) is limited by the diffusion rate only, the anodic process (18.3) takes place under conditions of stationary electrode voltammetry with linear potential change, and the recorded anodic current peaks depend on the polarization rate, rate constant k_1 and time t', measured from the beginning of the cathodic electrolysis to the start of the anodic oxidation.

The relation between the current function and the ratio k_1/a (where $a = nFV/RT$) at constant values of at' is shown in Fig. 18.1.

At small values of k_1/a the maximum value of the current function (measured from the extrapolation of the current–time curve) $\sqrt{\pi}\,\chi_p(at)$ always tends to 0.446 of the value characteristic of the diffusion process under consideration. At constant values of at', $\sqrt{\pi}\,\chi_p(at)$ decreases with increasing k_1/a, but the peak height is a function of k_1/a and at'.

In order to calculate the rate constants it is convenient to utilize working curves showing the dependence of, $\sqrt{\pi}\,\chi_p(at)$ on k_1/a, for various values of at'. Such curves can be plotted from the values given in Table 18.1 [2]. According to Lundquist and Nicholson this method is superior to that proposed by Schwarz and Shain.

The relationships resulting from the above considerations were used to determine the rate constants of the benzidine rearrangement of hydrazobenzene and *m*-hydrazotoluene.

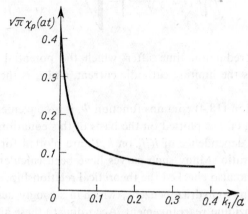

Fig. 18.1 Dependence of $\sqrt{\pi}\chi_p(at)$ on the ratio of chemical reaction rate constant to parameter a; $at' = 62.5$.

Christie [3] made a theoretical study of the application of double step chronocoulometry to the investigation of first-order chemical reactions following electrode reduction processes.

Expressions describing the time dependence of the charge were obtained by Schwarz and Shain [1] by integration of the current–time equations. They are fairly complex and therefore it is more convenient to use the working curve of the dependence of the ratio of charge in the second half-cycle Q_b [reaction (18.3)], to the charge in the first half-cycle Q_f [reaction (18.1)] on parameter $(k_1 t')^{1/2}$. For catalytic reactions Christie expressed the ratio of charges by the following equation:

$$\frac{Q_b}{Q_f} = \frac{Q(2t') - Q(t')}{Q(t')} = 2 - \frac{\left(2\beta^2 t' + \frac{1}{2}\right)\mathrm{erf}(2\beta^2 t')^{1/2} + \left(\frac{2\beta^2 t'}{\pi}\right)^{1/2} e^{-2\beta^2 t'}}{\left(\beta^2 t' + \frac{1}{2}\right)\mathrm{erf}(\beta^2 t')^{1/2} + \left(\frac{\beta^2 t'}{\pi}\right)^{1/2} e^{-\beta^2 t'}}$$

(18.5)

where $\beta^2 = k_1 C_Z^0$ (it is assumed that the excess of substance Z transforming Red into Ox is so large that its concentration is practically constant throughout the experiment).

In the derivation of equation (18.5) it was assumed that the diffusion coefficients of Ox and Red are equal.

18.1] Cyclic Chronoamperometry; Double Step Chronocoulometry

Table 18.1 Dependence of $\sqrt{\pi}\chi_p(at)$ on k_1/a and at' (from [2] by permission of the copyright holders, Elsevier Publishing Co.)

$at' = 0.5$		$at' = 2.5$		$at' = 7.5$		$at' = 20$		$at' = 37.5$		$at' = 62.5$	
k_1/a	$\sqrt{\pi}\chi_p(at)$	k_1/a	$\sqrt{\pi}\chi_p(at)$	k_1/a	$\sqrt{\pi}\chi_p(at)$	k_1/a	$\sqrt{\pi}\chi_p(at)$	k_1/a	$\sqrt{\pi}\chi_p(at)$	k_1/a	$\sqrt{\pi}\chi_p(at)$
0.0250	0.412	0.0100	0.426	0.0067	0.425	0.0010	0.442	0.0010	0.440	0.0010	0.430
0.0400	0.392	0.0200	0.410	0.0133	0.405	0.0030	0.438	0.0027	0.424	0.0024	0.415
0.0600	0.368	0.0400	0.377	0.0267	0.373	0.0050	0.415	0.0100	0.363	0.0050	0.382
0.1000	0.329	0.0600	0.347	0.0400	0.345	0.0100	0.393	0.0160	0.328	0.0075	0.355
0.1600	0.281	0.0800	0.325	0.0670	0.299	0.0200	0.347	0.0267	0.271	0.0100	0.324
0.2000	0.259	0.1200	0.286	0.1000	0.258	0.0300	0.313	0.0400	0.225	0.0150	0.283
0.2500	0.239	0.2000	0.230	0.1330	0.228	0.0500	0.260	0.0533	0.196	0.0250	0.226
0.4000	0.201	0.3200	0.193	0.2000	0.191	0.0750	0.216	0.0800	0.159	0.0500	0.155
0.5000	0.185	0.4000	0.183	0.2760	0.172	0.1400	0.160	0.1000	0.140	0.0800	0.121
0.7500	0.168	0.5000	0.175	0.4000	0.149	0.2500	0.128	0.2000	0.107	0.1500	0.094
—	—	—	—	—	—	0.4000	0.110	0.4000	0.087	0.4000	0.068

Values of function $\sqrt{\pi}\chi_p(at)$ calculated from equation:

$$\sqrt{\pi}\chi_p(at) = \frac{i_p}{i_{t'}(\pi at')^{1/2}}$$

where i_p is the experimental peak current and $i_{t'}$ is the current after time t'.

Two limiting cases of this equation are particularly simple:

(1) when $\beta^2 t' \to 0$, $\left|\dfrac{Q_b}{Q_f}\right| \to 2 - \sqrt{2}$; as expected, the ratio of charges tends to the value predicted for a diffusion controlled process when the chemical reaction rate decreases considerably;

(2) when $\beta^2 t' \to \infty$, $\left|\dfrac{Q_b}{Q_f}\right| \to 0$. In this case the catalytic reaction is very fast and, as a result, practically all substance Red is consumed in the chemical reaction. Hence the change in the oxidation half-cycle is practically equal to zero.

This theory was used by Lingane and Christie [4] in their studies of the catalytic oxidation of electrolytically generated titanium(III) with hydroxylamine. The rate constant, 43.4 ± 4.1 l. mole^{-1}s^{-1} obtained was in very good agreement with results by other methods.

Table 18.2 Dependence of $(k_1 t')^{1/2}$ on Q_b/Q_f in chronocoulometry with double potential jump for an electrode process followed by an irreversible chemical reaction of the (pseudo) first order (from [5] by permission of the copyright holders, Elsevier Publishing Co.)

$(k_1 t')^{1/2}$	Q_b/Q_f	$(k_1 t')^{1/2}$	Q_b/Q_f
0.00	0.5858	1.10	0.2398
0.10	0.5808	1.20	0.2085
0.20	0.5662	1.30	0.1809
0.30	0.5429	1.40	0.1569
0.40	0.5123	1.50	0.1362
0.50	0.4761	1.60	0.1186
0.60	0.4363	1.70	0.1037
0.70	0.3947	1.80	0.0911
0.80	0.3530	1.90	0.0805
0.90	0.3126	2.00	0.0716
1.00	0.2746		

Results of later work of Reilley and co-workers [5] have shown that Christie's result regarding the description of the working curve and the dependence of Q_b/Q_f on $(k_1 t')^{1/2}$ for an electrode process followed by an irreversible first-order chemical reaction is incorrect, particularly for $(k_1 t')^{1/2} > 0.5$. The correct results obtained by Reilley and co-workers are shown in Table 18.2.

In theoretical considerations of the discussed mechanism of EC the adsorption of substrate and of primary product of the electrode reaction has also been taken into account.

The correctness of the theoretical results has been proved by experimental investigation of electroreduction of azobenzene in acidic aqueous ethanolic media [6]. The resulting hydrazobenzene is largely adsorbed on the mercury and then rearranges to benzidine. The experimentally determined constant 4.27 s^{-1} for 1.18 M HClO$_4$ is in a very good agreement with that obtained earlier [1] by the chronoamperometric method.

Olmstead and Nicholson [7] have developed the theory of an electrode process followed by dimerization taking place under cyclic chronoamperometry conditions. The results of this work cannot be presented in analytical form but the tabulated data make it possible to determine rate constants of irreversible dimerizations on condition that the linearity of diffusion is maintained.

18.2 Cyclic Stationary Electrode Voltammetry

Cyclic stationary electrode voltammetry is as suitable as chronoamperometry for investigating the kinetics of consecutive reactions. First order reversible and irreversible consecutive processes have been studied theoretically by Nicholson and Shain [8]. When reduction is the primary electrode process the anodic current depends mainly on the rate of the chemical reaction, but the anodic current also depends on the potential at which the direction of polarization is reversed. For this reason the values of $\chi(at)$ are not quoted in the literature, although such functions were calculated by Nicholson and Shain.

The anodic current can be utilized in kinetic measurements by selecting the potential at which the direction of the current is reversed and preparing working curves of the dependence of i_{pa}/i_{pc} on $K\sqrt{a/(k_1+k_2)}$ for the potential selected.

The working curve for the potential of reversal of polarization direction, given by the equation $(E_z-E_{1/2})_n-(RT/F)\ln(K+1) = -90$ mV, is shown in Fig. 18.2; K is the equilibrium constant of the chemical reaction determined by the ratio of the rate constants k_1/k_2, k_1 is the rate constant of transformation of Red into A, k_2 is the rate constant of transformation of A into Red, and E_z is the polarization reversal potential.

Nicholson and Shain obtained a solution for the case of an irreversible chemical reaction, in the form of a relation between the ratio of the anodic peak current (i_{pa}) to cathodic peak current (i_{pc}) and $k\Delta t$, where Δt is the time (in seconds) during which the potential changes from $E_{1/2}$ to E_z at a given rate of polarization. This relationship is shown in Table 18.3.

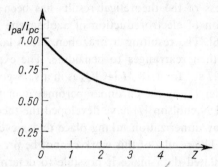

Fig. 18.2 The dependence of the ratio of anodic to cathodic currents peaks on parameter $Z = K\sqrt{a/(k_1+k_2)}$ in an electrode process followed by a reversible chemical reaction (from [8] by permission of the copyright holders, the American Chemical Society).

Table 18.3 RATIO OF ANODIC TO CATHODIC PEAK CURRENT IN A CYCLIC PROCESS WITH AN IRREVERSIBLE FIRST-ORDER CHEMICAL REACTION FOLLOWING THE CATHODIC REDUCTION (FROM [8] BY PERMISSION OF THE COPYRIGHT HOLDERS, THE AMERICAN CHEMICAL SOCIETY)

$k_1 \Delta t$	$\dfrac{i_{pa}}{i_{pc}}$	$k_1 \Delta t$	$\dfrac{i_{pa}}{i_{pc}}$
0.004	1.00	0.525	0.641
0.023	0.986	0.550	0.628
0.035	0.967	0.778	0.551
0.066	0.937	1.050	0.486
0.105	0.900	1.168	0.466
0.195	0.828	1.557	0.415
0.350	0.727		

The anodic current is measured from the extrapolation of the cathodic current line, as was described in the previous chapter. In this case the ratio i_{pa}/i_{pc} of the diffusion process currents is equal to one.

The solution, shown in Table 18.3, is much simpler than that obtained for reversible chemical reactions. When the chemical reaction is irreversible E_z can be selected fairly arbitrarily, since the ratio of peaks depends on Δt.

An interesting method of determining the current ratio for primary electrode processes, followed by chemical reactions, was proposed by Nicholson [9].

Kuempel and Schaap [10] used cyclic stationary electrode voltammetry in their studies on the rates of ligand exchange between cadmium and $Ca(EDTA)^{2-}$ ions. They interpreted their results using the theory deduced by Nicholson and Shain [8].

The theory of electrode processes followed by disproportionation was developed by Olmstead and Nicholson [11], who assumed irreversibility of the chemical reaction and rapid charge transfer. They considered both flat and spherical electrodes and expressed their results in tabular form (Table 18.4).

Table 18.4 RATIO OF ANODIC TO CATHODIC PEAK CURRENT FOR PLANAR AND SPHERICAL ELECTRODES* (FROM [11] BY PERMISSION OF THE COPYRIGHT HOLDERS, THE AMERICAN CHEMICAL SOCIETY)

γ	$\varrho \times 10^4$				
	0.0	6.25	25.0	56.25	100.0
0.02	0.998	0.998	0.998	0.998	0.998
0.04	0.995	0.995	0.995	0.995	0.995
0.06	0.992	0.992	0.992	0.992	0.992
0.10	0.985	0.986	0.986	0.986	0.987
0.16	0.976	0.977	0.977	0.978	0.978
0.25	0.962	0.964	0.965	0.966	0.967
0.40	0.940	0.943	0.945	0.947	0.949
0.60	0.913	0.917	0.921	0.924	0.928
0.90	0.877	0.883	0.889	0.895	0.900
1.30	0.835	0.845	0.854	0.862	0.870
1.90	0.784	0.798	0.812	0.824	0.835
2.70	0.732	0.752	0.770	0.786	0.801
4.00	0.672	0.700	0.724	0.746	0.766
6.00	0.613	0.649	0.681	0.710	0.736

* For $\varrho > 0$, $\gamma = k_1 C_{Ox}^0 \Delta t$ and $a\Delta t = 4$; $P = \dfrac{D}{r_0^2 C_{Ox}^0 k_1}$, $a\Delta t = \dfrac{nF}{RT}(E^0 - E_t')$ and $\log \gamma = \log(k_1 C_{Ox}^0 \Delta t) + 0.047(a\Delta t - 4)$; Δt has the previously defined meaning.

Using their equation and Table 18.4 it is possible to plot working curves i_{pa}/i_{pc} vs. $k_1 C_{Ox}^0 \Delta t$, from which the disproportionation rate constant k_1 can be determined. This constant can also be determined in the case of spherical electrodes, but here the condition $a\Delta t = 4$ must be fulfilled, since only then can the values from Table 18.4 be used.

Shuman [12] investigated the effect of following reactions other than first-order on the cyclic curves. In his model the product of the primary reversible electrode reaction was in equilibrium with the final product, which was an inactive substance.

Feldberg [13] and Feldberg and Jeftić [14] investigated the finer details of the ECE mechanism (see Chapter 15) complicated by chemical interactions between Ox and Red', or between Ox and Ox'.

The problem of irreversible dimerization following the primary process (18.1) was treated theoretically by Olmstead et al. [15].

18.3 Current Reversal Chronopotentiometry

Processes (18.1)–(18.3) taking place under conditions of chronopotentiometry with reversal of current were studied theoretically by three groups of investigators independently [16–18]. The basic equation relating the rate constant of an irreversible chemical reaction to the reduction time (t_{Red}) and transition time of the oxidation process τ_{Ox} is as follows:

$$(\gamma + \lambda)\,\mathrm{erf}[k_1 \tau_{Ox}]^{1/2} = \lambda\,\mathrm{erf}[k_1(t_{Red} + \tau_{Ox})]^{1/2} \tag{18.6}$$

where $\gamma = i_0/nFD$, $\lambda = i'_0/nFD$, i_0 is the reduction current density, and i'_0 is the oxidation current density.

When $i_0 = i'_0$ equation (18.6) becomes:

$$2\,\mathrm{erf}[k_1 \tau_{Ox}]^{1/2} = \mathrm{erf}[k_1(t_{Red} + \tau_{Ox})]^{1/2}. \tag{18.6a}$$

Various methods of treatment for the experimental results have been suggested, but that proposed by Testa and Reinmuth [18] is probably the simplest. They plotted the working curve (Fig. 18.3) of the dependence of τ_{Ox}/t_{Red} on $k_1 t_{Red}$, which makes it possible to determine $k_1 t_{Red}$ from the experimentally found value of τ_{Ox}/t_{Red}.

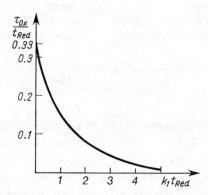

Fig. 18.3 Dependence of the ratio of oxidation transition time to reduction time on $k_1 t_{Red}$ (from [18] by permission of the copyright holders, the American Chemical Society).

18.4] Current Reversal Chronopotentiometry

For the primary reduction process the solution obtained by these authors can be represented by the equation

$$\tau_{Ox}/t_{Red} = \frac{\ln(1+u-ue^{-k_1 t_{Red}})}{k_1 t_{Red}} \tag{18.7}$$

where u is the ratio of current intensities in the reduction and oxidation cycle. The effect of partial dissolution of Red has also been considered.

The working curve of the dependence of τ_{Ox}/t_{Red} on $k_1 t_{Red}$ plotted on the basis of this equation has a shape identical with that of the working curve shown in Fig. 18.3, but the ratios τ_{Ox}/t_{Red} are approximately three times larger in a wide interval of $k_1 t_{Red}$ parameter values.

Dračka [19] investigated electrode processes followed by higher order irreversible reactions and derived the following equation for reactions of mth order:

$$\frac{2}{k_1^{m+1}} \frac{(1-m)}{D_{Red}^{m+1}} \left(\frac{i_0}{nF}\right)^{\frac{2(m-1)}{m+1}} \tau_{Ox} = f_m(u) \tag{18.8}$$

where u is the ratio of the reduction current to the oxidation current, and $f_m(u)$ is a function of this ratio.

When $m = 2$ equation (18.8) becomes:

$$k_1^{2/3} D_{Red}^{-1/3} \left(\frac{i_0}{nF}\right)^{2/3} \tau_{Ox} = f_2(u). \tag{18.8a}$$

When $u = 1$, the case usually observed in practice, $f_2(u) = 0.355$.

The order of the chemical process can be readily determined by plotting log τ_{Ox} against log i_0. This should be linear and the slope of the straight line should be $P = \Delta \log \tau_{Ox}/\Delta \log i_0$.

The reaction order m is connected with P according to the following simple equation:

$$m = \frac{2-P}{2+P}. \tag{18.9}$$

It follows from equation (18.7) that in the case of an irreversible following chemical reaction, the oxidation transition time does not depend on the reduction time (neglecting short reduction times), but it depends on the current density.

Feldberg used his methods to solve [20] various problems connected with first and second order following reactions in chronopotentiometry with reversal of current direction.

Herman and Bard [21, 22] developed the theory of cyclic chronopotentiometry for kinetically complicated systems taking into account consecutive and catalytic reactions and tabulated the values of Q_n for various values of n and $k_1 \tau_1$. These tables make it possible to determine the rate constants of the reactions investigated. In reported experimental studies they investigated the electroreduction of *p*-nitrosophenol and titanium(IV) in the presence of hydroxylamine.

Cyclic chronopotentiometry of kinetically complicated processes was also investigated by Vuković and Pravdić [23].

18.4 The Rotating Ring-Disc Electrode Method

Kinetics of chemical reactions can also be investigated by the rotating ring-disc electrode method.

In the case of systems (18.1)–(18.3) reaction (18.1) takes place on the disc electrode and some of the resulting Red is oxidized on the ring electrode. The efficiency of collection N_k in the case of kinetic processes is of course lower than it is in the case of diffusion processes of reduction on the disc, and oxidation on the ring, electrodes. Bruckenstein and Feldman [24] obtained an approximate solution of this problem for first-order irreversible chemical reactions, and a more exact solution was published by Albery and Bruckenstein [25].

When $k_1^{1/2}(0.51\omega^{3/2}v^{-1/2}D^{1/2})^{2/3} < 0.5$ the final equation becomes:

$$\frac{N}{N_k} = 1 + 1.28 \left(\frac{v}{D}\right)^{1/3} \frac{k_1}{\omega}. \tag{18.10}$$

According to Albery and Bruckenstein this method can be used for the investigation of reaction kinetics, when the rate constants are higher than 4×10^{-2} s^{-1}, but lower than 10^3 s^{-1}. Thus this method cannot be used for detecting intermediate products having a half-life period shorter than 10^{-3} s.

Albery and Bruckenstein have also considered the case of a reversible chemical reaction [26] and a second order chemical reaction [27].

The correctness of the theory elaborated for an electrode process followed by a chemical reaction has been checked [28] by Albery and co-workers on the examples of an earlier investigated bromination reaction of anisole with bromine generated on a disc and the reaction of VO_2^+ with iron(II) generated on a disc electrode in an acid medium. The agreement of the results with the values of constants obtained earlier by other methods was satisfactory.

Albery and co-workers [29] have also improved and further developed the theory of electrode processes followed by irreversible second-order reactions. All these studies as well as other investigations are discussed in Albery and Hitchman's monograph [30].

Prater and Bard [31] have presented theoretical considerations regarding ring-disc electrodes without the kinetic complications obtained in the numerical simulation method and applied this technique to the case in which a substance generated on a disc electrode participates in a first or second order chemical reaction [32]. Their results are in a good agreement with those obtained by less exact methods [25, 27–29]. These authors have elaborated the theory of a catalytic process accompanied by ECE reactions [33]. The chemical reactions involved in these processes were of the first or second order. The results have been presented in the form of working curves which make it possible to define the mechanism of the reactions examined and to determine their rate constants.

The theory of catalytic processes has been checked on the example of reaction of iron(II) generated on a disc electrode from iron(III) with hydrogen peroxide in hydrochloric acid, using carbon paste electrodes. The value of the constant obtained, 96 l. mole^{-1} s^{-1} is in good agreement with those reported in earlier publications.

Puglisi and Bard [34] have also applied the numerical simulation method to the investigation of electrode processes in which a substance generated on a disc electrode partly reaches the ring electrode and partly participates in a dimerization reaction or a second-order reaction with the substrate. They have considered two cases.

(1) The electrode reaction (18.1) is followed by chemical reaction

$$\text{Ox} + \text{Red} \xrightarrow{k} \text{C}.$$

Substance C is immediately reduced on the disc electrode. It is assumed that on the ring electrode only substance Red or substances Red and C are oxidized.

(2) Substance Red generated on the disc reacts with Ox and gives an electrode inactive substance Y:

$$\text{Ox} + \text{Red} \xrightarrow{k} \text{Y}.$$

On the ring electrode reaction (18.3) takes place.

The authors have given diagnostic criteria making it possible to differentiate between the mechanisms discussed. The working curves make it possible to calculate rate constants of chemical stages occurring in these processes.

References

[1] Schwarz, W. M. and Shain, I., *J. Phys. Chem.*, **69**, 30 (1965).
[2] Lundquist, J. T., Jr., and Nicholson, R. S., *J. Electroanal. Chem.* **16**, 445 (1968).
[3] Christie, J. H., *J. Electroanal. Chem.*, **13**, 79 (1967).
[4] Lingane, P. J. and Christie, J. H., *J. Electroanal. Chem.*, **13**, 227 (1967).
[5] Ridgway, T. H., Van Duyne, R. P. and Reilley, C. N., *J. Electroanal. Chem.*, **34**, 267 (1972).
[6] Van Duyne, R. P., Ridgway, T. H. and Reilley, C. N., *J. Electroanal. Chem.*, **34**, 283 (1972).
[7] Olmstead, M. L. and Nicholson, R. S., *Anal. Chem.*, **41**, 851 (1969).
[8] Nicholson, R. S. and Shain, I., *Anal. Chem.*, **36**, 706 (1964).
[9] Nicholson, R. S., *Anal. Chem.*, **38**, 1406 (1966).
[10] Kuempel, J. R. and Schaap, W. B., *Inorg. Chem.*, **7**, 2435 (1968).
[11] Olmstead, M. L. and Nicholson, R. S., *Anal. Chem.*, **41**, 862 (1969).
[12] Shuman, M. S., *Anal. Chem.*, **42**, 521 (1970).
[13] Feldberg, S. W., *J. Phys. Chem.*, **75**, 2377 (1971).
[14] Feldberg, S. W. and Jeftić, Lj., *J. Phys. Chem.*, **76**, 2349 (1972).
[15] Olmstead, M. L., Hamilton, R. G. and Nicholson, R. S., *Anal. Chem.*, **41**, 260 (1969).
[16] Dračka, O., *Coll. Czechoslov. Chem. Communs.*, **25**, 338 (1960).
[17] Furlani, C. and Morpurgo, G., *J. Electroanal. Chem.*, **1**, 351 (1960).
[18] Testa, A. C. and Reinmuth, W. H., *Anal. Chem.*, **32**, 1512 (1960).
[19] Dračka, O., *Coll. Czechoslov. Chem. Communs.*, **26**, 2144 (1961).
[20] Feldberg, S. W. and Auerbach, C., *Anal. Chem.*, **36**, 505 (1964).
[21] Herman, H. B. and Bard, A. J., *J. Phys. Chem.*, **70**, 396 (1966).
[22] Herman, H. B. and Bard, A. J., *J. Electrochem. Soc.*, **115**, 1028 (1968).
[23] Vuković, M. and Pravdić, V., *Croat. Chem. Acta*, **42**, 21 (1970).
[24] Bruckenstein, S. and Feldman, G. A., *J. Electroanal. Chem.*, **9**, 395 (1965).
[25] Albery, W. J. and Bruckenstein, S., *Trans. Faraday Soc.*, **62**, 1946 (1966).
[26] Albery, W. J. and Bruckenstein, S., *Trans. Faraday Soc.*, **62**, 2598 (1966).
[27] Albery, W. J. and Bruckenstein, S., *Trans. Faraday Soc.*, **62**, 2584 (1966).
[28] Albery, W. J., Hitchman, M. L. and Ulstrup, J., *Trans. Faraday Soc.*, **64**, 2831 (1968).
[29] Albery, W. J., Hitchman, M. L. and Ulstrup, J., *Trans. Faraday Soc.*, **65**, 1101 (1969).
[30] Albery, W. J. and Hitchman, M. L., "*Ring-Disc Electrodes*", Clarendon Press, Oxford 1971.
[31] Prater, K. B. and Bard, A. J., *J. Electrochem. Soc.*, **117**, 207 (1970).
[32] Prater, K. B. and Bard, A. J., *J. Electrochem. Soc.*, **117**, 335 (1970).
[33] Prater, K. B. and Bard, A. J., *J. Electrochem. Soc.*, **117**, 1517 (1970).
[34] Puglisi, V. J. and Bard, A. J., *J. Electrochem. Soc.*, **119**, 833 (1972).

Chapter **19**

Deviations from Normal Depolarizer Transport

In these discussions it has been assumed so far that electrode processes take place at definite transport rates, and therefore the current and transition time values can be predicted for any moment of the electrolysis time. However, in each method discussed the agreement between theoretical and experimentally determined values of limiting current and transition time is observed only in a certain range of transport rate values. Deviations from theoretical values are particularly pronounced at slow transport rates, as a result of the appearance of convection transport in chronoamperometry, chronopotentiometry and stationary electrode voltammetry, and as a result of intensification of diffusional transport in the rotating disc method. Sometimes excessive limiting currents are recorded even when optimal transport rates are used. This phenomenon is often observed in polarography where the abnormal current increases are called polarographic maxima. In order to elucidate these phenomena it is necessary to examine the conditions under which the agreement between observed and theoretical transport rates is satisfactory.

19.1 Polarography

Many authors have shown that the relation between the depolarizer concentration and the corresponding limiting current is linear, when the drop time is maintained between 1 and 8 seconds. This time interval t_1 corresponds to the most suitable transport rates in polarography, but under these conditions polarographic maxima may be observed. Usually, when the applied potential is increased a sudden decrease in the current to the theoretically predicted value occurs at a certain potential of the dropping mercury electrode.

Polarographic maxima are generally divided into two groups. Maxima of the first kind appear at the beginning of a polarographic wave and are most frequently observed when dilute solutions of the supporting electrolyte are used. In most cases they are in the form of sharp peaks, and the decrease of the current is much more rapid than its increase. Maxima of the second kind are also observed on limiting currents, but they usually appear when normal, or more concentrated solutions, of the supporting electrolyte are used. In this case the current increases slowly and, when the maximum of the second kind appears at the beginning of the limiting current and is not large, it can be difficult to observe.

The maxima of the third kind appear only in the presence of surface active substances. Their occurrence has been described in many publications [1–3]. It has been suggested that they can appear if adsorption equilibrium is not reached [1] or when there is a strong interaction between the adsorbed molecules.

Recently Frumkin and co-workers [4] investigated in detail the conditions of appearance of these maxima. They showed that the maxima are caused by turbulent movements of the dropping electrode surface which appear during adsorption of sparingly soluble surface active substances. At relatively high concentrations of such substances in the solution condensed surface layers are formed.

The potentials of appearance of the maxima of the third kind are similar to those of the maximum values of function $\partial \Gamma / \partial C$ (change of the surface concentration due to changing concentration of the surface active substance).

A typical example of a maximum of the first kind is that observed on the first wave of oxygen reduction in dilute (of the order of 10^{-3} M) solutions of supporting electrolyte [5, 6]. An oxygen wave with this maximum is shown in Fig. 19.1.

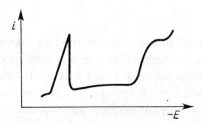

Fig. 19.1 Oxygen maximum in 0.001 M KCl solution.

Heyrovský [7] divided maxima of the first kind into two groups, depending on the potential of the dropping electrode with respect to the zero charge potential. The maxima observed at positive potentials are called positive and those observed at negative potentials are called negative. At the zero charge potential maxima of the first kind are not observed [8]. The heights of these maxima depend on the depolarizer concentration, but this dependence is not linear. When ordinary polarographic electrodes are used in dilute solutions the heights of the maxima increase linearly with $m^{2/3}$, provided that the drop time is constant. When the capillary yield is constant, maxima increase linearly with $t^{2/3}$ [9, 10]. All maxima, irrespective of their kind, decrease with increasing viscosity of the electrolyte solution.

Negative maxima may decrease with increasing concentration of cations in the solution and with increasing charge of these cations. The charge of anions does not effect negative maxima, but it has a pronounced effect on positive maxima, which do not depend on the charge of cations. Positive maxima are affected, not only by the charge of anions but also by their kind. Varasova [6] found that the effect of anions increases in the following order:

$$Cl^- < I^- < Br^- < BrO_3^- < NO_3^- < CO_3^{2-} < SO_4^{2-} < OH^-.$$

Maxima are also considerably affected by surface active substances such as pigments, alkaloids, colloids and high molecular weight aliphatic compounds, when they are present in the solution. In the case of high molecular weight hydrocarbons the potential at which the formation of the maximum is inhibited depends on the nature of the surface active agent. The formation of positive maxima is inhibited by acidic pigments and negatively charged colloids, and that of negative maxima by alkaline pigments, alkaloids, and positively charged colloids.

Uncharged substances and zwitterions (e.g. gelatine) inhibit the formation of maxima over a wide range of potentials. The potential ranges over which various substances have an inhibiting effect depend on the potential ranges in which these substances are adsorbed and, as a result, the inhibition range increases with increasing adsorption energy and with increasing concentration.

Obviously, at high concentrations of surface active agents the inhibiting effect on the maxima can be accompanied by a decrease in the electrode process rate.

The nature of polarographic maxima was first investigated by Heyrovský [5, 7]. In his opinion the current increases were due to adsorption of depolarizer on the electrode surface, this adsorption being due to heterogeneity of the electrical field around the mercury drop, caused by the flow of charging current. However, in 1938 Antweiler [11, 12] found that the presence of maxima is connected with streaming of the liquid in the neighbourhood of dropping electrodes. In the case of positive maxima the liquid moves from the neck of the drop downwards, while in the case of negative maxima it moves in the opposite direction (Fig. 19.2). The formation of the maxima is also connected with streaming of the mercury [11–14]. Maxima of the first kind are caused by movements of the solution, which are due

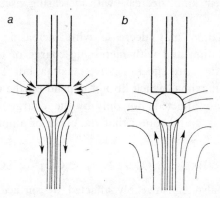

Fig. 19.2 Movement of liquid in the vicinity of the electrode in the case of maxima of the first kind; a—positive maximum; b—negative maximum.

to a potential difference between the neck of the drop at the end of the capillary and bottom of the drop. This potential difference is caused by the screening effect of the capillary glass on the electrical field of the drop. As a result of this screening effect, in the case of cathodic polarization, the potential of the neck of the drop is more positive than that of the bottom. When the electrode potential is positive with respect to the zero charge potential positive maxima are formed and the surface tension of the bottom of the drop is higher than that of the neck. Thus the surface of the drop moves downwards together with the adjoining solution layers.

When the potential reaches the values at which the concentration at the electrode surface is zero (the limiting current region), the surface tension becomes the same on the whole surface of the drop and the maximum disappears. Since the solution layer adjoining the electrode reappears when a new drop is formed, the decrease of the maximum current to the limiting current value can be very fast.

It is noteworthy that in 1934 Frumkin and Bruns [15] had already shown that current increases exceeding the diffusion current value are due to tangential movements of the mercury surface whereas the decreases of these abnormally high currents correspond to the disappearance of such movements.

In order to obtain an exact description of polarographic maxima Frumkin and Levich [16] carried out a rigorous investigation of the movements of mercury drops in solutions of electrolytes in electric fields. Since the tangential movements of the surface of falling mercury drops with respect to the drop centre are analogous with those of drops hanging at the capillary end, the equations obtained for falling drops can be used for describing the nature of polarographic maxima.

Frumkin and Levich [16] described the dependence of the rate of movement of drop surface on the potential difference across the surface. Later Levich [17] derived an equation for the instantaneous current due to depolarizer transport at a rate of tangential movement, V_r, on a drop electrode having radius r:

$$i_t = 10.7nFC^0D^{1/2}\overline{V}_r^{1/2}r^{3/2}. \tag{19.1}$$

In the case of an average current this equation becomes:

$$\bar{i} = 6.4nFC^0D^{1/2}\overline{V}_r^{1/2}r_1^{3/2}. \tag{19.2}$$

$\overline{V}_r$ and r_1 correspond to the moment of detachment of the drop.

Equation (19.1) indicates that the instantaneous current of the maximum depends on $t^{2/3}$. Results obtained by means of this equation are larger than those obtained experimentally.

The explanation that the non-homogeneous potential distribution is due to the screening of the drop by the capillary glass is unconvincing, since Antweiler [11] also observed whirls when the capillary end was very thin (the external diameter was 80 μm), when the screening of the electrode should be negligible. Flemming and Berg [18] confirmed this observation, using capillaries having extremely thin walls.

The theory of polarographic maxima is not yet completely worked out, and the non-uniform distribution of the potential on the drop surface has not been unambiguously explained. In the light of investigations carried out by Flemming and Berg [18] it appears unlikely that this distribution could be due to the screening effect alone, or to steric hindrance at the neck of the electrode, caused by the capillary glass, as a result of which the depolarizer cannot be transported freely to the electrode.

According to De Levie [19] the whirls are due to non-centric growth of drops, causing the movement of the mercury surface to be much faster on the bottom of the drop than on the neck. Therefore, the current density at the bottom of the drop differs considerably from that at the neck, so that the potential distribution on the solution side of the electrode–solution interface is not uniform.

Maxima of the second kind are due to movements of the mercury surface which cause transport of the electrolyte in the direction of the neck of the drop. These maxima have been described by Kryukova [20–22] and they are observed mainly at zero charge potentials.

According to Frumkin, Antweiler and von Stackelberg these maxima are due to movements of mercury inside the drops, which are shown schematically in Fig. 19.3. The movements of the mercury surface cause movements

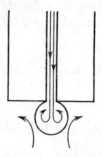

Fig. 19.3 Movements of mercury and solution in the case of a maximum of the second kind.

in adjoining solution layers. As a result charges are transferred to the neck of the drop and in this situation the formation of the whirl is prevented. At zero charge on the surface the opposition to whirl formation is weakest and for this reason whirls are mainly formed at electrocapillary zero potentials.

Maxima of the second kind decrease with increasing drop time.

The speed of the whirling movement of the liquid near the mercury electrode has also been investigated. Dvořak and Hermann [23] have determined this speed during the observation of oxygen maxima. For this purpose they recorded by means of a cine-camera the microscopic picture of moving carbon particles suspended in the solution.

O'Brien and Dieken [24] have determined the speed of the whirling movement at the maxima of copper(II) and cobalt(II) by means of laser interferometry.

Bauer and co-workers [25] have described in detail the procedure for determination of these speeds at maxima of the second kind. They have found that the speed decreases with decreasing capillary yield and is higher at the bottom of the drop than at its sides.

In the case of the maximum for cadmium(II) the speed of whirling movement was increased on the average by 2–3 times when 0.1 M KCl was replaced with saturated potassium chloride.

As could be expected, an addition of Triton X-100 caused a considerable decrease of the speed of this movement. This decrease was greater at the potential -0.80 V than at -1.20 V probably as a result of the fact that adsorption of Triton decreases as the potential becomes more negative.

The effect of depolarizer on maxima of the second kind has also been examined [26]. It has been found that this effect is a considerable one and the presence of depolarizer leads to an increase of the speed of movement of solution near the dropping electrode. This indicates that the speed does not depend only on the rate of outflow of mercury from the capillary.

Polarographic maxima and the methods of their elimination have been discussed in detail by Meites [27].

Recently Bauer [28] published a review of articles and views on the nature of the maxima and tried to reach general conclusions and Rangarajan published the paper on the theory of polarographic maxima [42].

The discrepancies between theoretical and observed limiting currents can sometimes be due to latent limiting currents. These are the result of interactions between depolarizers, or between the product of one depolarizer and another depolarizer. These currents were first described by Kemula and Michalski [29], who found that the hydrogen ion reduction current is much decreased by oxygen dissolved in the solution. This decrease in current is due to neutralization of hydrogen ions by hydroxide ions generated at the electrode surface during the reduction of oxygen.

In other cases hydroxide ions can decrease the reduction waves of metal ions [30] by formation of sparingly soluble hydroxides.

The theory of latent limiting currents has been given by Kemula and Grabowski [31, 32], who divided them into two groups according to the chemical reactions leading to their formation. Assuming that a solution contains two depolarizers Ox_1 and Ox_2 which can be reduced at potentials E_1 and E_2, respectively:

$$Ox_1 + n_1 e \rightarrow Red_1 \qquad (19.3)$$

$$Ox_2 + n_2 e \rightarrow Red_2 \qquad (19.4)$$

and assuming that E_2 is more negative than E_1, a latent limiting current of the first kind will be observed in the case of a fast chemical reaction:

$$p Red_1 + r Ox_2 \rightarrow (Red_1)_p (Ox_2)_r. \qquad (19.5)$$

As a result of reaction (19.5) the diffusion of Ox_2 to the electrode is decreased. In certain cases Ox_2 can be completely converted into compound $(Red_1)_p(Ox_2)_r$ in the neighbourhood of the electrode. In this case the latent limiting current value is determined by the difference between the limiting currents of Ox_1, observed in the presence and in the absence of Ox_2.

When the fast chemical reaction starts only as a result of the reduction of Ox_2 according to the following equation:

$$qOx_1 + mRed_2 \rightarrow (Ox_1)_q(Red_2)_m \qquad (19.6)$$

latent limiting currents of the second kind are observed. In this case a decrease or disappearance of the Ox_2 wave takes place.

The theory of Kemula and Grabowski leads to results which are in good agreement with experiment when water is the product of reactions (19.5) and (19.6). If these reactions yield insoluble hydroxides they can be adsorbed onto the electrode and the electrode process can be inhibited [33]. In such cases discrepancies between theoretical and experimental results are observed.

Latent limiting currents have also been investigated theoretically by other authors [34–36].

19.2 Stationary Electrode Voltammetry

Although whirls have been observed at hanging [11] and suspended [15] mercury drops and at large mercury electrodes [12, 37], they are seldom encountered in stationary electrode voltammetry. The corresponding maxima have been observed [38] in the case of pyridine complexes of cobalt(II) at hanging mercury electrodes. In this case the maximum was due to specific adsorption of the complex on the electrode surface. During the reduction of the complex, different potentials, and consequently different surface tensions, appeared on various parts of the electrode. In this situation movements of solution occurred. It should be stressed that in this case the maximum was formed only in the potential region corresponding to peak formation. When the concentration of the complex on the electrode surface decreased to zero, as a result of the increase of negative potential, the maximum disappeared and the current decreased to the value predicted on the basis of diffusion transport.

In stationary electrode voltammetry, deviations from the expected transport appear at low scan rates. When hanging drop mercury electrodes are used increased current values are observed at scan rates slower than 0.2 V/min, provided that the electrodes and solution are protected from shocks. For this reason the use of very low scan rates in stationary electrode voltammetry is not recommended, as it leads not only to an increase in peak current, but also to the appearance of convection transport, which causes a considerably more significant increase of the current after the peak, and as a result the recorded curves resemble polarographic waves.

19.3 Chronopotentiometry

It has already been mentioned that the use of transition times longer than one minute is not recommended, since the contribution of convection to the depolarizer transport becomes appreciable in such cases. Detailed studies of this problem for hanging drop mercury electrodes [39] showed that the errors are small when the transition times are shorter than 30 seconds. It might be expected that systems showing abnormally high current peaks in stationary electrode voltammetry would show abnormally long transition times in chronopotentiometry.

19.4 The Rotating Disc Method

In the case of the rotating disc method it is not advisable to use very low rates of electrode rotation, as the differences between the measured current and that calculated on the basis of the Nernst and Levich equation are then considerable. It follows from this equation that, when the rotation rate tends to zero, the limiting current also tends to zero. This conclusion is not in agreement with experimental results, since the current is also observed when the rotation rate is equal to zero. In this case the current is described by chronoamperometric equations.

Very fast electrode rotation rates should also not be used, since at such rates electrode vibrations in the plane perpendicular to its axis can appear. Levich [40] showed that when the Reynolds number Re is greater than 10^4–10^5 the flow is no longer laminar, whereas in the derivation of equations describing the rotating disc method laminar flow was assumed. Emery and Hintermann [41] showed that, in the case of small disc electrodes, the laminar flow is observed even at fast rotation rates. These authors used an electrode having a surface area of 0.0097 cm^2 and rotating at the rate of 400–36,100 r.p.m. They found that in this range of rotation rates the system investigated obeyed the Nernst and Levich equation.

References

[1] Doss, K. D. and Venkatesan, D., *Proc. Indian Acad. Sci.*, **49**, 129 (1959); Sathanarayana, S., *J. Electroanal. Chem.*, **10**, 56 (1965).
[2] Frumkin, A. N., Stenina, E. V. and Fedorovitch N. V., *Elektrokhimiya*, **6**, 1572 (1970).
[3] Frumkin, A. N., Stenina, E. V., Nikolayeva-Fedorovitch, N. V., Petukhova, G. N. and Yusupova, V. A., *ERev. Roumaine Chim.*, **17**, 155 (1972).
[4] Frumkin, A. N., Fedorovitch, N. V., Damaskin, B. B., Stenina, E. V. and Krylov, V. S., *J. Elektroanal. Chem.*, **50**, 103 (1974); Stenina, E. V., Frumkin, A. N., Nikolayeva-Fedorovitch, N, V. and Osipov, I. V., *J. Electroanal. Chem.*, **62**, 11 (1975); Nikolayeva-Fedorovitch, N. V., Stenina, E. V., Petukhova, G. N., Yusupova, V. A. and Shokhova, E. A., *Elektrokhimiya*, **9**, 157 (1972).
[5] Heyrovský, J. and Šimunek, R., *Phil. Mag.*, **7**, 951 (1929).
[6] Varasova, E., *Coll. Czechoslov. Chem. Communs.*, **2**, 8 (1930).
[7] Heyrovský, J., "*A polarographic study of the electrokinetic phenomena of adsorption electroreduction and overpotential displayed at the dropping mercury cathode*", Hermann et Cie, Paris 1934.
[8] Heyrovský, J. and Vascautzanu, E., *Coll. Czechoslov. Chem. Communs.*, **3**, 418 (1931).
[9] Vavruch, I., "*Polarograficka maxima w teorii a praxi*", SPICH, Prague 1949.
[10] Dvořak, J., *Proc. 1st Congress of Polarography, Prague 1952*, Vol. 3, page 418.
[11] Antweiler, H. J., *Z. Elektrochem.*, **44**, 719, 831, 888 (1938).
[12] von Stackelberg, M., Antweiler, H. J. and Kieselbach, L., *Z. Elektrochem.*, **44**, 663 (1938).
[13] von Stackelberg, M., *Z. Elektrochem.*, **45**, 466 (1939).
[14] von Stackelberg, M., *Fortschr. chem. Forsch.*, **2**, 229 (1951); *Proc. 1st Congress of Polarography, Prague 1951*, Vol. 1, page 359.
[15] Frumkin, A. N. and Bruns, B., *Acta physicochimica U.R.S.S.*, **1**, 232 (1934).
[16] Frumkin, A. N. and Levich, V. G., *Zhurn. Fiz. Khim.*, **19**, 573 (1945).
[17] Levich, V. G., *Zhurn. Fiz. Khim.*, **22**, 721 (1948).
[18] Flemming, J. and Berg, H., *J. Electroanal. Chem.*, **8**, 291 (1964).
[19] De Levie, R., *J. Electroanal. Chem.*, **9**, 311 (1965).
[20] Kryukova, T. A., *Zav. Lab.*, **9**, 691, 699 (1940).
[21] Kryukova, T. A., *Zhurn. Fiz. Khim.*, **20**, 1179 (1946).
[22] Kryukova, T. A., *Zav. Lab.*, **14**, 511, 639, 767 (1948).
[23] Dvořak, I. and Hermann, P., *Chem. Listy*, **46**, 565 (1952).
[24] O'Brien, R. N. and Dieken, F. B., *Can. J. Chem.*, **48**, 2651 (1970).
[25] Lal, S., Kumar, A. and Bauer, H. H., *J. Electroanal. Chem.*, **42**, 423 (1973).
[26] Lal, S., Holt, T. W. and Bauer, H. H., *J. Electroanal. Chem.*, **42**, 429 (1973).
[27] Meites, L., "*Polarographic Techniques*", 2nd Ed., Interscience, New York 1965, Chapter 6.
[28] Bauer, H. H., *Electrochim, Acta*, **18**, 355 (1973); Bauer, H. H., in "*Electroanalytical Chemistry*", Bard, A. J., Ed., Dekker, New York 1974, Vol. 8.
[29] Kemula, W. and Michalski, M., *Roczniki Chem.*, **16**, 533 (1936).
[30] Kryukova, T. A. and Kabanov, V. N., *Zhurn. Fiz. Khim.*, **13**, 1454 (1939).
[31] Kemula, W. and Grabowski, Z. R., *Sprawozdania Posiedzeń Tow. Nauk. Warszaw., Wydział III Nauk Mat. Fiz.*, **44**, 78 (1951).

[32] Kemula, W. and Grabowski, Z. R., *Roczniki Chem.*, **26**, 266 (1952).
[33] Behr, B. and Chodkowski, J., *Roczniki Chem.*, **30**, 1301 (1956).
[34] Kolthoff, I. M. and Lingane, J. J., *"Polarography"*, 2nd. Ed., Interscience, New York 1952, Chapter VI.
[35] Feoktistov, L. G. and Zhdanov, S. I., *Izv. Akad. Nauk SSSR, ser. khim.*, 45 (1963).
[36] Feoktistov, L. G., Tamilov, A. P. and Galdin, M. M., *Izv. Akad. Nauk SSSR, ser. khim.*, 352 (1963).
[37] Bruns, B., Frumkin, A., Jofa, S., Vanyukova, L. and Zolotarevska, S., *Acta Physicochimica U.R.S.S.*, **9**, 359 (1938).
[38] Janiszewska, L. and Galus, Z., *Roczniki Chem.*, **44**, 1107 (1970).
[39] Gumiński, C. and Galus, Z., *Roczniki Chem.*, **44** 1767 (1970).
[40] Levich, V. G., *"Physicochemical Hydrodynamics"*, Ed. Akad. Nauk USSR, Moscow 1952 (in Russian).
[41] Emery, C. A. and Hintermann, H. E., *Electrochim. Acta*, **13**, 127 (1968).
[42] Rangarajan, S. K., *J. Electroanal. Chem.*, **62**, 21 (1975).

Chapter 20

Some New Developments in Polarography

Since the third decade of this century polarography has been widely used as an analytical method and as a very useful tool for solving various physical and chemical problems. In recent years new problems have appeared, in which it is necessary to analyse solutions containing electroactive substances at concentrations lower than 10^{-6} mole per litre. This cannot be undertaken by means of classical polarography.

The efficiency of polarography, like that of any analytical method, depends on the ratio of measured to interfering signals (signal to noise ratio). When this ratio approaches unity or is lower than unity the accuracy of the method decreases rapidly. Provided that pure reagents and oxygen-free solutions are used, the polarographic interfering signals depend almost entirely on the capacitance current, which is connected with the charging of the drop electrode double layer.

In general the capacitance current i_c can be expressed as:

$$i_c = \frac{dq}{dt}. \tag{20.1}$$

Using the definition of integral capacitance

$$^cC = \frac{q}{E^m} \tag{20.2}$$

the following equation is obtained from expression (20.1):

$$i_c = E^m \frac{d^cC}{dt}, \tag{20.3}$$

since $dE^m/dt = 0$.

The rate of change of the double layer capacity is proportional to the electrode capacity and to the rate of increase of electrode surface area

$$\frac{d^cC}{dt} = {}^cC \frac{dA}{dt}. \tag{20.4}$$

By combining equations (20.3) and (20.4) the following expression is obtained:

$$i_c = {}^cCE^m \frac{dA}{dt}, \qquad (20.5)$$

and by introducing the equation describing the dependence of the dropping electrode surface on m and t:

$$i_c = 0.57{}^cCE^m m^{2/3} t^{-1/6}. \qquad (20.6)$$

In these equations E^m is the electrode potential referred to the zero charge potential, and q is the charge.

Polarographs are usually used for recording average currents. The following expression for the average capacitance current

$$\bar{i}_c = \frac{1}{t_1} \int_0^{t_1} i_c \, dt = 0.85{}^cCE^m m^{2/3} t_1^{-1/3} \qquad (20.7)$$

can thus be obtained from equation (20.6) describing the instantaneous capacitance current.

The approximate value of the average capacitance current at potentials differing by 0.5 V from the zero charge potential can be readily calculated. Assuming ${}^cC = 20 \ \mu\text{F/cm}^2$, $m = 2 \times 10^{-3}$ g/s and $t_1 = 2$ s, equation (20.7) gives a value of $i = 0.11 \ \mu\text{A}$. By substituting this value in the Ilkovič equation, assuming that $n = 2$, $D = 9 \times 10^{-6}$ cm^2/s, and retaining the previous values of m and t_1, the depolarizer concentration at which the Faraday current is equal to the capacitance current is obtained. This concentration is 1.4×10^{-5} mole per litre.

It follows from this simple calculation that analysis of solutions containing electroactive substances at concentrations lower than 10^{-5} M is difficult, even when compensation for the capacitance current is used. Analysis of solutions more dilute than 10^{-6} M is not possible by means of classical polarography.

In physical and chemical research polarography is widely used in studies of the kinetics of electrode processes, but this is possible only when these processes show deviations from reversibility, that is from a course of reaction controlled exclusively by the rate of depolarizer transport to the electrode.

On the basis of equations discussed in Chapter 3 it is possible to calculate that when $D = 9 \times 10^{-6}$ cm^2/s and $t_1 = 2$ s the average rate of depolarizer transport in polarography, V, is 2.4×10^{-3} cm/s. Therefore under such conditions electrode processes of redox systems, with standard rate constants exceeding 2×10^{-2} cm/s, are practically reversible, and as a result their kinetics cannot be studied by polarographic methods. Since many electrode processes behave in this way, the usefulness of classical polarography in kinetic studies is limited. Therefore many attempts have been made to modify polarographic methods in order to adapt them for analyses of very dilute solutions and for determinations of kinetic parameters of the above-mentioned electrode processes.

20.1 Principles of New Polarographic Methods

The limitations of polarography in the case of fast electrode processes could be overcome if it were possible to construct a polarographic system having a short drop time, since then the transport rate would be given by the following equation:

$$\overline{V} = \frac{2D^{1/2}}{(\pi t_1)^{1/2}}, \qquad (20.8)$$

i.e. it would be much faster than that observed in classical polarography. Assuming a drop time of 10^{-3} s and a diffusion coefficient of 9×10^{-6} cm^2/s the average transport rate $\overline{V}$ is 0.11 cm/s, i.e. 46 times faster than that obtained for $t_1 = 2$ s. Under such conditions it would be possible to study polarographically electrode systems having rate constants up to 1 cm/s.

In practice it is difficult to construct electrodes with such short drop times, although this has been achieved for special studies, by Reynolds and Powell [1]. Therefore other approaches to this problem have been employed.

A slowly and linearly changing potential applied to the reference electrode is modulated with alternating potentials of small amplitude. The modulation frequency is of the order of 100 Hz, and the measurements of the current are limited to that resulting from these short voltage impulses, which is the current formed at high transport rates. Thus the limiting effect of diffusion is shifted and electrode processes can be investigated, which are much faster than the fastest processes that can be studied by conventional polarography.

This is the principle [2, 3] of sinusoidal alternating current polarography which was developed by Breyer et al. [4, 5] and of alternating square-wave and pulse polarography introduced by Barker [6–9].

20.1.1 ALTERNATING CURRENT SINUSOIDAL POLAROGRAPHY

It is convenient to discuss sinusoidal polarography first, since this method is older than square-wave and pulse polarography. The slowly and linearly increasing potential applied to the electrodes is modulated with sinusoidal voltage according to the scheme shown in Fig. 20.1. The

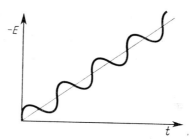

Fig. 20.1 Time dependence of voltage in a.c. sinusoidal polarography.

amplitude of the alternating voltage usually does not exceed 30 millivolts, and its frequency is usually equal to that of the mains current, 50–60 Hz. The current flowing in the circuit consists of three components; the diffusion current connected with the electrode reaction, taking place under the influence of the slowly changing applied potential, the current due to sinusoidal voltage changes connected with charging and discharging of the double layer and the current caused by sinusoidal voltage changes, but related to the electrode reaction. In this method the direct current component is rejected and only the current caused by the periodic voltage changes is measured. The general arrangement is shown in Fig. 20.2.

When the solution contains a substance which can be either oxidized or reduced at a suitable potential the polarographic waves observed at this potential are not those of conventional polarography. Instead a peak is obtained as shown in Fig. 20.3.

In the case of reversible electrode processes the peak potential is characteristic of the redox system being studied and does not depend on experimental conditions. It will be shown that it is equal to the polarographic half-wave potential. The reasons why the shape of this type of polarographic wave differs from those of normal polarography should be briefly considered. For this purpose it is convenient to divide Fig. 20.3 into three voltage regions, A, B and C. At point E_1 in region A the slowly and linearly increasing potential is insufficient, in classical polarography, to initiate the electrode reaction and commence the polarographic wave.

When E_1 is modulated with a sinusoidal alternating voltage of amplitude ΔE the resulting potential changes do not correspond to changes in the Faraday current, as the electrode reaction is not taking place. This argument applies to all potentials in region A, since they are more positive than E_2.

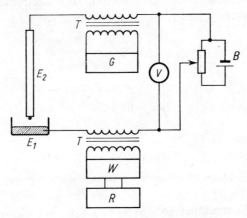

Fig. 20.2 Scheme of a simple a.c. polarograph. E_1—reference electrode; E_2—dropping electrode; T—transformer; G—oscillator; W—amplifier; R—a.c. recorder; B—EMF power source; V—voltmeter (from [5] by permission of the copyright holders, Interscience Publishers).

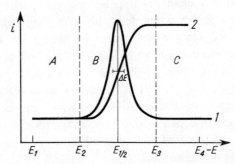

Fig. 20.3 Curve recorded in a.c. polarography on the background of classical polarography.

Consider region C, where voltage E_4 is greater than E_3. This is the limiting current region in classical polarography. The application of a sinusoidal alternating voltage of amplitude ΔE is not accompanied by changes in current. The electrode reaction is taking place at its maximum value, but it must be remembered that, in the polarographic limiting current region the concentration of the oxidized form must be zero at the electrode surface when the process is a reduction, since this is a condition for the existence of a limiting current. Substance Red generated on the electrode when the potential changes by ΔE cannot be oxidized in this potential region when the potential returns to its initial value. Therefore, under these conditions an alternating current does not flow and the only current flowing in the circuit is the direct current corresponding to potential E_4. However, since the direct current component is not included in the measurement, the recorded faradaic current is zero, as in region A.

Finally, potentials in region B in Fig. 20.3 should be considered, especially $E_{1/2}$, as related to the flow of direct current. When the potential becomes more negative by ΔE an additional concentration of Red is produced at the electrode surface. This will be re-oxidized when the potential returns to $E_{1/2}$, provided that the rate of electron exchange between Ox and Red is sufficiently fast for Red not to form the substrate of a subsequent reaction. Similarly, when the potential becomes less negative by an amount ΔE the concentration of Red is decreased. Therefore, when $E_{1/2}$ is modulated by a sinusoidal voltage ΔE, the small voltage changes give rise to an alternating current.

It follows that, when the direct potential applied to the electrode in region B is more positive than $E_{1/2}$ (i.e. in the initial part of the polarographic wave), modulation with the alternating voltage ΔE gives rise to an alternating current smaller than that obtaining at $E_{1/2}$. Similarly if the voltage is more negative than $E_{1/2}$, but not in the region of the limiting current the alternating current is again smaller.

Thus, from the above, and by reference to the classical polarographic wave in Fig. 20.3, it can be seen that the alternating current values are the smallest at the beginning and end of the wave and that the highest value occurs in the vicinity of the half-wave potential. The magnitude of the current increases with the degree of reversibility of the reaction studied.

In alternating current polarography only the alternating current caused by voltage changes is measured. However the capacity current is recorded in addition to the faradaic current and therefore this method is little more sensitive than classical polarography. It does not make possible the analysis of solutions much more dilute than about 10^{-5} M. Although alternating current polarography is not more sensitive than conventional polarography as an analytical method, it makes possible study of the kinetics of electrode reactions faster than those which can be studied by the conventional method. It also has analytical value because the peaked nature of the waves permits their resolution when they are too close together for resolution by normal polarography.

20.1.2 SQUARE-WAVE ALTERNATING CURRENT POLAROGRAPHY

The analytical sensitivity of sinusoidal polarography can be increased considerably if square-wave, instead of sinusoidal, alternating voltage is applied to the linearly increasing potential [6]. The time dependence of the potential in this case is shown in Fig. 20.4.

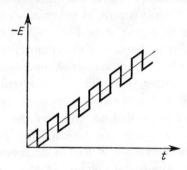

Fig. 20.4 Time dependence of the potential in square-wave a.c. polarography.

As before, the alternating current caused by changes of the square-wave potential, usually having a frequency of 225–250 Hz, is recorded. The shapes of the recorded curves are similar to the peak shown in Fig. 20.3 for the reasons already given in the case of sinusoidal potential changes.

The application of the alternating square-wave potential to the linearly increasing potential makes possible the significant elimination from the recorded current of the alternating current due to charging and discharging of the double layer, resulting from high frequency potential changes.

When the time constant of the circuit is not large this can be achieved in experimental systems. The capacitance current disappears very rapidly after the change of the electrode potential by ΔE, according to the following equation:

$$i_c = \frac{\Delta E}{R} \exp\left(-\frac{\tau}{RC}\right) \tag{20.9}$$

where R and C are the resistance and the capacity of the electrolytic system, respectively, and τ is the time elapsing since the last potential change. The faradaic current on the other hand disappears at a much slower rate and is approximately proportional to $t^{-1/2}$. These changes are shown schematically in Fig. 20.5.

The capacitance current is significant only at the beginning of each half-cycle. Therefore, if the current is measured only during the time intervals represented by the shaded areas in Fig. 20.5, the measured value will be practically equal to the Faraday current. This practice is employed in rectangular polarography, generally known as "square-wave polarography", by arranging for the current to be measured only during the latter part of

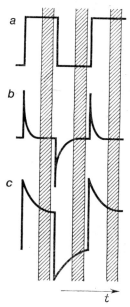

Fig. 20.5 Voltage (*a*), charging current (*b*) and Faraday current (*c*) changes in square-wave a.c. polarography. The current is measured in time intervals corresponding to the shaded areas.

each half-cycle. The elimination of capacitance current makes it possible to use this method for determinations of substances in solution at concentrations of the order of 10^{-8} M, or less [10].

20.1.3 Pulse Polarography

Pulse polarography was introduced by Barker and Gardner [8] in 1960. At the present time two variants of this method are used.

In the first variant one pulse of rectangular voltage, usually having a duration of 1/25 s and amplitude of about 30 mV, is applied to the slowly and linearly increasing potential after a definite time in the life of each mercury drop. The changes of the dropping electrode potential in this method are shown schematically in Fig. 20.6.

This method of application of the low amplitude pulse to the linearly increasing potential is characteristic of "derivative" pulse polarography. The current is again measured during the second half of the pulse, so that

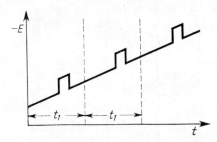

Fig. 20.6 Time dependence of electrode potential in derivative pulse polarography.

the capacity current is eliminated. This method is thus very useful for analytical purposes. In this method the difference between currents of individual pulses is measured and for this reason the recorded curves have the shape of peaks such as curve *1* in Fig. 20.3.

In the second variant of pulse polarography the electrode is polarized with a pulse having linearly increasing amplitude. The potential changes of the indicator electrode are shown in Fig. 20.7.

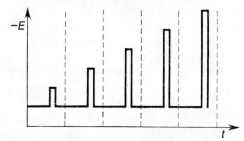

Fig. 20.7 Time dependence of the electrode potential in normal pulse polarography.

This method of polarization leads to polarograms identical in shape with those obtained in classical polarography. Since the time between the moment of application of the pulse to the moment at which the current is measured is much shorter than the drop time, the limiting value of the measured current is much higher than that observed in classical polarography. In contrast to d.c. polarography the potential is applied near the end of the drop life to diminish the capacitive current.

In sinusoidal, square-wave, and derivative pulse polarography the peak size depends on the degree of reversibility of the process although, in the case of derivative pulse polarography, the effect is much less. In normal pulse polarography, however, the degree of reversibility has only slight effect on the magnitude of the recorded limiting current.

20.1.4 Triangular-Wave Alternating Current Polarography

Sluyters and co-workers have proposed [11] another variant of alternating current polarography, in which a triangular alternating voltage of frequency ω is applied to the linearly increasing potential (Fig. 17.4, p. 428). In this case the capacity current is a square wave, which can be eliminated after rectification, followed by filtering the resulting direct current. Thus, at least theoretically, after filtering the direct current component it is possible to obtain the faradaic current peak on the alternating current polarogram in which the residual current curve is practically coincident with the zero current line.

Barker [12] has pointed out that this kind of polarization was first used and described in 1960 [13]. At that time Barker did not develop further this kind of polarography since in his opinion it did not offer the possibility of a wider practical application than the rectangular alternating current polarography. Later Barker [12] discussed conditions which should be observed in order to obtain satisfactory results by means of this kind of polarography and briefly outlined the theory of this method.

Sluyters and Sluyters-Rehbach [14] suggested that saw tooth voltage can be utilized in alternating current polarography.

Square-wave, pulse and triangular-wave polarography are superior to the sinusoidal technique, since in these methods the capacity current is practically eliminated from the measurements. For this reason these methods have rapidly become of importance in analytical chemistry.

In the case of sinusoidal polarography it is also possible to eliminate the capacity current from measurements [5, 15], since the capacity component precedes the applied voltage by $\pi/2$, while the faradaic component precedes it by $\pi/4$, provided that the depolarizer is not specifically adsorbed and reacts reversibly with the electrode. With the aid of a phase detector it is possible to record only the current which is in phase with the voltage, that is the faradaic current. Instruments are available which employ this principle.

The elimination of the capacity current from the total alternating current has no significant effect on the recorded curves; their shapes are similar to that shown in Fig. 20.3.

The analytical sensitivity of sinusoidal polarography based on the principle described above is, in theory, similar to that of square-wave polarography.

20.2 Theoretical Aspects of New Polarographic Methods

In order to outline the theory of alternating current polarography it is necessary to consider the phenomena accompanying the application of a low amplitude alternating voltage to the electrode. This problem has been investigated theoretically by many authors. Important contributions to the knowledge of these phenomena have been made by Randles [16], Ershler [17], Gerischer [18] and Grahame [19].

The following reasoning is based on the work of Delahay [20] and Damaskin [21]. Assume that the electrode is in a solution containing the components of a redox system involved in the following electrode reaction:

$$Ox + ne \rightleftharpoons Red. \qquad (20.10)$$

When the alternating voltage amplitude is small it can be assumed that in the small region of potential change the potential dependence of the current is linear (Fig. 20.8), and therefore the current corresponding to the

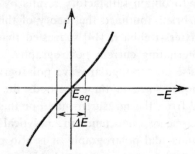

Fig. 20.8 Linearity of the voltage dependence of the current in the region of small values of the amplitude of alternating voltage ΔE.

potential changes is sinusoidal and its frequency is equal to that of the voltage. In order to describe the value of the alternating current theoretically, it is necessary to obtain an expression describing the electrode resistance. At the frequencies commonly employed the electrode behaves like a combination of capacity and resistance and therefore the problem of electrode impedance has to be considered.

The current flowing under such conditions consists of the faradaic current due to the reaction (20.10) and the capacity current due to charging and discharging of the double layer.

The two components of the current correspond to two components of electrode impedance: the faradaic impedance Z_f and the double layer impedance Z_p. Thus the current results from voltage changes on these impedances. Presenting the electrolytic system used in alternating current polarography in the form of an electric circuit, the two component impedances would have to be connected in parallel. The impedance of the double layer could be represented in the form of differential capacity of the double layer C_d. The faradaic impedance is usually represented in the form of reaction resistance R_s and capacity C_s connected in series. Thus during the flow of the alternating current the electrode behaves like the electrical circuit shown in Fig. 20.9.

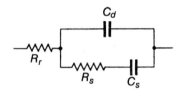

Fig. 20.9 Electrical diagram of electrolytic cell.

Since the current depends on the impedance of the system an expression describing R_s and C_s must be found. The resistance of the solution R_r and the differential capacity of the double layer can be readily determined by the examination of a solution containing no depolarizer.

The alternating current flowing through the faradaic impedance can be expressed as follows:

$$\Delta i = I \sin \omega t \qquad (20.11)$$

where I is the amplitude of the current and ω is described by the relationship $\omega = 2\pi f$ where f is the frequency of voltage change.

Using the circuit shown in Fig. 20.9 the potential changes on the faradaic impedance can be described by the following expression:

$$E = R_s i + \frac{1}{C_s} \int i \, dt. \qquad (20.12)$$

Differentiating equation (20.12) with respect to time and combining the result with equation (20.11) gives:

$$\frac{dE}{dt} = \omega I R_s \cos \omega t + \frac{I}{C_s} \sin \omega t. \qquad (20.13)$$

In order to describe R_s and C_s the expression for dE/dt must be found on the basis of the properties of faradaic impedance.

Since the alternating current is a function of the potential and concentrations of components of reaction (20.10) on the electrode surface of the system, the potential is also determined by the values of current and concentrations of the redox system components according to the following general equation:

$$E = f[i, C_{\text{Ox}}(0, t), C_{\text{Red}}(0, t)]. \tag{20.14}$$

From this equation:

$$\frac{dE}{dt} = \frac{\partial E}{\partial i}\frac{di}{dt} + \frac{\partial E}{\partial C_{\text{Ox}}(0, t)}\frac{dC_{\text{Ox}}(0, t)}{dt} + \frac{\partial E}{\partial C_{\text{Red}}(0, t)}\frac{dC_{\text{Red}}(0, t)}{dt}. \tag{20.15}$$

Putting:

$$\varkappa = \frac{\partial E}{\partial i}, \tag{20.16}$$

$$\beta_{\text{Ox}} = \frac{\partial E}{\partial C_{\text{Ox}}(0, t)}, \tag{20.17}$$

$$\beta_{\text{Red}} = \frac{\partial E}{\partial C_{\text{Red}}(0, t)}, \tag{20.18}$$

equation (20.15) can be expressed in the following form:

$$\frac{dE}{dt} = \varkappa\frac{di}{dt} + \beta_{\text{Ox}}\frac{dC_{\text{Ox}}(0, t)}{dt} + \beta_{\text{Red}}\frac{dC_{\text{Red}}(0, t)}{dt}. \tag{20.19}$$

Differentiating equation (20.11) with respect to time and combining it with equation (20.19) gives:

$$\frac{dE}{dt} = \omega I \varkappa \cos\omega t + \beta_{\text{Ox}}\frac{dC_{\text{Ox}}(0, t)}{dt} + \beta_{\text{Red}}\frac{dC_{\text{Red}}(0, t)}{dt}. \tag{20.20}$$

Since the right-hand sides of equations (20.13) and (20.20) must be equal a combined equation can be written:

$$\omega I R_s \cos\omega t + \frac{I}{C_s}\sin\omega t = \omega I \varkappa \cos\omega t + \beta_{\text{Ox}}\frac{dC_{\text{Ox}}(0, t)}{dt} + \beta_{\text{Red}}\frac{dC_{\text{Red}}(0, t)}{dt}. \tag{20.21}$$

In order to describe expressions $dC_{\text{Ox}}(0, t)/dt$ and $dC_{\text{Red}}(0, t)/dt$ appearing in the above equation it is necessary to introduce expressions describing changes in concentrations of the redox system, taking place on the electrode surface during the electrolysis. These can be obtained by solving the Fick equations for substances Ox and Red:

$$\frac{\partial C_{Ox}}{\partial t} = D_{Ox} \frac{\partial^2 C_{Ox}}{\partial x^2}, \qquad (20.22)$$

$$\frac{\partial C_{Red}}{\partial t} = D_{Red} \frac{\partial^2 C_{Red}}{\partial x^2}, \qquad (20.23)$$

with the following initial conditions:

$$t = 0, \quad x \geqslant 0, \quad C_{Ox} = C_{Ox}^0, \qquad (20.24)$$

$$C_{Red} = C_{Red}^0, \qquad (20.25)$$

and the following boundary conditions:

$$t > 0, \quad x \to \infty, \quad C_{Ox} \to C_{Ox}^0, \qquad (20.26)$$

$$C_{Red} \to C_{Red}^0, \qquad (20.27)$$

$$t > 0, \quad x = 0, \quad nFAD_{Ox}\left(\frac{\partial C_{Ox}}{\partial x}\right)_{x=0} = i_s + I\sin\omega t, \qquad (20.28)$$

$$D_{Ox}\left(\frac{\partial C_{Ox}}{\partial x}\right)_{x=0} + D_{Red}\left(\frac{\partial C_{Red}}{\partial x}\right)_{x=0} = 0, \qquad (20.29)$$

where D_{Ox} and D_{Red} are the diffusion coefficients of substances Ox and Red, respectively, C_{Ox}^0 and C_{Red}^0 are the initial concentrations of these substances, and i_s is the direct current component.

The general solution of equation (20.22) for the given conditions is:

$$C_{Ox}(x, t) = f(x, t) + \frac{I}{nFA}\left(\frac{1}{2\omega D_{Ox}}\right)^{1/2} \exp\left[-\left(\frac{\omega}{2D_{Ox}}\right)^{1/2} x\right] \times$$
$$\left\{\sin\left[+\omega t - \left(\frac{\omega}{2D_{Ox}}\right)^{1/2} x\right] - \cos\left[\omega t - \left(\frac{\omega}{2D_{Ox}}\right)^{1/2} x\right]\right\} \qquad (20.30)$$

where $f(x, t)$ describes concentration changes connected with the direct current component i_s.

The concentration at the electrode surface is described by the following simple expression:

$$C_{Ox}(0, t) = f(0, t) + \frac{I}{nFA}\left(\frac{1}{2\omega D_{Ox}}\right)^{1/2}(\sin\omega t - \cos\omega t) \qquad (20.31)$$

from which the following expression for $dC_{Ox}(0, t)/dt$ can be readily obtained by differentiation with respect to time:

$$\frac{dC_{Ox}(0,t)}{dt} = \frac{df(0,t)}{dt} + \frac{I}{nFA}\left(\frac{\omega}{2D_{Ox}}\right)^{1/2}(\cos\omega t + \sin\omega t). \quad (20.32)$$

The expression describing $dC_{Red}(0,t)/dt$ is similar to equation (20.32), but it contains the diffusion coefficient of the reduced form D_{Red} instead of D_{Ox} and the second term of the right-hand side is preceded by a minus sign:

$$\frac{dC_{Red}(0,t)}{dt} = \frac{df(0,t)}{dt} - \frac{I}{nFA}\left(\frac{\omega}{2D_{Red}}\right)^{1/2}(\cos\omega t + \sin\omega t). \quad (20.33)$$

Combining equations (20.32) and (20.33) with equation (20.21):

$$\omega I R_s \cos\omega t + \frac{I}{C_s}\sin\omega t = \omega I\varkappa\cos\omega t + \sigma I\omega^{1/2}(\sin\omega t + \cos\omega t) \quad (20.34)$$

where σ is described by the expression:

$$\sigma = \frac{1}{2^{1/2}nFA}\left[\frac{\beta_{Ox}}{D_{Ox}^{1/2}} - \frac{\beta_{Red}}{D_{Red}^{1/2}}\right]. \quad (20.35)$$

By comparing the coefficients of $\sin\omega t$ and $\cos\omega t$ in equation (20.34) expressions for R_s and C_s are obtained:

$$R_s = \varkappa + \frac{\sigma}{\omega^{1/2}}, \quad (20.36)$$

$$C_s = \frac{1}{\sigma\omega^{1/2}}. \quad (20.37)$$

Assuming that $\varkappa = 0$ the impedance of the system can be expressed in general terms using R_s and C_s, as follows:

$$Z = \left[R_s^2 + \frac{1}{\omega^2 C_s^2}\right]^{1/2}. \quad (20.38)$$

The assumption that $\varkappa = 0$ is equivalent to the assumption that the electrode process is reversible, i.e. that it is diffusion controlled. Utilizing equations (20.36) and (20.37) the impedance can be expressed by the relationship:

$$Z = \sigma\left(\frac{2}{\omega}\right)^{1/2}. \quad (20.39)$$

On the basis of this relationship the amplitude of the alternating current flowing through the faradaic impedance can be expressed by means of

20.2] Theoretical Aspects of New Polarographic Methods

the equation:

$$\Delta i = \frac{\Delta E \omega^{1/2} A}{2^{1/2} \sigma}. \qquad (20.40)$$

In order to expand this equation σ must be defined. This problem is equivalent to description of parameters β_{Ox} and β_{Red} given by equations (20.17) and (20.18).

For a reversible electrode process (20.10) it follows from the Nernst equation that:

$$\beta_{Ox} = \frac{\partial E}{\partial C_{Ox}(0, t)} = \frac{RT}{nFC'_{Ox}(0, t)}, \qquad (20.41)$$

$$\beta_{Red} = \frac{\partial E}{\partial C_{Red}(0, t)} = \frac{RT}{nFC'_{Red}(0, t)}, \qquad (20.42)$$

where $C'_{Ox}(0, t)$ and $C'_{Red}(0, t)$ are the concentrations on the electrode surface when the alternating current does not flow through the circuit.

Using the equation of reversible polarographic waves the following expressions are obtained:

$$C'_{Ox}(0, t) = C^0_{Ox} \frac{i_g - i}{i_g} = \frac{\Theta}{(1+\Theta)} C^0_{Ox}, \qquad (20.43)$$

$$C'_{Red}(0, t) = C^0_{Ox} \left(\frac{D_{Ox}}{D_{Red}}\right)^{1/2} \frac{i}{i_g} = C^0_{Ox} \left(\frac{D_{Ox}}{D_{Red}}\right)^{1/2} \frac{1}{(1+\Theta)}. \qquad (20.44)$$

In these equations

$$\Theta = \exp\left[\frac{(E - E_{1/2})nF}{RT}\right] \qquad (20.45)$$

and C^0_{Ox} is the concentration of Ox in the bulk of the solution.

Introducing equations (20.43) and (20.44) into equations (20.41) and (20.42), respectively, and combining the results with equation (20.35) gives:

$$\sigma = \frac{RT}{n^2 F^2 C^0_{Ox}(2D_{Ox})^{1/2} A} \frac{(1+\Theta)^2}{\Theta}. \qquad (20.46)$$

The dependence on potential of the alternating current in sinusoidal alternating current polarography Δi^s is described by the equation obtained by introduction of (20.46) into equation (20.40):

$$\Delta i^s = \frac{n^2 F^2 C^0_{Ox} \Delta E (D_{Ox} \omega)^{1/2} A}{RT} \frac{\Theta}{(1+\Theta)^2}. \qquad (20.47)$$

It follows from this relationship that the current reaches the maximum value when $\Theta = 1$:

$$\Delta i^s_{\max} = \frac{n^2 F^2 C^0_{Ox} \Delta E (D_{Ox}\omega)^{1/2} A}{4RT}. \qquad (20.48)$$

This equation describes the maximum current attained by alternating current sinusoidal polarography for a reversible electrode process, i.e. it is equivalent to the Ilkovič equation used in classical polarography. It follows from this equation that the recorded peak current is proportional to the depolarizer concentration in the bulk of solution, to the electrode surface area, to the amplitude of the alternating voltage (up to several millivolts) and to the square root of frequency.

It follows from equation (20.45) that, since the current reaches the maximum value when $\Theta = 1$, the corresponding potential is equal to $E_{1/2}$.

The theory of alternating current polarography has been worked out by Kambara [22] and by Senda and Tachi [23]. Koutecký [24] first derived the general equation of curves recorded in a.c. polarography for reversible processes. In this work he took into account the effect of growth of the dropping electrode. This was found to be small when commonly used electrodes were employed. Under these conditions the effect of spherical diffusion was also small [25].

Matsuda [26] propounded the general theory of a.c. polarographic curves for small amplitudes of polarizing voltage, which indicates that in *quasi*-reversible processes the growth of the drop affects the recorded current peaks.

The further development of the theory of this method has been largely due to the work of Smith and co-workers.

Delmastro and Smith [27] published a critical discussion of the approximate theories and compared them with exact treatments. They also considered the effect of the amplitude of alternating voltage on the recorded curves, and in further work they evaluated the effect of the growth of the dropping electrode and its geometry [28]. In several papers Smith has described the investigations of other theoretical aspects of a.c. polarography. They are also discussed in a review article [29].

An equation similar to (20.47) has been derived by Barker [30] for a.c. square-wave polarography. It also describes the dependence of the recorded alternating current Δi^k on the electrode potential for reversible electrode processes:

$$\Delta i^k = \frac{n^2 F^2 \Delta E C_{Ox}^0 D_{Ox}^{1/2} A}{RT\pi^{1/2} t_k^{1/2}} \frac{\Theta}{(1+\Theta)^2} \sum_{m=0}^{\infty} (-1)^m \frac{1}{\left[m+\dfrac{t}{t_k}\right]^{1/2}} \quad (20.49)$$

where t_k is the half-cycle time, and t/t_k is the part of this time after which the current is measured.

Expression (20.47) can be written in the following abridged form:

$$\Delta i^k = \frac{n^2 F^2 \Delta E C_{Ox}^0 D_{Ox}^{1/2} A}{RT\pi^{1/2} t_k^{1/2}} \frac{\Theta H}{(1+\Theta)^2} \quad (20.50)$$

where H is a constant, characteristic of the apparatus. By comparing expressions (20.49) and (20.50) this constant is defined as follows:

$$H = \sum_{m=0}^{\infty} (-1)^m \frac{1}{\left(m+\dfrac{t}{t_k}\right)^{1/2}}. \quad (20.51)$$

It follows from equation (20.50) that the current reaches a maximum value when $\Theta = 1$, and that the corresponding potential E_{max}^k is equal to the polarographic half-wave potential $E_{1/2}$.

Assuming that $\Theta = 1$, the expression describing the maximum current is obtained from equation (20.50):

$$\Delta i_{max}^k = \frac{n^2 F^2 \Delta E C_{Ox}^0 D_{Ox}^{1/2} HA}{4RT\pi^{1/2} t_k^{1/2}}. \quad (20.52)$$

By expressing the wave equation for conventional polarography in the following form:

$$i = i_g \left[\frac{1}{1+\exp(E-E_{1/2})\dfrac{nF}{RT}} \right], \quad (20.53)$$

differentiating this equation and expressing the diffusion current by the Cottrell equation:

$$i_g = \frac{nFAC_{Ox}^0 D_{Ox}^{1/2}}{\pi^{1/2} t^{1/2}}, \quad (20.54)$$

an expression is obtained for the current which is observed when a potential pulse having amplitude ΔE is applied to the electrode:

$$\Delta i^p = \frac{n^2 F^2 \Delta E C_{Ox}^0 D_{Ox}^{1/2} A}{RT\pi^{1/2} t_p^{1/2}} \frac{\Theta}{(1+\Theta)^2}. \quad (20.55)$$

This is the equation of curves recorded in derivative pulse polarography, but since its derivation is based on the equation of a reversible polarographic wave, equation (20.55) is valid for reversible processes only and for small voltage amplitudes. Also in this case it can be seen from equation (20.55) that when $\Theta = 1$, $\Delta i^p = \Delta i^p_{max}$ and the current reaches the maximum value; when $\Theta = 1$, $E^p_{max} = E_{1/2}$, where E^p_{max} is the potential at which the current in pulse polarography reaches the maximum value.

Putting $\Theta = 1$ in equation (20.55) an equation for the maximum current is obtained:

$$\Delta i^p_{max} = \frac{n^2 F^2 \Delta E C^0_{Ox} D^{1/2}_{Ox} A}{4RT\pi^{1/2} t_p^{1/2}}. \tag{20.56}$$

Equations (20.48), (20.52) and (20.56) describing the maximum currents of sinusoidal pulse and square-wave polarography are very similar. Equation (20.48) differs slightly from the remaining two, but by using the relationship:

$$\omega = \frac{\pi}{t_s} \tag{20.57}$$

where t_s is the half-cycle time, equation (20.48) can be expressed in the following form:

$$\Delta i^s_{max} = \frac{n^2 F^2 \Delta E C^0_{Ox} D^{1/2}_{Ox} A \pi^{1/2}}{4RT t_s^{1/2}}. \tag{20.58}$$

It is convenient to introduce the concept of the kinetic parameter in the treatment of these methods, as was done in the case of polarography, chronopotentiometry, stationary electrode voltammetry and the rotating disc method. In the case of a.c. polarography the kinetic parameter is the duration of disturbed electrochemical equilibrium, i.e. in sinusoidal and square-wave polarography it is the half-cycle time t_s and t_k, while in pulse polarography it is the time t_p elapsing from the moment of application of the pulse to that of the measurement of the current.

Calling the kinetic parameter X, equations (20.52), (20.56) and (20.58) can be expressed in the form of the following general equation:

$$\Delta i^{s,p,k}_{max} = \frac{Kn^2 F^2 \Delta E C^0_{Ox} D^{1/2}_{Ox} A}{RT X^{1/2}} \tag{20.59}$$

where K is a constant, characteristic of a given method.

Using equation (20.59), the maximum current can be discussed in general terms.

In all the methods considered, the maximum current is proportional to the depolarizer concentration in the bulk of solution, to the square root of the diffusion coefficient, to the pulse amplitude, and to the electrode surface area, but it is inversely proportional to the square root of the kinetic parameter.

Normal pulse polarography differs from the other methods, including derivative pulse polarography, as it gives curves resembling classical polarographic waves [31].

Selecting the initial potential such that in the case of an electroreduction it is in the region of the formation of classical polarograms, and assuming linearity of diffusion, the curve describing a reversible process can be expressed by the following equation:

$$i = nFAC_{Ox}^0 \left(\frac{D_{Ox}}{\pi t_p}\right)^{1/2} \frac{1}{(1+\Theta)}. \qquad (20.60)$$

When the potential becomes sufficiently negative and Θ tends to zero the following expression for the limiting current in normal pulse polarography is obtained from equation (20.60):

$$i_g = nFAC_{Ox}^0 \left(\frac{D_{Ox}}{\pi t_p}\right)^{1/2}. \qquad (20.61)$$

A detailed treatment of this problem was put forward in 1964 by Brinkman and Los [32], who took into account the conditions of transport to the growing dropping electrode. Space does not permit their work to be discussed here.

It follows from equation (20.60) that at the polarographic half-wave potential the current reaches half of its limiting value.

The discussion has, so far, been confined to the current of processes controlled by diffusion rate alone, i.e. reversible processes. Therefore, the question arises which criteria of reversibility should we use? When the standard rate constant of the electrode process of the system under consideration is known, the reversibility of the process can be estimated by calculating the transport rates and comparing them with the rate constant. When the process is investigated simultaneously by d.c. and a.c. polarography the agreement between $E_{1/2}$ and E_{max} potentials can be regarded as a proof of reversibility of the process.

Another criterion of reversibility, specific for a.c. polarography, is the strictly defined width of the current peaks observed in the case of reversible processes. The recorded current maximum is symmetrical with respect to the potential peak. Since the potential dependence is very similar in all the methods, with the exception of normal pulse polarography (the difference is limited to the value of the constant), the width of the current peaks should be also similar. Usually the peak width is measured in the middle of the peak and the measured value is quoted in millivolts.

The peak width S can be obtained from equations (20.47), (20.50) and (20.55) in the form of the following expression

$$S = 3.52 \frac{RT}{nF}, \tag{20.62}$$

and at 25°C:

$$S = \frac{90.5}{n} \text{mV}. \tag{20.62a}$$

These equations are valid for small amplitude pulses only. In derivative pulse polarography the reversibility can be estimated on the basis of an equation identical to that derived by Tomeš.

Irreversible processes should now be considered. It can be expected that, in equations of processes controlled by electron exchange rate alone, the electrode process rate constant will appear instead of the transport rate, which was the determining factor in reversible processes.

The potential dependence of the current in a.c. sinusoidal polarography has been described by several authors [5] who obtained linear increase of the alternating current with increasing standard rate constant. This result was challenged in 1967 by Timmer et al. [33, 34] and then by Smith and McCord [35], who derived an equation showing that in irreversible processes the peak current does not depend on the rate constant of the electrode process, but has a small constant value independent of frequency:

$$_n\Delta i^s_{max} = \left(\frac{7}{3\pi t}\right)^{1/2} \frac{\alpha n^2 F^2 A C^0_{Ox} D^{1/2}_{Ox} \Delta E}{RT}, \tag{20.63}$$

where $_n\Delta i^s_{max}$ is the maximum alternating current observed in a.c. sinusoidal polarography, in the case of irreversible electrode processes, and α is the transfer coefficient.

This equation was derived by Timmer et al.; the equation derived by Smith and McCord is slightly more accurate, but also more complex.

The curves of irreversible processes have the shape of peaks, but the current peak potential can differ considerably from the reversible polarographic half-wave potential. The relation between these potentials was found by Smith and McCord to have the following form:

$$E^s_{max} = E_{1/2} + \frac{RT}{\alpha nF} \ln\left(\frac{1.349 k_s t^{1/2}}{D_{Ox}^{1/2}}\right) - \frac{RT}{2\alpha nF} \ln 1.907(\omega t)^{1/2}. \quad (20.64)$$

In square-wave a.c. polarography the current–voltage curves of fully irreversible processes are described by an equation derived by Barker [7]:

$$_n\Delta i^k = \frac{n^2 F^2}{RT} A \Delta E C_{Ox} k_{fh} \sum_{m=0}^{\infty} (-1)^m \exp N^2\left(m + \frac{t}{t_k}\right) \operatorname{erfc} N\left(m + \frac{t}{t_k}\right)^{1/2}$$

(20.65)

where C_{Ox} is the depolarizer concentration on the electrode surface and N is defined by the equation:

$$N = k_{fh}\left(\frac{t_k}{D_{Ox}}\right)^{1/2}. \quad (20.66)$$

Function $\exp \lambda^2 \operatorname{erfc} \lambda$ and constant k_{fh} have been defined earlier in this book.

The principles of the theory of simple electrode processes controlled by diffusion or by charge transfer rate have been only briefly discussed here. Space does not permit a full treatment of electrode processes connected with chemical reactions. These will be treated schematically only, later in this chapter.

20.3 Potential Usefulness of New Polarographic Methods

A.c. polarography is used mainly in studies of kinetics and mechanisms of electrode processes and the accompanying chemical reactions, in the analysis of trace elements (which is perhaps of greater importance), and in studies on the double layer structure.

Particular cases of the application of a.c. polarography will now be discussed here, but advantages and disadvantages will be shown in comparison with other electrochemical techniques. For the sake of clarity the application of the methods in kinetics, analysis and studies of double layer structure will be considered separately.

20.3.1 NEW POLAROGRAPHIC METHODS OF INVESTIGATION OF THE KINETICS OF ELECTRODE PROCESSES AND ACCOMPANYING CHEMICAL REACTIONS

In the discussion of the potential usefulness of classical polarography, chronopotentiometry, stationary electrode voltammetry and the rotating disc technique in kinetic studies the author has introduced the concept of the kinetic parameter as the principal factor determining the rate of depolarizer transport to the electrode. This concept and the derivation of equations for the transport rate has made it possible to compare the potential usefulness of the four methods. The a.c. polarographic methods will be discussed similarly.

The expressions describing the rates of depolarizer transport to the electrode can be derived from equations (20.52), (20.56) and (20.58), since the recorded currents are proportional to the depolarizer concentration, to the electrode surface area and to the transport rate. By removing the first two terms from these equations the following expression for the transport rate V are obtained; for sinusoidal polarography:

$$\bar{V}_s = \frac{nF\Delta E D_{Ox}^{1/2} \pi^{1/2}}{4RTt_s^{1/2}}, \tag{20.67}$$

for square-wave polarography:

$$\bar{V}_k = \frac{nF\Delta E D_{Ox}^{1/2} H}{4RT\pi^{1/2}t_k^{1/2}}, \tag{20.68}$$

for derivative pulse polarography:

$$\bar{V}_p = \frac{nF\Delta E D_{Ox}^{1/2}}{4RT\pi^{1/2}t_p^{1/2}}, \tag{20.69}$$

Using the kinetic parameter X (in a.c. polarography X is equal to t_s, t_p and t_k), equations (20.67)–(20.69) can be expressed in general terms by means of the following equation:

$$\bar{V}_{s,k,p} = \frac{K'nF\Delta E D_{Ox}^{1/2}}{RTX^{1/2}}. \tag{20.70}$$

It is assumed that the depolarizer studied has a diffusion coefficient of 9×10^{-6} cm^2/s, $n = 1$ and $\Delta E = RT/nF$. In sinusoidal polarography the frequency of the alternating voltage is usually equal to 50 Hz, so that $t_s = 10^{-2}$ s. Using these values and equation (20.67) the value $\bar{V}_s = 1.3 \times 10^{-2}$ cm/s is obtained.

In calculations of transport rates characteristic of square-wave polarography it will be assumed that $H = 1.0$. Since the frequency is usually 225 Hz, the half-cycle time is 4.7×10^{-2} s. According to previous assumptions regarding D_{Ox} and ΔE, $\bar{V}_k = 1.2 \times 10^{-2}$ cm/s is obtained from equation (20.68).

In pulse polarography the pulse time is usually 1/25 s, and the current is measured during the second half of the pulse, when the capacity current is insignificant. Thus it can be assumed that $t_p = 1/50$ s, and, from equation (20.69), the value $\bar{V}_p = 3 \times 10^{-3}$ cm/s is obtained.

From equation (20.61) the expression describing the depolarizer transport rate in normal pulse polarography can easily be obtained:

$$\bar{V}_n = \left(\frac{D_{Ox}}{\pi t_p}\right)^{1/2}. \tag{20.71}$$

Assuming that in normal pulse polarography $t_p = 0.02$ s, and introducing this value into equation (20.71) together with the previously assumed value of D_{Ox}, the value $\bar{V}_n = 1.2 \times 10^{-2}$ cm/s is obtained.

The comparison of transport rates is least favourable with derivative pulse polarography, but the differences between the transport rates are not large. It can be assumed that processes with standard rate constants ten times smaller than the transport rates are virtually controlled by the charge transfer rate and therefore are irreversible, whereas redox systems having standard rate constants exceeding ten times the transport rate are reversible.

It can therefore be assumed, somewhat arbitrarily, that when the disturbance time is not longer than the times assumed in this discussion (and usually observed in practice) the kinetics of systems with standard rate constants not exceeding 0.1 cm/s can be studied by a.c. polarographic methods. This does not mean that the investigation of the kinetics of systems with rate constants lower than the quoted values is simple. It is not very complex when the systems are completely irreversible, i.e. when the rate constants are very small.

The above calculations indicate that, at the assumed times of disturbed equilibrium, the systems with standard rate constants lower than 10^{-3} cm/s are controlled by the charge transfer rate only. When the standard potential is known, it is possible to obtain the standard rate constant from the peak current by means of equation (20.66).

The kinetics of systems having standard rate constants between 10^{-3} and 10^{-1} cm/s can be studied by a.c. polarography, but in this case the calculations are usually fairly tedious. These difficulties can be readily understood by considering the equation derived by Barker [7] for the potential dependence of the current in square-wave polarography of electrode processes controlled simultaneously by diffusion and charge transfer kinetics.

When the system is almost reversible, i.e. when the standard rate constant is close to 0.1 cm/s, the potential dependence of the alternating current is expressed by the following equation:

$$\Delta i = \frac{n^2 F^2 C^0_{Ox} \Delta E k_{1/2}}{RT(\Theta^{-\alpha}+\Theta^{\beta})} \sum_{m=0}^{\infty} (-1)^m \exp M^2\left(m+\frac{t}{t_k}\right) \operatorname{erfc} M\left(m+\frac{t}{t_k}\right)^{1/2}$$
(10.72)

where

$$M = (\Theta^{-\alpha}+\Theta^{\beta})k_{1/2}\left(\frac{\tau}{D_{Ox}}\right)^{1/2},$$
(20.73)

α and β are the transfer coefficients of the cathodic and anodic processes, respectively, and $k_{1/2}$ is the reduction rate constant, measured at $E = E_{1/2}$.

When the degree of irreversibility is higher, but the process is still controlled by diffusion and charge transfer rates, the equation becomes:

$$\Delta i = \frac{n^2 F^2 \Delta E}{RT}(\alpha k_{fh} C_{Ox} + \beta k_{bh} C_{Red}) \sum_{m=0}^{\infty} (-1)^m \times$$

$$\exp N^2\left(m+\frac{t}{t_k}\right) \operatorname{erfc} N\left(m+\frac{t}{t_k}\right)^{1/2}$$
(20.74)

where C_{Ox} and C_{Red} are surface concentrations of Ox and Red, respectively. N is given by the equation:

$$N = \left(\frac{k_{fh}}{D_{Ox}^{1/2}} + \frac{k_{bh}}{D_{Red}^{1/2}}\right)\tau^{1/2}$$
(20.75)

and k_{fh} and k_{bh} are the heterogeneous rate constants of the reduction and oxidation processes, respectively.

It follows from these equations that the determination of the rate constants on the basis of current measurement data is not easy, since these constants appear in the term following the plus sign. For this reason the method based on measurements of the phase angle between the faradaic alternating current and the alternating potential applied to the electrode is at present more widely used than the other methods. It is possible that in the near future normal pulse polarography will find wider practical applications.

It is appropriate now to discuss briefly the phase angle method. On the basis of Matsuda's [26] theoretical study of *quasi*-reversible processes, Tamamushi and Tanaka [36] showed that the phase angle Φ can be described by the following equation:

$$\cot \Phi = 1 + \frac{(2D\omega)^{1/2}}{k_s[\exp(-\alpha L)+\exp(\beta L)]}$$
(20.76)

where
$$L = \frac{nF(E-E_{1/2})}{RT} \tag{20.77}$$

and $D = D_{Ox}^{\beta} D_{Red}^{\alpha}$.

Equation (20.76) is valid when the amplitude of the applied alternating voltage is smaller than $8/n$ millivolts.

It follows from this equation that when $E = E_{1/2}$ the dependence of $\cot \Phi$ on $\omega^{1/2}$ is described by the equation:

$$\cot \Phi = 1 + \frac{(D\omega)^{1/2}}{k_s 2^{1/2}}. \tag{20.78}$$

From the linear relation between $\cot \Phi$ and $\omega^{1/2}$ the slope of the straight line $\Delta \cot \Phi / \Delta \omega^{1/2}$ can be obtained. When D is known, the rate constant can be calculated from the expression:

$$k_s = \frac{D^{1/2}}{2^{1/2} \frac{\Delta \cot \Phi}{\Delta \omega^{1/2}}}. \tag{20.79}$$

However, this procedure does not lead to the evaluation of the transfer coefficients. They can be obtained by comparing the relation between $\cot \Phi$ and $\omega^{1/2}$ at two different potential values. As a result the following relationship is obtained:

$$\alpha = \frac{2.303}{(L_2 - L_1)} \left\{ \log \frac{(1+e^{L_2})}{(1+e^{L_1})} - \log \frac{\left(\frac{\Delta \cot \Phi}{\Delta \omega^{1/2}}\right)_1}{\left(\frac{\Delta \cot \Phi}{\Delta \omega^{1/2}}\right)_2} \right\}, \tag{20.80}$$

where subscripts "1" and "2" correspond to potentials E_1 and E_2.

In normal pulse polarography the current–potential curves of electrode processes controlled by both diffusion and charge transfer rates are described by the equation [31, 37]:

$$i = \frac{nFAC_{Ox}^0 k_s D_{Ox}^{\alpha/2}}{D_{Red}^{\alpha/2}} \exp(-\alpha L) \exp(\lambda^2 t) \operatorname{erfc}(\lambda t^{1/2}), \tag{20.81}$$

where

$$\lambda = \frac{k_s}{D_{Red}^{\alpha/2} D_{Ox}^{\beta/2}} \exp(-\alpha L)(1+\exp L). \tag{20.82}$$

When $E - E_{1/2} \leqslant 0$ equation (20.81) is simplified to equation (20.61).

Dividing equation (20.81) by equation (20.61):

$$\frac{i}{i_g}[1+\exp L] = \pi^{1/2}\lambda t^{1/2}\exp(\lambda^2 t)\operatorname{erfc}(\lambda t^{1/2}). \quad (20.83)$$

On the basis of this relationship the value of $\lambda t^{1/2}$ corresponding to the values of i/i_g can be determined from a graph of $\pi^{1/2}\lambda t^{1/2}\exp(\lambda t^2)\operatorname{erfc}(\lambda t^{1/2})$ vs. $\lambda t^{1/2}$.

Using equation (20.82) and the values of $\lambda t^{1/2}$ the following expression can be obtained:

$$k_{fh}t^{1/2} = \frac{k_s t^{1/2}}{D_{\text{Red}}^{\alpha/2} D_{\text{Ox}}^{-\beta/2}} \exp(-\alpha L) = \lambda t^{1/2}(1+\exp L)^{-1}. \quad (20.84)$$

Plotting the values of k_{fh} against the potential, a straight line having the slope $-\alpha nF/2.303RT$ is obtained from which the transfer coefficient α can be determined. It follows from equation (20.84) that the value of k at $E = E_{1/2}$ is connected with the standard rate constant by means of the following equation:

$$\log k_{1/2} = \log k_s - \frac{\alpha}{2}\log D_{\text{Red}} + \frac{\beta}{2}\log D_{\text{Ox}}. \quad (20.85)$$

Thus, normal pulse polarography makes it possible to carry out a complete kinetic analysis of moderately fast electrode processes. This method was applied by Oldham and Parry [38] in their kinetic studies of irreversible electrode processes.

It is probable that with the aid of computers it will be possible to determine rate constants by means of the equations describing the alternating current in a.c. polarography.

A.c. polarography also plays an important part in kinetic studies of chemical reactions accompanying electrode processes. The theory of these, with particular reference to sinusoidal polarography, has been mainly worked out by Smith et al. [29].

In this brief review of a.c. polarography it is not possible to consider the details. The discussion must be limited to an assessment of advantages which these techniques offer in comparison with classical polarography. The fast processes which can be studied by means of these methods will also be defined.

It is already known that a.c. polarography allows the investigation of chemical processes faster then those possible with classical polarography, since in the former technique the transport rates are faster. In order to carry out quantitative comparison of the methods the equations of the transport rates can be used, and also the relation between the heterogeneous k_{fh} and the homogeneous k_1 rate constants:

$$k_{fh} = k_1 \mu \qquad (20.86)$$

where μ is the thickness of the reaction layer described by the equation:

$$\mu = \sqrt{\frac{D}{k_2}}. \qquad (20.87)$$

By combining equations (20.86) and (20.87) the following equation is obtained:

$$k_{fh} = k_1 k_2^{-1/2} D^{1/2} \qquad (20.88)$$

and by introducing the chemical equilibrium constant $K = k_1/k_2$:

$$k_{fh} = K k_2^{1/2} D^{1/2}. \qquad (20.89)$$

The maximum rate constants k_{fh}^m which can still be measured by the techniques discussed exceed by tenfold the transport rate $\overline{V}$:

$$\overline{V} = \frac{k_{fh}^m}{10}. \qquad (20.90)$$

Combining equations (20.90), (20.89) and (20.70):

$$(K k_2^{1/2})^m = \frac{10 K' n F \Delta E}{RT X^{1/2}} \qquad (20.91)$$

where $(K k_2^{1/2})^m$ is the maximum value which is still measurable in a.c. polarography.

Since the transport rates characteristic of a.c. polarographic methods have already been evaluated, it is convenient to use the formula obtained from equations (20.90) and (20.89):

$$(K k_2^{1/2})^m = \frac{10 \overline{V}}{D^{1/2}}. \qquad (20.92)$$

Except for derivative pulse polarography it can be assumed that the transport rate in these methods is equal to 1.2×10^{-2} cm/s. Assuming further that $D = 9 \times 10^{-6}$ cm²/s, $(K k_2^{1/2})^m = 40$ s$^{-1/2}$ is obtained.

Using equation (20.92), the maximum value of $(K k_2^{1/2})^m$ determinable by classical polarography can be calculated by introducing into this equation the previously calculated transport rate for classical polarography equal to 1.7×10^{-3} cm/s, $(K k_2^{1/2})^m = 5.7$ s$^{-1/2}$ is obtained.

It follows from this discussion that a.c. polarography makes it possible to investigate the kinetics of chemical reactions approximately one order of magnitude faster than those accessible to classical polarography. Obviously in this case, as in the case of electrode processes, the time of disturbed equilibrium can theoretically be shortened and, as a result, the transport rates can be increased. Although this would increase the advantages offered by a.c. polarography in kinetic studies, such modifications are not always advantageous for practical reasons [7].

20.3.2 New Polarographic Techniques as Analytical Methods

It has already been mentioned that a.c. polarography was developed as a result of the search for improved analytical methods, since certain analytical problems, in particular those of determination of trace elements, could not be solved by classical polarographic methods.

The discussion of the usefulness of a.c. polarography in analytical chemistry should commence be considering the general equation for the maximum current:

$$\Delta i_{max} = nFAC^0 \bar{V} \qquad (20.93)$$

which indicates that, at a constant concentration, the current increases with increasing transport rate $\bar{V}$. Therefore in this discussion the transport rates described by equations (20.67), (20.68), (20.69) and (20.71) will be compared. The transport rates from these equations have already been calculated and found to be 1.3×10^{-2}, 1.2×10^{-2}, 3×10^{-3} and 2×10^{-2} cm/s for sinusoidal, square-wave, derivative pulse and normal pulse polarography, respectively. These values compare favourably with the average transport rate of classical polarography (1.7×10^{-3} cm/s), since the recorded alternating currents are several times larger than the limiting current recorded in classical polarography at the same depolarizer concentration. This is one of the reasons why a.c. polarographic methods are particularly suitable for analyses of traces. This advantage is the least pronounced in sinusoidal polarography in which the analytically accessible concentration region is only slightly better than that of classical polarography, although the transport rates in the former case are much faster. This situation is obviously due to fairly large capacity currents which, in the sinusoidal method, are recorded together with faradaic currents.

The square-wave and derivative pulse polarographic techniques can be successfully used in analyses of more dilute solutions (down to 10^{-8} M)

since in these cases the capacity component is not included in the measured alternating current.

The conclusions reached in this discussion are valid when the electrode processes of analysed components are reversible under the experimental conditions.

The analytical efficiency of the a.c. polarographic methods depends on the nature of the analysed substances. Thus in the case of reversible processes the sensitivity limit is in the region of 10^{-8} M solutions, whereas in the case of substances reacting irreversibly with the electrode the limit is usually in the region of 10^{-6} M; in the case of square-wave polarography and derivative pulse polarography it is 5×10^{-8} M.

The maximum sensitivity of the square-wave a.c. polarographic method is not limited by the noise level of the electric circuit but it is limited by the instability of the signal due to the tendency of the solution to penetrate up the dropping electrode capillary (capillary response). When the potential of this electrode is changed rapidly a small slowly disappearing current and a larger rapidly disappearing current due to the charging of the electrode double layer are observed.

Since the disappearance of the current due to the capillary response is proportional to $1/t^n$ (where $n > 1/2$) and the faradaic current of a reversible reaction is proportional to $1/t^{1/2}$ it is obvious that for a large ratio of the signal of electrode reaction to the signal interfering in a.c. polarography, the frequency should be as low as possible.

It should be mentioned that in square-wave polarography at a frequency of 225 Hz difficulties also appear since the signal connected with the capillary response can change from drop to drop.

This signal is more reproducible when capillaries are dilated near the tip whereas the tip itself is narrow. Barker [6] showed that the use of such capillaries leads to more reproducible polarograms with much lower residual currents.

In sinusoidal a.c. polarography the use of frequencies exceeding 100 Hz is not recommended in view of the unfavourable ratio of faradaic current to the capacity current. In an analytical work Bond [39] recommends the use of considerable amplitudes of alternating voltage (from 10 to 50 mV) since then the noise level is relatively low.

The importance and possibilities of a.c. polarography in analysis and its comparison with other polarographic techniques have been discussed by Bond [39, 40].

In normal pulse polarography the limiting current is proportional to the transport rate irrespective of the irreversibility of the electrode process.

In analyses of irreversibly reacting systems pulse polarography is the most sensitive and the sinusoidal method is the least sensitive. In the latter case the faradaic currents of irreversible systems are usually so small that they cannot be detected in the background of the large capacity currents, which are not eliminated in this method. Therefore sinusoidal polarography is, in practice, suitable only for analysis of depolarizers reacting reversibly at the electrode.

In the introduction to this chapter it was briefly mentioned that the possibility exists of improving sinusoidal a.c. polarography as the analytical method, by separation of the faradaic alternating current from the capacity current. This can be done by measuring the current which is in phase with the voltage. When the alternating voltage changes according to the expression:

$$\Delta V = \Delta E \sin\omega t, \qquad (20.94)$$

the faradaic component of the alternating current of a reversible electrode reaction can be expressed in the following form:

$$\Delta i_f = I_f \sin\left(\omega t + \frac{\pi}{4}\right) = I_f \frac{\sqrt{2}}{2}\sin\omega t + I_f \frac{\sqrt{2}}{2}\cos\omega t, \qquad (20.95)$$

where I_f is the amplitude of the faradaic current. When the process is controlled by both diffusion kinetics and charge transfer the phase angle depends on the rate constant of the electrode process.

The capacity component of the alternating current can be expressed by means of the following relationship:

$$\Delta i_c = I_c \sin\left(\omega t + \frac{\pi}{2}\right) = I_c \cos\omega t. \qquad (20.96)$$

The capacity current has a phase shift of one quarter phase with respect to the applied alternating voltage. When the voltage reaches the maximum value the capacity current is zero.

It follows from equations (20.94) and (20.95) that the faradaic current of a reversible electrode process has a component in phase with the applied alternating voltage, equal to the value of the component shifted in phase by 90°. Using an apparatus measuring the current component which is in phase with the alternating voltage, the value proportional to the faradaic current is obtained, with the capacity current excluded. The potential dependence of the current recorded by means of this apparatus (called a vectorial polarograph), has the shape of a peak, as in the case of square-wave polarography in which the capacity component of the current is also eliminated.

The importance of these techniques in analytical chemistry should again be emphasised. A combination of square-wave polarography with the method based on accumulation of mercury-soluble metals in a hanging mercury drop allows [41–43] the analysis of solutions at concentration levels of the order of 10^{-9}–10^{-10} M. A.c. polarographic methods are convenient analytical tools, and the reproducibility of their results is at least equal to those obtained by classical polarography.

A further advantage of these methods, with the exception of the normal pulse method, is the considerably higher resolving power in the analysis of mixtures, compared with classical polarography. In the case of classical polarography mixtures of substances can be analysed provided that the differences between the half-wave potentials of the components exceed $200/n$ mV, since below this value serious overlapping of waves occurs. In a.c. polarography distinct and separate peaks are observed even when the difference between their potentials is only about 40 mV.

In analyses of mixtures it is possible to utilize the phenomenon mentioned earlier, that in sinusoidal polarography the currents due to substances reacting irreversibly with the electrode are negligible. In square-wave and pulse polarography these currents are much smaller than those due to reversible reactions. By combining classical polarography, by which it is possible to determine the sum of two components with similar half-wave potentials, and a.c. polarography, which determines only the reversibly reacting component, it is often possible to analyse mixtures of substances, reacting reversibly and irreversibly with the electrode. Leake and Reynolds employed this procedure for the analysis of lead styphnate [44]. The large styphnate reduction, comprising up to three waves (depending on pH), which precedes the lead wave, was measured first by normal polarography. The determination was then repeated using a sinusoidal a.c. polarograph. The irreversible styphnate wave disappeared and the lead was determined without interference.

20.3.3 A.C. Polarography as a Method for Investigation of Double Layer Structure

A.c. polarographic methods are also very useful in studies of double layer structure and adsorption of various substances on mercury electrodes, since large currents flow in the circuit as a result of the high frequency voltage changes. These currents are related to the charging and discharging of the double layer:

$$i_c = C_t \frac{dE}{dt} \tag{20.97}$$

where C_t is the differential capacity of the double layer.

The rate of change of applied potential dE/dt can be held constant for any experiment, since this is a feature of most types of polarograph. Therefore, the current recorded in the absence of the faradaic process is proportional to the differential capacity of the double layer.

In sinusoidal polarography this capacity current makes it difficult to investigate the oxidation and reduction of substances at concentrations lower than 10^{-5} M. Thus, when the sensitivity of the apparatus is sufficiently high the sinusoidal a.c. polarogram of e.g. pure 0.5 M K_2SO_4 is a curve resembling that of the potential dependence of the differential capacity of the electrode double layer. This curve is shown in Fig. 20.10. By recording

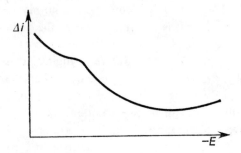

Fig. 20.10 A.c. current recorded by a.c. polarography of 0.5 M potassium sulphate solution.

such a relationship for a system of known capacity it is possible to calculate approximately the capacity from the recorded current.

The phenomenon preventing analytical utilization of the method can thus be employed in studies of the double layer. It is evident that the accuracy of these measurements is lower than that of capacity determinations by means of an alternating current bridge, carried out by the method described by Grahame [45], but the a.c. polarographic method is very convenient and rapid, since it allows the whole curve of the potential dependence of capacity to be obtained from one measurement.

It has been mentioned that in square-wave and derivative pulse polarography the measuring devices are so constructed that it is possible to measure the current a certain time after the potential increment, thus eliminating the capacity current from the measurement. In certain devices the square-wave half-pulse time is divided into a definite number of time intervals. The faradaic current is measured in the intervals near the termination of the drop but the polarograph can be so adjusted that the current is measured in the time intervals soon after the potential change. Then the current measured is practically limited to that connected with the charging and discharging of the double layer, particularly when the solution contains the supporting electrolyte only.

A.c. polarography, and particularly the sinusoidal method, is a useful tool in the study of adsorption on electrodes. This method of investigation of adsorption phenomena was first used independently by Breyer and Hacobian [46] and by Doss and Kalyanasundaram [47]. The former authors gave the name tensammetry to this method of investigation of adsorption; an abbreviation of the words surface tension and amperometry. It employs the measurements of the potential dependence of the alternating current of a solution of surface active agent. The resulting curves are shown schematically in Fig. 20.11. They resemble the curves of potential dependence of

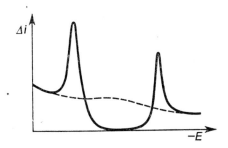

Fig. 20.11 Tensammetric curves. Full line shows a.c. current changes in the presence of a specifically adsorbable substance.

the double layer differential capacity, which are recorded for solutions of adsorbable compounds.

The two current maxima observed in Fig. 20.11 correspond to adsorption and desorption of the surface active substance, which occur when the potential changes to the negative values. They are often called the tensammetric maxima. As might be expected, the current magnitude of these maxima is proportional to the square root of the alternating current frequency. Their concentration dependence is not linear, and they decrease with increasing temperature. Since, under these conditions, the adsorption also decreases, the maxima are shifted in the direction of the electrocapillary zero, which indicates that the adsorption potential region decreases.

The tensammetric method is inferior to that based on the a.c. bridge, since it gives only approximate results, but it is often used, particularly when the determinations have to be carried out rapidly.

During investigations of a new system it is sometimes difficult to differentiate between tensammetric maxima and the a.c. current peaks due to faradaic processes. This problem can be readily solved also by examining the system by means of classical polarography; the a.c. peaks corresponding to polarographic waves are due to electrode reactions, whereas tensammetric maxima have no equivalents in classical polarography.

In recent years Jehring and co-workers [48–52] have used extensively the tensammetric method for investigation the of adsorption of various substances on mercury electrodes and their effect on certain electrode processes. Several publications dealing with tensammetry have been discussed or mentioned by Jehring [53] in a review article.

A.c. polarographic methods and in particular their analytical applications have also been discussed in books [54–56].

20.4 A Brief Discussion of Some Other Electroanalytical Methods

In addition to the new methods discussed above other techniques were developed. We will mention some of them here, in particular those used in chemical analysis. One of these methods is polarography at radio frequencies, introduced by Barker [7]. It is based on Faraday rectification phenomenon caused by the non-linear character of electrode processes (see e.g. [57]).

It can be shown that when an electrode in the equilibrium state is polarized by sinusoidal alternating current a change of the mean electrode potential is observed. It can be expressed by the equation

$$\Psi = (\Delta E)^2 \frac{nF}{RT} \left\{ \frac{2\alpha - 1}{4} + \left[\frac{(1-\alpha)\sqrt{D_{Ox}}\, C^0_{Ox} - \alpha \sqrt{D_{Red}}\, C^0_{Red}}{2 C^0_{Red} \sqrt{D_{Red}} + 2 C^0_{Ox} \sqrt{D_{Ox}}} \right] \times \left[\frac{\gamma(k_e + \sqrt{2}\, k_e^2 \gamma)}{\sqrt{2}\,(1 + \sqrt{2}\, k_e \gamma + k_e^2 \gamma^2)} \right] \right\} \quad (20.98)$$

where

$$\gamma = \frac{C^0_{Red} \sqrt{D_{Red}} + C^0_{Ox} \sqrt{D_{Ox}}}{C^0_{Red} \sqrt{\omega C^0_{Red} D_{Red}}},$$

ΔE is the amplitude of the variable component of the potential in the interface, α is the transfer coefficient of the cathodic process and k_e is the equilibrium rate constant of this process. Equation (20.98) is valid when $\Delta E \ll \dfrac{RT}{nF}$. The problems of Faraday rectification have been studied by many authors.

In practice it is more convenient to measure not the mean potential changes when alternating current is applied to the electrode system, but the alternating current which should be applied to the system in order to keep the mean electrode potential constant when it is polarized with a current having radio frequency with modulated amplitude. Let us assume that in this case the mean electrode potential is kept at the value E_0. In time intervals during which the current having the radio frequency does not flow, the electrode potential will have this value.

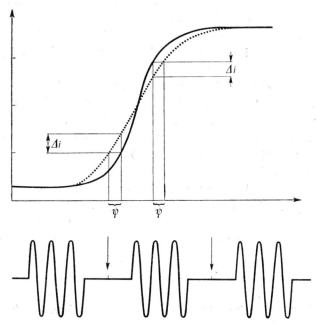

Fig. 20.12 Scheme illustrating the principles of radio frequency polarography (from [58] by permission of the copyright holders, Aeademic Press).

In Fig. 20.12 these time intervals are shown by arrows. During the flow of the current the potential has the tendency to change by the value of Ψ given by equation (20.98), but is kept constant as a result of current

Δi which is recorded by means of the polarograph used in square-wave alternating current polarography. Thus the measurements of low frequency alternating current replace the measurements of voltage Ψ. In analytical work ΔE can reach 25 mV. The sensitivity of the method increases, but its resolving power decreases with increasing ΔE. In kinetic studies ΔE is $\leqslant 5$ mV. The most suitable frequency is 400 kHz. The use of much higher frequencies is usually undesirable, since it can cause the appearance of an additional signal due to the heating of the solution surrounding the drop of mercury. In this method the signal due to the penetration of the solution to the capillary is unimportant [7]. In the case of depolarizers reacting reversibly with the electrode this method is about twice as sensitive as square-wave alternating current polarography and in the case of depolarizers reacting irreversibly the increase of the sensitivity is even greater, but the principal advantage of this method is a greater reproducibility of the recorded signals. The principles of this method were also described by Schmidt and von Stackelberg [58].

In recent years a new method, which could be called coulostatic polarography, was developed. It is based on the coulostatic technique and its analytical applications have been described by Delahay and co-workers [59–65]. A new variant of this method, its theoretical principles [67, 68], apparatus [66], and experimental details [69, 70] have been described by Astruc and co-workers [66–72]. Another variant of this method has also been considered [71–73]. Later a method based on the same principles was described by Kudirka, Abel and Enke [74]. In this method the dropping mercury electrode is maintained at such a potential that the faradaic current practically does not flow. At the end of the drop life time a small charge is rapidly injected into the electrode. This causes a change of the electrode potential which in the potential region in which the electrode reaction is limited exclusively by the transport rate can be expressed by the equation

$$E = \pm \frac{2nFC^0 D^{1/2} t^{1/2}}{\pi^{1/2} C_t} \qquad (20.99)$$

where C_t is the capacity of the double layer formed on the electrode used in the investigations.

The concentration of the substance participating in the electrode reaction can be determined from the slope of the plot of $\Delta E - t^{1/2}$. In this method the limitations resulting from the charging current of the double layer become insignificant when the polarography is used for analysis of low concentrations.

In the method proposed by Kudirka *et al.*, after the introduction of the charge a small digital computer calculates the slope of the initial plot of E–$t^{1/2}$ and the intercept (E for $t = 0$). The experiment is repeated but a slightly larger charge is introduced at the end of the life time of each successive drop. The increase of the charge leads to an increase of the slope and when the charge becomes so large that it brings the electrode potential to the range of the polarographic wave plateau the slope becomes independent of the potentials. Thus the potential dependence of the slope at $t = 0$ resembles the polarograms recorded in classical polarography and can be used for qualitative and quantitative analysis of solutions.

Using this method the authors [74] obtained well formed waves for zinc(II) and cadmium(II) in concentrations of 5×10^{-6} M. Astruc, Del Rey and Bonastre [69, 70] reported that the sensitivity limit of this method is $5 \times 10^{-8} M$.

It should be mentioned that this method is particularly suitable for determinations carried out at very low concentrations of the base electrolyte. Successful determinations have been carried out at 10^{-4} M concentrations of the base electrolyte.

During the 1950's a variant of polarography appeared which differs from the classical polarography only by the method of recording the current. In the classical polarography the mean current is usually measured by means of a strongly damped polarograph. The faradaic current considered during the life time of the drop changes with $t^{1/6}$ whereas the capacity current of charging of the double layer is proportional to $t^{-1/3}$. This means that the ratio of the faradaic current to the capacity current is small in the initial period of the drop life and it reaches the maximum values at the end of the drop life.

Since in analytical measurements we are interested in elimination of the capacity component of the measured current, the measurements of the current only during the final stages of the drop life brings us closer to the ideal situation. Such measurements were first carried out by Wåhlin and Bresle [75] who measured the current during the last second of the drop life, which was of the order of 5 seconds.

A polarograph working on this basis was described by Kronenberger and co-workers [76]. Circuits making possible recordings of polarographs according to the principles described above were also developed by Nash [77] and Yasumori [78]. They differ considerably from the circuits of classical polarographs and for this reason it is difficult to adapt the latter to recording of currents at the end of the drop life.

Kane [79] described a simple circuit which does not require changes in the polarizing circuit of a classical polarograph but only a simple change in the recording circuit.

According to Schmidt and von Stackelberg [58] this method has the following advantages: the sensitivity is higher than that of classical polarography because the current is recorded during the periods of the drop life in which the capacity current is small in comparison with the faradaic current, and it gives the possibility of recording polarograms without suppressing the oscillations, which are slight. The suppression of oscillations can alter the shape and sometimes also the size of the recorded polarograms. This method of recording the currents is also utilized in pulse polarography.

In electroanalytical practice methods which are generally called oscillopolarography are also used. They are also based on electrolysis. Changes of various parameters connected with the electrode reaction are observed on the oscillograph screen, usually as functions of electrolysis time or electrode potential. Stationary electrode voltammetry is sometimes included in this group of methods when the scan rate is fast and the current–potential curve is recorded on an oscillograph screen.

Heyrovský [80–82] developed and used an oscillopolarographic method based on recording of potential–time curves. In this method the electrolysis vessel is connected in series with a large resistance of the order of 10^5 ohms and to this system is applied 120–220 V sinusoidal voltage of mains frequency. The working electrode is polarized to a suitable potential by a constant voltage applied to the electrodes.

Since under such conditions the resistance of the electrolytic cell is small in comparison with the resistance introduced into the circuit, and the electrode potential changes are small in comparison with the changes of total potential, the current is in principle determined by external voltage and constant resistance.

The time dependence of potential difference is recorded by means of an oscillograph, the time basis of which is synchronized with the variable voltage. The potential–time curves contain steps at potentials corresponding to reduction and oxidation potentials of the substances examined. When the process examined is reversible the steps appear at practically identical potentials. In the case of slow electrode processes the difference between these potentials changes, depending on the degree of reversibility. Thus this method can be used for evaluation of reversibility of electrode reactions. The extent of the dip at a given potential is proportional to the depolarizer concentration. When the electrode potential range is of the order of 2 V, at potentials close to zero the independence of potential on time is due to oxidation and reduction of mercury whereas the plateau at negative potentials is connected with the discharge of cations of the base electrolyte (Fig. 20.13).

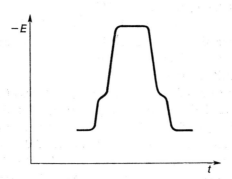

Fig. 20.13 Potential–time curve obtained in the oscillopolarography method introduced by Heyrovský.

In a solution containing only the base electrolyte the slope of potential–time curves is determined only by the capacity of the double layer.

Obviously in the case when the solution contains several depolarizers additional steps appear on the curve at the corresponding potentials.

Another method [83] in which the derivative of potential with respect to time is recorded as a function of time has also been used. This method is particularly suitable for qualitative investigation of adsorption of substances on electrodes.

In another method developed by Heyrovský [82], which has been fairly widely used, $\frac{dE}{dt}-E$ curves are recorded.

These methods have been described in detail in several extensive works [84–86]. We may remember that the limitations in analysis of low concentrations, because of the capacity currents, have been minimized as a result of the different time dependences of faradaic and capacity currents when a potential is applied to the electrode or when the potential is changed (square-wave polarography) or as a result of utilization of the different phase-angles of faradaic and capacity currents with respect to the applied voltage.

There is yet another possibility of a partial elimination of the capacity current from the total current measured.

It is known that electrodes at which redox processes take place behave like non-linear electrical elements. For this reason, after the application of the variable voltage the contribution of higher harmonics to the measured alternating current becomes considerable. However since the capacity impedance of the double layer behaves like an almost perfectly linear electrical element the measurement of higher harmonics of the current can give information about the electrode process which is more valuable for analytical purposes than that derived from the measurements of the basic frequency current, since in the former case the interfering effect of the double layer capacity is negligible.

Bauer [87] was one of the first authors who showed that this procedure can be useful in analytical chemistry. Analysing 2.5×10^{-4} M Cd^{2+} solution in 0.5 M $HClO_4$ he found that the charging current constitutes about 75% of the total alternating current in the case of the normal alternating current polarography developed by Breyer *at al.* but only 5% in the case when the second harmonic is measured. Analytical applications of this method have also been considered by Neeb [88].

Van Cakenberghe [89] carried out the first experimental studies on the second harmonic present in the alternating current flowing through an electrolytic system in which the electrodes are polarized by variable sinusoidal voltage superimposed on a constant voltage.

Hungarian investigators [90] have considered analytical applications of the higher harmonics of alternating current and reviewed the theoretical principles of this method. They have also studied [91] applications of second harmonic alternating current polarography to the analysis of systems containing several depolarizers.

References

[1] Reynolds, G. F. and Powell, K. G., *Nature*, **205**, 695 (1965).
[2] Müller, R. H., Garman, R. L., Droz, M. E. and Petras, J., *Ind. Eng. Chem., Anal. Ed.*, **10**, 339 (1938).
[3] Boeke, J. and van Suchtelen, H., *Z. Elektrochem.*, **45**, 753 (1939).
[4] Breyer, B. and Gutman, F., *Trans. Faraday Soc.*, **42**, 650 (1946).
[5] Breyer, B. and Bauer, H. H., "*Alternating Current Polarography and Tensammetry*", Interscience, New York 1963.
[6] Barker, G. C. and Jenkins, I. L., *Analyst*, **77**, 685 (1952).
[7] Barker, G. C., *Anal. Chim. Acta*, **18**, 118 (1958).
[8] Barker, G. C. and Gardner, A. W., *Z. anal. Chem.*, **173**, 79 (1960).

References

[9] Barker, G. C., in *"Progress in Polarography"* Zuman, P., Ed., Interscience, New York 1962, Vol. 2, page 411.
[10] Ferrett, D. J., Milner, G. W. C., Shalgosky, H. J. and Slee, L. J., *Analyst*, **81**, 506 (1956).
[11] Sluyters, J. H., Brenkel, J. S. M. C. and Sluyters-Rehbach, M., *J. Electroanal. Chem.*, **31**, 201 (1971).
[12] Barker, G. C., *J. Electroanal. Chem.*, **14**, 95 (1973).
[13] Barker, G. C. and Faircloth, R. L., in *"Advances in Polarography"*, Longmuir, J. S., Ed., Pergamon, Oxford 1960, page 313.
[14] Sluyters, J. H. and Sluyters-Rehbach, M., *J. Electroanal. Chem.*, **34**, 542 (1972).
[15] Smith, D. E., *Anal. Chem.*, **35**, 1811 (1963); Niki, E., *Rev. Polarogr. Kyoto*, **3**, 41 (1955); Takahashi, T. and Niki, E., *Talanta*, **1**, 245 (1958).
[16] Randles, J. E. B., *Disc. Faraday Soc.*, **1**, 11 (1947).
[17] Ershler, B. B., *Disc. Faraday Soc.*, **1**, 269 (1947); *Zhurn. Fiz. Khim.*, **22**, 683 (1948).
[18] Gerischer H., *Z. Elektrochem.*, **55**, 98 (1951).
[19] Grahame, D. C., *J. Electrochem. Soc.*, **99**, C370 (1952).
[20] Delahay, P., *"New Instrumental Methods in Electrochemistry"*, Interscience, New York 1954, Chap. 7.
[21] Damaskin, B. B., *"The Principles of Current Methods for the Study of Electrochemical Reactions"*, McGraw-Hill, New York 1967
[22] Kambara, T., *Z. physik. Chem., Frankfurt*, **5**, 52 (1955).
[23] Senda, M. and Tachi, I., *Bull. Chem. Soc. Japan*, **28**, 632 (1955).
[24] Koutecký, J., *Coll. Czechoslov. Chem. Communs.*, **21**, 433 (1956).
[25] Gerischer, H., *Z. physik. Chem., Leipzig*, **198**, 286 (1951).
[26] Matsuda, H., *Z. Elektrochem.*, **62**, 977 (1958).
[27] Delmastro, J. R. and Smith, D. E., *J. Electroanal. Chem.*, **14**, 261 (1967).
[28] Delmastro, J. R. and Smith, D. E., *J. Electroanal. Chem.*, **9**, 192 (1965).
[29] Smith, D. E., in *"Electroanalytical Chemistry"*, Bard, A. J., Ed., Vol. 1, Dekker, New York 1966.
[30] Barker, G. C., Faircloth, R. L. and Gardner, A. W., AERE C/R 1786, Harwell (1956).
[31] Barker, G. C. and Gardner, A. W., AERE C/R 2297, Harwell (1958).
[32] Brinkman, A. A. and Los, J. M., *J. Electroanal. Chem.*, **7**, 171 (1964).
[33] Timmer, B., Sluyters-Rehbach, M. and Sluyters, J. H., *J. Electroanal. Chem.*, **14**, 169 (1967).
[34] Timmer, B., Sluyters-Rehbach, M. and Sluyters, J. H., *J. Electroanal. Chem.*, **14**, 181 (1967).
[35] Smith, D. E. and McCord, T. G., *Anal. Chem.*, **40**, 474 (1968).
[36] Tamamushi, R. and Tanaka, N., *Z. physik. Chem., Frankfurt*, **21**, 89 (1959).
[37] Christie, J. H., Parry, E. P. and Osteryoung, R. A., *Electrochim. Acta*, **11**, 1525 (1966).
[38] Oldham, K. B. and Parry, E. P., *Anal. Chem.*, **40**, 65 (1968).
[39] Bond, A. M., *Anal. Chem.*, **44**, 315 (1972).
[40] Bond, A. M. and Canterford, D. R., *Anal. Chem.*, **44**, 721 (1972).
[41] Von Sturm, F. and Ressel, M., *Z. anal. Chem.*, **186**, 63 (1962).
[42] Kaplan, Ya. B. and Sorokovskaya, I. A., *Zav. Lab.*, **30**, 1177 (1964).
[43] Krause, M. S., Jr. and Ramaley, L., *Anal. Chem.*, **41**, 1362, 1365 (1969).

[44] Leake, L. R. and Reynolds, G. F., *Anal. Chim. Acta*, **21**, 255 (1959).
[45] Grahame, D. C., *J. Am. Chem. Soc.*, **68**, 301 (1946).
[46] Breyer, B. and Hacobian, S., *Australian J. Sci. Res.*, **A5**, 500 (1952).
[47] Doss, K. S. G. and Kalyanasundaram, A., *Proc. Indian Acad. Sci.*, **35A**, 27 (1952).
[48] Jehring, H. in *"Polarography 1964"*, Hills, G. J., Ed., Macmillan, London 1966, Vol. 1, page 349.
[49] Jehring, H., Horn, E., Reklat, A. and Stolle, W., *Coll. Czechoslov. Chem. Communs.*, **33**, 1038 (1968).
[50] Jehring, H. and Stolle, W., *Coll. Czechoslov. Chem. Communs.*, **33**, 1670 (1968).
[51] Jacobasch, H. J., Jehring, H. and Schuman, U., *Faserforsch. u. Textiltechnik*, **23**, 42 (1972).
[52] Horn, E. and Jehring, H., *Faserforsch. u. Textiltechnik*, **24**, 240 (1973).
[53] Jehring, H., *J. Electroanal. Chem.*, **21**, 77 (1969).
[54] Geissler, M. and Kuhnhardt C., *"Square-wave-Polarographie"*, VEB Deutscher Verlag für Grundstoffindustrie, Leipzig 1970.
[55] Pats, P. G. and Vasilyeva, L. N., *"Analytical methods using a.c. polarography"*, Ed. Metallurgiya, Moscow 1967 (in Russian).
[56] Zaretsky, L. C., *"Impulse polarographic concentratometer"*, Ed. Energiya, Moscow 1970 (in Russian).
[57] Delahay, P., *Adv. Electrochem., Electrochem. Eng.*, **1**, 233 (1961).
[58] Schmidt, H. and von Stackelberg, M., *"Modern Polarographic Methods"*, Academic Press, New York, London 1963.
[59] Delahay, P., *Anal. Chim. Acta*, **27**, 90 (1962).
[60] Delahay, P., *Anal. Chim. Acta*, **27**, 400 (1962).
[61] Delahay, P., *Anal. Chem.*, **34**, 1267 (1962).
[62] Delahay, P. and Ide, Y., *Anal. Chem.*, **34**, 1580 (1962).
[63] Delahay, P., *Anal. Chem.*, **34**, 1662 (1962).
[64] Aramata, A. and Delahay, P., *Anal. Chem.*, **35**, 1117 (1963).
[65] Delahay, P. and Ide, Y., *Anal. Chem.*, **35**, 1119 (1963).
[66] Bonastre, J., Astruc, M. and Lelu, J., *Electron. Ind.*, **99**, 763 (1966).
[67] Astruc, M., Bonastre, J. and Royer, R., *J. Electroanal. Chem.*, **34**, 211 (1972).
[68] Astruc, M. and Bonastre, J., *J. Electroanal. Chem.*, **36**, 435 (1972).
[69] Astruc, M., Del Rey, F. and Bonastre, J., *J. Electroanal. Chem.*, **43**, 113 (1973).
[70] Astruc, M., Del Rey, F. and Bonastre, J., *J. Electroanal. Chem.*, **43**, 125 (1973).
[71] Astruc, M. and Bonastre, J., *J. Electroanal. Chem.*, **40**, 311 (1972).
[72] Bonastre, J., Astruc, M. and Bentata, J. L., *Chim. anal., Paris*, **50**, 113 (1968).
[73] Astruc, M. and Bonastre, J., *Anal. Chem.*, **45**, 421 (1973).
[74] Kudirka, J. M., Abel, R. and Enke, C. G., *Anal. Chem.*, **44**, 425 (1972).
[75] Wåhlin, E., *Radiometer Polarogr.*, **1**, 113 (1952); Wåhlin, E. and Bresle, Å., *Acta Chem. Scand.*, **10**, 935 (1956).
[76] Kronenberger, K., Strehlow, H. and Ebel, A. W., *Polarogr. Ber.*, **5**, 62 (1957).
[77] Nash, L. F., *Brit. Patent* 850,078, 28/9/1960.
[78] Yasumori, Y., *Japan Analyst*, **7**, 354 (1958).
[79] Kane, P. O., *J. Electroanal. Chem.*, **11**, 276 (1966).
[80] Heyrovský, J., *Chem. Listy*, **35**, 155 (1941).
[81] Heyrovský, J., *Anal. Chim. Acta*, **2**, 533 (1948).
[82] Heyrovský, J., *Anal. Chim. Acta*, **8**, 283 (1953).

[83] Heyrovský, J. and Forejt, J., *Z. physik. Chem., Leipzig*, **193**, 77 (1943).
[84] Heyrovský, J. and Forejt J., *"Oscillographic polarography"*, Prague 1953 (in Czech).
[85] Heyrovský, J. and Kalvoda, R., *"Oscillographische Polarographie mit Wechselstrom"*, Akad. Verlag, Berlin 1960.
[86] Heyrovský, J. and Micka, K., in *"Electroanalytical Chemistry"*, Bard, A. J., Ed., Dekker, New York 1967, Vol. 2, page 193.
[87] Bauer, H. H., *J. Electroanal. Chem.*, **1**, 256 (1959/60).
[88] Neeb, R., *Z. anal. Chem.*, **188**, 401 (1962).
[89] van Cakenberghe, J., *Bull. Soc. Chim. Belge*, **60**, 3 (1951).
[90] Dèvay, J., Garai, T., Mészáros, L. and Pungor, E., *Hung. Scient. Instruments*, No. 12, 1 (1968).
[91] Dèvay, J., Garai, T., Mészáros, L. and Palágyi-Fényes, B., *Hung. Scient. Instruments*, No. 15, 1 (1969).

Index

acetylacetone complexes, 375
adsorption effects, 396
  catalytic, 413
  chronoamperometry, 397
  chronocoulometry, 397
  chronopotentiometry, 411
  inhibitory, 416
  polarography, 400
  stationary electrode voltammetry, 407
aldoses, bromination, 258
amalgams, activity of metal in, 221
p-aminophenol, 308
ammonium ions, 340
n-amyl alcohol, inhibitory effect, 416
anisole, bromination, 456
aromatic carbonyls, dimerization of reaction products, 340
ascorbic acid, 296
  iron complexes, 328
azobenzene, 447, 451

benzidine rearrangement, 447, 448, 451
p-benzoquinoneimine, hydrolysis, 308
Bockris isotherm, 20
Boltzmann equation, 2, 11
n-butanol, inhibitory effect, 417

cadmium, 507
  amalgam, 308
  cyanide complexes, 288
  EDTA complex, 373
  exchange with calcium EDTA complex, 453
  chloride complex, 434
  pyrazole complexes, 364
  thiocyanate complexes, 367
calcium EDTA complex, ligand exchange with cadium, 453
capacitance current, 470, 481
capillary efficiency, 28, 29, 148
catalytic reactions, 311, 315, 321, 325, 328
charge transfer rates, 55, 56
chronoamperometry, 24, 25, 97, 139, 156, 157, 164, 185, 262, 313, 383, 397
  catalytic reactions, 313, 325
  current, 137, 140, 158, 165
  cyclic, 421, 422, 425, 446
  cylindrical electrodes, 157
  flat electrodes, 140
  irreversible reactions, 185, 205
  limited diffusion, 164
  linear diffusion, 97, 185
  spherical electrodes, 140
chronocoulometry, 24, 25, 97, 104, 230, 383, 397
  double potential step, 423, 446
  for measuring kinetic parameters, 230
chronopotentiometry, 24, 43, 105, 153, 156, 160, 169, 173, 175, 193, 208, 239, 275, 288, 321, 336, 347, 350, 357, 391, 411, 434, 467
  catalytic reactions, 321, 325
  circuitry, 45
  current, 137, 154, 155
    reversal, 453
  cyclic, 46, 421, 438, 455
  electrodes, 46
    boron carbide, 46
    carbon paste, 46
    cylindrical, 160
    gold, 46
    hanging drop, 46
    mercury film, 46
    platinum, 46
    spherical, 154
  irreversible processes, 193, 208, 242
  kinetic parameters, 239
  multi-stage processes, 173
  resolution, 175
  reversible processes, 239
  transition times, 45, 58, 107, 108, 153, 155, 170, 173, 175, 288
  transport rates, 57, 59, 60, 196
cobalt(III), 329
  complexes, 379
complexes, electroanalytical investigation, 360
  DeFord and Hume method, 364
  irreversibly reducible, 369

mechanism of reduction, 375
reversibly reducible, 360
(see also individual compounds)
convolution potential sweep voltammetry, 24, 116
copper(I), anodic oxidation, 443
copper(II), 443
Cottrell equation, 25, 99, 116, 423
cyclic methods, 46, 420, 421, 422, 425, 427, 431, 446, 451
   diffusion processes, 420, 431
   kinetic processes, 446, 451
cyclic polarization, 36

DeFord and Hume method, 364
depolarizer transport, 492
   deviations, 459
diffusion coefficients of metals in mercury, 165
diffusion processes, 420
diffusion to the electrode, 78
   convective to rotating electrode, 92
   cylindrically symmetrical, 86, 157, 209, 235, 437
   limited field, 164
   linear, 79, 82, 119, 184, 431, 436
      to expanding drop, 88
   spherically symmetrical, 83, 138, 224, 234, 237, 263, 266, 436
dimerization rate constants, 342
p-dimethylaminobenzaldehyde, 291
double layer, adsorption, of alkali metals, 13
   of caesium, 16
   of halides, 13
   of thallium, 16
   of water, 11, 12
   study of, 501, 503
charge density, 3, 4
charging, in chronopotentiometry, 108
differential capacity, 8
diffuse layer capacity, 7, 8
distribution of potential, 2, 5, 6, 7
electric field intensity, 5, 11
Ershler model, 16

Gouy and Chapman theory, 6, 8, 9, 11, 12
Grahame model, 16
Helmholtz theory, 2, 7
minimum capacity, 8, 9
Quincke theory, 2, 7
rigid layer capacity, 7, 8
specific adsorption, 13
statistical theory, 12
Stern model, 16
Stern theory, 7, 8
structure, 1
   study of, 501
dropping electrode, 26
   drop area, 27, 30
   drop growth, 28, 148
   drop life, 29
   drop shape, 148
   mercury flow-rate, 29, 148, 149
   streaming, 462

ECC processes, 382, 385
ECE processes, 382, 453, 457
   chronoamperometry, 383
   chronocoulometry, 383
   chronopotentiometry, 391
   polarography, 386
   rotating disc methods, 393
   stationary electrode voltammetry, 388
EDTA (Complexone III) complexes, 308, 373, 453
electrochemical affinity, 54
electrochemical potentials, 53
electrode processes, 61
   adsorption effects on, 396
   catalytic, 311
      chronoamperometry, 313, 325
      chronopotentiometry, 321, 325
      examples, 328
      polarography, 322, 325
      rotating disc method, 324, 325
      stationary electrode voltammetry, 315, 325
   controlled by linear diffusion, 95
   controlled by transport rate, 95
   followed by dimerization of product, 340

chronoamperometry, 341
chronopotentiometry, 347, 350
polarography, 341, 350
rotating disc method, 348, 350
stationary electrode voltammetry, 344, 350
followed by disproportionation, 353, 453
chronopotentiometry, 357
polarography, 353
rotating disc method, 357
stationary electrode voltammetry, 354
followed by first-order reaction, 296
chronopotentiometry, 303, 307
polarography, 297, 307
rotating disc method, 305, 307
stationary electrode voltammetry, 299, 307
irreversible, 222, 228, 230, 235, 242, 247, 319, 490
kinetic parameters, 211, 433, 488
chronocoulometry, 230
chronopotentiometry, 239, 243
polarography, 224
rotating disc method, 245, 249
kinetic studies, 471, 491, 493
utility, 292
kinetics, effect of double layer, 288
of charge exchange, 432
order, determination of, 455
preceded by first-order reaction, 255
chronoamperometry, 262
chronopotentiometry, 275, 283
polarography, 264, 283
rotating disc method, 278, 283
stationary electrode voltammetry, 268, 283
preceded by higher-order reactions, 331
chronopotentiometry, 336
polarography, 332
rotating disc method, 337
stationary electrode voltammetry, 335
preceded by pseudo first-order reaction, 285
preceded by two first-order reactions, 286
quasi-reversible, 211, 225, 228, 229, 230, 238, 239, 248, 251, 494

rate constants, 225, 238, 239, 244, 245, 250
and half-wave potential, 227
reversible, 212, 229, 231, 239, 245, 315, 485
electrodes for stationary electrode voltammetry, 34
boron carbide, 35
carbon paste, 34
gold, 34
graphite, 34
hanging drop, 34, 38, 167
maxima with, 466
mercury film, 35, 167
platinum, 34
spherical, 151
enediols, 308
erf, 102
erfc, 102
Ershler model, 16
Esin and Markov coefficient, 15
Esin and Markov effect, 14
exchange current, 63

faradaic current, 480
ferrocyanide /ferricyanide system, 251
Fick's laws, 81, 83, 95, 96, 138
Flory–Huggins isotherm, 20
formaldehyde, electroreduction, 256

Gibbs equation, 52

half-wave potentials, 213, 220
interpretation, 219
hanging drop electrode, 34, 38, 167, 466
harmonic alternating current polarography, 510
Helmholtz theory, 2, 7
Henry isotherm, 10
hydrazobenzene, 448
m-hydrazotoluene, 448
hydrodynamic voltammetry, 136
hydrogen peroxide, as catalyst, 328, 329, 457

Index

Ilkovič constant, 123
Ilkovič equation, 31, 57, 104, 122, 123, 141, 148, 212, 226
   modifications, 141
imines, dimerization of reaction products, 340
immonium ions, dimerization of reaction products, 340
indium, anodic oxidation, 444
iodine, catalytic reduction, 328
iron(III), catalytic reduction, 328
   complexes, 328, 364
   EDTA complex, 373
irreversibility, 52
   total, 56

Kalousek method, 425
kinetic current, 256, 267, 288
kinetic parameters, 211, 224, 433, 448, 492
kinetic processes, 446
kinetic studies, 471, 491, 493

Langmuir isotherm, 19
lead styphnate determination, 501
leucoriboflavin, inhibitory effect, 417
Levich equation, 41, 57, 58, 249
Levich and Krylov isotherm, 20
Levine isotherm, 20
limiting current, 31, 99, 123
Lippmann equation, 6

manganese(II), 308
   EDTA complex, 373
maximum current, 145, 489
mercury EDTA complex, 308, 373
methylene blue, adsorption, 410
molybdates, catalytic reduction, 328

Nernst equation, 44
Nernst theory, 40, 41
nickel, 288
nickel complexes, 371
nitrilotriacetic acid complexes, 372
o-nitrophenol, 382
p-nitrosophenol, 455

Oldham's method, 104, 105, 116
oscillopolarography, 508
overvoltage, 54, 55, 64, 65
oxalate complexes, of iron(III), 328, 364
   of titanium, 328
oxygen reduction, 444

phase angle method, 494
polarography, 24, 26, 118, 141, 172, 174, 196, 224, 264, 332, 341, 350, 353, 386, 400 459
   alternating square wave, 472, 476, 479, 486, 492
   anodic limiting current, 149
   applications to analysis, 498
   catalytic reactions, 322, 325
   coulostatic, 506
   current distribution on drop, 147
   derivative pulse, 478, 490, 492
   diffusion current constant, 124
   drop area, 27, 30
   drop growth, 28, 146, 148
   drop life, 29
   drop shape, 148
   drop surface movements, 147, 462
   half-wave potential, 213, 220
   harmonic alternating current, 510
   impoverishment effect, 144, 145
   instantaneous current, 143, 144
   irreversible processes, 197, 222, 228, 230, 407
   kinetic currents, 256
   limiting current, 123
      and mercury head, 123
      latent, 465
   maxima, 459
   maximum current, 145
   mean current, 122, 142, 197, 212
   multi-stage processes, 172
   pulse, 472, 477, 479, 489, 493, 495
   quasi-reversible, 211, 225, 228, 229, 230
   radiofrequency, 504
   resolution, 174
   reversibility, 203, 217
   reversible processes, 212, 229, 400
   saw-tooth, 479

sinusoidal alternating current, 472, 473, 479, 490, 492
transport rates, 57, 59, 60
triangular-wave alternating current, 479
waves, 31
pyridine derivatives, dimerization of reaction products, 340
pyrocatechol complexes of iron (III), 328
pyrogallol complexes of iron (III), 328
pyruvic acid, 257

quasi-reversibility, 56
Quincke theory, 2, 7

Randles–Ševčik equation, 37, 57, 114, 116, 135, 152
rate constants, 66, 225, 238, 239, 244, 245, 250
  table, 67
reversibility, 51, 489
rotating disc voltammetry, 38, 124, 172, 175, 199, 245, 278, 337, 348, 350, 357, 393, 467
  catalytic reactions, 324, 325
  constant current, 131
  constant potential, 125
  current, 135, 137, 200
    and sweep rate, 133
    Levich equation, 41, 57, 58
  electrodes, 39
    carbon paste, 39
    copper, 40
    hydrodynamic, 136
    platinum, 39, 136
    porous, 136
  errors, 467
  irreversible processes, 199
  kinetic parameters, 245, 249
  multi-stage processes, 172
  reversibility, 201
  transport rates, 57, 59, 60
rotating ring-disc electrode method, 440, 455

Sand's equation, 44, 57, 107
silver cyanide complexes, catalytic reduction, 328
stationary electrode voltammetry, 33, 110, 150, 156, 159, 166, 172, 174, 188, 206, 231, 268, 315, 335, 344, 350, 354, 388, 407, 466
  catalytic reactions, 315, 325
  controlled by electron transfer and diffusion rate, 190
  current, 115, 117, 137, 152
    maxima, 457, 466
  cyclic, 421, 427, 451
    and kinetics, 431
    and linear diffusion, 431
  irreversible reactions, 188, 206, 235
  kinetic parameters, 231, 238
  multi-stage processes, 172
  peak current, 115, 150, 152, 159, 166
  quasi-reversible processes, 238, 239
  resolution, 175
  reversible processes, 231
  transport rates, 57, 59, 60
Stern model, 16
Stern theory, 7, 8

Tafel equations, 65, 71
tensammetric maxima, 503, 504
tensammetry, 503
thymol, inhibitory effect, 417
titanium (III), catalytic oxidation, 450
titanium (IV), 455
  oxalate complexes, catalytic reduction, 328
Tomeš criterion, 217, 241, 490
transfer coefficients, measurement, 62, 69, 224, 237, 238, 239, 244, 245
transport rates, 55, 56
  and kinetic parameters, 57
triethanolamine complexes of iron (III), 328
triphenylamine and derivatives, dimerization of reaction products, 340, 343, 394
tungstates, catalytic reduction, 328

uranium (V) disproportionation, 357
uranyl system, 251, 328, 432

vanadates, catalytic reduction, 328
vanadium (IV), catalytic reduction, 328
　reaction with iron (II), 456
vanadium, EDTA complexes, 373

Wicke and Eigen equation, 11

zero charge potentials, 13, 14, 15, 16
zinc, 507
　ammonia complexes, 364
　EDTA complex, 373
　fluoride complexes, 375
　hydroxide complexes, 378
　thiocyanate complexes, 367

3 2311 00067 682 8

/543.087G181F>C1/

543.087
G181f

Galus, Zbigniew.
  Fundamentals of electrochemical
analysis.

A. C. BUEHLER LIBRARY
ELMHURST COLLEGE
Elmhurst, Illinois 60126